Derivative matrices (see page 757)

For $f(x, y)$ at (x_0, y_0):

- **Gradient:** $\nabla f(x_0, y_0) = [\, f_x(x_0, y_0), f_y(x_0, y_0)\,]$
- **Hessian matrix:** $H_f(x_0, y_0) = \begin{bmatrix} f_{xx}(x_0, y_0) & f_{xy}(x_0, y_0) \\ f_{yx}(x_0, y_0) & f_{yy}(x_0, y_0) \end{bmatrix}$

For $\mathbf{f}(x, y) = (u, v)$ at (x_0, y_0):

- **Jacobian matrix:** $\mathbf{f}'(x_0, y_0) = \begin{bmatrix} u_x(x_0, y_0) & u_y(x_0, y_0) \\ v_x(x_0, y_0) & v_y(x_0, y_0) \end{bmatrix} = \begin{bmatrix} \dfrac{\partial u}{\partial x} & \dfrac{\partial u}{\partial y} \\ \dfrac{\partial v}{\partial x} & \dfrac{\partial v}{\partial y} \end{bmatrix}_{(x,y)=(x_0, y_0)}$

Linear and quadratic approximation
(see page 724)

For $f(x, y)$ near (x_0, y_0):

- **Linear:** $L(x, y) = f(x_0, y_0) + f_x(x_0, y_0)(x - x_0) + f_y(x_0, y_0)(y - y_0)$

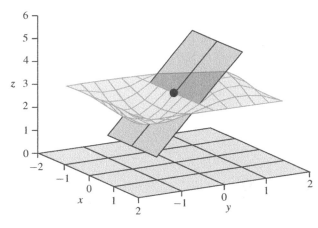

The gray plane is tangent to the surface at the black dot.

- **Quadratic:** $Q(x, y) = f(x_0, y_0) + f_x(x_0, y_0)(x - x_0) + f_y(x_0, y_0)(y - y_0) + \dfrac{f_{xx}(x_0, y_0)}{2}(x - x_0)^2$

$$+ f_{xy}(x_0, y_0)(x - x_0)(y - y_0) + \dfrac{f_{yy}(x_0, y_0)}{2}(y - y_0)^2$$

Multivariate chain rule (see page 760)

For $\mathbf{f}$ and $\mathbf{g}$ differentiable functions, with derivative matrices $\mathbf{f}'$ and $\mathbf{g}'$:

$$(\mathbf{f} \circ \mathbf{g})'(\mathbf{X_0}) = \mathbf{f}'(\mathbf{g}(\mathbf{X_0})) \cdot \mathbf{g}'(\mathbf{X_0}) \quad \text{(dot represents matrix multiplication)}$$

Multivariable
Calculus

From Graphical, Numerical, and Symbolic Points of View

SECOND EDITION

Arnold Ostebee
St. Olaf College

Paul Zorn
St. Olaf College

FREEMAN
Custom Publishing

ISBN-13: 978-1-4292-7178-3
ISBN-10: 1-4292-7178-7

Printed in the United States of America

W. H. Freeman Custom Publishing
W. H. Freeman and Company
41 Madison Avenue
New York, NY 10010

www.whfreeman.com/custompub

About This Book:
Notes For Instructors

This book aims to do what its title suggests: present multivariable calculus from graphical, numerical, and symbolic points of view. In doing so, this work continues the philosophy and viewpoints embodied in our two-volume single-variable text, *Calculus from Graphical, Numerical, and Symbolic Points of View*, 2nd edition. For more details on philosophy, strategy, use of technology, and other issues, see either those volumes or our Web site:

www.stolaf.edu/people/zorn/ozcalc/mvcindex.html

Audience and prerequisites

The text addresses a general mathematical audience: mathematics majors, science and engineering majors, and non-science majors. We assume a little more mathematical maturity than for single-variable calculus, but the presentation is not rigorous in the sense of mathematical analysis. We want students to encounter, understand, and use the main concepts and methods of multivariable calculus and to see how they extend the simpler objects and ideas of elementary calculus. We believe that a fully rigorous logical development belongs later in a student's mathematical education.

We assume that students have had the "usual" one-year, single-variable calculus preparation but little or nothing more than that. A basic familiarity with numerical integration techniques (such as the midpoint rule) is helpful, but it could be developed enroute, if necessary. (We do *not* assume that students have studied single-variable calculus from our own text!)

Although linear functions and linear approximation are stressed, we do not assume that students have had formal experience with linear algebra. Vectors are used often but are introduced from scratch. Matrices appear only occasionally but are important when they do appear. Students unfamiliar with matrices, or who need basic review, should find Appendix A useful. It can either be covered "officially" or left to student reading.

Main themes and strategies

We aim to focus on the main *concepts* of multivariable calculus: the derivative and integral in their higher-dimensional versions, linear approximation, parametrization, vector fields and vector operations, the multivariable analogues of the fundamental theorem of calculus, and a few geometric and physical applications. As in our treatment of single-variable calculus, the key strategy for improving conceptual understanding is to combine, compare, and move among multiple viewpoints—graphical, numerical, and symbolic.

To these ends, we emphasize several themes:

- **Linearity and linear approximation**   These crucial ideas, generally familiar from single-variable calculus, have natural, but more complicated, analogues in the multivariate setting. We try to help students see objects such as gradients, tangent planes, and Jacobian matrices as natural generalizations of their single-variable counterparts.

- **Explicit parametrization**   Multivariate objects—curves, lines, planes, surfaces, and others—are best and most concretely understood, we believe, when students themselves produce and manipulate them through explicit parametrization and calculation. Technology is crucial; using it, students can parametrize objects directly and see at a glance the results, correct or incorrect, of their work.

- **Varying views of functions: on beyond surfaces**   Students find multivariate functions —even of two variables—far harder to visualize and reason about than functions of one variable. With help from technology we offer a variety of graphical and numerical views of such functions, including not only the usual surfaces in space but also numerical tables and contour plots—unglamorous but perhaps underappreciated representations.

Changes in this edition

This edition is substantially improved and expanded from its predecessor. Nearly every section has been substantially revised to clarify explanations, add examples and detail in calculations, improve figures, and increase the quantity and variety of exercises. In addition:

- **Interludes**   The text now includes a selection of "Interludes"—brief, project-oriented expositions designed for independent student work—addressing topics or questions that are "optional" or out of that chapter's main stream of development.

- **Improper integrals**   An entirely new section (Section 15.5) treats improper multiple integrals.

- **Chapter reviews**   Brief chapter summaries and extensive review exercise sets have been added to Chapters 12–14.

- **A look at theory**   A new, brief appendix (Appendix B) offers samples of the analytic theory of multivariable calculus.

- **"Basic" and "Further" exercises**   Each section has exercises of two types: "Basic" and "Further." Typical "Basic" exercises are relatively straightforward and focus on a single important idea. All students should aim to master most of these exercises. "Further" exercises are a little more ambitious; they may require the synthesis of several ideas, deeper or more sophisticated understanding of basic concepts, or better symbol manipulation skills.

Technology

Technology is an important tool for illustrating and comparing graphical, numerical, and symbolic viewpoints in calculus—especially in multivariable calculus, where calculations can be messy and geometric intuition is harder to come by. Although we refer occasionally to computations done with *Maple*, other programs (*Mathematica*, *Derive*, and some symbol-manipulating calculators) would do as well.

In any event, we strongly recommend that students have access to (and use) *some* capable and flexible technology—especially for graphical representations. In particular, some exercises effectively require technology. Less tangibly, but just as important, technology helps foster an experiment-oriented, hands-on, concrete approach to the subject.

Annotated Table of Contents

Brief chapter-by-chapter information follows. (See the Web sites for information about Volumes 1 and 2.)

Chapter 11: Infinite Series   *Note. This chapter appears in this volume to accommodate institutions that may treat this material in a third semester.* Unlike some "reform" text authors, we do treat convergence and divergence of numerical series, but in a somewhat nontraditional way. We stress (i) the analogy with improper integrals, (ii) concrete (sometimes graphical) treatment of partial sums, and (iii) numerical estimation of limits. We think these strategies help make this difficult subject more concrete and accessible than it often is when the principal concern is the abstract question of convergence or divergence. The chapter ends with power series.

Chapter 12: Curves and Vectors   This chapter introduces curves and vectors and their properties, first in the relatively simple context of the xy-plane. Explicit parametrization of various objects—curves, lines, and planes—is stressed throughout. Physical motion is the most important physical application.

Chapter 13: Derivatives   The idea of derivative has various incarnations in several variables. We develop a variety of them here together with some standard applications. The concepts of linearity and local linearity are key: differentiable functions in any number of variables are "almost linear" in an appropriate sense.

Chapter 14: Integrals   We consider the idea, meaning, and various applications of multiple integrals in various coordinate systems. Numerical and graphical as well as symbolic views are represented. The general change-of-variable formula unifies several earlier ideas.

Chapter 15: Other Topics   The chapter is a sampler of extensions and applications of ideas developed earlier; some are presented as student projects. All material here is independent of the sequel, and topics may be covered at an instructor's option.

Chapter 16: Vector Calculus   We introduce the basic objects (especially vector fields and line and surface integrals) and theorems of vector calculus. Special emphasis is laid on Green's theorem—the most accessible vector form of the general fundamental theorem of calculus. Surfaces, surface integrals, and theorems relating them are treated at the end of the chapter. The final section collects and relates many versions of the fundamental theorem of calculus.

Supplements for the instructor

Multivariable Calculus from Graphical, Numerical, and Symbolic Points of View has a support package for the instructor that includes the following:

Instructor's Solutions Manual with Test Bank   The *Instructor's Solutions Manual with Test Bank* offers worked-out solutions to all the exercises in each exercise set. A *Printed Test Bank* is also available in the manual. The *Printed Test Bank* provides a printout of one example of each of the algorithmic items in the HM Testing 6.0 (see the description of HM Testing 6.0 below, under HM ClassPrep with HM Testing 6.0 CD–ROM).

HM ClassPrep with HM Testing CD–ROM   This CD–ROM is a combination of two course management tools.

- HM Testing 6.0 computerized testing software provides instructors with an array of algorithmic test items, allowing for the creation of an unlimited number of tests for each chapter, including cumulative tests and final exams. HM Testing also offers online testing via a Local Area Network (LAN) or the Internet as well as a grade book function.
- HM ClassPrep features supplements and text-specific resources such as ready-to-use Chapter Tests (two formats: free response and multiple choice), PowerPoint® slides, and *Maple* and *Mathematica* activities.

Instructor text-specific Web site   The companion Web site provides additional teaching resources such as ready-to-use Chapter Tests (two formats: free response and multiple choice), PowerPoint® slides, and *Maple* and *Mathematica* activities. Visit `math.college.hmco.com/instructors` and choose *Multivariable Calculus from Graphical, Numerical, and Symbolic Points of View 2e* from the list provided on the site. Appropriate items will be password protected. Instructors have access to the student Web site as well.

Navigating Calculus CD–ROM   The *Navigating Calculus* CD–ROM authored by Jason Brown of Dalhousie University in Nova Scotia and by Arnold Ostebee and Paul Zorn is keyed closely to the *Calculus from Graphical, Numerical, and Symbolic Points of View* (volumes 1 and 2) table of contents and covers both single-variable and multivariable material. *Navigating Calculus* contains a variety of useful activities, tools, and resources, including a powerful graphing calculator utility, a glossary with examples, and many interactive activities that deepen students' understanding of calculus fundamentals. This learning aid is accompanied by the *Navigating Calculus Workbook* written by Stephen Kokoska of Bloomsburg University in Pennsylvania. This workbook is designed to help both instructors and students fully utilize *Navigating Calculus* by offering guided instruction through the workings of the CD–ROM and providing additional examples and exercises.

Supplements for the student

Multivariable Calculus from Graphical, Numerical, and Symbolic Points of View 2e has a support package for the student that includes the following:

Student's Solutions Manual   The *Student's Solutions Manual*, prepared by the authors contains complete worked-out solutions to all odd-numbered exercises. (Brief answers to odd-numbered exercises are available in the back of the text.)

Student text-specific Web Site   This textbook has a companion Web site that provides additional learning resources for the student. Visit `math.college.hmco.com/students` and choose *Multivariable Calculus from Graphical, Numerical, and Symbolic Points of View* from the list provided on the site.

SMARTTHINKING™ Live, On-line Tutoring   Houghton Mifflin has partnered with SMARTTHINKING™ to provide an easy-to-use, effective, on-line tutorial service. Through state-of-the-art tools and a two-way whiteboard, students communicate in real time with qualified e-structors who can help the students understand difficult concepts and guide them through the problem-solving process while studying or completing homework.

Four levels of service are offered to the students.

- **Live, on-line tutoring support** is available Sunday–Thursday 2 P.M.–5 P.M. and 9 P.M.–1 A.M eastern standard time (hours are subject to change).

- **Question submission** allows students to submit questions to the tutor outside the scheduled hours and receive a response within 24 hours.

- **Prescheduled time** allows students to schedule tutoring with an e-structor in advance.

- **Review past on-line sessions** allows students to access and review their progress from previous sessions on a personal academic home page.

Advice from you

Our Web site (the address is given in the opening paragraph) offers various resources and information (*Maple* worksheets, information on obtaining review copies, etc.) that instructors may find useful. We also appreciate hearing your suggestions, comments, and advice.

Arnold Ostebee and Paul Zorn
Department of Mathematics
St. Olaf College
1520 St. Olaf Avenue
Northfield, Minnesota   55057-1098

e-mail: **ostebee@stolaf.edu**   **zorn@stolaf.edu**

ACKNOWLEDGMENTS

This text owes its existence to (literally) countless professors, students, publishing company professionals, friends, advisors, critics, "competitors," family, and others. (These categories are not mutually exclusive!) It is a pleasure to acknowledge by name some—but, necessarily, only some—of the people who attended this book through its long gestation, birth, and publication.

We are indebted, for many reasons, to members of Houghton Mifflin's staff, including (alphabetically) Kathryn Dinovo, Senior Project Editor; Lisa Pettinato, Assistant Editor; and Lauren Schultz, Sponsoring Editor. Zachary Dorsey of TechBooks handled the book's physical production efficiently and effectively. We also thank Alexa Epstein and Leslie Lahr, formerly of Harcourt College Publishers, for many forms of help and advice over the years.

We owe thanks to many professional colleagues for useful suggestions, criticism, and advice. We took some—but not all—of their good advice; all errors of omission and commission are ours alone. It is impossible to list all our creditors and our specific debts to them, but Matthew Bloss, St. Olaf College; Caren Diefenderfer, Hollins College; Ruth Dover, Illinois Mathematics and Science Academy; Doreen Hamilton, St. Olaf College; Bruce Hanson, St. Olaf College; Reg Laursen, Luther College; Richard Mercer, Wright State University; Edward Nichols, Chattanooga State Technical Community College; Sharon Robbert, Trinity Christian College; Joanne Snow, Saint Mary's College; and Douglas Swan, Morningside College, all deserve special thanks, as do our students and theirs. Participants in several summer workshops on multivariable calculus also taught us at least as much as we taught them.

We also thank the following colleagues for various forms of useful editorial advice and manuscript reviews: Nazanin Azarnia, Santa Fe Community College; William Barnier, Sonoma State University; Russell Blyth, Saint Louis University; Philip K. Hotchkiss, Westfield State College; Glenn Ledder, University of Nebraska–Lincoln; Steven G. Krantz, Washington University in St. Louis; Linda McGuire, Muhlenberg College; Javad Namazi, Fairleigh Dickinson University; Dr. Cornelius Nelan, Quinnipiac University; Edward Nichols, Chattanooga State Technical Community College; Todd D. Oberg, Illinois College; Sharon K. Robbert, Trinity Christian College; Elyn Rykken, Muhlenberg College; Joanne E. Snow, Saint Mary's College; Thomas Stohmer, University of California–Davis; Karel Stroethoff, University of Montana; and Amy K.C.S. Vanderbilt, Xavier University.

We thank St. Olaf College in general, and our departmental colleagues in particular, for their advice, support, good humor, and (sometimes) forbearance during the many years of this project's development and progress to a second edition. Countless students, here and at other institutions, also offered generous advice, praise, and criticism—all of it useful.

Our families, finally, deserve our deepest thanks. They have coped cheerfully with peculiar hours, extended absences, mental distraction, blizzards of paper, missed meals, and every other vagary that such a project entails. Without their love and sacrifice, we would never have begun—let alone completed—this project.

Special acknowledgment   This text was developed with support from the National Science Foundation (Grant DUE-9450765).

April 2003

How to Use This Book:
Notes for Students

All authors want their books to be used: read, studied, thought about, puzzled over, reread, underlined, disputed, understood, and, ultimately, enjoyed. So do we.

That might go without saying for some books—beach novels, user manuals, field guides, and others—but it may need repeating for a calculus textbook. We know as teachers, and remember as students, that mathematics textbooks are too often read *backwards*: faced with Exercise 231(b) on page 1638, we have all shuffled backwards through the pages in search of something similar. (Too often, moreover, our searches were rewarded.)

A textbook is not a novel. It is a peculiar hybrid of encyclopedia, dictionary, atlas, anthology, daily newspaper, shop manual, *and* novel—not exactly light reading, but essential reading nevertheless. Ideally, a calculus book should be read in *all* directions: left to right, top to bottom, back to front, and even front to back. That's a tall order. Here are some suggestions for coping with it.

Read the narrative   Each section's narrative is designed to be read from beginning to end. The examples, in particular, are supposed to illustrate ideas and make them concrete—not just serve as templates for homework exercises.

Read the examples   Examples are, if anything, more important than theorems, remarks, and other "talk." We use examples to show already familiar ideas "in action" and to set the stage for new ideas.

Read the pictures   We're serious about the "graphical points of view" mentioned in our title. The pictures in this book are not "illustrations" or "decorations." They are an important part of the language of calculus. An ability to think "pictorially"—as well as symbolically and numerically—about mathematical ideas may be the most important benefit calculus can offer.

Read the language   Mathematics is not a "natural language" like English or French, but it has its own vocabulary and usage rules. Calculus, especially, relies on careful use of technical language. Words and phrases like **partial derivative, gradient, linear approximation, tangent plane, Jacobian matrix, stationary point,** and **vector field** have precise, agreed-upon mathematical meanings. Understanding such words goes a long way toward understanding the mathematics they convey; misunderstanding the words leads inevitably to confusion. When in doubt, consult the index.

Read the instructors' preface (if you like)   Get a jump on your teacher.

In short: *Read the book.* Read it actively with paper and pencil at hand and, if possible, with technology at your elbow. Do the calculations for yourself. Plot some curves and surfaces for yourself. You bought the book—do whatever you can to make it your own.

A last note

Why study calculus at all? There are plenty of good practical and "educational" reasons: because it's good for applications, because higher mathematics requires it, because it's good mental training, because other majors require it, and because jobs require it. The ideas and methods of multivariable calculus, in particular, are even more powerful and flexible than those of single-variable calculus in modeling the physical and human worlds in all their higher-dimensional richness.

There is another, different, better reason to study the subject: Calculus is among our species' deepest, richest, farthest reaching, and most beautiful intellectual achievements. We hope this book will help you see it in that spirit.

A last request   Last, a request. We sincerely appreciate—and take very seriously—students' opinions, suggestions, and advice on this book. We invite you to offer your advice either through your teacher or by writing us directly.

Arnold Ostebee and Paul Zorn
Department of Mathematics
St. Olaf College
1520 St. Olaf Avenue
Northfield, Minnesota 55057-1098

Contents

Multivariable

Calculus

WHAT IS MULTIVARIABLE CALCULUS?

What is this book about?

A rough, short answer is suggested by the name itself: Multivariable calculus resembles ordinary calculus but allows more variables. The basic object of ordinary calculus is a function that accepts *one* number as input and produces *one* number as output. The squaring function, for example, can be described using variables—one for inputs and another for outputs—by the equation

$$y = f(x) = x^2.$$

The equation mentions two variables in all, and so its graph is a curve (a parabola) in *two*-dimensional xy-space. Using standard tools and operations of beginning calculus, such as derivatives and integrals, we can calculate such quantities as slopes and areas determined by the graph of f.

A simple multivariable analogue of the function f is the function g defined by

$$z = g(x, y) = x^2 + y^2.$$

This equation mentions three variables—two for inputs and one for outputs. The graph of g turns out, therefore, to be a surface (called a *paraboloid*) in *three*-dimensional xyz-space. Figure 8, page 616, offers one possible view of the graph.

This new graph is a little more complicated, but a lot more interesting, than its counterpart in xy-space. We might ask, for instance, how *steep* the surface would seem to an ant walking along it, and how the answer depends both on the ant's position and direction of motion. We might also ask about the *volume* enclosed by some part of the surface or about the *surface area* of some part of the graph. These questions, like their simpler analogues mentioned earlier, can all be answered using derivatives and integrals, but only after "derivative," "integral," and other standard calculus objects and processes have been defined and understood in ways that extend usefully to higher-dimensional settings.

Extending basic ideas and methods of elementary calculus to new and more general settings is the main theme of multivariable calculus. Doing so takes work and care, but the rewards are real. Multivariable calculus shows the power and generality of calculus ideas, not only in mathematics itself but also in modeling our multidimensional world.

INFINITE SERIES

11.1 SEQUENCES AND THEIR LIMITS

This section, on infinite *sequences*, prepares the ground for the next topic—infinite *series*. Convergent series are defined in terms of the simpler, more basic idea of convergent sequences. We start with a brief introduction to sequences—what they are, what it means for them to converge or diverge, and how to find their limits.

Terminology and basic examples

A **sequence** is an infinite list of numbers, of the general form

$$a_1, a_2, a_3, a_4, \ldots, a_k, a_{k+1}, \ldots.$$

Read "a sub three" and "a sub k."

Individual entries are called the **terms** of the sequence; a_3 and a_k, ◄ for instance, are the third term and the kth term, respectively. The full sequence is, technically speaking, an **ordered set**; the standard notation

$$\{a_k\}_{k=1}^{\infty}$$

(or simply $\{a_k\}$) uses set (wiggly) brackets to emphasize this view.

Our main interest in sequences is in their **limits**. For the simplest sequences, limits (or the lack thereof) are evident at a glance. The next three examples are of this type.

EXAMPLE 1 Discuss the sequence $\{a_k\}_{k=1}^{\infty}$ defined by the formula $a_k = 1/k$. Does this sequence have a limit?

Solution Sampling some terms—

$$\frac{1}{1}, \frac{1}{2}, \frac{1}{3}, \ldots, \frac{1}{10}, \frac{1}{11}, \ldots, \frac{1}{100}, \frac{1}{101}, \ldots$$

shows (to nobody's surprise) that the sequence **converges** to zero: As k increases, the terms a_k approach zero arbitrarily closely. In symbols,

$$\lim_{k \to \infty} a_k = \lim_{k \to \infty} \frac{1}{k} = 0.$$

■

E X A M P L E 2   Suppose that $\{b_j\}_{j=1}^{\infty}$ is defined by

$$b_j = \frac{(-1)^j}{j}, \quad j = 1, 2, 3, \ldots.$$

(We used j, not k, as our **index variable**—the choice is up to us.) What's $\lim_{j \to \infty} b_j$?

Solution   Writing out terms shows a pattern similar to that in Example 1:

$$-\frac{1}{1}, \frac{1}{2}, -\frac{1}{3}, \ldots, \frac{1}{10}, -\frac{1}{11}, \ldots, \frac{1}{100}, -\frac{1}{101}, \ldots$$

Although the terms oscillate in sign, they approach zero more and more closely as j increases. Eventually, all the terms—positive or negative—remain within any specified distance from zero. (All terms past b_{1000}, for instance, are within 0.001 of zero.) Hence, the sequence $\{b_j\}$ **converges** to zero:

$$\lim_{j \to \infty} b_j = \lim_{j \to \infty} \frac{(-1)^j}{j} = 0. \qquad \blacksquare$$

E X A M P L E 3   Does the sequence $\{c_k\}_{k=0}^{\infty}$ with general term $c_k = (-1)^k$ converge?

Solution   No; it **diverges**. Successive terms have the pattern

$$1, -1, 1, -1, 1, \ldots,$$

never settling on a single limit. $\qquad \blacksquare$

E X A M P L E 4   Discuss the **Fibonacci sequence**, defined by the rules

$$F_1 = 1; \quad F_2 = 1; \quad F_{n+2} = F_n + F_{n+1};$$

each term is the sum of its two predecessors. ➧ Such definitions are called **recursive**: Each term is defined by means of earlier terms.

The sequence is named for the Italian mathematician Leonardo Fibonacci (1170–1250), who related it to a rabbit population explosion under certain conditions.

Solution   The first few terms of this sequence are

$$1, 1, 2, 3, 5, 8, 13, 21, 34, 55, \ldots.$$

As the pattern suggests, the sequence **diverges** to infinity, and we write $\lim_{n \to \infty} F_n = \infty$. $\quad \blacksquare$

Lessons from the examples

Sequences have their own notational quirks and conventions. Here are several to watch for:

- **Where to start?**   The sequence in Example 3 began with c_0, not c_1. Other starting points, such as a_2 or even b_{-3}, occasionally arise. In practice, such differences are unimportant. What matters for sequences is their long-run behavior, not the presence or absence of a few initial terms.

- **Index variable names don't matter**   We can define the squaring function by writing either $f(t) = t^2$ or $f(x) = x^2$; the *variable name* makes no difference. In the same way, a sequence's *index name* is arbitrary: $\{a_k\}_{k=1}^{\infty}$ and $\{a_j\}_{j=1}^{\infty}$ mean exactly the same thing.

- **Reindexing** The sequence

$$\frac{1}{1}, \frac{1}{2}, \frac{1}{3}, \dots, \frac{1}{10}, \frac{1}{11}, \dots$$

of Example 1 can be described symbolically in various different-looking but still equivalent ways. Here are two:

$$a_k = \frac{1}{k}, \quad k = 1, 2, \dots \qquad \text{or} \qquad a_j = \frac{1}{j+1}, \quad j = 0, 1, 2, \dots.$$

One description or the other may be preferable, depending on the situation.

Sequences as functions Sequences are closely related to functions, as expressions such as

$$a_k = \frac{\sin k}{k} \qquad \text{and} \qquad f(x) = \frac{\sin x}{x}$$

illustrate. The formal definition makes this connection precise.

DEFINITION An **infinite sequence** is a real-valued function that is defined for *positive integer inputs.*

The definition and the preceding examples give us several useful ways to think of a sequence:

- **As a list** As an infinite list of numbers: $a_1, a_2, a_3, \dots$.
- **As a function** As a function $a(n)$, where n takes only positive integer values. (The rule for a may or may not make sense for other inputs.) Thus, $a(1) = a_1$, $a(2) = a_2$, $a(3) = a_3$, and so on.
- **As a discrete sample** As a "discrete sample" of values of an ordinary calculus-style function $f(x)$ defined for real $x \geq 1$. The function $f(x) = 1/x$, for example, produces the sequence $a_k = 1/k$ in Example 1.

Graphs of sequences, graphs of functions The graph of any function f consists of the points $(x, f(x))$ for x in the domain of f. The graph of a sequence $\{a_k\}$ is therefore the set of points (k, a_k) as k runs through positive integer values.

EXAMPLE 5 Let a function f and a sequence $\{a_k\}$ be defined by

$$f(x) = \frac{\sin x}{x}; \qquad a_k = \frac{\sin k}{k}.$$

Plot graphs of both. What do the graphs say about limits?

Solution   Figure 1 shows the graphs.

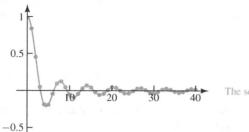

The sequence is plotted with dots.

FIGURE 1

Graphs of $f(x) = \dfrac{\sin x}{x}$ **and** $a_k = \dfrac{\sin k}{k}$

The graphs illustrate how the sequence $\{a_k\}$ is a "discrete sample" of the continuous function f. They show, too, that as x (or k) tends to infinity, this sequence and function, although oscillating in sign, tend to zero.   ◼

  Not every sequence comes naturally from a familiar calculus function. Nevertheless, graphs or tables often suggest limits.

EXAMPLE 6   A sequence $\{b_j\}$ has general term

$$b_j = 1 \cdot 2 \cdot 3 \cdot 4 \cdots (j-1) \cdot j = j!$$

(In words: b_j is j **factorial**.) Tabulate $\{b_j\}$; find its limit, if any.

Solution   As the table suggests, $\{b_j\}$ diverges—quickly—to infinity.

As $k \to \infty, k! \to \infty$: explosive numerical evidence						
k	1	2	4	8	16	32	64
$k!$	1	2	24	40,320	2.092×10^{13}	2.631×10^{35}	1.269×10^{89}

Can any doubt remain? ➡

◼ *Graphs work poorly for this sequence—most windows are too small.*

Sequence limits defined

Sequences are special sorts of functions; limits of sequences, therefore, are mild variants on limits at infinity. An informal definition will suffice. ➡

A formal definition describes more precisely what "approaches" means.

> **DEFINITION (Limit of a sequence)**   Let $\{a_k\}$ be a sequence and L a real number. If a_k approaches L to within any desired tolerance as k increases without bound, then the sequence **converges** to L. In symbols,
>
> $$\lim_{k\to\infty} a_k = L.$$
>
> Otherwise, the sequence **diverges**.

Notice:

- **Divergence to infinity** If either $a_k \to \infty$ or $a_k \to -\infty$ as k increases without bound, then the sequence diverges to (positive or negative) infinity. We write, for instance,

$$\lim_{k \to \infty} k! = \infty.$$

- **Other divergence behavior** Example 3 shows that a divergent sequence need not "blow up"; other patterns of "wandering" behavior (or no pattern at all) are possible.

- **Asymptotes** A sequence, like a function, converges to a finite limit L if and only if $y = L$ is a horizontal asymptote of its graph.

Sequences, functions, and limits For sequences that are "discrete samples" of familiar functions, we can often use what we know of the underlying function to find limits.

> **FACT** Let f be a function defined for $x \geq 1$. If $\lim_{x \to \infty} f(x) = L$ and $a_k = f(k)$ for all $k \geq 1$, then $\lim_{k \to \infty} a_k = L$.

Example 5 illustrates this Fact:

$$\lim_{k \to \infty} \frac{\sin k}{k} = \lim_{x \to \infty} \frac{\sin x}{x} = 0.$$

EXAMPLE 7 Does the sequence with general term $a_k = \sin k$ converge or diverge?

Solution Because $\lim_{x \to \infty} \sin x$ does not exist, the preceding Fact does not help. Instead, let's look at the graph (Figure 2).

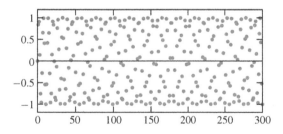

FIGURE 2
The sequence $a_k = \sin k$

Although full of interesting shapes, the graph never settles on a single limit, and so the sequence diverges. ∎

Finding limits of sequences

Many of the tools we have developed for finding limits of functions will help us find limits of sequences as well.

We studied l'Hôpital's rule in Chapter 4.

Using l'Hôpital's rule If a sequence has a nice symbolic formula, l'Hôpital's rule can sometimes be applied to find the limit. ◄

EXAMPLE 8 Numerical evidence suggests that if $a_k = 2^k/k^2$, then $\{a_k\}$ diverges to infinity. Use l'Hôpital's rule to show this result symbolically.

Solution l'Hôpital's rule applied to $h(x) = 2^x/x^2$ gives

$$\lim_{x\to\infty} \frac{2^x}{x^2} = \lim_{x\to\infty} \frac{2^x \ln 2}{2x} \quad \text{(differentiate top and bottom separately)}$$

$$= \lim_{x\to\infty} \frac{2^x \ln 2 \cdot \ln 2}{2} \quad \text{(apply l'Hôpital's rule again)}$$

$$= \frac{(\ln 2)^2}{2} \lim_{x\to\infty} 2^x = \infty.$$

Because $h(x) \to \infty$, $a_k \to \infty$ too; the sequence diverges. ∎

EXAMPLE 9 Show that $\lim_{n\to\infty} n^{1/n} = 1$.

Solution l'Hôpital's rule doesn't apply directly, but applying the natural logarithm produces a fraction:

$$a_n = n^{1/n} \implies \ln(a_n) = \frac{\ln n}{n}.$$

l'Hôpital's rule *does* apply to the last quantity:

$$\lim_{n\to\infty} \frac{\ln n}{n} = \lim_{n\to\infty} \frac{1/n}{1} = 0.$$

We have shown that $\ln(a_n) \to 0$ as $n \to \infty$; it follows that $a_n \to 1$, as desired. ∎

New sequence limits from old Limits of sequences, like limits of functions, can be combined in various symbolic ways to give new limits. We will need some *known* limits, of course, to get started. Here are several useful and important ones:

$$\lim_{n\to\infty} n^{1/n} = 1$$

$$\lim_{n\to\infty} x^{1/n} = 1 \quad \text{(for all } x > 0\text{)}$$

$$\lim_{n\to\infty} \frac{1}{n^k} = 0 \quad \text{(for all } k > 0\text{)}$$

$$\lim_{n\to\infty} r^n = 0 \quad \text{(if } -1 < r < 1\text{)}.$$

(We derived the first of these limits in Example 9, using l'Hôpital's rule. The other limits are easier.)

Calculating with limits Plausible-seeming calculations such as

$$\lim_{k\to\infty} \left(\frac{1}{k} + \frac{3k}{k+1} \right) = \lim_{k\to\infty} \frac{1}{k} + 3 \lim_{k\to\infty} \frac{k}{k+1} = 0 + 3 = 3$$

rely implicitly on the following theorem. We have already seen it—in almost identical form—for functions.

> **THEOREM 1 (Algebra with limits)** Suppose that
> $$a_k \to L \quad \text{and} \quad b_k \to M$$
> as $k \to \infty$, where L and M are finite numbers. Let c be any real constant. Then
> $$ca_k \to cL, \quad a_k \pm b_k \to L \pm M, \quad \text{and} \quad a_k b_k \to LM.$$
> If $M \neq 0$, then $\dfrac{a_k}{b_k} \to \dfrac{L}{M}$.

Squeezing limits For sequences, as for functions, unknown limits can sometimes be found by "squeezing" them between known limits.

> **THEOREM 2 (The squeeze principle)** Suppose that
> $$a_k \leq b_k \leq c_k \quad \text{for all } k \geq 1$$
> and that
> $$\lim_{k \to \infty} a_k = \lim_{k \to \infty} c_k = L.$$
> Then $\displaystyle\lim_{k \to \infty} b_k = L$.

EXAMPLE 10 We have seen graphically that $\displaystyle\lim_{k \to \infty} \frac{\sin k}{k} = 0$. Use squeezing to show this legalistically.

Solution The squeeze inequality

$$-\frac{1}{k} \leq \frac{\sin k}{k} \leq \frac{1}{k}$$

holds for all integers $k > 0$. Because both the left and the right sides tend to 0 as k tends to infinity, the middle expression must do so too. ∎

Guaranteeing convergence: an existence theorem

The best way to show that a sequence converges is to find a limit. But doing so is sometimes hard. The following example illustrates the difficulty—and also suggests a remedy.

EXAMPLE 11 Does the sequence $a_n = \left(1 + \dfrac{1}{n}\right)^n$ converge?

Solution The answer is not obvious from the formula, so let's tabulate some numerical values:

Values of $(1 + 1/n)^n$							
n	16	32	64	128	256	512	1024
$(1 + 1/n)^n$	2.63793	2.67699	2.69734	2.70774	2.71299	2.71563	2.71696

As $n \to \infty$, the terms a_n seem to increase but not to blow up. Thus, apparently, the sequence converges to *some* limit. (The number $e \approx 2.71828$ is a tempting guess.) ∎

A convergence theorem   The sequence in Example 11 appears from the table of numbers to be **nondecreasing**:

$$a_1 \leq a_2 \leq \cdots \leq a_k \leq a_{k+1} \leq \ldots .$$

(For this sequence, *strict* inequalities happen to hold, and so we could use the stronger adjective **increasing**.) Common sense suggests that any nondecreasing sequence should either (i) converge to a limit; or (ii) diverge to ∞. Because the sequence in Example 11 appears to be bounded above (by 3, for example), it is reasonable to guess that it converges. Theorem 3 makes these commonsense impressions precise. It applies to any **monotone** sequence (i.e., any sequence that is either nondecreasing on nonincreasing).

> **THEOREM 3**   Suppose that the sequence $\{a_k\}$ is nondecreasing and bounded above by a number A. That is,
>
> $$a_1 \leq a_2 \leq a_3 \leq \cdots a_k \leq a_{k+1} \leq \cdots \leq A.$$
>
> Then $\{a_k\}$ converges to some finite limit a with $a \leq A$.
>   Similarly, if $\{b_k\}$ is nonincreasing and bounded below by a number B, then $\{b_k\}$ converges to a finite limit $b \geq B$.

EXAMPLE 12   Consider the increasing sequence $\{a_k\}$ defined recursively by

$$a_1 = 0; \quad a_{k+1} = \sqrt{6 + a_k} \quad \text{if } k \geq 1.$$

Show that the sequence $\{a_k\}$ converges and that $\lim\limits_{k \to \infty} a_k \leq 3$.

Solution   The first few terms suggest that the sequence is increasing: ➡

$$a_1 = 0; \quad a_2 = \sqrt{6} \approx 2.45; \quad a_3 \approx 2.91; \quad a_4 \approx 2.98; \quad a_5 \approx 2.997.$$

We'll take this for granted, though it's not hard to show.

The sequence also seems to be bounded above by 3. This is easy to show. If $a_k < 3$, then, by definition,

$$a_{k+1} = \sqrt{6 + a_k} < \sqrt{6 + 3} = 3.$$

Thus, *no* term can exceed 3. It follows from the theorem, therefore, that the sequence converges to a limit no greater than 3. ➡

The limit is 3, but that requires further proof.

BASIC EXERCISES

In Exercises 1–17, find the limit of the sequence or explain why the limit does not exist.

1. $a_k = (-3/2)^k$

2. $a_k = (-0.8)^k$

3. $a_k = (1.1)^k$

4. $a_k = \left(\sqrt{26}/17\right)^k$

5. $a_k = (1/k)^k$

6. $a_k = \sin k$

7. $a_k = \arctan k$

8. $a_k = \cos(1/k)$

9. $a_k = \sin(k\pi)$

10. $a_k = \cos(k\pi)$

11. $a_k = \dfrac{k^2}{k^2 + 3}$

12. $a_k = \sqrt{\dfrac{2k}{k+3}}$

13. $a_m = m^2 e^{-m}$

14. $a_j = \dfrac{\ln j}{\sqrt[3]{j}}$

15. $a_k = \dfrac{k!}{(k+1)!}$

16. $a_k = \ln\left(\dfrac{k}{k+1}\right)$

17. $a_k = 3^{1/k}$

18. (a) For which values of x does $\lim\limits_{n \to \infty} x^n$ exist?

   (b) Find all values of x for which $\lim\limits_{n \to \infty} x^n = 0$.

   (c) Are there any values of x for which $\lim\limits_{n \to \infty} x^n = L \neq 0$? For each such x, find the limit L.

In Exercises 19–24, give an example of a sequence that is

19. convergent but not monotone.

20. bounded but not monotone.

21. monotone but not convergent.

22. nonincreasing and unbounded.

23. nonincreasing and convergent.

24. unbounded but not monotone.

25. Use Theorem 3 to show that the sequence 0.7, 0.77, 0.777, 0.7777, 0.77777, 0.777777, 0.7777777, ... has a limit.

26. Suppose that $\lim_{k \to \infty} a_k = L$, where L is a finite number, and that the terms of the sequence $\{b_k\}$ are defined by $b_k = L - a_k$. Explain why $\lim_{k \to \infty} b_k = 0$.

FURTHER EXERCISES

In Exercises 27–32, find the limit of the sequence or explain why the limit does not exist. [HINT: l'Hôpital's rule may be useful.]

27. $a_n = n \sin(1/n)$

28. $a_k = (2^k + 3^k)^{1/k}$

29. $a_k = \int_0^k e^{-x} \, dx$

30. $a_k = \int_k^\infty \dfrac{dx}{1 + x^2}$

31. $a_k = \sqrt{k^2 + 1} - k$

32. $a_k = \sqrt{k^2 + k} - k$

In Exercises 33–36, determine the values of x for which the sequence converges as $k \to \infty$. Evaluate $\lim_{k \to \infty} a_k$ for these values of x.

33. $a_k = e^{kx}$

34. $a_k = (\ln x)^k$

35. $a_k = (\arcsin x)^k$

36. $a_k = 2^{-k}(\arctan x)^k$

37. Show that $\lim_{n \to \infty} x^{1/n} = 1$ for all $x > 0$.

38. Let $a_k = \left(1 + \dfrac{x}{k}\right)^k$, where x is a real number.

(a) Show that $\lim_{k \to \infty} \ln(a_k) = x$.

(b) Use part (a) to evaluate $\lim_{k \to \infty} a_k$.

39. Evaluate $\lim_{n \to \infty} \left(1 - \dfrac{1}{2n}\right)^n$.

40. Let $a_n = \displaystyle\sum_{k=1}^{n} \dfrac{k}{n^2}$.

(a) Evaluate a_{10}.

(b) Explain why $\lim_{n \to \infty} a_n = \displaystyle\int_0^1 x \, dx = 1/2$.
[HINT: $k/n^2 = (k/n) \cdot (1/n)$.]

41. Let $a_n = \displaystyle\sum_{k=1}^{n} \dfrac{1}{n + k}$.

(a) Show that this sequence is increasing (i.e., $a_{n+1} > a_n$).

(b) Show that $a_n \le \dfrac{n}{n+1} < 1$.

(c) What do parts (a) and (b) imply about $\lim_{n \to \infty} a_n$?

(d) Explain why $\lim_{n \to \infty} a_n > 1/2$. [HINT: $a_1 = 1/2$.]

(e) Show that $\lim_{n \to \infty} a_n = \ln 2$.

[HINT: $\dfrac{1}{n+k} = \dfrac{1}{1 + k/n} \cdot \dfrac{1}{n}$, and a_n is a Riemann-sum approximation to an integral.]

42. Let $a_n = \displaystyle\sum_{k=1}^{n} k^2/n^3$. Evaluate $\lim_{n \to \infty} a_n$. [HINT: Think of a_n as a Riemann-sum approximation to an integral.]

43. Let $a_n = \dfrac{\sqrt[n]{n!}}{n}$.

(a) Show that $\ln a_n = \dfrac{1}{n} \displaystyle\sum_{k=1}^{n} \ln k - \dfrac{1}{n} \displaystyle\sum_{k=1}^{n} \ln n$.

(b) Use part (a) to show that $\ln a_n$ is a right-sum approximation to $\displaystyle\int_0^1 \ln x \, dx$.

(c) Use part (b) to show that $\lim_{n \to \infty} a_n = e^{-1}$.

44. Let $a_n = 4^n/n!$.

(a) Find a number N such that $a_{n+1} \le a_n$ for all $n \ge N$.

(b) Use part (a) to explain why $\lim_{n \to \infty} a_n$ exists.

(c) Evaluate $\lim_{n \to \infty} a_n$.

45. Suppose that $\{a_n\}$ is a sequence with the property $|a_{n+1}/a_n| \le (n+3)/(2n+1)$ for all $n \ge 1$. Show that $\lim_{n \to \infty} a_n = 0$.
[HINT: Start by showing that $|a_{n+1}/a_3| \le (6/7)^{n-2}$ for all $n \ge 3$.]

In Exercises 46–49, let $a_n = \cos 1 \cdot \cos 2 \cdot \cos 3 \cdot \cos 4 \cdots \cos n$.

46. Is the sequence $\{a_n\}$ bounded? Justify your answer.

47. Is the sequence $\{a_n\}$ monotone? Justify your answer.

48. Is the sequence $\{|a_n|\}$ bounded? Justify your answer.

49. Is the sequence $\{|a_n|\}$ monotone? Justify your answer.

50. Let $a_n = \dfrac{1 \cdot 3 \cdot 5 \cdots (2n-1)}{2 \cdot 4 \cdot 6 \cdots (2n)}$. Use Theorem 3 to show that $\lim_{n \to \infty} a_n$ exists.

51. Show that $\lim_{n \to \infty} \sin\left(\dfrac{\pi}{2^2}\right) \cdot \sin\left(\dfrac{\pi}{3^2}\right) \cdots \sin\left(\dfrac{\pi}{n^2}\right) = 0$.

[HINT: $0 < \sin x < x$ when $0 < x < 1$.]

52. Does the sequence defined by $a_1 = 1$, $a_{n+1} = 1 - a_n$ converge? Justify your answer.

53. Does the sequence defined by $a_1 = 1$, $a_{n+1} = a_n/2$ converge? Justify your answer.

54. Consider the sequence defined by $a_1 = 1$, $a_{n+1} = \left(\dfrac{n}{n+1}\right) a_n$.

(a) Show that the sequence converges.

(b) Find the limit. [HINT: Write out the first few terms and look for a pattern.]

55. For which values of $x \ge 0$ does the sequence defined by $a_1 = x$, $a_{n+1} = \sqrt{a_n}$ converge? Justify your answer.

11.2 INFINITE SERIES, CONVERGENCE, AND DIVERGENCE

An **infinite series** (just **series** for short) is an expression of the form

$$\sum_{k=1}^{\infty} a_k = a_1 + a_2 + a_3 + a_4 + \cdots + a_k + a_{k+1} + \cdots .$$

(Notice the **sigma notation** on the left; we used it earlier for approximating sums for integrals.) A series results from *adding* the terms of a sequence $a_1, a_2, a_3, \ldots$. If, say, $a_k = 1/k^2$, then

$$\sum_{k=1}^{\infty} a_k = \sum_{k=1}^{\infty} \frac{1}{k^2} = \frac{1}{1} + \frac{1}{4} + \frac{1}{9} + \frac{1}{16} + \cdots .$$

If $a_k = k$, then

$$\sum_{k=1}^{\infty} a_k = \sum_{k=1}^{\infty} k = 1 + 2 + 3 + \cdots .$$

Natural questions   The notion of an infinite sum raises natural questions:

- What does it *mean* to add infinitely many numbers?
- Which series add up to a finite number? Which series blow up?
- If a series has a finite sum, how can we find (or estimate) it?
- What good are infinite series?

The rest of this chapter addresses these questions.

Improper sums vs. improper integrals   Standard examples of infinite series include

$$\text{(i)} \quad \sum_{k=1}^{\infty} \frac{1}{k^2}; \quad \text{(ii)} \quad \sum_{k=1}^{\infty} \frac{1}{k}; \quad \text{(iii)} \quad \sum_{k=1}^{\infty} \frac{1}{\sqrt{k}}; \quad \text{(iv)} \quad \sum_{k=1}^{\infty} \frac{1}{2^k}.$$

Notice the close typographical resemblance to improper integrals:

$$\text{(i)} \quad \int_{1}^{\infty} \frac{dx}{x^2}; \quad \text{(ii)} \quad \int_{1}^{\infty} \frac{dx}{x}; \quad \text{(iii)} \quad \int_{1}^{\infty} \frac{dx}{\sqrt{x}}; \quad \text{(iv)} \quad \int_{1}^{\infty} \frac{dx}{2^x}.$$

This is no accident; infinite series and improper integrals are closely analogous; we will often exploit the connection. We'll see, in fact, that items (i) and (iv) of *both* preceding lists converge, whereas (ii) and (iii) diverge.

Why series matter: a look ahead

Understanding series takes some work. To preview why the work is worthwhile, consider the fact (we'll see later why it's true) that for any real number x,

$$\cos x = 1 - \frac{x^2}{2!} + \frac{x^4}{4!} - \frac{x^6}{6!} + \frac{x^8}{8!} - \cdots . \tag{1}$$

If, say, $x = 1$, then

$$\cos 1 = 1 - \frac{1}{2!} + \frac{1}{4!} - \frac{1}{6!} + \frac{1}{8!} - \cdots . \tag{2}$$

So what? Why would we write something familiar—the cosine function—in terms of something exotic—a series?

One good answer is that the cosine function has no algebraic formula, and so finding accurate numerical values of $\cos x$ for arbitrary inputs x is a genuine problem. ← Infinite series help solve this problem. Although not quite a formula in the ordinary sense (ordinary formulas don't include dots ...), Equation (1) gives a concrete, computable recipe for *approximating* $\cos x$: Given an input x, calculate, as far out as practically possible, the "infinite polynomial"

Finding $\cos x$ is easy for a few special values of x, such as $x = \pi/4$.

$$1 - \frac{x^2}{2!} + \frac{x^4}{4!} - \frac{x^6}{6!} + \frac{x^8}{8!} - \cdots .$$

With any luck, the result should closely approximate the "true" value of $\cos x$.

Equation (2) shows how to approximate $\cos 1$. A calculator readily gives, to seven decimals,

$$1 - \frac{1}{2!} = 0.5000000;$$

$$1 - \frac{1}{2!} + \frac{1}{4!} = 0.5416667;$$

$$1 - \frac{1}{2!} + \frac{1}{4!} - \frac{1}{6!} = 0.5402778;$$

$$1 - \frac{1}{2!} + \frac{1}{4!} - \frac{1}{6!} + \frac{1}{8!} = 0.5403026.$$

The results converge with gratifying speed to the "right" answer—the true value of $\cos 1$ (≈ 0.5403023).

Good questions   These calculations raise good questions:

> *Where did Equation (1) come from? What do all the "dots" really mean? Are similar equations available for other functions—sine, arctangent, logarithmic, and so on? How many terms are needed to guarantee accuracy to, say, five decimals?*

We will answer all of these questions in this chapter.

Definitions and terminology

Working successfully with series requires some up-front investment in definitions and technical language. After stating terms and definitions, we show by example why they are reasonable.

Series language   Let

$$\sum_{k=1}^{\infty} a_k = a_1 + a_2 + a_3 + \cdots + a_k + a_{k+1} + \cdots$$

be an infinite series. (In most cases the index variable starts at 1, but sometimes it's convenient to start k at 0 or elsewhere.) ➤ The summand a_k is called the kth **term** of the series, and the nth **partial sum**, usually denoted by S_n, is the (finite) sum of all terms *through index n*:

Sometimes we'll use other names, such as j, for the index variable.

$$S_n = a_1 + a_2 + a_3 + \cdots + a_{n-1} + a_n = \sum_{k=1}^{n} a_k. \tag{3}$$

The nth **tail**, denoted by R_n, is the (infinite) sum of all terms *beyond index n*:

$$R_n = a_{n+1} + a_{n+2} + a_{n+3} + \cdots = \sum_{k=n+1}^{\infty} a_k.$$

As the notation R_n suggests, the nth tail is a *remainder*—what's left after adding terms through index n. In symbols,

$$\sum_{k=1}^{\infty} a_k = \sum_{k=1}^{n} a_k + \sum_{k=n+1}^{\infty} a_k = S_n + R_n.$$

(We did the same thing in Chapter 10 with improper integrals:

$$\int_{1}^{\infty} a(x)\,dx = \int_{1}^{n} a(x)\,dx + \int_{n}^{\infty} a(x)\,dx,$$

where the last integral is another type of upper tail.)

The crucial definition of convergence involves the partial sums S_n:

DEFINITION If $\lim_{n\to\infty} S_n = S$, for some finite number S, then the series $\sum_{k=1}^{\infty} a_k$ **converges** to the limit S. (S is also called the **sum** of the series.) Otherwise, the series **diverges**.

Notice the following aspects of the definition:

- **Divergent series**   A divergent series is one for which the sequence of partial sums does *not* converge to a finite limit S. One possibility is that the partial sums S_n blow up to infinity. Another possibility is that the partial sums remain bounded but never settle on a specific limit.

- **Improper integrals, improper sums**   The definition says, in symbols, that

$$\sum_{k=1}^{\infty} a_k = \lim_{n\to\infty} \sum_{k=1}^{n} a_k$$

if the limit exists. Convergence for improper integrals means much the same thing:

$$\int_{x=1}^{\infty} f(x)\,dx = \lim_{n\to\infty} \int_{x=1}^{n} f(x)\,dx$$

if *this* limit exists. ➤ An infinite series is an improper sum in exactly the sense that an integral may be improper.

In each case we take the limit of something proper.

- **Convergence, partial sums, and tails** To say that $\sum a_k$ converges to the sum S means that $S_n \to S$ as $n \to \infty$. Since for all n, $S = S_n + R_n$, it follows that $R_n \to 0$ as $n \to \infty$.

- **Two sequences—keep them straight** Every series $\sum a_k$ involves *two* sequences: the sequence $\{a_k\}$ of *terms* and the sequence $\{S_n\}$ of *partial sums*. Keeping these related but different sequences separate is essential. We will take special care to do so in the following examples.

EXAMPLE 1 Does the series $\displaystyle\sum_{k=0}^{\infty} \frac{1}{2^k} = 1 + \frac{1}{2} + \frac{1}{4} + \frac{1}{8} + \cdots$ converge? If so, to what limit?

Solution Because the index k begins at 0, not 1, so does the sequence of partial sums. Direct calculation yields

$$S_0 = 1 = 2 - 1;$$

$$S_1 = 1 + \frac{1}{2} = \frac{3}{2} = 2 - \frac{1}{2};$$

$$S_2 = 1 + \frac{1}{2} + \frac{1}{4} = \frac{7}{4} = 2 - \frac{1}{4};$$

$$S_3 = 1 + \frac{1}{2} + \frac{1}{4} + \frac{1}{8} = \frac{15}{8} = 2 - \frac{1}{8};$$

$$S_4 = 1 + \frac{1}{2} + \frac{1}{4} + \frac{1}{8} + \frac{1}{16} = \frac{31}{16} = 2 - \frac{1}{16}.$$

The pattern is easy to see: For any $n \geq 1$,

$$S_n = 2 - \frac{1}{2^n}.$$

This explicit formula for S_n lets us answer the question posed above: Because $S_n \to 2$ as $n \to \infty$, the series converges to 2.

$R_n = 2 - S_n$

A numerical table of partial sums and tails supports this conclusion: ◂

Partial sums and tails of $1 + \frac{1}{2} + \frac{1}{4} + \cdots$										
n	0	1	2	3	4	5	6	7	8	9	10
S_n	1	1.5	1.75	1.875	1.938	1.969	1.984	1.992	1.996	1.998	1.999
R_n	1	0.5	0.25	0.125	0.063	0.031	0.016	0.008	0.004	0.002	0.001

As the numbers show, the partial sums S_n converge to 2 while the tails R_n tend to 0. ∎

EXAMPLE 2 Does $\displaystyle\sum_{k=1}^{\infty} (-1)^k = -1 + 1 - 1 + 1 - \cdots$ converge?

Solution No. Successive partial sums are $-1, 0, -1, 0, \ldots$. Thus, the sequence $\{S_n\}$ diverges and so does the series. ∎

E X A M P L E 3   Does the **harmonic series** $\sum_{k=1}^{\infty} \frac{1}{k}$ converge? ➡

The harmonic series will be one of our most important examples.

S o l u t i o n   The answer depends on the partial sums S_n. By definition,

$$S_n = 1 + \frac{1}{2} + \frac{1}{3} + \frac{1}{4} + \frac{1}{5} + \frac{1}{6} + \cdots + \frac{1}{n}.$$

Because no simple formula for S_n comes to mind (in fact, there *is* no such simple formula), we will investigate the partial sums numerically. Here are some results, computed to three decimal places:

$$S_1 = 1; \quad S_{20} = 3.598; \quad S_{50} = 4.499; \quad S_{100} = 5.187; \quad S_{1000} = 7.485.$$

The numerical evidence is ambiguous; the S_n's seem to keep growing, although slowly. Whether the sequence of partial sums $\{S_n\}$ converges or diverges is not yet clear. ➡ ■

In fact, the series diverges. We'll show this fact soon by comparing the series to an integral.

When does a series (or an integral) converge?

Whether a given infinite series converges or diverges is a delicate question. ➡ Such series as

So was the analogous question for improper integrals

$$\sum_{k=1}^{\infty} \frac{1}{k} = 1 + \frac{1}{2} + \frac{1}{3} + \cdots \quad \text{and} \quad \sum_{k=1}^{\infty} \frac{1}{k^2} = 1 + \frac{1}{4} + \frac{1}{9} + \cdots$$

pose the same puzzle: Although successive terms of both series tend to zero, the number of terms is infinite. Convergence or divergence hinges on which of these conflicting tendencies "wins" in the long run.

The same dilemma arose for the improper *integrals*

$$\int_1^{\infty} \frac{1}{x}\, dx \quad \text{and} \quad \int_1^{\infty} \frac{1}{x^2}\, dx,$$

and it is worth recalling exactly what happened. In each case, the "partial integral" $I(t)$ is easily found. For the first integral, we found

$$I(t) = \int_1^t \frac{dx}{x} = \ln t \to \infty.$$

For the second,

$$I(t) = \int_1^t \frac{dx}{x^2} = 1 - \frac{1}{t} \to 1.$$

Thus, the two similar-looking integrals led to opposite results: The first integral diverges to infinity, and the second converges to 1. Deciding whether the corresponding series converge or diverge is a bit harder because no convenient formulas for S_n are available.

Convergence and divergence: graphical views   For a series $\sum a_k$, plotting both the terms $\{a_k\}$ and the partial sums $\{S_n\}$ on the same axes illustrates the connection between the two—and sometimes suggests whether the series converges or diverges. The two series $\sum 1/k$ and $\sum 1/k^2$ generate Figures 1(a) and 1(b), respectively.

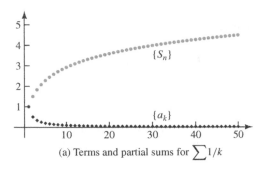 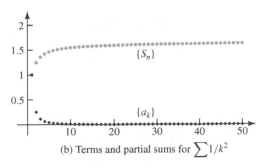

(a) Terms and partial sums for $\sum 1/k$ (b) Terms and partial sums for $\sum 1/k^2$

FIGURE 1

Terms and partial sums for two series

The two pictures give different impressions. The pictured behavior of partial sums suggests (but doesn't prove) that the first series diverges; the second series appears to converge because the partial sums seem to level off, perhaps approaching a limit near 1.7.

Geometric series: the nicest kind

The most important family of infinite series are the **geometric series**. They have the form

$$\sum_{k=0}^{\infty} ar^k = a + ar + ar^2 + ar^3 + ar^4 + \cdots;$$

here a is called the **leading term**, and r is the **ratio** because each term is r times the preceding term. An especially simple example is

$$\sum_{k=0}^{\infty} \left(\frac{1}{2}\right)^k = 1 + \frac{1}{2} + \frac{1}{4} + \frac{1}{8} + \cdots$$

with $a = 1$ and $r = 1/2$. We saw in Example 1, page 558, that this series converges to 2.

Partial sums of geometric series   Geometric series have a great advantage: It is easy to tell whether they converge and, if so, to find their sums. This is because geometric series—unlike many others—have a simple, explicit formula for the partial sums S_n. The formula depends on a beautiful algebraic fact. If $r \neq 1$ and $n \geq 0$, then

$$1 + r + r^2 + r^3 + \cdots + r^n = \frac{1 - r^{n+1}}{1 - r}. \tag{4}$$

Multiplying Equation (4) by a gives a formula for S_n:

$$S_n = a + ar + ar^2 + ar^3 + \cdots + ar^n = a\frac{1 - r^{n+1}}{1 - r}. \tag{5}$$

From Formula (5) follows the whole story of convergence and divergence for geometric series. As always, the question is how the sequence $\{S_n\}$ of partial sums behaves. In this case, the issue has to do entirely with the power r^{n+1}. We will need the following facts:

$$\lim_{m \to \infty} r^m = \begin{cases} 0 & \text{if } |r| < 1 \\ 1 & \text{if } r = 1 \\ \text{does not exist} & \text{if } |r| > 1 \text{ or } r = -1. \end{cases}$$

The conclusion for geometric series follows, and it's worth emphasizing:

THEOREM 4 (Convergence and divergence of geometric series) If $|r| < 1$, the geometric series

$$\sum_{k=0}^{\infty} ar^k = a + ar + ar^2 + ar^3 + \cdots$$

converges to $\dfrac{a}{1-r}$. If $a \neq 0$ and $|r| \geq 1$, the series diverges.

EXAMPLE 4   The series $\dfrac{1}{3} - \dfrac{1}{6} + \dfrac{1}{12} - \dfrac{1}{24} + \cdots$ converges. To what limit?

Solution   The series is geometric with $a = 1/3$ and $r = -1/2$. By Theorem 4, it converges to

$$\frac{a}{1-r} = \frac{1/3}{1+1/2} = \frac{2}{9} \approx 0.2222222.$$

A numerical look at partial sums and tails supports this computation. ➤

$R_n = \dfrac{2}{9} - S_n.$

Partial sums and tails of $\dfrac{1}{3} - \dfrac{1}{6} + \dfrac{1}{12} - \dfrac{1}{24} + \cdots$							
n	0	1	2	3	4	5	...	10
S_n	0.3333	0.1667	0.2500	0.2083	0.2292	0.2188	...	0.2223
R_n	−0.1111	0.0555	−0.0278	0.0139	−0.0069	0.0035	...	−0.0001

Telescoping series   Most series *other* than the geometric variety do not admit explicit formulas for partial sums. Telescoping series are among the unusual (and pleasant) exceptions. The next example illustrates how telescoping series work and explains the name.

EXAMPLE 5   Show that $\displaystyle\sum_{k=1}^{\infty} \frac{1}{k(k+1)}$ converges, and find its limit.

Solution   A little algebra inside the summation sign lets us rewrite the series in a more convenient form: ➤

$$\sum_{k=1}^{\infty} \frac{1}{k(k+1)} = \sum_{k=1}^{\infty} \left(\frac{1}{k} - \frac{1}{k+1} \right).$$

To convince yourself that the two sides are equal, find a common denominator on the right side.

Writing out some terms shows the "telescoping" pattern: ➤

$$S_n = \sum_{k=1}^{n} \left(\frac{1}{k} - \frac{1}{k+1} \right) = \left(\frac{1}{1} - \frac{1}{2} \right) + \left(\frac{1}{2} - \frac{1}{3} \right) + \left(\frac{1}{3} - \frac{1}{4} \right) + \cdots + \left(\frac{1}{n} - \frac{1}{n+1} \right)$$

$$= 1 - \frac{1}{n+1}.$$

The summands collapse like an old-fashioned spyglass.

Now it's clear that, as $n \to \infty$, $S_n \to 1$; that's the sum. Again the numbers agree: ➤

Watch the tails go to zero.

Partial sums and tails of $\displaystyle\sum_{k=1}^{\infty} \dfrac{1}{k^2+k}$							
n	1	2	3	4	5	6	...	11
S_n	0.5000	0.6667	0.7500	0.8000	0.8333	0.8571	...	0.9167
R_n	0.5000	0.3333	0.2500	0.2000	0.1667	0.1429	...	0.0833

Algebra with series

As with functions and sequences, combining series algebraically produces new series. Combining convergent series produces new convergent series, with limits related in the expected way.

THEOREM 5   Suppose that $\sum a_k$ converges to S and that $\sum b_k$ converges to T. Let c be any constant. Then

$$\sum_{k=1}^{\infty} (a_k + b_k) \quad \text{converges to } S + T,$$

and

$$\sum c a_k \quad \text{converges to } cS.$$

In short:

$$\sum (a_k + b_k) = \sum a_k + \sum b_k \quad \text{and} \quad \sum c a_k = c \sum a_k.$$

After all, the limit of a series is defined as the limit of the sequence of partial sums.

These reasonable-looking properties of convergent *series* follow directly from the analogous properties of convergent *sequences*. ◂

A little series algebra, cleverly applied, can immensely simplify finding the limits of certain series.

EXAMPLE 6   Evaluate $\displaystyle\sum_{k=0}^{\infty} \frac{4 + 2^k}{3^k}$.

Solution   The series is the sum of two (convergent) geometric series. Applying Theorems 4 and 5 gives

$$\sum_{k=0}^{\infty} \frac{4 + 2^k}{3^k} = \sum_{k=0}^{\infty} \frac{4}{3^k} + \sum_{k=0}^{\infty} \left(\frac{2}{3}\right)^k = 4 \cdot \frac{3}{2} + 3 = 9.$$ ∎

EXAMPLE 7   Calculate the tail R_{10} for the geometric series $\displaystyle\sum_{k=0}^{\infty} \frac{3}{2^k}$.

Solution   A little algebra is all we need:

$$R_{10} = \sum_{k=11}^{\infty} \frac{3}{2^k} = \frac{3}{2^{11}} + \frac{3}{2^{12}} + \frac{3}{2^{13}} + \cdots$$

$$= \frac{3}{2^{11}} \left(1 + \frac{1}{2} + \frac{1}{2^2} + \frac{1}{2^3} + \cdots\right) \quad \text{(factoring out the constant)}$$

$$= \frac{3}{2^{11}} \cdot 2 = \frac{3}{2^{10}} = \frac{3}{1024}. \quad \text{(summing the geometric series)}$$ ∎

Detecting divergent series

We are mainly interested in convergent series and their limits. Theorem 5, for instance, applies only to convergent series, and so it is important to recognize divergence when it occurs. One strategy is to use Theorem 5 indirectly.

EXAMPLE 8 Suppose that $\sum a_k$ diverges, and let $c \neq 0$ be a constant. Explain why $\sum c\,a_k$ diverges, too.

Solution If $\sum c\,a_k$ were convergent, then, by Theorem 5, the series

$$\sum \frac{1}{c}\,c\,a_k = \sum a_k$$

must also converge, which contradicts our assumption. ∎

The *n*th term test The following theorem describes another useful divergence detector:

> **THEOREM 6 (The *n*th term test for divergence)**   If $\lim\limits_{n \to \infty} a_n \neq 0$, then $\sum a_n$ diverges.

The theorem holds because, for a series $\sum a_k$ to converge, the partial sums S_n must converge to a limit. For this to occur the difference $S_n - S_{n-1}$ between *successive* partial sums must tend to zero. ➡ But this difference is just the nth term:

Otherwise the partial sum sequence wouldn't "level off."

$$S_n - S_{n-1} = (a_1 + a_2 + \cdots + a_{n-1} + a_n) - (a_1 + a_2 + \cdots + a_{n-1}) = a_n.$$

Thus, the terms of a convergent series must tend to zero, as the theorem says.

What the theorem does not say It is especially important to notice that the nth term test does *not* guarantee that $\sum a_n$ converges whenever $a_n \to 0$. For example, the harmonic series $\sum 1/k$ "passes" the nth term test, but it diverges. ➡

We've said this several times, and we'll show it rigorously in the next section.

In practice, the nth term test is an effective but rather blunt instrument—it sometimes detects divergence, but it *never* detects convergence. We will develop tools for that purpose in the next section.

EXAMPLE 9 Does $\displaystyle\sum_{k=1}^{\infty} \frac{k}{k + 1000}$ converge?

Solution No. Because $a_k = \dfrac{k}{k + 1000} \to 1$ as $k \to \infty$, the series diverges. ∎

EXAMPLE 10 Assuming that $\displaystyle\sum_{k=0}^{\infty} \frac{3^k}{k!}$ converges (as it really does), find $\lim\limits_{k \to \infty} \dfrac{3^k}{k!}$.

Solution The limit is zero: By Theorem 6, the terms of *every* convergent series must tend to zero. ∎

BASIC EXERCISES

1. Consider the series $\sum_{k=0}^{\infty} a_k = \sum_{k=0}^{\infty} \frac{1}{5^k}$.

   (a) Evaluate $a_1, a_2, a_5, a_{10}, S_1, S_2, S_5$, and S_{10}.

   (b) Show that the sequence $\{a_k\}$ is decreasing and bounded below.

   (c) Show that the sequence $\{S_n\}$ is increasing and bounded above. What does this imply about the sequence of partial sums?

   (d) Find the sum of the series (i.e., $\lim_{n \to \infty} S_n$).

   (e) Evaluate R_1, R_2, R_5, and R_{10}.

   (f) Show that $\{R_n\}$ is decreasing and bounded below.

   (g) Evaluate $\lim_{n \to \infty} R_n$.

2. Consider the series $\sum_{k=0}^{\infty} a_k = \sum_{k=0}^{\infty} (-0.8)^k$.

   (a) Evaluate $a_1, a_2, a_5, a_{10}, S_1, S_2, S_5$, and S_{10}.

   (b) Find the sum of the series (i.e., $\lim_{n \to \infty} S_n$).

   (c) Evaluate R_1, R_2, R_5, and R_{10}.

   (d) Is the sequence $\{a_k\}$ decreasing? Is it bounded?

   (e) Is the sequence $\{S_n\}$ increasing? Justify your answer.

   (f) Show that the sequence $\{R_n\}$ is neither increasing nor decreasing.

   (g) Show that the sequence $\{|R_n|\}$ is decreasing.

   (h) Evaluate $\lim_{n \to \infty} R_n$.

3. Consider the series $\sum_{k=0}^{\infty} a_k = \sum_{k=0}^{\infty} \frac{1}{k + 2^k}$.

   (a) Show that the sequence $\{a_k\}$ is decreasing.

   (b) Explain why $a_k \le 2^{-k}$ for all $k \ge 0$.

   (c) Show that the sequence $\{S_n\}$ is increasing.

   (d) Use part (b) to show that $S_n \le 2 - 2^{-n} < 2$. [HINT: $\sum_{k=0}^{n} 2^{-k}$ is a geometric series.]

   (e) Show that $\sum_{k=0}^{\infty} a_k$ converges.

4. Consider the series $\sum_{j=0}^{\infty} a_j = \sum_{j=0}^{\infty} \frac{1}{2 + 3^j}$.

   (a) Evaluate S_1, S_2, S_5, and S_{10}.

   (b) Show that the sequence $\{S_n\}$ is increasing and bounded above.

   (c) Does $\sum_{j=0}^{\infty} a_j$ converge? Justify your answer.

5. Consider the series $\sum_{k=0}^{\infty} \frac{1}{k!}$.

   (a) Explain why $\frac{1}{k!} \le \frac{1}{2^{k-1}}$ if $k \ge 1$. [HINT: Explain why $1/3! < 1/(2 \cdot 2)$ and $1/4! < 1/(2 \cdot 2 \cdot 2)$.]

   (b) Show that the sequence of partial sums $\{S_n\}$ is increasing.

   (c) Use parts (a) and (b) to show that the series converges.

6. The series $\sum_{k=0}^{\infty} a_k = \sum_{k=0}^{\infty} \frac{1}{k!}$ converges to $e \approx 2.718282$.

   (a) Evaluate $a_1, a_2, a_5, a_{10}, S_1, S_2, S_5$, and S_{10}.

   (b) Show that $\{a_k\}$ is a decreasing sequence.

   (c) Show that $\{S_n\}$ is an increasing sequence.

   (d) Show that $R_n > 0$ for all $n \ge 0$.

   (e) Show that $\{R_n\}$ is a decreasing sequence.

   (f) Find a value of n for which S_n differs from e by less than 0.001.

   (g) Find a value of n for which S_n differs from e by less than 10^{-5}.

   (h) Use parts (e) and (g) to show that $R_{50} < 10^{-5}$.

In Exercises 7 and 8, use the fact that $\sum_{m=1}^{\infty} \frac{1}{m^4} = \frac{\pi^4}{90}$ to evaluate the series.

7. $\sum_{i=0}^{\infty} \frac{1}{(i+1)^4}$ 8. $\sum_{k=3}^{\infty} \frac{1}{k^4}$

9. This exercise is about partial sums of the geometric series $\sum_{k=0}^{\infty} ar^k$.

   (a) Find a formula for S_n when $r = 1$. Explain why the formula for S_n (Equation 5) does not hold in this case.

   (b) Show that $S_n - rS_n = (1 - r)S_n$ for any r.

   (c) Use part (b) to show that Equation 5 holds if $r \ne 1$.

10. Use Equation 5 to evaluate $3 + 6 + 12 + 24 + 48 + 96 + \cdots + 3072$. [HINT: $3072 = 3 \cdot 2^{10}$.]

In Exercises 11–18, find the limit of the series.

11. $\frac{1}{16} + \frac{1}{32} + \frac{1}{64} + \frac{1}{128} + \cdots + \frac{1}{2^{i+4}} + \cdots$

12. $2 - 5 + 9 + \frac{1}{3} + \frac{1}{9} + \frac{1}{27} + \frac{1}{81} + \cdots + \frac{1}{3^n} + \cdots$

13. $\sum_{n=0}^{\infty} e^{-n}$ 14. $\sum_{k=3}^{\infty} \left(\frac{e}{\pi}\right)^k$

15. $\sum_{m=2}^{\infty} (\arctan 1)^m$ 16. $\sum_{i=10}^{\infty} \left(\frac{2}{3}\right)^i$

17. $\sum_{j=5}^{\infty} \left(-\frac{1}{2}\right)^j$ 18. $\sum_{j=0}^{\infty} \frac{3^j + 4^j}{5^j}$

19. Show that $\sum_{k=1}^{\infty} \frac{1}{2 + \sin k}$ diverges.

20. Does $\sum_{k=0}^{\infty} (-1)^k$ converge? Justify your answer.

In Exercises 21–24, find an expression for the partial sum S_n of the series. Use this expression to determine whether the series converges and, if so, to find its limit.

21. $\sum_{k=0}^{\infty} (\arctan(k+1) - \arctan k)$

22. $\displaystyle\sum_{j=1}^{\infty} \frac{j}{(j+1)!}$

23. $\displaystyle\sum_{m=1}^{\infty} \left(\frac{1}{\sqrt{m}} - \frac{1}{\sqrt{m+2}} \right)$

24. $\displaystyle\sum_{j=1}^{\infty} \ln\left(1 + \frac{1}{j}\right)$ [HINT: $\displaystyle\sum_{j=1}^{\infty}(\ln(j+1) - \ln j)$.]

25. Suppose that the partial sums of the series $\displaystyle\sum_{k=1}^{\infty} a_k$ are
$$S_n = \sum_{k=1}^{n} a_k = 5 - \frac{3}{n}.$$

  (a) Evaluate $\displaystyle S_{100} = \sum_{k=1}^{100} a_k$.

  (b) Evaluate $\displaystyle\sum_{k=1}^{\infty} a_k$.

(c) Evaluate $\displaystyle\lim_{k\to\infty} a_k$.

(d) Show that $a_k > 0$ for all $k \geq 1$. [HINT: $a_{n+1} = S_{n+1} - S_n$.]

26. Let $\displaystyle H_n = \sum_{k=1}^{n} \frac{1}{k}$ and let $\displaystyle S_n = \sum_{k=0}^{n} \frac{1}{2k+1}$.

  (a) Explain why $\displaystyle\lim_{n\to\infty} H_n = \infty$.

  (b) Show that $S_n \geq \frac{1}{2} H_n$.

  (c) What do the results in parts (a) and (b) imply about $\displaystyle\sum_{k=0}^{\infty} 1/(2k+1)$? Justify your answer.

FURTHER EXERCISES

In Exercises 27–29, use the fact that $\displaystyle\sum_{i=1}^{\infty} \frac{1}{i^2} = 1 + \frac{1}{4} + \frac{1}{9} + \frac{1}{16} + \frac{1}{25} + \cdots = \frac{\pi^2}{6}$ *to find the limit of the series.*

27. $\displaystyle\sum_{j=1}^{\infty} \frac{1}{(2j)^2}$

28. $\displaystyle\sum_{k=0}^{\infty} \frac{1}{(2k+1)^2}$ [HINT: See the previous exercise.]

29. $\displaystyle\sum_{m=1}^{\infty} \frac{(-1)^{m+1}}{m^2} = 1 - \frac{1}{4} + \frac{1}{9} - \frac{1}{16} + \frac{1}{25} - \frac{1}{36} + \cdots$

30. Express $\displaystyle\sum_{m=3}^{\infty} \frac{2^{m+4}}{5^m}$ as a rational number.

For each of the series in Exercises 31–36, find all values of x for which the series converges; then state the limit as a simple expression involving x. (Assume that $x^0 = 1$ for all x.)

31. $\displaystyle\sum_{k=0}^{\infty} x^k$

32. $\displaystyle\sum_{m=2}^{\infty} \left(\frac{x}{5}\right)^m$

33. $\displaystyle\sum_{j=5}^{\infty} x^{2j}$

34. $\displaystyle\sum_{k=1}^{\infty} x^{-k}$

35. $\displaystyle\sum_{n=3}^{\infty} (1+x)^n$

36. $\displaystyle\sum_{j=4}^{\infty} \frac{1}{(1-x)^j}$

37. Find the limit of the sequence defined by $S_1 = 1$, $S_{n+1} = S_n + 1/3^n$. [HINT: Write out the first few terms to see the pattern.]

38. Find the limit of the sequence defined by $a_1 = 4$, $a_{n+1} = a_n - 1/2^n$.

In Exercises 39–52, determine whether the series converges or diverges. If a series converges, find its limit. Justify your answers.

39. $\displaystyle\sum_{n=0}^{\infty} \frac{n+1}{2n+1}$

40. $\displaystyle\sum_{j=0}^{\infty} (\ln 2)^j$

41. $\displaystyle\sum_{n=2}^{\infty} \frac{2}{n^2 - 1}$

42. $\displaystyle\sum_{n=1}^{\infty} \left(1 + \frac{1}{n}\right)^n$

43. $\displaystyle\sum_{j=1}^{\infty} \sqrt[j]{\pi}$

44. $\displaystyle\sum_{k=1}^{\infty} \frac{1}{\ln(10^k)}$

45. $\displaystyle\sum_{j=2}^{\infty} \frac{3^j}{4^{j+1}}$

46. $1 - \dfrac{1}{2} - \dfrac{1}{3} - \dfrac{1}{4} - \dfrac{1}{5} - \cdots$

47. $\dfrac{1}{100} + \dfrac{1}{200} + \dfrac{1}{300} + \cdots$

48. $2 - 2 + 2 - 2 + 2 - 2 + \cdots$

49. $\dfrac{3}{10} - \dfrac{3}{20} + \dfrac{3}{40} - \dfrac{3}{80} + \dfrac{3}{160} - \dfrac{3}{320} + \cdots$

50. $1 - \dfrac{1}{2} + \dfrac{1}{2} - \dfrac{1}{3} + \dfrac{1}{3} - \cdots$

51. $1 - 1 + 2 - 1 - 1 + 3 - 1 - 1 - 1 + 4 - 1 - 1 - 1 - 1 + \cdots$

52. $\dfrac{4}{7^{10}} + \dfrac{4}{7^{12}} + \dfrac{4}{7^{14}} + \dfrac{4}{7^{16}} + \dfrac{4}{7^{18}} + \cdots$

53. A rubber ball rebounds to two-thirds the height from which it falls. If it is dropped from a height of 4 feet and is allowed to continue bouncing indefinitely, what is the total distance it travels?

54. Let $\displaystyle S_n = \sum_{k=1}^{n} \frac{1}{\sqrt{k}}$.

  (a) Evaluate $\displaystyle\lim_{k\to\infty} \frac{1}{\sqrt{k}}$.

  (b) Show that $S_n \geq \dfrac{n}{\sqrt{n}} = \sqrt{n}$ for all $n \geq 1$. [HINT: If $k \leq n$, then $1/k \geq 1/n$.]

  (c) Use part (b) to show that $\displaystyle\sum_{k=1}^{\infty} \frac{1}{\sqrt{k}}$ diverges.

55. Use the previous exercise and the fact that $\ln x \leq \sqrt{x}$ for all $x \geq 1$ to show that $\sum_{k=2}^{\infty} \frac{1}{\ln k}$ diverges.

56. Let $\{a_k\}$ be an increasing sequence such that $a_1 > 0$ and $a_k \leq 100$ for all $k \geq 1$.

(a) Does $\lim_{k \to \infty} a_k$ exist? Justify your answer.

(b) Show that $\sum_{k=1}^{\infty} a_k$ diverges.

57. Let $\{a_k\}$ be a sequence of positive terms such that $\sum_{k=1}^{n} a_k \leq 100$ for all $n \geq 1$. Explain why $\lim_{k \to \infty} a_k = 0$ must be true.

58. Suppose that the partial sums of the series $\sum_{j=1}^{\infty} b_j$ are
$$S_n = \sum_{j=1}^{n} b_j = \ln\left(\frac{2n+3}{n+1}\right).$$

(a) Evaluate $\lim_{n \to \infty} S_n$.

(b) Does the series converge? Justify your answer.

(c) Show that $b_j < 0$ for all $j \geq 1$.

59. Suppose that the partial sums of the series $\sum_{k=1}^{\infty} a_k$ satisfy the inequality $\frac{6 \ln n}{\ln(n^2+1)} < S_n < 3 + ne^{-n}$ for all $n \geq 100$.

(a) Does the series converge? If so, to what limit? Justify your answers.

(b) What, if anything, can be said about $\lim_{k \to \infty} a_k$? Explain.

60. Let $S_n = \sum_{k=1}^{n} a_k$, and suppose that $0 \leq S_n \leq 100$ for all $n \geq 1$.

(a) Give an example of a sequence $\{a_k\}$ that satisfies these conditions but $\sum_{k=1}^{\infty} a_k$ diverges.

(b) Show that if $a_k > 0$ for all $k \geq 1$, then $\sum_{k=1}^{\infty} a_k$ converges.

(c) Show that if $a_k > 0$ for all $k \geq 10^6$, then $\sum_{k=1}^{\infty} a_k$ converges.

61. Suppose that $\sum_{k=1}^{\infty} a_k$ diverges.

(a) Explain why $a_k > 0$ for all $k \geq 1$ implies that $\lim_{n \to \infty} S_n = \infty$.

(b) Give an example of a divergent series for which $\lim_{n \to \infty} S_n$ does not exist.

62. Let $H_n = \sum_{k=1}^{n} \frac{1}{k}$, and let $I_n = \int_1^{n+1} \frac{dx}{x}$.

(a) Let L_n be the left Riemann-sum approximation, with n equal subdivisions, to I_n. Show that $L_n = H_n$.

(b) Use part (a) to show that the harmonic series diverges. [HINT: Start by comparing L_n and I_n.]

63. Let $H_n = \sum_{k=1}^{n} \frac{1}{k}$, and let $a_m = \sum_{j=1}^{2^{m-1}} \frac{1}{2^{m-1}+j}$. Then $H_{2^n} = 1 + \sum_{m=1}^{n} a_m$. [NOTE: a_m is the sum of a "block" of 2^{m-1} consecutive terms of the harmonic series—those from $n = 2^{m-1}+1$ through $n = 2^m$.]

(a) Show that $a_1 = 1/2$, $a_2 = 7/12$, and $a_3 = 533/840$.

(b) Show that $H_8 = 1 + a_1 + a_2 + a_3 = 761/280$.

(c) Show that $a_k \geq 1/2$ for all $k \geq 1$.
[HINT: $\frac{1}{2^{m-1}+j} \geq \frac{1}{2^{m-1}+2^{m-1}}$ if $1 \leq j \leq 2^{m-1}$.]

(d) Use part (c) to show that $\lim_{n \to \infty} H_n = \infty$ (i.e., the harmonic series diverges).

64. Consider the series $\sum_{k=1}^{\infty} \frac{1}{k^p}$ with $p > 1$. This exercise outlines a proof that this series converges.

(a) Let $S_n = \sum_{k=1}^{n} \frac{1}{k^p}$. Show that the sequence of partial sums $\{S_n\}$ is increasing.

(b) Show that $S_{2m+1} = 1 + \sum_{k=1}^{m} \frac{1}{(2k)^p} + \sum_{k=1}^{m} \frac{1}{(2k+1)^p}$.

(c) Explain why $S_{2m+1} < 1 + 2 \sum_{k=1}^{m} \frac{1}{(2k)^p}$.
[HINT: $1/(x+1) < 1/x$ if $x > 0$.]

(d) Show that $S_{2m+1} < 1 + 2^{1-p} S_{2m+1}$.
[HINT: First show that $S_{2m+1} < 1 + 2^{1-p} S_m$.]

(e) Show that $\{S_n\}$ is bounded above.

11.3 TESTING FOR CONVERGENCE; ESTIMATING LIMITS

In theory, the question of convergence is simple: The series $\sum a_k$ converges to the sum S if the sequence $\{S_n\}$ of partial sums tends to S. The trouble, in practice, is that a simple, explicit formula for S_n is often unavailable. It might seem, then, that testing for convergence—let alone finding a limit—would be difficult or impossible. Surprisingly, that isn't so. All the convergence tests of this section and the next (comparison test, integral test, ratio test, and so on) offer clever, indirect ways of testing whether $\{S_n\}$ converges.

Nonnegative series   A series $\sum a_k$ for which $a_k \geq 0$ for all k is called **nonnegative**. For such a series, the sequence $\{S_n\}$ of partial sums is nondecreasing: ➡

Convince yourself; it's easy but important.

$$S_1 \leq S_2 \leq S_3 \leq \cdots \leq S_n \leq S_{n+1} \leq \ldots .$$

Nondecreasing sequences are easier to study than others for one main reason: A nondecreasing sequence either (i) converges or (ii) blows up to infinity. ➡ (An arbitrary sequence can diverge without blowing up.) This means, in practice, that the question of convergence or divergence for a *nonnegative* series $\sum a_k$ boils down to the following simpler question:

Theorem 3, page 553, says so.

> *Do the partial sums S_n blow up or remain bounded as $n \to \infty$?*

The answer may be obvious at a glance if a simple formula for S_n is available. ➡ If not, our best strategy may be to *compare* the given series to something that is better understood (another series, an integral—whatever works). Sometimes an obvious comparison suggests itself.

We have such a formula for geometric series.

> **EXAMPLE 1**   Does $\displaystyle\sum_{k=0}^{\infty} \frac{1}{2^k + 1}$ converge?
>
> **Solution**   We saw in the preceding section that the geometric series $\sum_{k=0}^{\infty} 1/2^k$ converges to 2. For all k, it is clear that
>
> $$\frac{1}{2^k + 1} < \frac{1}{2^k} ,$$
>
> and so each partial sum of $\sum_{k=0}^{\infty} 1/(2^k + 1)$ is less than the corresponding partial sum of $\sum_{k=0}^{\infty} 1/2^k$. ➡ Therefore, the partial sums of the original series cannot blow up; they must tend to some limit less than 2. Numerical evidence agrees. For the original series, calculation gives
>
> $$S_{10} \approx 1.263523536; \quad S_{20} \approx 1.264498827; \quad S_{100} \approx 1.264499780. \quad \blacksquare$$

Adding smaller summands gives a smaller sum.

The comparison test: one series vs. another

In Example 1 we showed that one series converges by comparing it to another series that is *known* to converge. The following theorem makes this idea precise.

> **THEOREM 7 (Comparison test for nonnegative series)**   Consider two series $\sum a_k$ and $\sum b_k$, with $0 \leq a_k \leq b_k$ for all k.
> - If $\sum b_k$ *converges*, so does $\sum a_k$, and $\sum a_k \leq \sum b_k$.
> - If $\sum a_k$ *diverges*, so does $\sum b_k$.

Observe:

- **What we'd expect**   The theorem's assertions should seem reasonable. A formal proof uses the fact that every partial sum of $\sum a_k$ is less than the corresponding partial sum of $\sum b_k$.

Sequences are "discrete" versions of functions; series are "discrete" versions of integrals.

- **Comparing integrals, comparing series** An almost identical comparison theorem applies to improper integrals. ← Theorem 1, Section 10.2, says that if $0 \le a(x) \le b(x)$ for all $x \ge 1$, then

$$\int_1^\infty a(x)\,dx \le \int_1^\infty b(x)\,dx.$$

We did this explicitly in Example 1.

- **Successful comparisons** In a "successful" comparison, either both series converge or both diverge. To use the comparison test, therefore, it's necessary first to *guess* whether the series in question converges or diverges. ←

Comparing tails Theorem 7 says that if $0 \le a_k \le b_k$ for all $k \ge 1$ and $\sum b_k$ converges, then $\sum a_k$ converges too, and

$$\sum_{k=1}^\infty a_k \le \sum_{k=1}^\infty b_k.$$

A similar inequality holds for upper tails of convergent series. If N is any positive integer, $0 \le a_k \le b_k$ for all $k \ge N$, and $\sum b_k$ converges, then $\sum a_k$ converges too, and

$$\sum_{k=N+1}^\infty a_k \le \sum_{k=N+1}^\infty b_k.$$

We will "compare tails" in the following example to estimate the numerical value of a limit.

EXAMPLE 2 We have seen (using comparison) that $\sum_{k=0}^\infty \dfrac{1}{2^k+1}$ converges, and a calculator gives $S_{100} \approx 1.264499781$. How closely does S_{100} approximate the *true* sum S?

Solution The error in using S_{100} to estimate the limit comes from ignoring the tail R_{100}. In symbols,

$$S = \sum_{k=0}^{100} a_k + \sum_{k=101}^\infty a_k = S_{100} + R_{100}.$$

Clearly, $\dfrac{1}{2^k+1} < \dfrac{1}{2^k}$ for all k; in particular, the inequality holds for $k \ge 101$. Now we compare tails:

$$R_{100} = \sum_{k=101}^\infty \frac{1}{2^k+1} < \sum_{k=101}^\infty \frac{1}{2^k}.$$

The comparison is worthwhile because the right side is a *geometric* series, which we can calculate exactly:

$$\sum_{k=101}^\infty \frac{1}{2^k} = \frac{1}{2^{101}} + \frac{1}{2^{102}} + \frac{1}{2^{103}} + \cdots = \frac{1}{2^{101}}\left(1 + \frac{1}{2} + \frac{1}{2^2} + \cdots\right) = \frac{1}{2^{101}} \cdot 2 = \frac{1}{2^{100}}.$$

This shows that we commit *very* little error by ignoring the tail. The error, R_{100}, is less than $1/2^{100} \approx 8 \times 10^{-31}$. ∎

The integral test: series vs. integrals

The *idea* of comparison is easy. Harder, in practice, is deciding what to compare a series *to*. Our only reliable "benchmark" series, so far, are geometric series. The **integral test** enlarges our stock of benchmark series considerably.

Comparing areas Thinking of the terms of a nonnegative *series* $\sum_{k=1}^{\infty} a_k$ as rectangular areas, each with base 1, helps clarify the close connection with the *integral* $\int_1^{\infty} a(x)\,dx$. Figure 1 shows two ways of doing so.

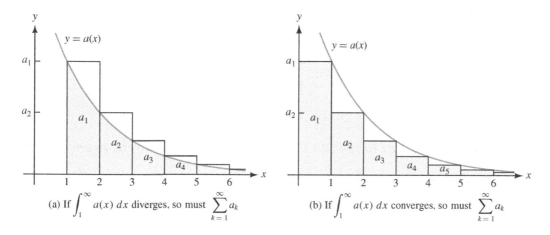

(a) If $\displaystyle\int_1^{\infty} a(x)\,dx$ diverges, so must $\displaystyle\sum_{k=1}^{\infty} a_k$ (b) If $\displaystyle\int_1^{\infty} a(x)\,dx$ converges, so must $\displaystyle\sum_{k=1}^{\infty} a_k$

FIGURE 1
Relating improper integrals to infinite series

Look closely:

- **Total areas** The successive rectangles have heights $a(1) = a_1$, $a(2) = a_2$, $a(3) = a_3$, and so on; each *base* is 1. The respective areas, therefore, are a_1, a_2, a_3, and so on. Thus,

    The series $a_1 + a_2 + a_3 + a_4 + \cdots$ represents the total "left-rule" rectangular area from 1 to ∞.

- **An important inequality** The shaded area in Figure 1(a) represents the integral $\int_1^{\infty} a(x)\,dx$. Thus, if the integral diverges, so must the series. Here is the message of Figure 1(a) in inequality form:

$$a_1 + a_2 + a_3 + \cdots \geq \int_1^{\infty} a(x)\,dx. \tag{1}$$

    If the right side diverges to infinity, then so must the left.

- **A decreasing integrand** The reasoning that led to Inequality (1) requires that $a(x)$ be *decreasing* for $x \geq 1$, as shown in the pictures. (We collect such technical hypotheses carefully in the following theorem.)

- **Bounding a series from above**   Figure 1(b) shows an integral bounding a series from *above*. This time, comparing areas gives

$$a_2 + a_3 + \cdots \leq \int_1^\infty a(x)\, dx,$$

or, equivalently,

$$a_1 + a_2 + a_3 + \cdots \leq a_1 + \int_1^\infty a(x)\, dx. \tag{2}$$

If the right side converges, so must the left.

Combining Inequalities (1) and (2) gives upper and lower bounds for the series:

$$\int_1^\infty a(x)\, dx \; \leq \; \sum_{k=1}^\infty a_k \; \leq a_1 + \int_1^\infty a(x)\, dx. \tag{3}$$

In particular:

The integral $\int_1^\infty a(x)\, dx$ and the series $\sum_{k=1}^\infty a_k$ either both converge or both diverge.

The final picture, Figure 2, relates the tails of an integral and of a series.

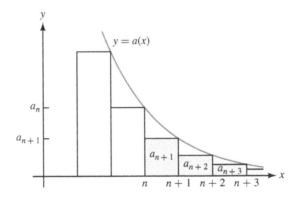

FIGURE 2

Comparing tails: Why $\displaystyle\sum_{k=n+1}^\infty a_k \leq \int_n^\infty a(x)\, dx$

Comparing areas in Figure 2 shows that, for all n, the tail R_n satisfies

$$R_n = \sum_{k=n+1}^\infty a_k \leq \int_n^\infty a(x)\, dx.$$

Lessons from the pictures   Figures 1 and 2 show a particular function $a(x)$, but the conclusions we drew hold for *any* function $a(x)$ that is both positive and decreasing. It is time to collect our observations in a theorem.

> **THEOREM 8 (Integral test for positive series)** Suppose that, for all $x \geq 1$, the function $a(x)$ is continuous, positive, and decreasing. Consider the series and the integral
>
> $$\sum_{k=1}^{\infty} a_k \quad \text{and} \quad \int_{1}^{\infty} a(x)\,dx,$$
>
> where $a_k = a(k)$ for integers $k \geq 1$.
>
> - If either diverges, so does the other.
> - If either converges, so does the other. In this case, we have
>
> $$\int_{1}^{\infty} a(x)\,dx \leq \sum_{k=1}^{\infty} a_k \leq a_1 + \int_{1}^{\infty} a(x)\,dx.$$
>
> and
>
> $$R_n = \sum_{k=n+1}^{\infty} a_k \leq \int_{n}^{\infty} a(x)\,dx.$$

Convergent and divergent p-series Series of the form $\sum_{k=1}^{\infty} 1/k^p$ are called **p-series**. They form an important family of examples.

EXAMPLE 3 For which p does the p-series $\sum_{k=1}^{\infty} \dfrac{1}{k^p}$ converge?

Solution We saw in Chapter 10 that the improper integral $\int_{1}^{\infty} dx/x^p$ converges if and only if $p > 1$. By the integral test, the same is true for the series. In particular, the harmonic series $\sum 1/k$ diverges. ■

This useful result deserves special mention:

> **FACT (The p-test for series)** The p-series $\sum_{k=1}^{\infty} \dfrac{1}{k^p}$ converges if and only if $p > 1$.

Harmonic divergence The fact that the harmonic series

$$1 + \frac{1}{2} + \frac{1}{3} + \frac{1}{4} + \frac{1}{5} + \cdots$$

diverges to infinity—even though the terms themselves tend to zero—has fascinated mathematicians for many centuries. One early proof (unrelated to integrals) is attributed to Nicole Oresme, a 14th-century French bishop, scientist, and mathematician.

EXAMPLE 4 The p-series $\sum_{k=1}^{\infty} 1/k^3$ converges to some limit S. (The integral test says so.) How large must n be to ensure that S_n differs from S by less than 0.0001?

Solution   We need to choose n so that the tail R_n is less than 0.0001. By the last inequality in Theorem 8,

$$R_n \le \int_n^\infty \frac{dx}{x^3} = \frac{1}{2n^2}.$$

(The last step is an easy calculation.) This quantity (and hence also R_n) are less than 0.0001 if $n \ge 71$. Hence, $S_{71} \approx 1.20196$ differs from the true limit by less than 0.0001.

EXAMPLE 5   Does $\displaystyle\sum_{k=1}^\infty \frac{1}{10k+1}$ converge?

Solution   The integral test says no.

$$\int_1^\infty \frac{1}{10x+1}\,dx = \lim_{n\to\infty} \int_1^n \frac{1}{10x+1}\,dx = \lim_{n\to\infty} \left.\frac{\ln(10x+1)}{10}\right]_1^n = \infty.$$

The comparison test *also* says no. Since $10k+1 \le 11k$ for all $k \ge 1$, we have

$$\frac{1}{10k+1} \ge \frac{1}{11k} \implies \sum_{k=1}^\infty \frac{1}{10k+1} \ge \sum_{k=1}^\infty \frac{1}{11k} = \frac{1}{11}\sum_{k=1}^\infty \frac{1}{k}.$$

Because the last series diverges to infinity, so must the first.

The ratio test: comparison with a geometric series

In a geometric series $a + ar + ar^2 + ar^3 + ar^4 + \cdots$, the ratio of successive terms is r; by Theorem 4, page 561, the series converges if and only if $|r| < 1$.

The **ratio test** is also based on ratios of successive terms; it amounts to a lightly disguised form of comparison with a geometric series.

THEOREM 9 (Ratio test for positive series)   Suppose that $a_k > 0$ for all k and that

$$\lim_{k\to\infty} \frac{a_{k+1}}{a_k} = L.$$

- If $L < 1$, then $\sum a_k$ converges.

- If $L > 1$, then $\sum a_k$ diverges.

- If $L = 1$, the test is inconclusive—either convergence or divergence is possible.

Two points deserve emphasis:

The limit is 1 for every p-series.

- **Best cases**   The ratio test works best for series like $\sum 1/k!$, $\sum r^k$, and $\sum 1/(2^k+3)$, in which the index k appears in an exponent or a factorial.

- **Other cases**   For many series, unfortunately, the ratio a_{k+1}/a_k either tends to 1 or has *no* limit. ← In such cases, the ratio test tells us nothing.

EXAMPLE 6   Show that $\displaystyle\sum_{k=0}^{\infty} \frac{1}{k!}$ converges. Guess its limit.

Solution   The ratio test works nicely. Since

$$\lim_{k\to\infty} \frac{a_{k+1}}{a_k} = \lim_{k\to\infty} \frac{k!}{(k+1)!} = \lim_{k\to\infty} \frac{1}{k+1} = 0,$$

the series converges. In fact, it converges very, very fast. Here are some representative partial sums:

$$S_5 \approx 2.716667; \quad S_{10} \approx 2.718282; \quad S_{30} \approx 2.718281828459045235360287471354.$$

Is e involved somehow? It certainly seems so—S_{30} agrees with e in all decimal places shown. In fact, this series can be shown to converge to e. We explore this phenomenon further in later sections.   ■

EXAMPLE 7   Does $\displaystyle\sum_{k=0}^{\infty} \frac{100^k}{k!}$ converge?

Solution   Yes, by the ratio test:

$$\lim_{k\to\infty} \frac{a_{k+1}}{a_k} = \lim_{k\to\infty} \frac{100^{k+1}}{(k+1)!} \cdot \frac{k!}{100^k} = \lim_{k\to\infty} \frac{100}{k+1} = 0.$$

Notice what the result means: Even though 100^k grows very fast, $k!$ grows even faster.   ■

Why the ratio test works   To illustrate the connection between the ratio test and geometric series, and to give the idea of a proof, let's suppose, for instance, that

$$\lim_{k\to\infty} \frac{a_{k+1}}{a_k} = \frac{1}{2}.$$

Why must $\sum a_k$ converge?

The idea is that $a_{k+1} \approx a_k/2$ for large k, and so the given series should be comparable to a geometric series. Suppose, for instance, that $a_{k+1} < 0.6 a_k$ for all $k \geq 1000$. ➡ Then

$$a_{1001} < (0.6)a_{1000}, \quad a_{1002} < (0.6)a_{1001} < (0.6)^2 a_{1000}, \quad a_{1003} < (0.6)^3 a_{1000}, \ldots.$$

Such an inequality must hold for large k because the ratio limit is $1/2$.

Therefore,

$$a_{1000} + a_{1001} + a_{1002} + \cdots < a_{1000}\left(1 + (0.6) + (0.6)^2 + (0.6)^3 + \cdots\right).$$

The preceding inequality is the point; it shows that an upper tail of the original series $\sum a_k$ converges by comparison with the geometric series in the last line. (The divergence part of the ratio test can be proved using similar ideas.)

BASIC EXERCISES

1. Consider the series $\displaystyle\sum_{k=0}^{\infty} a_k$, where $a_k = \dfrac{1}{k+2^k}$.

   (a) Use the comparison test to show that the series converges.

   (b) Show that $0 \leq R_{10} \leq 2^{-10}$.

   (c) Compute an estimate of the limit of the series that is guaranteed to be within 0.001 of the exact value.

   (d) Is your estimate in part (c) an overestimate? Justify your answer.

2. Consider the series $\sum_{j=0}^{\infty} \frac{1}{2+3^j}$.

   (a) Show that the series converges.

   (b) Estimate the limit of the series within 0.01.

   (c) Is your estimate in part (b) an overestimate? Justify your answer.

In Exercises 3 and 4, suppose that $a(x)$ is continuous, positive, and decreasing for all $x \geq 1$ and that $a_k = a(k)$ for all integers $k \geq 1$.

3. Rank the values $\int_1^n a(x)\,dx$, $\sum_{k=1}^{n-1} a_k$, and $\sum_{k=2}^{n} a_k$ in increasing order. [HINT: Draw a picture.]

4. Rank the values $\int_n^\infty a(x)\,dx$, $\sum_{k=n+1}^{\infty} a_k$, and $\int_{n+1}^\infty a(x)\,dx$ in increasing order.

Suppose that $a(x)$ is continuous, positive, and decreasing for all $x \geq 1$ and that $a_k = a(k)$ for all integers $k \geq 1$. In Exercises 5–8, draw a carefully annotated picure that shows that

5. $\int_1^{n+1} a(x)\,dx \leq \sum_{k=1}^{n} a_k$

6. $\sum_{k=2}^{n} a_k \leq \int_1^n a(x)\,dx$

7. $\sum_{k=n+1}^{\infty} a_k \leq a_{n+1} + \int_{n+1}^\infty a(x)\,dx$

8. $\sum_{k=n+1}^{\infty} a_k \leq \int_n^\infty a(x)\,dx$

In Exercises 9–12, use the integral test to find upper and lower bounds on the limit of the series.

9. $\sum_{k=1}^{\infty} \frac{1}{k^3}$

10. $\sum_{k=1}^{\infty} \frac{1}{k\sqrt{k}}$

11. $\sum_{j=1}^{\infty} je^{-j}$

12. $\sum_{k=0}^{\infty} \frac{1}{k^2+1}$

13. Consider the series $\sum_{k=1}^{\infty} \frac{e^{\sin k}}{k^2}$.

    (a) $\int_1^\infty \frac{e^{\sin k}}{k^2}$ is a convergent improper integral. Does it follow from this fact and Theorem 8 that the series converges? Justify your answer.

    (b) Show that this series converges.

14. Use the fact that $\ln x \leq \sqrt{x}$ for all $x \geq 1$ to show that $\sum_{n=2}^{\infty} \frac{1}{(\ln n)^2}$ diverges.

15. Use the fact that $\ln x \leq x$ for all $x \geq 1$ to show that $\sum_{n=2}^{\infty} \frac{1}{(\ln n)^2}$ diverges. [HINT: $x \ln x \geq (\ln x)^2$ if $x \geq 1$.]

16. Show that $\sum_{k=3}^{\infty} \frac{1}{(\ln k)^k}$ converges. [HINT: $\ln 3 \approx 1.0986$.]

In Exercises 17–20, use the comparison test to show that the series converges. Then find upper and lower bounds on the limit of the series.

17. $\sum_{n=1}^{\infty} \frac{1}{n^2+\sqrt{n}}$

18. $\sum_{j=0}^{\infty} \frac{1}{j+e^j}$

19. $\sum_{m=1}^{\infty} \frac{1}{m\sqrt{1+m^2}}$

20. $\sum_{k=1}^{\infty} \frac{k}{(k^2+1)^2}$

In Exercises 21–24, use the ratio test to show that the series converges.

21. $\sum_{j=0}^{\infty} \frac{j^2}{j!}$

22. $\sum_{k=1}^{\infty} \frac{2^k}{k!}$

23. $\sum_{n=1}^{\infty} \frac{n^2}{2^n}$

24. $\sum_{m=1}^{\infty} \frac{m!}{(2m)!}$

25. Consider the series $\sum_{k=1}^{\infty} a_k = \sum_{k=1}^{\infty} \frac{\ln k}{k}$.

    (a) Use the integral test to show that the series diverges. [HINT: Be careful, the integrand is not monotone.]

    (b) Use the comparison test to show that the series diverges. [HINT: $1 - x^{-1} \leq \ln x$ for all $x > 0$.]

    (c) Can the ratio test be used to show that the series diverges? Explain.

26. Consider the series $\sum_{k=1}^{\infty} a_k = \frac{1}{2} + \frac{1}{3} + \frac{1}{2^2} + \frac{1}{3^2} + \frac{1}{2^3} + \frac{1}{3^3} + \cdots$.

    (a) Explain why $\lim_{k\to\infty} a_{k+1}/a_k$ does not exist.

    (b) What, if anything, does the ratio test say about the convergence of $\sum_{k=1}^{\infty} a_k$?

    (c) Show that the series converges, and evaluate its limit. [HINT: Rewrite the series as the sum of two convergent series.]

27. (a) What, if anything, does the ratio test say about the convergence of the series $\frac{1}{2} + \frac{1}{2} + \frac{1}{4} + \frac{1}{4} + \frac{1}{8} + \frac{1}{8} + \cdots$?

    (b) Does the series in part (a) converge or diverge? Justify your answer.

28. Give an example of a divergent series $\sum_{k=1}^{\infty} a_k$ such that $a_k > 0$ and $a_{k+1}/a_k < 1$ for all $k \geq 1$.

29. Use the ratio test to show that $\sum_{n=1}^{\infty} n^{-n}$ converges.

30. Use the ratio test to show that the series $\sum_{n=1}^{\infty} \frac{n^n}{n!}$ diverges.

FURTHER EXERCISES

In Exercises 31–36, determine whether the series converges or diverges. If the series converges, find a number N such that the partial sum S_N approximates the sum of the series within 0.001. If the series diverges, find a number N such that $S_N \geq 1000$.

31. $\displaystyle\sum_{k=0}^{\infty} \frac{1}{k^2+3}$

32. $\displaystyle\sum_{m=1}^{\infty} \frac{\arctan m}{m}$

33. $\displaystyle\sum_{k=0}^{\infty} \frac{1}{2+\cos k}$

34. $\displaystyle\sum_{m=2}^{\infty} \frac{\ln m}{m^3}$

35. $\displaystyle\sum_{k=0}^{\infty} \frac{k}{k^6+17}$

36. $\displaystyle\sum_{k=2}^{\infty} \frac{1}{k\,(\ln k)^5}$

In Exercises 37 and 38, let $S_n = \displaystyle\sum_{k=1}^{n} a_k$, $T_n = \displaystyle\sum_{k=1}^{n} b_k$, and assume that $0 \leq a_k \leq b_k$ for all $k \geq 1$.

37. (a) Suppose that $\displaystyle\sum_{k=1}^{\infty} b_k$ converges. Explain why there is a number M such that $S_n \leq T_n \leq M$ for all $n \geq 1$.

(b) Explain why $\{S_n\}$ is an increasing sequence.

(c) Explain why parts (a) and (b) together imply that $\displaystyle\sum_{k=1}^{\infty} a_k$ converges.

38. (a) Suppose that $\displaystyle\sum_{k=1}^{\infty} a_k$ diverges. Explain why $\displaystyle\lim_{n \to \infty} S_n = \infty$.

(b) Suppose that $\displaystyle\sum_{k=1}^{\infty} a_k$ diverges. Use part (a) to show that $\displaystyle\sum_{k=1}^{\infty} b_k$ diverges.

39. Suppose that $a(x)$ is continuous, positive, and decreasing for all $x \geq 1$, that $a_k = a(k)$ for all integers $k \geq 1$, and that $\displaystyle\int_1^{\infty} a(x)\,dx$ converges.

(a) Explain why the sequence of partial sums $\{S_n\}$ is an increasing sequence.

(b) Explain why $\displaystyle\int_1^{n} a(x)\,dx \leq \int_1^{\infty} a(x)\,dx$.

(c) Use parts (a) and (b) to show that the sequence of partial sums $\{S_n\}$ converges.

40. (a) Where in the proof of the integral test (Theorem 8) is the assumption that $a(x)$ is a decreasing function used?

(b) Suppose that the requirement that $a(x)$ be decreasing for all $x \geq 1$ is replaced by the "weaker" requirement that $a(x)$ be decreasing for all $x \geq 10$. How does this change in assumptions affect the conclusions of Theorem 8?

41. Does the series $1 + \dfrac{1}{1 \cdot 3} + \dfrac{1}{1 \cdot 3 \cdot 5} + \dfrac{1}{1 \cdot 3 \cdot 5 \cdot 7} + \cdots + \dfrac{1}{1 \cdot 3 \cdot 5 \cdot 7 \cdots (2k+1)} + \cdots$ converge? Justify your answer.

42. Does the series $\dfrac{1}{2} + \dfrac{1}{2 \cdot 4} + \dfrac{1}{2 \cdot 4 \cdot 6} + \cdots + \dfrac{1}{2 \cdot 4 \cdot 6 \cdots (2k)} + \cdots$ converge? Justify your answer.

43. Show that $\displaystyle\sum_{k=3}^{\infty} \frac{1}{(\ln k)^{\ln k}}$ converges.

[HINT: $\ln k > e^2$ if $k > 1619$.]

44. Suppose that $a_n \geq 0$ for all $n \geq 1$ and that $\displaystyle\sum_{n=1}^{\infty} a_n$ converges. Show that $\displaystyle\sum_{n=1}^{\infty} |\sin a_n|$ converges.

[HINT: $|\sin x| \leq |x|$ for all x.]

In Exercises 45–52, determine whether the series converges. If it converges, find upper and lower bounds on its limit. Justify your answers.

45. $\displaystyle\sum_{n=1}^{\infty} \frac{\arctan n}{1+n^2}$

46. $\displaystyle\sum_{m=1}^{\infty} \frac{m^3}{m^5+3}$

47. $\displaystyle\sum_{j=1}^{\infty} \frac{1}{100+5j}$

48. $\displaystyle\sum_{k=2}^{\infty} \frac{1}{k \ln k}$

49. $\displaystyle\sum_{n=1}^{\infty} \frac{1}{n\,3^n}$

50. $\displaystyle\sum_{n=2}^{\infty} \frac{1}{\sqrt[3]{n^2-1}}$

51. $\displaystyle\sum_{j=0}^{\infty} \frac{j!}{(j+2)!}$

52. $\displaystyle\sum_{n=0}^{\infty} \frac{n!}{(2n)!}$

53. Suppose that $\{a_k\}$ is a sequence of positive terms and that r is a constant such that $a_{k+1}/a_k \leq r < 1$ for all $k \geq 1$.

(a) Show that $a_2 \leq a_1 r$.

(b) Show that $a_{k+1} \leq a_1 r^k$ for every $k \geq 1$.

(c) Use part (b) to show that $\displaystyle\sum_{k=1}^{\infty} a_k$ converges.

[HINT: Use the comparison test and the formula for the sum of geometric series.]

(d) Show that $R_n = \displaystyle\sum_{k=n+1}^{\infty} a_k \leq \frac{a_{n+1}}{1-r}$.

54. Use Exercise 53 to find an integer N such that the partial sum S_N approximates the sum of the series $\displaystyle\sum_{n=1}^{\infty} n^2/2^n$ within 0.0005.

55. Use Exercise 53 to find an integer N such that the partial sum S_N approximates the sum of the series $\displaystyle\sum_{n=1}^{\infty} (n!)^2/(2n)!$ within 0.0005.

11.4 ABSOLUTE CONVERGENCE; ALTERNATING SERIES

Not-necessarily-positive series   The integral, comparison, and ratio tests apply only to *nonnegative* series. Some interesting series, however, have both positive and negative terms.

> **EXAMPLE 1**   Does the **alternating harmonic series**
>
> $$\sum_{k=1}^{\infty} \frac{(-1)^{k+1}}{k} = 1 - \frac{1}{2} + \frac{1}{3} - \frac{1}{4} + \frac{1}{5} - \cdots$$
>
> converge or diverge? To what limit?
>
> **Solution**   As always, the question is how partial sums behave. Tabulating some of them shows a pattern.

Partial sums of $1 - \frac{1}{2} + \frac{1}{3} - \frac{1}{4} + \frac{1}{5} - \cdots$										
n	1	2	3	4	5	6	7	8	9	10	...
S_n	1	0.500	0.833	0.583	0.783	0.617	0.760	0.635	0.746	0.646	...
n	51	52	53	54	55	56	57	58	59	60	...
S_n	0.703	0.684	0.702	0.684	0.702	0.684	0.702	0.685	0.702	0.685	...

Successive partial sums seem to hop back and forth across some limiting value. Plots of partial sums and terms (Figure 1) exhibit the same pattern:

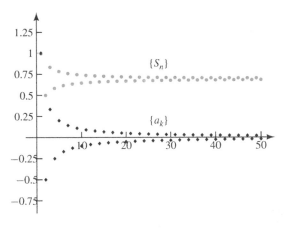

Blue dots show partial sums; black diamonds show terms.

FIGURE 1

Terms and partial sums for $1 - \frac{1}{2} + \frac{1}{3} - \frac{1}{4} + \frac{1}{5} - \cdots$

Because the terms alternate in sign, the partial sums successively rise and fall, alternately overshooting and undershooting the limiting value, which is apparently around 0.69. (It can be shown—with considerable effort—that the exact limit is $\ln 2 \approx 0.69315$.)

Absolute vs. conditional convergence

The alternating harmonic series illustrates the phenomenon of **conditional convergence**. Although

$$1 - \frac{1}{2} + \frac{1}{3} - \frac{1}{4} + \frac{1}{5} - \cdots$$

converges (as Example 1 suggests), the *ordinary* harmonic series

$$1 + \frac{1}{2} + \frac{1}{3} + \frac{1}{4} + \frac{1}{5} + \cdots$$

diverges, as we saw from the integral test.

EXAMPLE 2   Does $\displaystyle\sum_{k=1}^{\infty} \frac{\sin k}{k^2}$ converge? Does $\displaystyle\sum_{k=1}^{\infty} \frac{|\sin k|}{k^2}$? Estimate limits.

Solution   The first series, like the alternating harmonic series, has both negative and positive terms—although in no regular order this time. Plotting terms and partial sums (Figure 2) suggests (but doesn't prove) that this series converges.

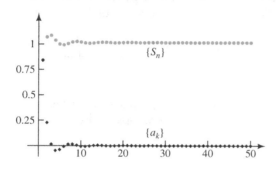

FIGURE 2

Terms and partial sums for $\displaystyle\sum_{k=1}^{\infty} \frac{\sin k}{k^2}$

The partial sums wander up and down just slightly but still appear to approach a horizontal asymptote, perhaps near $y = 1$.

The second series also seems to converge, as Figure 3 suggests. This time the limit appears to be near 1.25.

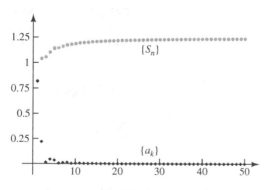

FIGURE 3

Terms and partial sums for $\displaystyle\sum_{k=1}^{\infty} \frac{|\sin k|}{k^2}$

If $\sum a_k$ is the first series, $\sum |a_k|$ is the second.

The second series comes from the first by taking the absolute value of each term. ← The new series is nonnegative, and so the comparison test applies. Because

$$0 \le \frac{|\sin k|}{k^2} \le \frac{1}{k^2}$$

for all $k \ge 1$ and $\sum_{k=1}^{\infty} 1/k^2$ converges, so must $\sum_{k=1}^{\infty} |\sin k|/k^2$. ■

Examples 1 and 2 illustrate the phenomena of **conditional convergence** and **absolute convergence**, respectively. The formal definitions are as follows:

DEFINITION (Absolute and conditional convergence)   Let $\sum a_k$ be any series.

- If $\sum |a_k|$ converges, then $\sum a_k$ **converges absolutely**.
- If $\sum |a_k|$ diverges but $\sum a_k$ converges, then $\sum a_k$ **converges conditionally**.

The wacky world of conditional convergence   Conditionally convergent series have some surprising properties. Here is one of the oddest:

Let $\sum a_k$ be conditionally convergent, and let L be any real number. Then the terms of $\sum a_k$ can be reordered in such a way that the resulting series converges to L.

(For more details, see your instructor.) Notice how drastically this peculiar property of conditionally convergent series upsets the naive hope that addition is commutative even for infinite sums.

Pluses and minuses of pluses and minuses   Let $\sum_{k=1}^{\infty} a_k$ be any series. If it happens that $a_k \ge 0$ for all k, the advantage is simplicity: The partial sums are nondecreasing. The disadvantage, as the harmonic series shows, is that the partial sums may tend to infinity.

Mixing positive and negative terms may cost something in simplicity, but it is an advantage for convergence. As the alternating harmonic series shows, positive and negative terms can offset each other, thus helping the cause of convergence.

Absolute convergence implies ordinary convergence   We saw in Chapter 10 that if $\int_1^{\infty} |f(x)|\,dx$ converges, then so must $\int_1^{\infty} f(x)\,dx$, and $\left| \int_1^{\infty} f(x)\,dx \right| \le \int_1^{\infty} |f(x)|\,dx$. (See Theorem 2, page 534.) The same principle applies to infinite series:

THEOREM 10   If $\sum_{k=1}^{\infty} |a_k|$ converges, then so does $\sum_{k=1}^{\infty} a_k$, and

$$\left| \sum_{k=1}^{\infty} a_k \right| \le \sum_{k=1}^{\infty} |a_k|.$$

The theorem's inequality should be believable—it is the infinite version of the fact that the absolute value of a sum can not exceed the sum of the absolute values. The idea of a rigorous proof is to write the original series as a sum of two new series, one entirely positive and the other entirely negative. Using the comparison test, one can show that each of these new series converges.

The theorem shows that, as Figure 3 suggested, the series $\sum_{k=1}^{\infty} \sin k / k^2$ of Example 2 does indeed converge because $\sum_{k=1}^{\infty} |\sin k| / k^2$ does. Our limit estimates from Example 2 are also consistent with the theorem:

$$1 \approx \left| \sum_{k=1}^{\infty} \frac{\sin k}{k^2} \right| \leq \sum_{k=1}^{\infty} \frac{|\sin k|}{k^2} \approx 1.25.$$

EXAMPLE 3 For which values of x does the **power series**

$$\sum_{k=1}^{\infty} kx^k = x + 2x^2 + 3x^3 + 4x^4 + \cdots$$

converge? (A power series is something like an "infinite polynomial"; we discuss power series carefully in the next section.)

Solution First we use the ratio test to check for absolute convergence:

$$\lim_{k \to \infty} \left| \frac{(k+1)x^{k+1}}{kx^k} \right| = \lim_{k \to \infty} \left| \frac{k+1}{k} \right| \cdot |x| = 1 \cdot |x| = |x|.$$

If $|x| < 1$, the original series converges absolutely. By Theorem 10, it also converges *without* absolute value signs.

If $|x| \geq 1$, the series fails the nth term test and so diverges. ■

Estimating limits

Estimating a limit for any series—nonnegative or not—depends on keeping the upper tail small. Theorem 10, combined with earlier estimates, can help.

EXAMPLE 4 For the series $\sum_{k=1}^{\infty} (-1)^{k+1}/k^3$, we find (using technology) that $S_{100} \approx 0.901542$. How closely does S_{100} approximate S, the true sum of the series?

Solution Because

$$S = \sum_{k=1}^{100} \frac{(-1)^{k+1}}{k^3} + \sum_{k=101}^{\infty} \frac{(-1)^{k+1}}{k^3} = S_{100} + R_{100},$$

we need only estimate R_{100}, as follows:

$$|R_{100}| - \left| \sum_{k=101}^{\infty} \frac{(-1)^{k+1}}{k^3} \right| \leq \sum_{k=101}^{\infty} \frac{1}{k^3} \qquad \text{(by Theorem 10)}$$

$$< \int_{100}^{\infty} \frac{1}{x^3} dx. \qquad \text{(by the integral test)}$$

The last integral is easy to calculate:

$$\int_{100}^{\infty} \frac{1}{x^3} dx = \frac{-1}{2x^2} \Big]_{100}^{\infty} = \frac{1}{20,000} = 0.00005,$$

and so the estimate $S \approx S_{100} \approx 0.901542$ is good to at least four decimal places. ■

Alternating series

For series with both positive and negative terms, testing for absolute convergence is usually the best option. In the special (but surprisingly useful) case that the terms alternate in sign, we can sometimes do better.

> **DEFINITION**   An **alternating series** is one whose terms alternate in sign, that is, a series of the form
>
> $$c_1 - c_2 + c_3 - c_4 + c_5 - c_6 + \cdots ,$$
>
> where each c_i is positive.

The **alternating harmonic series**

$$1 - \frac{1}{2} + \frac{1}{3} - \frac{1}{4} + \frac{1}{5} - \cdots$$

See Example 1, especially Figure 1.

illustrates the best possibility. ← Because successive terms alternate in sign and decrease in size, successive partial sums straddle smaller and smaller intervals. If the terms also tend to zero, then the partial sums narrow down on a limit. The following theorem makes these observations formal; it also gives a convenient error bound.

> **THEOREM 11 (Alternating series test)**   Consider the series
>
> $$\sum_{k=1}^{\infty} (-1)^{k+1} c_k = c_1 - c_2 + c_3 - c_4 + \cdots ,$$
>
> where
>
> (i) $c_1 \geq c_2 \geq c_3 \geq \cdots \geq 0;$ and   (ii) $\lim_{k \to \infty} c_k = 0.$
>
> Then the series converges, and its sum S lies *between* any two successive partial sums S_n and S_{n+1}. In particular,
>
> $$|S - S_n| < c_{n+1}$$
>
> for all $n \geq 1$.

A formal proof is slightly tricky, but the underlying idea is simple: Because the limit S lies *between* successive partial sums, adding another term to any partial sum always "overshoots" the limit, which explains the final inequality.

Using the theorems: miscellaneous examples

Combining Theorems 10 and 11 with results from earlier sections enables us to handle many not-necessarily-positive series, detecting convergence or divergence and estimating limits. The following examples illustrate some useful tricks of this trade.

EXAMPLE 5   (An alternating p-series: another look)   What does Theorem 11 say about $\sum_{k=1}^{\infty} (-1)^{k+1}/k^3$ and its 100th partial sum $S_{100} \approx 0.9015422$?

Solution In this context, $c_k = 1/k^3$. Now Theorem 11 says not only that the series converges—which we already knew—but also that

$$|S - S_{100}| < c_{101} = \frac{1}{101^3} \approx 0.000001.$$

Thus, $S_{100} \approx 0.9015422$ lies within 0.000001 of the true limit S. Equivalently, S lies between $S_{100} \approx 0.9015422$ and $S_{101} \approx 0.9015432$. ∎

EXAMPLE 6 Does $\sum_{j=1}^{\infty}(-1)^j \dfrac{j}{j+1}$ converge or diverge? Why?

Solution The alternating series test looks tempting at first glance. But it does not apply because

$$\lim_{j\to\infty}\frac{j}{j+1} = \lim_{j\to\infty}\frac{1}{1+1/j} = 1,$$

not zero as Theorem 11 requires. The nth term test does apply, however. (Maybe we should call it the "jth term test" here.) Since

$$\frac{j}{j+1} \to 1 \quad \text{as} \quad j \to \infty,$$

it follows that the jth term has no limit as $j \to \infty$. ➡ Therefore, the given series diverges. ∎

Successive terms are alternately near 1 and −1, and so the terms diverge.

EXAMPLE 7 Does the series

$$1 + 2 + 3 + 4 + 5 - \frac{1}{6} + \frac{1}{7} - \frac{1}{8} + \frac{1}{9} - \cdots$$

converge? If so, find or estimate the limit.

Solution The alternating series test does not apply right out of the box because the first five terms break the desired pattern. The problem isn't fatal, however. Basic series algebra lets us group our terms into two blocks as follows:

$$(1 + 2 + 3 + 4 + 5) - \left(\frac{1}{6} - \frac{1}{7} + \frac{1}{8} - \frac{1}{9} + \cdots\right).$$

The first block is finite, and so convergence isn't an issue; its sum is 15. The second block clearly satisfies all hypotheses of the alternating series test and so converges to some limit L. Any partial sum of the second block, moreover, differs from L by less than the magnitude of the next term (by the last line of Theorem 11).

The entire series therefore converges to $S = 15 - L$, and any partial sum differs from S by no more than the next term. The partial sum $S_9 = 1 + 2 + \cdots + 1/9 \approx 14.962$, for instance, overshoots the true limit by less than 1/10. In other words, the exact limit S satisfies $14.862 \le S \le 14.962$. ∎

EXAMPLE 8 Does $\sum_{n=1}^{\infty} \dfrac{\sin n}{n^3 + n^2 + n + 1 + \cos n}$ converge or diverge? Why?

Solution The problem is easier than it looks. Hoping for absolute convergence, we start by taking absolute values:

$$\left|\frac{\sin n}{n^3 + n^2 + n + 1 + \cos n}\right| = \frac{|\sin n|}{n^3 + n^2 + n + 1 + \cos n}.$$

The general appearance of numerator and denominator suggests a comparison. A simple inequality make the job much easier:

$$\frac{|\sin n|}{n^3 + n^2 + n + 1 + \cos n} \le \frac{1}{n^3}.$$

Because the p-series $\sum 1/n^3$ converges, so must the absolute-value version of the given series. Finally, Theorem 10 guarantees that the original series also converges. ∎

BASIC EXERCISES

Exercises 1–4 are about the series in Example 7.

1. Does the series converge conditionally or absolutely? Justify your answer.

2. Does the partial sum S_{15} overestimate the limit of the series? Justify your answer.

3. $S_{60} \approx 14.902$. Use this result to find upper and lower bounds for S.

4. The alternating harmonic series can be shown to converge to $\ln 2$. Use this fact to find the limit of the series exactly.

5. Consider the series in Example 8.

  (a) Compute S_{50}.

  (b) Explain why $|R_{50}| \le \int_{50}^{\infty} \frac{dx}{x^3}$.

  (c) Use parts (a) and (b) to find upper and lower bounds for the limit of the series.

6. Suppose that $\sum_{k=1}^{\infty} a_k$ converges absolutely. Show that $\sum_{k=1}^{\infty} \frac{a_k}{k}$ also converges.

7. Suppose that $\sum_{k=1}^{\infty} \frac{a_k}{k}$ converges. Must $\sum_{k=1}^{\infty} a_k$ also converge? Justify your answer.

8. Show that $\sum_{k=2}^{\infty} (-1)^k \frac{k}{k^2 - 1}$ converges conditionally.

In Exercises 9–12, show that the series converges. Then compute an estimate of the limit that is guaranteed to be in error by no more than 0.005.

9. $\sum_{k=1}^{\infty} \frac{(-1)^k}{k^4}$

10. $\sum_{k=1}^{\infty} \frac{(-1)^k}{k^2 + 2^k}$

11. $\sum_{k=0}^{\infty} \frac{(-3)^k}{(k^2)!}$

12. $\sum_{k=5}^{\infty} (-1)^k \frac{k^{10}}{10^k}$

13. Suppose that $a_k \ge 0$ for all $k \ge 1$. Is it possible that $\sum_{k=1}^{\infty} a_k$ converges conditionally? Justify your answer.

14. Does $\sum_{n=2}^{\infty} (-1)^n \frac{n}{2n - 1}$ converge? Justify your answer.

FURTHER EXERCISES

15. For which values of p, if any, does $\sum_{k=2}^{\infty} \frac{\ln k}{k^p}$ converge? Justify your answer.

16. For which values of p, if any, does $\sum_{k=2}^{\infty} (-1)^k \frac{\ln k}{k^p}$ converge? Justify your answer.

17. For which values of p, if any, does $\sum_{k=2}^{\infty} (-1)^k \frac{\ln k}{k^p}$ converge absolutely? Justify your answer.

18. For which values of p, if any, does $\sum_{k=2}^{\infty} (-1)^k \frac{\ln k}{k^p}$ converge conditionally? Justify your answer.

In Exercises 19–26, determine whether the series converges absolutely, converges conditionally, or diverges. If the series converges, find upper and lower bounds on its limit. Justify your answers.

19. $\sum_{j=1}^{\infty} \frac{(-1)^{j+1}}{j^2}$

20. $\sum_{k=1}^{\infty} \frac{(-1)^k}{\sqrt{k}}$

21. $\sum_{n=1}^{\infty} \frac{(-3)^n}{n^3}$

22. $\sum_{k=4}^{\infty} (-1)^k \frac{\ln k}{k}$

23. $\displaystyle\sum_{k=0}^{\infty}(-1)^k\frac{k}{2k+1}$

24. $\displaystyle\sum_{m=0}^{\infty}(-1)^m\frac{m^3}{2^m}$

25. $\displaystyle\sum_{j=0}^{\infty}(-1)^j\frac{j!}{(j^2)!}$

26. $\displaystyle\sum_{n=1}^{\infty}(-1)^{n+1}\frac{\arctan n}{n}$

27. Consider the series $\displaystyle\sum_{k=1}^{\infty}(-1)^{k+1}a_k$. Suppose that the terms of the sequence $\{a_k\}$ are positive and decreasing for all $k \geq 10^9$ and $\lim_{k\to\infty} a_k = 0$ but that $a_{10^9} > a_1$. Explain why the series converges. [HINT: Theorem 11 does not apply directly.]

28. Does $1-\dfrac{1}{2^3}+\dfrac{1}{3^2}-\dfrac{1}{4^3}+\dfrac{1}{5^2}-\dfrac{1}{6^3}+\dfrac{1}{7^2}-\dfrac{1}{8^3}+\cdots$ converge? Justify your answer.

29. Suppose that $\displaystyle\sum_{j=1}^{\infty}b_j$ converges to a number S and that $b_j \geq 0$ for all $j \geq 1$. Show that $\displaystyle\sum_{j=1}^{\infty}(-1)^{j+1}b_j$ also converges.

30. Suppose that $0 \leq b_{j+1} \leq b_j$ for all $j \geq 1$ and that the partial sum $\displaystyle\sum_{j=1}^{100}b_j$ approximates $S = \displaystyle\sum_{j=1}^{\infty}b_j$ within 0.005. Explain why $0 \leq \displaystyle\sum_{j=1}^{\infty}(-1)^{j+1}b_j - \sum_{j=1}^{100}(-1)^{j+1}b_j \leq 0.005$.

31. Give an example of a convergent series $\displaystyle\sum_{k=1}^{\infty}a_k$ with the property that $\displaystyle\sum_{k=1}^{\infty}(a_k)^2$ diverges.

32. Suppose that $\displaystyle\sum_{k=1}^{\infty}a_k$ diverges. Is it possible that $\displaystyle\sum_{k=1}^{\infty}|a_k|$ converges? Justify your answer.

33. **A proof of the alternating series test.** Let $S_n = \displaystyle\sum_{k=1}^{n}(-1)^{k+1}c_k$ denote the partial sum of the first n terms of a series satisfying the hypotheses of the alternating series test (Theorem 11).
  (a) Show that the sequence of even partial sums, $S_2, S_4, S_6, S_8, \ldots$ is increasing.
  (b) Show that the sequence of odd partial sums, $S_1, S_3, S_5, S_7, \ldots$, is decreasing.
  (c) Show that $S_{2m} \leq S_{2m-1}$ for any integer $m \geq 1$.
  (d) Use part (c) to show that the sequence of even partial sums and the sequence of odd partial sums both converge. [NOTE: Although both sequences converge, we must still show that they converge to the same limit.]
  (e) Show that $\lim_{m\to\infty}(S_{2m+1} - S_{2m}) = 0$. From this it follows that there is a real number S such that $\lim_{n\to\infty} S_n = S$.
  (f) Explain why $0 < S - S_{2m} < c_{2m+1}$ and $0 < S_{2m+1} - S < c_{2m+2}$.

11.5 POWER SERIES

A **power series** is a series of the form

$$a_0 + a_1(x-x_0) + a_2(x-x_0)^2 + a_3(x-x_0)^3 + \cdots + a_n(x-x_0)^n + \cdots = \sum_{k=0}^{\infty} a_k(x-x_0)^k.$$

Here x is a variable, the constants a_k are called the **coefficients**, and the constant x_0 is called the **base point**. For convenience we'll often use $x_0 = 0$; in this case the appearance is slightly simpler:

$$a_0 + a_1 x + a_2 x^2 + a_3 x^3 + \cdots + a_n x^n + \cdots = \sum_{k=0}^{\infty} a_k x^k.$$

A power series may converge for some values of x and diverge for others. We illustrate these words and ideas with examples.

EXAMPLE 1 (**A geometric power series**) One of the simplest and most useful power series is

$$S(x) = 1 + x + x^2 + x^3 + x^4 + x^5 + \cdots = \sum_{k=0}^{\infty} x^k.$$

(In this case, $x_0 = 0$ and $a_k = 1$ for all $k \geq 0$.) For which real numbers x does $S(x)$ converge?

Solution Setting $x = 1$ gives the series

$$S(1) = 1 + 1 + 1^2 + 1^3 + 1^4 + \cdots = 1 + 1 + 1 + 1 + 1 + \cdots,$$

which clearly diverges. If, instead, $x = 1/2$, then the series converges to 2:

$$S(1/2) = 1 + \frac{1}{2} + \frac{1}{4} + \frac{1}{8} + \cdots + \frac{1}{2^n} + \cdots = 2.$$

Indeed, $S(x)$ is a geometric series (in powers of x) for *any* value of x, and so the series converges if and only if $|x| < 1$. We even know the limit:

$$\text{If } |x| < 1, \text{ then } \quad 1 + x + x^2 + x^3 + x^4 + \cdots \quad \text{converges to} \quad \frac{1}{1-x}.$$

Thus, $S(x)$ converges if $-1 < x < 1$ and diverges otherwise. ∎

EXAMPLE 2 We will show in the next section that, for any number x,

$$e^x = 1 + x + \frac{x^2}{2!} + \frac{x^3}{3!} + \frac{x^4}{4!} + \cdots.$$

Interpret the right side in the language of power series. For which x does the series converge?

Solution Writing the series in the form

$$1 + x + \frac{x^2}{2!} + \frac{x^3}{3!} + \cdots = \sum_{k=0}^{\infty} \frac{1}{k!} x^k$$

shows the pattern of coefficients. (Here $x_0 = 0$ and $a_k = 1/k!$.)

We will use the ratio test to decide where the series converges. (Since the ratio test applies only to *positive* series, we will check for *absolute* convergence.) For any input x,

$$\lim_{k \to \infty} \frac{\left| a_{k+1} x^{k+1} \right|}{\left| a_k x^k \right|} = \lim_{k \to \infty} \frac{|x|^{k+1}}{(k+1)!} \cdot \frac{k!}{|x|^k} = \lim_{k \to \infty} \frac{|x|}{k+1} = 0.$$

Recall: If a series converges with absolute value signs, then it converges without.

Because $0 < 1$, the ratio test guarantees that this series converges absolutely—and therefore also in the ordinary sense—for all values of x. ◂ ∎

Power series and polynomials Power series are, roughly speaking, "infinite-degree" polynomials. Notice these points of similarity:

- **The terms are power functions** For both polynomials and power series, each summand is of the form $a_k x^k$ with k a nonnegative integer.

- **Partial sums are ordinary polynomials** Every partial sum S_n of the power series $\sum_{k=0}^{\infty} a_k x^k$ has the form

$$S_n = a_0 + a_1 x + a_2 x^2 + a_3 x^3 + \cdots + a_n x^n,$$

that is, an ordinary polynomial of degree n.

- **Easy to use** Polynomials are easy to differentiate and integrate, term by term. So are power series—if due care is taken for convergence. ◂ We'll return soon to this theme and see why it matters.

An important proviso!

Choosing base points   The polynomial functions

$$p(x) = 1 + 2x + 3x^2 \quad \text{and} \quad q(x) = 1 + 2(x - 5) + 3(x - 5)^2$$

are not much different—the q-graph is found by sliding the p-graph 5 units to the right. The differences between p and q have to do with different choices of the **base point**: q is said to be **expanded** about $x = 5$, whereas p is expanded about $x = 0$.

The same notion of base point applies to power series. For example, the two power series

$$\sum_{k=0}^{\infty} 2^k x^k \quad \text{and} \quad \sum_{k=0}^{\infty} 2^k (x - 1)^k$$

are written in powers of $(x - 0)$ and powers of $(x - 1)$, and so their respective base points are $x = 0$ and $x = 1$. As with polynomials, the mathematical difference is small: The second series is the result of shifting the first series one unit to the right.

Power series as functions

Any power series

$$S(x) = \sum_{k=0}^{\infty} a_k (x - x_0)^k = a_0 + a_1(x - x_0) + a_2(x - x_0)^2 + a_3(x - x_0)^3 + \cdots$$

defines, in a natural way, a *function* of x. For a given input x, $S(x)$ is the limit—if one exists—of the power series.

Domains of power series   Any function given by a "formula" in x has a natural domain: the set of x for which the formula makes sense. Power series are no different. The domain of a power series function $S(x)$ is the set of inputs x for which the series converges; this set is also called the **interval of convergence**. We have seen, for instance, that $\sum_{k=0}^{\infty} x^k$ converges for x in $(-1, 1)$, whereas $\sum_{k=0}^{\infty} x^k/k!$ converges for x in $(-\infty, \infty)$.

EXAMPLE 3   A function $S(x)$ is defined by the power series

$$S(x) = \sum_{k=0}^{\infty} 2^k x^k = 1 + 2x + 4x^2 + 8x^3 + \cdots .$$

What is the domain of S? Is a simpler formula available for S?

Solution   We *could* use the ratio test. Instead, let's think of $S(x)$ as a geometric series $1 + r + r^2 + r^3 + \cdots$, with $r = 2x$:

$$S(x) = 1 + 2x + (2x)^2 + (2x)^3 + \cdots .$$

A geometric series converges if $|r| < 1$; this one converges, therefore, if $|2x| < 1$, that is, if $|x| < 1/2$. In that case the limit is $1/(1 - r) = 1/(1 - 2x)$. To summarize: The power series $S(x)$ has interval of convergence $(-1/2, 1/2)$; for x in that interval,

$$S(x) = 1 + 2x + (2x)^2 + (2x)^3 + \cdots = \frac{1}{1 - 2x}.$$

Thus, the function S has the simple formula $S(x) = 1/(1 - 2x)$, but the formula holds only if $|x| < 1/2$.

Finding the interval of convergence Given a power series, the first task is to find its interval of convergence. For many series, the ratio test is all that is needed. We illustrate with several (important) examples.

EXAMPLE 4 Where does $S(x) = \sum_{k=0}^{\infty}(x-5)^k$ converge?

Solution The series is geometric in powers of $r = (x-5)$:

$$S(x) = 1 + (x-5) + (x-5)^2 + (x-5)^3 + \cdots ,$$

and so it converges if (and only if) $|r| = |x-5| < 1$. Thus, the convergence interval is $(4, 6)$—the result of shifting $(-1, 1)$ five units to the right.   ∎

EXAMPLE 5 Show that $1 + 2x + 3x^2 + 4x^3 + 5x^4 + \cdots = \sum_{k=1}^{\infty} kx^{k-1}$ converges only for x in $(-1, 1)$. Guess a limit.

Solution We will use the ratio test to check for absolute convergence. The ratio of successive terms is

$$\frac{\left|a_{k+1}x^{k+1}\right|}{\left|a_k x^k\right|} = \frac{(k+1)|x|^k}{k|x|^{k-1}} = |x| \cdot \frac{k+1}{k}.$$

This ratio tends to $|x|$ as $k \to \infty$, and so the series converges absolutely if $|x| < 1$.

If $|x| = 1$, the ratio test is inconclusive. But it is easy to see that if $x = \pm 1$, then

$$\left|kx^{k-1}\right| = |k| \to \infty \quad \text{as} \quad k \to \infty.$$

Thus, by the nth term test, the series *diverges* if $x = \pm 1$. (It diverges for the same reason if $|x| > 1$.) The interval of convergence is therefore $(-1, 1)$, as claimed.

To guess a limit, let's recall from Example 1 that the equation

$$S(x) = 1 + x + x^2 + x^3 + x^4 + \cdots = \frac{1}{1-x} \tag{1}$$

holds for $|x| < 1$. Differentiating all three quantities in Equation (1) suggests a natural guess for the limit of the series at hand:

$$S'(x) = 1 + 2x + 3x^2 + 4x^3 + \cdots = \frac{1}{(1-x)^2}.$$

With $x = 1/2$, for example, our guess is that

$$\sum_{k=1}^{\infty} k\left(\frac{1}{2}\right)^{k-1} = 1 + \frac{2}{2} + \frac{3}{4} + \frac{4}{8} + \cdots = \frac{1}{(1-1/2)^2} = 4.$$

Numerical evidence is promising: For the preceding series, $S_{20} \approx 3.999958038$.   ∎

Good questions Our guess in Example 5 is reasonable (and correct), but it raises some good questions:

- Is it legitimate to differentiate a series term by term?
- On what interval does the resulting series converge?

- Is $S'(x)$ the correct derivative of $S(x)$?
- Can we antidifferentiate a series function term by term?

We explore these questions informally in the following examples and return to them more formally in the next section.

EXAMPLE 6 **(Antidifferentiating a power series)** Antidifferentiating the geometric series

$$S(x) = 1 + x + x^2 + x^3 + \cdots = \sum_{k=1}^{\infty} x^{k-1}$$

term by term gives the new series

$$T(x) = x + \frac{x^2}{2} + \frac{x^3}{3} + \frac{x^4}{4} + \frac{x^5}{5} + \cdots = \sum_{k=1}^{\infty} \frac{x^k}{k}.$$

Where does $T(x)$ converge? Guess a limit.

Solution Because $S(x)$ converges for $|x| < 1$, we might expect the same of $T(x)$. The ratio test supports this guess:

$$\lim_{k \to \infty} \frac{|a_{k+1} x^{k+1}|}{|a_k x^k|} = \lim_{k \to \infty} |x| \cdot \frac{k}{k+1} = |x|,$$

and so $T(x)$ converges (absolutely) on the interval $(-1, 1)$.

What happens at the endpoints? Setting $x = \pm 1$ in $T(x)$ produces two series we have seen before:

$$T(1) = \sum_{k=1}^{\infty} \frac{1}{k}; \qquad T(-1) = \sum_{k=1}^{\infty} \frac{(-1)^k}{k}.$$

As we saw earlier, the first series diverges; the second converges conditionally (by the alternating-series theorem). Thus, T converges for x in $[-1, 1)$.

Because $T(x)$ came from $S(x)$ by antidifferentiation, it is reasonable to guess a similar relationship for limits:

$$1 + x + x^2 + x^3 + \cdots = \frac{1}{1-x} \implies x + \frac{x^2}{2} + \frac{x^3}{3} + \cdots = -\ln(1-x).$$

Numerical evidence suggests that we are right. If $x = 1/2$, the series gives $S_{20} \approx 0.69314714$, which is not far from $-\ln 1/2 \approx 0.69314718$. ∎

EXAMPLE 7 Where does the power series $\sum_{k=0}^{\infty} k! \, x^k$ converge?

Solution Every power series converges at its base point—in this case, $x = 0$. But if $x \neq 0$, the ratio test (applied to absolute values) gives

$$\lim_{k \to \infty} \frac{|a_{k+1} x^{k+1}|}{|a_k x^k|} = \lim_{k \to \infty} \frac{(k+1)! \, |x|^{k+1}}{k! \, |x|^k} = \lim_{k \to \infty} (k+1) \cdot |x| = \infty,$$

and so the series diverges for *all* $x \neq 0$. This power series has, in a sense, the smallest possible domain: It converges *only* if $x = 0$. ∎

Lessons from the examples

The preceding examples illustrate several useful properties of power series and their convergence sets.

In Example 7 the power series converged only at the base point $x = 0$. We'll call the set $\{0\}$ an interval of radius 0.

- **Domains are intervals** In each example we have seen, the convergence set turned out to be an *interval*, centered at the base point. ← This is no accident—*every* power series converges on an interval centered at the base point. The radius of this interval is called the **radius of convergence**; it can be zero, finite, or infinite.

The ratio test shows this.

- **Any radius of convergence is possible** Every positive number R is a possible radius of convergence for a power series. Indeed, the series $\sum_{k=0}^{\infty} x^k / R^k$ has radius of convergence precisely R. ←

- **At endpoints, anything can happen** The series in Example 1 converges on the *open* interval $(-1, 1)$; the series in Example 6 converges on the "half-open" interval $[-1, 1)$. In fact, an interval of convergence may include either, both, or neither of its endpoints—any combination is possible. (In practice, what matters most is the radius of convergence; what happens at the endpoints can be interesting, but it is usually less important.)

The following theorem summarizes our observations.

THEOREM 12 Let $S(x) = \displaystyle\sum_{k=0}^{\infty} a_k (x - x_0)^k$ be a power series. The set of x for which $S(x)$ converges is an interval centered at x_0; endpoints may or may not be contained in the interval. The radius of convergence may be zero, finite, or infinite.

The idea of proof The theorem says that if $S(x)$ converges for $x = C$, then $S(x)$ also converges for $|x| < |C|$. Suppose, for instance, that $\sum_{k=0}^{\infty} a_k x^k$ converges for $x = C = 1$. Then $\sum_{k=0}^{\infty} a_k$ converges. By the nth term test, $a_k \to 0$ as $k \to \infty$, and so the a_k's must be bounded in absolute value. This means that there is a number $M > 0$ such that $|a_k| \le M$ for all k, which implies that

$$\left| a_k x^k \right| \le M |x|^k$$

for all k. Now if $|x| < 1$, then the geometric series $\sum_{k=0}^{\infty} M |x|^k$ converges; by comparison, so does $\sum_{k=0}^{\infty} \left| a_k x^k \right|$. (A general proof for any value of C is not much harder.)

Power series convergence, graphically The nth partial sum of the power series $S(x) = \sum_{k=0}^{\infty} a_k x^k$ is the polynomial

$$p_n(x) = a_0 + a_1 x + a_2 x^2 + a_3 x^3 + \cdots + a_n x^n.$$

To say that the power series converges for x in $(-R, R)$ means that, for any x in that interval, there is a number $S(x)$ such that $p_n(x) \to S(x)$ as $n \to \infty$.

Sorting out exactly what this means and when it happens is a worthy challenge. Figure 1 gives a graphical sense of the situation for the geometric power series $S(x) = \sum_{k=0}^{\infty} x^k$.

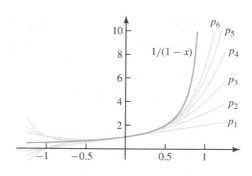

FIGURE 1

On $(-1, 1)$, $\displaystyle\sum_{k=0}^{\infty} x^k$ converges to $\dfrac{1}{1-x}$

The graphs labeled p_1 through p_6 represent the first six partial sums. Over the interval $(-1, 1)$, they appear to approach the graph of the limiting function (shown darker) more and more closely. Outside the interval $(-1, 1)$ the polynomial graphs seem to diverge, failing to approach any common limiting function.

BASIC EXERCISES

1. The power series $\displaystyle\sum_{k=1}^{\infty} x^k/k$ has radius of convergence 1. Plot the partial sum polynomials of degree 1, 2, 4, 6, 8, and 10 over the interval $[-2, 2]$. Is the interval of convergence of the series apparent? Explain.

2. The power series $\displaystyle\sum_{k=1}^{\infty} (x-3)^k/k$ has radius of convergence 1. Plot the partial sum polynomials of degree 1, 2, 4, 6, 8, and 10 over the interval $[1, 5]$. Is the interval of convergence of the series apparent? Explain.

In Exercises 3–6, find the radius of convergence of the power series.

3. $\displaystyle\sum_{j=1}^{\infty} \left(\dfrac{x}{2}\right)^j$

4. $\displaystyle\sum_{k=1}^{\infty} \dfrac{x^k}{k\,3^k}$

5. $\displaystyle\sum_{k=1}^{\infty} \dfrac{x^k}{\sqrt{k}}$

6. $\displaystyle\sum_{n=0}^{\infty} \dfrac{x^n}{n!+n}$

In Exercises 7–10, find the radius and the interval of convergence of the power series.

7. $\displaystyle\sum_{n=0}^{\infty} (x-2)^n$

8. $\displaystyle\sum_{n=2}^{\infty} \dfrac{(x-3)^{2n}}{n^4}$

9. $\displaystyle\sum_{n=2}^{\infty} \dfrac{(x+5)^n}{n \ln n}$

10. $\displaystyle\sum_{n=1}^{\infty} \dfrac{(x+1)^n}{n}$

FURTHER EXERCISES

11. Show that $\displaystyle\sum_{k=0}^{\infty} \dfrac{x^k}{R^k}$ converges on $(-R, R)$, where $R > 0$ is a constant.

12. Show that $\displaystyle\sum_{k=1}^{\infty} \dfrac{x^k}{kR^k}$ converges on $[-R, R)$, where $R > 0$ is a constant.

13. Show that $\displaystyle\sum_{k=1}^{\infty} \dfrac{x^k}{k^2 R^k}$ converges on $[-R, R]$, where $R > 0$ is a constant.

14. Concoct a power series that converges on $(-R, R]$, where $R > 0$ is a constant.

In Exercises 15–20, give an example of a power series that has the given interval as its interval of convergence.
[HINT: See Exercises 11–14.]

15. $[-4, 4)$

16. $(-3, 3]$

17. $[-1, 5]$

18. $(-4, 0)$

19. $(8, 16]$

20. $[-11, -3)$

In Exercises 21–24, suppose that the power series $\sum\limits_{k=0}^{\infty} a_k x^k$ converges only if $-2 < x \le 2$.

21. Explain why the radius of convergence of the power series is 2.

22. Explain why the power series $\sum\limits_{k=0}^{\infty} a_k(x-1)^k$ has radius of convergence 2.

23. Show that the interval of convergence of $\sum\limits_{k=0}^{\infty} a_k(x-3)^k$ is $(1, 5]$.

24. Find the interval of convergence of $\sum\limits_{k=0}^{\infty} a_k(x+1)^k$.

25. Suppose that $\sum\limits_{k=0}^{\infty} a_k(x-b)^k$ converges only if $-11 \le x < 17$.

 (a) What is the radius of convergence of the power series?

 (b) Determine the value of b.

26. Suppose that $\sum\limits_{n=1}^{\infty} a_n$ converges and that $|x| < 1$. Show that $\sum\limits_{n=1}^{\infty} a_n x^n$ converges absolutely.

Suppose that the power series $\sum\limits_{k=0}^{\infty} a_k x^k$ converges if $x = -3$ and diverges if $x = 7$. In Exercises 27–32, indicate whether the statement must be true, cannot be true, or may be true. Justify your answers.

27. The power series converges if $x = -10$.

28. The power series diverges if $x = 3$.

29. The power series converges if $x = 6$.

30. The power series diverges if $x = 2$.

31. The power series diverges if $x = -7$.

32. The power series converges if $x = -4$.

Suppose that the power series $\sum\limits_{k=0}^{\infty} a_k (x+2)^k$ converges if $x = -7$ and diverges if $x = 7$. In Exercises 33–38, indicate whether the statement must be true, cannot be true, or may be true. Justify your answers.

33. The power series converges if $x = -8$.

34. The power series converges if $x = 1$.

35. The power series converges if $x = 3$.

36. The power series diverges if $x = -11$.

37. The power series diverges if $x = 5$.

38. The power series diverges if $x = -5$.

In Exercises 39–42, let $f(x) = \sum\limits_{n=0}^{\infty} \dfrac{2x^n}{3^n + 5}$.

39. Explain why 10 is not in the domain of f.

40. Which of the numbers $0.5, 1.5, 3$, and 6 are in the domain of f? Justify your answer.

41. Estimate $f(1)$ within 0.01 of its exact value.

42. Estimate $f(-1.5)$ within 0.01 of its exact value.

In Exercises 43–46, let $h(x) = \sum\limits_{k=0}^{\infty} \dfrac{(x-2)^k}{k! + k^3}$.

43. What is the domain of h? Justify your answer.

44. Estimate $h(0)$ within 0.005 of its exact value.

45. Estimate $h(1)$ within 0.005 of its exact value.

46. Estimate $h(3)$ within 0.005 of its exact value.

In Exercises 47–50, let $g(x) = \sum\limits_{n=1}^{\infty} \dfrac{(x+4)^n}{n^3 5^n}$.

47. What is the domain of g?

48. Estimate $g(0)$ within 0.005 of its exact value.

49. Estimate $g(1)$ within 0.005 of its exact value.

50. Estimate $g(-5)$ within 0.005 of its exact value.

11.6   POWER SERIES AS FUNCTIONS

Any power series

$$S(x) = \sum_{k=0}^{\infty} a_k (x - x_0)^k = a_0 + a_1(x - x_0) + a_2(x - x_0)^2 + a_3(x - x_0)^3 + \cdots$$

can be thought of as a function of x; its domain is the series' interval of convergence—an interval centered at x_0. In this section we explore the remarkable—and useful—properties of functions defined by power series. (For simplicity we will mainly use $x_0 = 0$.)

Calculus with power series

Given *any* power series $S(x)$, convergent or divergent, it is easy to differentiate or antidifferentiate term by term to produce *new* series we will call $D(x)$ and $A(x)$:

$$S(x) = \sum_{k=0}^{\infty} a_k x^k = a_0 + a_1 x + a_2 x^2 + a_3 x^3 + \cdots \ ;$$

$$D(x) = \sum_{k=1}^{\infty} k a_k x^{k-1} = a_1 + 2a_2 x + 3a_3 x^2 + \cdots \ ;$$

$$A(x) = \sum_{k=0}^{\infty} a_k \frac{x^{k+1}}{k+1} = a_0 x + a_1 \frac{x^2}{2} + a_2 \frac{x^3}{3} + a_3 \frac{x^4}{4} + \cdots \ .$$

Important questions arise:

- If the series S has radius of convergence r, is the same true of D and A?
- Even if we assume that S, D, and A all converge on $(-r, r)$, must D be the derivative of S in the ordinary calculus sense? Must A be an antiderivative of S?

The following theorem answers all these questions in the affirmative. Conveniently, everything works just as we would hope.

> **THEOREM 13 (Derivatives and antiderivatives of power series)** Let $S(x)$ be a power series with radius of convergence $r > 0$. Let $D(x)$ and $A(x)$ be defined as above.
>
> - Both D and A have radius of convergence r.
> - If $|x - x_0| < r$, $D(x) = S'(x)$.
> - If $|x - x_0| < r$, $A'(x) = S(x)$.

The theorem says, among other things, that a function S given by a power series is differentiable and that its derivative is another power series S' with the same radius of convergence as S. The same principle applies to S', to S'', and so on to show that S has infinitely many derivatives—all available by repeated term-by-term differentiation.

EXAMPLE 1 For any x, $e^x = 1 + x + \dfrac{x^2}{2!} + \dfrac{x^3}{3!} + \cdots$. Explain why.

Solution Let $S(x)$ represent the preceding series. We saw in the last section that $S(x)$ converges for *all* x. By Theorem 13, S' can be found by differentiating S term by term. In this case, differentiation produces a curious result:

Differentiating S term by term leaves S unchanged.

But *every* differentiable function S for which $S' = S$ has the form $S(x) = Ce^x$. Since $S(0) = 1$, it follows that $C = 1$; thus, $S(x) = e^x$, as claimed. ■

Writing known functions as power series

In earlier examples we have written several functions, including $1/(1-x)$ and e^x, in power series form. Can *other* familiar functions be "represented" as power series? If so, how?

Theorem 13 suggests some answers. But first let's address an even more basic question.

Why bother? Examples with the sine function   Why write a function as a power series? What good, for instance, is the equation

$$\sin x = x - \frac{x^3}{3!} + \frac{x^5}{5!} - \frac{x^7}{7!} + \frac{x^9}{9!} - \cdots, \tag{1}$$

We'll explain in the next section why the equation holds.

which holds for all real numbers x? ◄ The next two examples suggest some answers.

Trigonometric functions lack finite algebraic formulas; power series are the next best thing. Using them lets us find approximate—but very accurate—values of trigonometric (and other transcendental) functions.

E X A M P L E 2 **(Transcendental family values)**   Use the series expression (1) to approximate $\sin 1$ accurately.

Solution   Substituting $x = 1$ into Equation (1) gives

$$\sin 1 = 1 - \frac{1}{3!} + \frac{1}{5!} - \frac{1}{7!} + \frac{1}{9!} - \frac{1}{11!} + \cdots,$$

an alternating series with terms decreasing (rapidly!) to zero. By the alternating series theorem, each partial sum S_n of such a series differs from the limit by no more than the size of the next term, a_{n+1}. In particular, the partial sum

$$1 - \frac{1}{3!} + \frac{1}{5!} - \cdots - \frac{1}{11!} \approx 0.8414709846$$

differs from $\sin 1$ by no more than $1/13! \approx 2 \times 10^{-10}$. In other words, our estimate is *guaranteed* accurate to at least nine decimal places—not bad for so little work!   ∎

As we have seen often, many integrals cannot be calculated in "closed form," by elementary antidifferentiation. Numerical methods—such as the midpoint rule—may help. Infinite series, being easy to integrate, offer another way to transcend our troubles.

E X A M P L E 3 **(Hard integrals made easy)**   Find, in series form, an antiderivative for $\sin(x^2)$. Use it to estimate $I = \int_0^1 \sin(x^2)\, dx$. (For comparison, the midpoint rule applied to I gives $M_{50} \approx 0.31025$.)

Solution   The function $\sin(x^2)$ has no *elementary* antiderivative, but it is easy to find an antiderivative in *series* form. To do so, we replace x with x^2 in the sine series (1) to produce the new series

$$\sin(x^2) = x^2 - \frac{x^6}{3!} + \frac{x^{10}}{5!} - \frac{x^{14}}{7!} + \cdots.$$

(Because the original series converges for all x, so does this one.)

It, too, converges for all x.

Antidifferentiating term by term gives yet another power series: ◄

$$\int_0^x \sin(t^2)\, dt = \frac{x^3}{3} - \frac{x^7}{7 \cdot 3!} + \frac{x^{11}}{11 \cdot 5!} - \frac{x^{15}}{15 \cdot 7!} + \cdots.$$

The result is not an elementary function, but it is an honest antiderivative for $\sin(x^2)$, and so we can find our definite integral in the obvious way:

$$\int_0^1 \sin(x^2)\, dx = \frac{x^3}{3} - \frac{x^7}{7 \cdot 3!} + \frac{x^{11}}{11 \cdot 5!} - \frac{x^{15}}{15 \cdot 7!} + \cdots \Bigg]_0^1$$

$$= \frac{1}{3} - \frac{1}{7 \cdot 3!} + \frac{1}{11 \cdot 5!} - \frac{1}{15 \cdot 7!} + \cdots.$$

The alternating series theorem applies to the last series, and so the estimate

$$\int_0^1 \sin(x^2)\,dx \approx \frac{1}{3} - \frac{1}{7 \cdot 3!} + \frac{1}{11 \cdot 5!} - \frac{1}{15 \cdot 7!} \approx 0.3102681578$$

is in error by no more than $1/(19 \cdot 9!) \approx 1.5 \times 10^{-7}$ (the size of the next term). The midpoint rule estimate agrees with the result through four decimal places.   ▪

New series from old: Help from algebra and calculus Power series can be found for many familiar functions by applying simple algebra or calculus operations to a few standard known series. Differentiating the sine series, for instance, gives the new series

$$\cos x = 1 - \frac{x^2}{2!} + \frac{x^4}{4!} - \frac{x^6}{6!} + \cdots,$$

which, like the old one, converges for all x.

We can also start with another famous series,

$$\frac{1}{1-x} = 1 + x + x^2 + x^3 + x^4 + x^5 + \cdots, \tag{2}$$

which converges for $|x| < 1$. With a little algebraic ingenuity we can produce many other useful series, all converging on the same set. Replacing x with $-x$ in Equation (2) gives the alternating series

$$\frac{1}{1+x} = 1 - x + x^2 - x^3 + x^4 - x^5 + \cdots. \tag{3}$$

Replacing x with x^2 in Equation (3) gives still another alternating series:

$$\frac{1}{1+x^2} = 1 - x^2 + x^4 - x^6 + x^8 - x^{10} + \cdots.$$

Integrating *this* series term by term gives another striking result:

$$\arctan x = x - \frac{x^3}{3} + \frac{x^5}{5} - \frac{x^7}{7} + \frac{x^9}{9} - \cdots.$$

Setting $x = 1$ in this last series yields a *really* remarkable result:

$$\frac{\pi}{4} = 1 - \frac{1}{3} + \frac{1}{5} - \frac{1}{7} + \frac{1}{9} - \cdots.$$

(Caution is needed—these simple arguments show only that the series for $\arctan x$ is valid if $-1 < x < 1$. Showing carefully that the series converges to $\pi/4$ at the endpoint $x = 1$ requires further argument.)

Multiplying power series Convergent power series can be multiplied together, something like polynomials, to form new convergent series. As always with series, convergence is a question. Here is the answer: The product of two power series converges wherever both factors converge.

EXAMPLE 4 We have seen already that

$$\frac{1}{1-x} = 1 + x + x^2 + x^3 + \cdots, \quad \text{and} \quad \frac{1}{1+x} = 1 - x + x^2 - x^3 + \cdots,$$

Multiply these series. Where does the new series converge? What familiar function does the result represent?

Solution   Symbolically, the problem looks like this:

$$(1+x+x^2+x^3+\cdots)\cdot(1-x+x^2-x^3+\cdots)=a_0+a_1x+a_2x^2+a_3x^3+\cdots.$$

We want numerical values for the constants on the right.

Both factors have infinitely many summands, and so ordinary expansion quickly gets out of hand. To avoid this, we collect like powers right from the start. It is clear, for instance, that $a_0=1\cdot1=1$; no other combination of factors yields a constant result. Similarly, tracking the first and second powers of x gives

$$a_1=1\cdot(-1)+1\cdot1=0;\qquad a_2=1\cdot1+1\cdot(-1)+1\cdot1=1.$$

Continuing this process reveals a simple pattern:

$$(1+x+x^2+x^3+\cdots)\cdot(1-x+x^2-x^3+\cdots)=1+x^2+x^4+x^6+\cdots.$$

The result is a geometric series in powers of x^2; it converges for $|x^2|<1$, that is, if $-1<x<1$.

What familiar function does the product series represent? Because the two factors represent the functions $1/(1-x)$ and $1/(1+x)$, it follows that the product *series* represents the product *function* $1/(1-x^2)$. ∎

EXAMPLE 5   Find a power series for $\ln(1+x)$; use it to estimate $\ln(1.5)$ with error guaranteed to be less than 0.0001.

Solution   Integrating both sides of Equation (3) gives

$$\ln(1+x)=\int\frac{1}{1+x}\,dx=\int\left(1-x+x^2-x^3+\cdots\right)\,dx$$

$$=x-\frac{x^2}{2}+\frac{x^3}{3}-\frac{x^4}{4}+\cdots.$$

(We used $C=0$ as the constant of integration because our "target" function $\ln(1+x)$ has the value 0 when $x=0$.) The new series, like the old, converges for x in $(-1,1)$. To estimate $\ln(1.5)$, we plug in $x=0.5$:

$$\ln(1.5)=0.5-\frac{0.5^2}{2}+\frac{0.5^3}{3}-\frac{0.5^4}{4}+\cdots=\sum_{k=1}^{\infty}(-1)^{k+1}\frac{0.5^k}{k}.$$

Now the alternating series theorem applies. To achieve our target accuracy, we can use any partial sum S_n for which

$$c_{n+1}=\frac{0.5^{n+1}}{n+1}<0.0001.$$

It is easy to see that $n=10$ works, with room to spare. In fact, $S_{10}\approx0.405435$; this compares closely with the value $\ln1.5\approx0.405465$ reported by a calculator. ∎

A brief atlas of power series   For easy reference, we collect a short list of "standard" power series for basic calculus functions. In the next section we will see how to show rigorously that the series really converge to the stated limits.

A power series sampler		
Function	**Series**	**Convergence interval**
$\sin x$	$x - \dfrac{x^3}{3!} + \dfrac{x^5}{5!} - \dfrac{x^7}{7!} + \dfrac{x^9}{9!} - \cdots$	$(-\infty, \infty)$
$\cos x$	$1 - \dfrac{x^2}{2!} + \dfrac{x^4}{4!} - \dfrac{x^6}{6!} + \dfrac{x^8}{8!} - \cdots$	$(-\infty, \infty)$
e^x	$1 + x + \dfrac{x^2}{2!} + \dfrac{x^3}{3!} + \dfrac{x^4}{4!} + \dfrac{x^5}{5!} + \cdots$	$(-\infty, \infty)$
$\dfrac{1}{1-x}$	$1 + x + x^2 + x^3 + x^4 + x^5 + \cdots$	$(-1, 1)$
$\dfrac{1}{1+x}$	$1 - x + x^2 - x^3 + x^4 - x^5 + \cdots$	$(-1, 1)$
$\dfrac{1}{1+x^2}$	$1 - x^2 + x^4 - x^6 + x^8 - x^{10} + \cdots$	$(-1, 1)$
$\arctan x$	$x - \dfrac{x^3}{3} + \dfrac{x^5}{5} - \dfrac{x^7}{7} + \dfrac{x^9}{9} - \cdots$	$[-1, 1]$

What's next? A power series for any function

As we have seen, knowing a power series expression for one function can lead, via various manipulations, to power series for related functions. A good question remains:

> Given any function f, how can we find a power series "from scratch," without knowing a related series to begin with?

We answer this question in the next section.

BASIC EXERCISES

In Exercises 1–3, $f(x) = \displaystyle\sum_{k=0}^{\infty} \left(\dfrac{x}{2}\right)^k$.

1. What is the radius of convergence of the power series for f?

2. According to Theorem 13, $f'(x) = \displaystyle\sum_{k=1}^{\infty} \dfrac{k\,x^{k-1}}{2^k}$. What is the radius of convergence of the series for f'?

3. According to Theorem 13, $F(x) = \displaystyle\sum_{k=0}^{\infty} \dfrac{x^{k+1}}{(k+1)\,2^k}$ is an antiderivative of f. What is the radius of convergence of the series for F?

4. Let $f(x) = \ln(1+x)$. Show that the power series for f and f' have the same radius of convergence but not the same interval of convergence.

In Exercises 5–8, use the power series representation of $1/(1-x)$ to produce a power series representation of the function f.

5. $f(x) = \dfrac{x^2}{1+x}$

6. $f(x) = \dfrac{1}{1-x^2}$

7. $f(x) = \dfrac{1}{(1+x)^2}$

8. $f(x) = \dfrac{x}{1-x^4}$

In Exercises 9–12, find a power series representation of the function and the radius of convergence of the power series. Then plot the function and the fifth-order polynomial that is a partial sum of the power series on the same axes. [HINT: Write out the first few terms of the series before trying to find the form of the general term.]

9. $f(x) = \arctan(2x)$

10. $f(x) = \cos(x^2)$

11. $f(x) = x^2 \sin x$

12. $f(x) = \ln\left(1 + x^3\right)$

13. Use the partial sum of a series to estimate $1/\sqrt{e}$ with an error less than 0.005.

14. Use the partial sum of a series to estimate $\displaystyle\int_0^{0.2} xe^{-x^3}\,dx$ with an error less than 10^{-5}.

15. Use power series to show that $\lim\limits_{x \to 0} \dfrac{(\sin x - x)^3}{x(1-\cos x)^4} = -\dfrac{2}{27}$.

16. Show that $\lim\limits_{x \to 0^+} \dfrac{x - \sin x}{(x \sin x)^{3/2}} = \dfrac{1}{6}$.

FURTHER EXERCISES

In Exercises 17–22, find a power-series representation of the function and the radius of convergence of the power series.

17. $f(x) = \dfrac{1}{2+x}$

18. $f(x) = \sin\left(x^3\right)$

19. $f(x) = \sin x + \cos x$

20. $f(x) = \ln\left(1+x^2\right)$

21. $f(x) = (x^2 - 1)\sin x$

22. $f(x) = \ln\left(\dfrac{1+x}{1-x}\right)$

23. Use power series to show that $y = e^x$ is a solution of the DE $y' = y$.

24. Use power series to show that $y = 2e^x$ is the solution of the IVP $y' = y$, $y(0) = 2$.

25. Use power series to show that $y = e^{3x}$ is the solution of the IVP $y' = 3y$, $y(0) = 1$.

26. Use power series to show that $y = \sin x$ is a solution of the DE $y'' = -y$.

27. Use power series to show that $y = 1/(1-x)$ is the solution of the IVP $y' = y^2$, $y(0) = 1$.

28. (a) Show that $f(x) = \tan x$ is the solution of the IVP $f'(x) = 1 + (f(x))^2$, $f(0) = 0$.

(b) Use part (a) to find the first four nonzero terms in the power series representation of $\tan x$.

In Exercises 29–34, use power series to evaluate the limit. Check your answer using l'Hôpital's rule.

29. $\lim\limits_{x \to 0} \dfrac{\sin x}{x}$

30. $\lim\limits_{x \to 0} \dfrac{e^x - 1}{x}$

31. $\lim\limits_{x \to 0} \dfrac{1 - \cos x}{x^2}$

32. $\lim\limits_{x \to 0} \dfrac{\arctan x}{x}$

33. $\lim\limits_{x \to 0} \dfrac{\ln(1+x) - x}{x^2}$

34. $\lim\limits_{x \to 0} \dfrac{1 - \cos^2 x}{x}$ [HINT: $1 - \cos^2 x = (1 - \cos(2x))/2$.]

35. Evaluate $\sum\limits_{n=1}^{\infty} \dfrac{n}{2^n}$ exactly. [HINT: If $f(x) = \sum\limits_{n=0}^{\infty} x^n$, then $f'(x) = \sum\limits_{n=1}^{\infty} n x^{n-1}$.]

36. Show that $\dfrac{1}{x-1} = \sum\limits_{k=1}^{\infty} \dfrac{1}{x^k}$ if $|x| > 1$.

[HINT: $1/(x-1) = x/(x-1) - 1$.]

37. (a) Use the formula for the sum of a geometric series to show that

$$\sum_{k=1}^{\infty} \dfrac{x^k}{k} = -\ln|1-x|.$$

(b) What is the interval of convergence of the series in part (a)?

(c) Show that if $N \geq 1$, then

$$0 < \ln 2 - \sum_{k=1}^{N} \dfrac{1}{k \, 2^k} \leq \dfrac{1}{(N+1)2^N}.$$

[HINT: $-\ln(1/2) = \ln 2$.]

38. Use a power series to show that $x - x^2/2 < \ln(1+x) < x$ for all x in the interval $(0, 1)$.

39. Use power series to show that $1 - \cos x < \ln(1+x) < \sin x$ if $0 < x < 1$.

40. Let $f(x) = \dfrac{1}{1+x^4}$.

(a) Find a power series representation of f.

(b) What is the interval of convergence of the series in part (a)?

(c) Use the series found in part (a) to evaluate $\int_0^{0.5} f(x)\,dx$ with an error no greater than 0.001.

41. (a) Find the power series representation of an antiderivative of e^{-x^2}.

(b) Use the result from part (a) to estimate $\int_0^1 e^{-x^2}\,dx$ within 0.005 of its exact value.

42. Estimate $\int_0^1 \cos(x^2)\,dx$ with an error no greater than 0.005.

43. Estimate $\int_0^1 \sqrt{x}\sin x\,dx$ with an error no greater than 0.001.

44. Estimate $\int_0^{1000} \dfrac{e^{-10x}\sin x}{x}\,dx$ with an error no greater than 5×10^{-5}.

In Exercises 45–50, find the first four nonzero terms in the power series representation of the function.

45. $f(x) = e^{2x} \ln(1 + x^3)$

46. $f(x) = \arctan x \sin(4x)$

47. $f(x) = \dfrac{e^x}{1 - x}$

48. $f(x) = \dfrac{\cos x}{1 + x^2}$

49. $f(x) = e^{\sin x}$

50. $f(x) = \ln(1 + \sin x)$

In Exercises 51–54, find the elementary function represented by the power series by manipulating a more familiar power series (e.g., the series for $\cos x$, $\sin x$, $(1 - x)^{-1}$).

51. $\displaystyle\sum_{k=1}^{\infty} kx^{k-1}$

52. $\displaystyle\sum_{k=0}^{\infty} \dfrac{x^k}{(k+1)!}$

53. $\displaystyle\sum_{k=1}^{\infty} (-1)^{k+1} x^k$

54. $\displaystyle\sum_{k=1}^{\infty} \dfrac{(2x)^k}{k}$

11.7 TAYLOR SERIES

In the preceding section we saw some of the practical advantages of writing a function as a power series. We also saw ways of using a power series for one function to derive power series for other functions. In this section we show, given a suitable function, how to find its power series "from scratch" by matching derivatives. The idea is not new—we did the same thing in Section 9.1, where we found Taylor polynomials. Here we extend these ideas to find **Taylor series**—"infinite Taylor polynomials."

Does every function have a power series? Mathematical life would be simpler if *every* function $f(x)$ could be written as a power series—ideally, a series that converges to $f(x)$ for all x. Alas, it isn't so. At least two things can go wrong:

- **Smaller domains** A power series may have a smaller domain than the function it represents. For instance, the series equation

$$\frac{1}{1 + x^2} = 1 - x^2 + x^4 - x^6 + x^8 - x^{10} + \cdots$$

  holds if—but only if—$|x| < 1$, even though the left side is defined (and well behaved) for all real numbers x.

- **No series at all** A function may have no series at all. As Theorem 13 (page 591) says, every power series can be differentiated repeatedly on its interval of convergence. Thus, a function that *has* a power series with base point $x = x_0$ must itself be repeatedly differentiable at $x = x_0$. This means, for instance, that $f(x) = |x|$ (which is not differentiable at $x = 0$) has *no* power series with base point $x = 0$.

Each derivative is another power series.

Coefficients and derivatives at the base point

A simple equation relates coefficients and derivatives of a power series. (We saw the same principle at work for Taylor polynomials.) If

$$S(x) = a_0 + a_1(x - x_0) + a_2(x - x_0)^2 + a_3(x - x_0)^3 + a_4(x - x_0)^4 + \cdots,$$

then, clearly, $S(x_0) = a_0$. Differentiating S repeatedly gives

$$S'(x) = a_1 + 2a_2(x - x_0) + 3a_3(x - x_0)^2 + 4a_4(x - x_0)^3 + \cdots;$$
$$S''(x) = 2a_2 + 6a_3(x - x_0) + 12a_4(x - x_0)^2 + \cdots;$$
$$S'''(x) = 6a_3 + 24a_4(x - x_0) + \cdots;$$
$$\vdots$$

These equations show that

$$S(x_0) = a_0; \quad S'(x_0) = a_1; \quad S''(x_0) = 2a_2; \quad S'''(x_0) = 6a_3; \quad \ldots.$$

The following Fact captures the general pattern:

> **FACT**   If $S(x) = \sum_{k=0}^{\infty} a_k(x - x_0)^k$, then $a_k = \dfrac{S^{(k)}(x_0)}{k!}$ for all $k \geq 0$.

This Fact amounts to a recipe for cooking up a power series for a given function expanded about a given base point x_0.

EXAMPLE 1   Assuming that $f(x) = \sin x$ *has* a power series based at $x = 0$, find it.

Solution   To find the desired series

$$f(x) = a_0 + a_1 x + a_2 x^2 + a_3 x^3 + \cdots,$$

we need "only" find the coefficients $a_0, a_1, a_2, a_3, \ldots$. Finding infinitely many of *anything* sounds difficult, but in this case it's easier than it seems. For the sine function (and for many functions of interest) the coefficients follow a simple pattern. The first few derivatives of $f(x) = \sin x$ reveal the pattern:

$$f(0), f'(0), f''(0), f'''(0), f^{(4)}(0), \ldots = 0, 1, 0, -1, 0, \ldots.$$

By the preceding Fact, the series coefficients have a similar pattern:

$$a_0, a_1, a_2, a_3, a_4, a_5, a_6, \ldots = 0, 1, 0, -\frac{1}{3!}, 0, \frac{1}{5!}, 0, \ldots.$$

Thus, the sine series is the one we have seen before:

$$\sin x = x - \frac{x^3}{3!} + \frac{x^5}{5!} - \frac{x^7}{7!} + \cdots.$$

It is not hard (using the ratio test) to show that this series converges for *all* x, as we had hoped.   ∎

Taylor series

Using the preceding Fact as a recipe for the coefficients, we can write a power series for *any* function f that has repeated derivatives at $x = x_0$.

> **DEFINITION (Taylor series)**   Let f be any function with infinitely many derivatives at $x = x_0$. The **Taylor series** for f is the series $\sum_{k=0}^{\infty} a_k(x - x_0)^k$ with coefficients given by
>
> $$a_k = \frac{f^{(k)}(x_0)}{k!}, \quad k = 0, 1, 2, \ldots.$$

If $x_0 = 0$, then the series looks a little simpler; such series are called **Maclaurin series** after the 17th-century Scottish mathematician Colin Maclaurin.

EXAMPLE 2 The function $f(x) = \ln x$ is not defined at $x = 0$, and so there is no Maclaurin series. Expand f about $x = 1$.

Solution Derivatives of $f(x) = \ln x$ are easy for either a human or a machine to calculate. Here are several:

$$f'(x) = \frac{1}{x}; \quad f''(x) = -\frac{1}{x^2}; \quad f'''(x) = \frac{2}{x^3}; \quad f^{(4)}(x) = -\frac{6}{x^4}.$$

At $x = 1$, therefore, we have

$$f(1) = 0; \quad f'(1) = 1; \quad f''(1) = -1; \quad f'''(1) = 2; \quad f^{(4)}(1) = -6, \quad \ldots,$$

and so the Taylor coefficients are

$$a_0 = 0; \quad a_1 = 1; \quad a_2 = -\frac{1}{2}; \quad a_3 = \frac{1}{3}; \quad a_4 = -\frac{1}{4}; \quad \ldots.$$

The Taylor series is therefore

$$(x-1) - \frac{(x-1)^2}{2} + \frac{(x-1)^3}{3} - \frac{(x-1)^4}{4} + \cdots = \sum_{k=1}^{\infty} \frac{(-1)^{k+1}}{k}(x-1)^k.$$ ∎

Finding Taylor series: help from technology Computing the necessary derivatives to find a Taylor series by hand can be tedious and error prone. Technology can help. The following, for instance, is how one computer algebra system finds Taylor polynomials (partial sums of Taylor series) in the cases illustrated in the preceding examples:

```
> taylorpoly( sin(x), x=0, 7 );
            3          5            7
     x - 1/6 x + 1/120 x - 1/5040 x
> taylorpoly( ln(x), x=1, 7 );
                  2            3              4
    x - 1 - 1/2 (x - 1) + 1/3 (x - 1) - 1/4 (x - 1)
                  5            6              7
      + 1/5 (x - 1) - 1/6 (x - 1) + 1/7 (x - 1)
```

Converging to the right place: Taylor's theorem

Any function f that is infinitely differentiable at $x = x_0$ has a Taylor series—ideally, one with a large radius of convergence. One possible problem remains: The series might converge at x but perhaps to a limit other than $f(x)$.

Taylor's theorem guarantees that this unfortunate event seldom occurs. We met Taylor's theorem in Section 9.2 (see Theorem 2, page 504); we used it there to predict how closely a Taylor polynomial approximates its "target" function f. That question is relevant here because Taylor *polynomials* are partial sums of Taylor *series*.

THEOREM 14 (Taylor's theorem) Suppose that f is repeatedly differentiable on an interval I containing x_0 and that

$$P_n(x) = a_0 + a_1(x - x_0) + a_2(x - x_0)^2 + \cdots + a_n(x - x_0)^n$$

is the nth-order Taylor polynomial based at x_0. Suppose that for all x in I,

$$\left| f^{(n+1)}(x) \right| \leq K_{n+1}.$$

Then $\left| f(x) - P_n(x) \right| \leq \dfrac{K_{n+1}}{(n+1)!} |x - x_0|^{n+1}.$

Observe:

- **Estimating difference**   Taylor's theorem estimates the difference between $f(x)$ and $P_n(x)$. Unless K_{n+1} grows very quickly with n, this difference tends to 0 as $n \to \infty$, and so the series converges to $f(x)$.

- **Why does it hold?**   We discussed the theorem's proof at length in Section 9.2; an important ingredient is the mean value theorem.

EXAMPLE 3   For $f(x) = \sin x$, show that the Maclaurin series converges to $\sin x$ for *every* value of x.

Solution   For $f(x) = \sin x$, derivatives of all orders are sines or cosines or their opposites, and so the inequality

$$\left| f^{(n+1)}(x) \right| \le 1$$

holds for all x and for all n. Thus, we can use $K_{n+1} = 1$ for all n in Taylor's theorem:

$$\left| \sin x - P_n(x) \right| \le \frac{1 \cdot |x|^{n+1}}{(n+1)!}.$$

The last quantity tends to 0 as $n \to \infty$, regardless of the value of x. Hence, the series converges, for all x, to $\sin x$. The graphs in Figure 1 suggest what this convergence means geometrically. ◄

The same figure appears, with additional comments, on page 500.

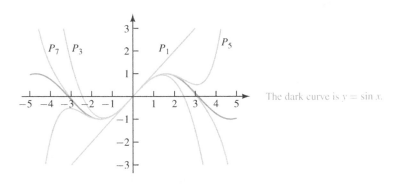

FIGURE 1
Maclaurin polynomials converging to $f(x) = \sin x$

Notice how the higher-order Maclaurin polynomials approximate the sine function better than lower-order ones. For instance, P_7, "sticks with" the sine curve much farther than P_1 or P_3.

BASIC EXERCISES

1. Let $f(x) = \displaystyle\int_3^x \sqrt{t}e^{-t} \, dt$.

(a) Show that if $x \approx 3$, then

$$f(x) \approx \sqrt{3}e^{-3}(x-3) - \frac{5}{12}\sqrt{3}e^{-3}(x-3)^2 + \frac{23}{216}\sqrt{3}e^{-3}(x-3)^3.$$

(b) Use Theorem 14 to bound the error made when $f(3.5)$ is estimated using the polynomial in part (a).

2. Let $f(x) = \sqrt{1+x}$.

(a) Find the first three nonzero terms in the Maclaurin series for f.

(b) Use Theorem 14 to bound the approximation error made if the Maclaurin polynomial from part (a) is used to estimate $f(1)$.

3. Let $f(x) = e^{2x}$. What is the coefficient of x^{100} in the Maclaurin series representation of f?

4. Let $f(x) = \dfrac{x}{1-x^3}$.

   (a) Find the Maclaurin series for $f(x)$.

   (b) What is the interval of convergence of the series in part (a)?

   (c) Use part (a) to find a power series for $f''(x)$.

   (d) Use part (a) to find the Maclaurin series for $\displaystyle\int_0^x f(t)\,dt$.

5. **(a)** Find the Maclaurin series representation of $f(x) = \dfrac{1}{2+x}$. [HINT: $\dfrac{1}{2+x} = \dfrac{1}{2} \cdot \dfrac{1}{1+(x/2)}$.]

   (b) Find $f^{(259)}(0)$ exactly.

6. Suppose that f is a function such that $f(1) = 1$, $f'(1) = 2$, and $f''(x) = 1/(1+x^3)$ for $x > -1$.

   (a) Estimate $f(1.5)$ using a quadratic Taylor polynomial.

   (b) Find an upper bound on the approximation error made in part (a).

FURTHER EXERCISES

7. Use Theorem 14 to show that the Maclaurin series for e^x converges to e^x for all x.

8. Use Theorem 14 to show that the Maclaurin series for $1/(1+x)$ converges to $1/(1+x)$ if $-1/2 < x < 1$. [HINT: Consider the cases $-1/2 < x < 0$ and $0 \le x < 1$ separately.]

9. Suppose that f is a function such that $f^{(n)}$ exists for all $n \ge 1$.

   (a) Explain why the Maclaurin series for f converges to f if $\left| f^{(n)}(x) \right| \le n$ for all $n \ge 1$ and all x.

   (b) Does Theorem 13 guarantee that the Maclaurin series for f converges to f if $\left| f^{(n)}(x) \right| \le 2^n$ for all $n \ge 1$ and all x?

10. Let $f(x) = \begin{cases} \dfrac{1-\cos x}{x^2} & \text{if } x \ne 0, \\ \dfrac{1}{2} & \text{if } x = 0. \end{cases}$ Evaluate $f^{(100)}(0)$.

11. Let $f(x) = \begin{cases} x^{-1}\sin x & \text{if } x \ne 0, \\ 1 & \text{if } x = 0. \end{cases}$

   (a) Find the Maclaurin series representation of f.

   (b) What is the interval of convergence of the power series found in part (a)?

   (c) Use the series in part (a) to estimate $f'''(1)$ with an error no greater than 0.005.

12. Let $f(x) = \begin{cases} e^{-1/x^2} & \text{if } x \ne 0, \\ 0 & \text{if } x = 0. \end{cases}$

   (a) Use the definition of the derivative (and l'Hôpital's rule) to show that $f'(0) = 0$.

   (b) Using methods similar to those in part (a), one can show that $f^{(k)}(0) = 0$ for all integers $k \ge 0$. Use this fact to find the Maclaurin series for f.

   (c) What is the radius of convergence of the series in part (b)?

   (d) For which values of x does the series in part (b) converge to $f(x)$?

Sequences Infinite series are formed by adding—in a special sense and with due care for convergence—the terms of an infinite **sequence**. To get started, we defined and studied sequences and their limits in their own right.

Series and convergence An infinite series **converges** if its sequence of partial sums, S_n, formed from only finitely many terms, converge to a limit S. For a given series, the sequence of partial sums may be hard to understand or handle directly. For **geometric series**, however, partial sums (and therefore limits) are easy to calculate. Geometric series are our simplest and most important examples.

Convergence tests The first question to ask about a series is whether it converges or diverges. As with improper integrals, the question can be subtle. To answer it, we developed several tests for convergence and divergence, including the **comparison test**, the **integral test**, and the **ratio test**. All of these tests apply only to series of **nonnegative** terms.

Absolute convergence Series that contain both positive and negative terms need special care; they raise the notions of **absolute** versus **conditional** convergence. The **alternating series test** handles certain series with terms that alternate in sign.

Estimating limits   Many convergent series are difficult to sum exactly. With help from technology, we can calculate specific partial sums and then estimate the error committed in ignoring the **upper tail**.

Series as functions   "Infinite polynomials"—**power series**—are useful and convenient; many standard calculus functions can be written in power series form. Power series, their **intervals of convergence**, and their uses are the subjects of the last two sections of the chapter.

Taylor series; Taylor's theorem   Taylor series, constructed by matching derivatives of a target function to those of a power series, give useful "formulas" for many calculus functions, including transcendental ones such as $\sin x$ and e^x. **Taylor's theorem** guarantees that, given appropriate conditions, the Taylor series of a function converges to the "right place"—the value $f(x)$.

REVIEW EXERCISES

In Exercises 1–8, find the limit of the sequence or explain why the limit does not exist.

1. $a_k = \left(\dfrac{\pi}{e}\right)^k$

2. $a_k = e^{-k}$

3. $a_k = (\arcsin 1)^k$

4. $a_k = \dfrac{k}{\sqrt{k}+10}$

5. $a_n = \dfrac{n+2}{n^3+4}$

6. $a_k = \dfrac{\ln(1+k^2)}{\ln(4+3k)}$

7. $a_k = \dfrac{\cos k}{\ln(k+1)}$

8. $a_k = e^{-k}\sin k$

In Exercises 9–34, determine whether the series converges absolutely, converges conditionally, or diverges. If the series converges, find an upper bound on its limit. If the series diverges, explain why.

9. $\displaystyle\sum_{k=0}^{\infty}\left(\dfrac{1}{k!}\right)^2$

10. $\displaystyle\sum_{n=3}^{\infty}\dfrac{1}{n(\ln n)^3}$

11. $\displaystyle\sum_{n=1}^{\infty}\left(\sum_{k=1}^{n}k^{-1}\right)$

12. $1-\dfrac{1}{2}+\dfrac{1}{2}-\dfrac{1}{4}+\dfrac{1}{3}-\dfrac{1}{6}+\dfrac{1}{4}-\dfrac{1}{8}+\dfrac{1}{5}-\dfrac{1}{10}+\cdots$

13. $\displaystyle\sum_{j=1}^{\infty}\dfrac{j}{5^j}$

14. $\displaystyle\sum_{j=1}^{\infty}\dfrac{j}{j^4+j-1}$

15. $\displaystyle\sum_{m=0}^{\infty}e^{-m^2}$

16. $\displaystyle\sum_{m=1}^{\infty}\dfrac{m^3}{m^4-7}$

17. $\displaystyle\sum_{k=1}^{\infty}\dfrac{k!}{(k+1)!-1}$

18. $\displaystyle\sum_{j=2}^{\infty}\dfrac{\ln j}{j^2}$

19. $\displaystyle\sum_{k=1}^{\infty}\dfrac{\sqrt{k}}{k^2+1}$

20. $\displaystyle\sum_{k=0}^{\infty}\dfrac{(-2)^k}{7^k+k}$

21. $\displaystyle\sum_{k=0}^{\infty}\dfrac{(-1)^k}{(k+1)\,2^k}$

22. $\displaystyle\sum_{m=8}^{\infty}\dfrac{\sin m}{m^3}$

23. $\displaystyle\sum_{n=1}^{\infty}\dfrac{\cos(n\pi)}{n}$

24. $\displaystyle\sum_{k=0}^{\infty}\dfrac{(-3)^k}{k^3+3^k}$

25. $\displaystyle\sum_{k=1}^{\infty}\dfrac{k^\pi}{k^e}$

26. $\displaystyle\sum_{m=2}^{\infty}\dfrac{1}{(\ln 3)^m}$

27. $\displaystyle\sum_{j=0}^{\infty}\left(\dfrac{1}{2^j}+\dfrac{1}{3^j}\right)^2$

28. $\displaystyle\sum_{k=1}^{\infty}\left(\int_k^{k+1}\dfrac{dx}{x^2}\right)$

29. $\displaystyle\sum_{m=0}^{\infty}\left(\int_0^m e^{-x^2}\,dx\right)$

30. $\displaystyle\sum_{n=1}^{\infty}\dfrac{\ln n}{\ln(3+n^2)}$

31. $\displaystyle\sum_{n=1}^{\infty}\sin(1/n)$

32. $\displaystyle\sum_{n=1}^{\infty}\dfrac{\sin(1/n)}{n}$

33. $\displaystyle\sum_{n=1}^{\infty}e^{-1/n}$

34. $\displaystyle\sum_{n=1}^{\infty}\left(1-e^{-1/n}\right)$

35. (a) Estimate a lower bound for $n!$ by comparing $\ln(n!)$ and $\displaystyle\int_1^n \ln x\,dx$.

   (b) Let b be a positive number. Use part (a) to find an integer N such that $b^N/N! < 1/2$.

36. Let $a_k = \displaystyle\int_k^{\infty}\dfrac{dx}{2x^2-1}$.

   (a) Evaluate $\displaystyle\lim_{k\to\infty}a_k$.   [HINT: $1/2x^2 \le 1/(2x^2-1) \le 1/x^2$ if $x \ge 1$.]

   (b) Does $\displaystyle\sum_{k=1}^{\infty}(-1)^{k+1}a_k$ converge absolutely? Justify your answer.

   (c) Does $\displaystyle\sum_{k=1}^{\infty}(-1)^{k+1}a_k$ converge? Justify your answer.

In Exercises 37–48, find the interval of convergence (endpoint behavior too!) of the power series.

37. $\displaystyle\sum_{m=1}^{\infty} \frac{x^m}{m^2+1}$

38. $\displaystyle\sum_{n=1}^{\infty} n^n x^n$

39. $\displaystyle\sum_{k=1}^{\infty} (3x)^k$

40. $\displaystyle\sum_{m=0}^{\infty} \frac{(3x)^m}{m!}$

41. $\displaystyle\sum_{n=1}^{\infty} \frac{(3x)^n}{n}$

42. $\displaystyle\sum_{j=1}^{\infty} \frac{(3x)^j}{j^2}$

43. $\displaystyle\sum_{m=0}^{\infty} \left(\frac{x-3}{2}\right)^m$

44. $\displaystyle\sum_{j=0}^{\infty} \frac{(x-2)^j}{j!}$

45. $\displaystyle\sum_{k=1}^{\infty} \frac{(x-1)^k}{k4^k}$

46. $\displaystyle\sum_{n=1}^{\infty} \frac{(x-1)^n}{\sqrt{n}}$

47. $\displaystyle\sum_{i=1}^{\infty} \frac{(x+5)^i}{i(i+1)}$

48. $\displaystyle\sum_{m=1}^{\infty} \frac{2^m (x-1)^m}{m}$

Consider a power series of the form $\displaystyle\sum_{k=0}^{\infty} a_k(x-1)^k$. In Exercises 49–53, indicate whether the statement must be true, cannot be true, or may be true. Justify your answers.

49. The power series converges only if $|x| > 2$.

50. The power series converges for all values of x.

51. If the radius of convergence of the power series is 3, the power series converges if $-2 < x < 4$.

52. The interval of convergence of the power series is $[-5, 5]$.

53. If the interval of convergence of the power series is $(-7, 9)$, the radius of convergence is 7.

54. **(a)** Evaluate $\displaystyle\lim_{x\to 1^-} \sum_{k=0}^{\infty} (-1)^k x^k$.

(b) Explain why the result in part (a) does *not* mean that $\displaystyle\sum_{k=0}^{\infty} (-1)^k$ converges.

In Exercises 55–58, use power series to evaluate the limit. Check your answers using l'Hôpital's rule.

55. $\displaystyle\lim_{x\to 0} \frac{1-\cos x}{x}$

56. $\displaystyle\lim_{x\to 0} \frac{e^x - e^{-x}}{x}$

57. $\displaystyle\lim_{x\to 0} \frac{x - \arctan x}{x^3}$

58. $\displaystyle\lim_{x\to 1} \frac{\ln x}{x-1}$

In Exercises 59–62, find the Maclaurin series for f and determine its radius of convergence.

59. $f(x) = 2^x = e^{x \ln 2}$

60. $f(x) = \cos^2 x = \frac{1}{2}(1 + \cos(2x))$

61. $f(x) = \dfrac{5+x}{x^2+x-2} = \dfrac{2}{x-1} - \dfrac{1}{x+2}$

62. $f(x) = \sin^3(x) = \frac{1}{4}(3\sin x - \sin(3x))$

63. Determine the coefficients a_k such that $\dfrac{1}{1-x} = \displaystyle\sum_{k=0}^{\infty} a_k(x-2)^k$.

[HINT: $\dfrac{1}{1-x} = -\dfrac{1}{1+(x-2)}$.]

64. Is $1 - x + \dfrac{x^2}{2} - \dfrac{x^4}{8} + \dfrac{x^5}{15} - \dfrac{x^6}{240} + \cdots$ the Maclaurin series representation of the function f shown in the graph below? Justify your answer.

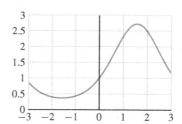

65. Suppose that f is a function that is positive, increasing, and concave down on the interval $[-2, 2]$.

(a) Is the coefficient of x^2 in the Maclaurin series representation of f positive? Justify your answer.

(b) Let $g(x) = 1/\sqrt{1 + f(x)}$. Is the coefficient of x^2 in the Maclaurin series representation of g positive? Justify your answer.

66. Let $H_n = \displaystyle\sum_{k=1}^{n} \frac{1}{k}$ be the nth partial sum of the harmonic series, let $a_n = H_n - \ln n$, and let $f(x) = \ln(x+1) - \ln x - \dfrac{1}{x+1}$.

(a) Show that $\ln(n+1) < H_n < 1 + \ln n$.
[HINT: Use the integral test.]

(b) Show that the sequence $\{a_n\}$ is decreasing.
[HINT: Explain why $\ln(n+1) - \ln n = \displaystyle\int_n^{n+1} x^{-1}\, dx > (n+1)^{-1}$.]

(c) Use part (c) to show that $\displaystyle\lim_{n\to\infty} a_n$ exists. (This limit, denoted by γ, is called **Euler's constant**; $\gamma \approx 0.57722$.)

(d) Show that $f(x) = -\displaystyle\int_x^{\infty} f'(t)\, dt$.

(e) Use part (d) to show that $f(x) > \dfrac{1}{2(x+1)^2}$.
[HINT: If $x > 0$, then $-f'(x) = \dfrac{1}{x(x+1)^2} > \dfrac{1}{(x+1)^3}$.]

(f) Show that $a_n - \gamma = \displaystyle\sum_{k=n}^{\infty} (a_k - a_{k+1})$.

(g) Use parts (e) and (f) to show that $\dfrac{1}{2(n+1)} < a_n - \gamma < \dfrac{1}{2n}$.
[HINT: $f(k) = a_k - a_{k+1}$.]

*Exercises 67–76 explore some properties of the **binomial series**:*

$$(1+x)^r = 1 + \sum_{n=1}^{\infty} \frac{r(r-1)(r-2)\cdots(r-n+1)}{n!} x^n,$$

where r is a constant.

67. Show that the binomial series converges if $|x| < 1$.

68. Show that $\sqrt{1+x} \approx 1 + \dfrac{1}{2}x - \dfrac{1}{8}x^2 + \dfrac{1}{16}x^3 - \dfrac{5}{128}x^4 + \dfrac{7}{256}x^5 \mp \cdots.$

In Exercises 69–72, find the first four nonzero terms of a power series representation of the function.

69. $f(x) = (1+x^4)^3$

70. $f(x) = \sqrt[3]{1-x^2}$

71. $f(x) = (1+x^2)^{-3/2}$

72. $f(x) = \arcsin x.$

73. Estimate $\displaystyle\int_0^{0.4} \sqrt{1+x^3}\, dx$ with an error less than 5×10^{-4}.

74. Let $f(x) = (1+x)^r$.

   (a) Show that $(1+x)f'(x) = rf(x)$.

   (b) Let $g(x) = (1+x)^{-r}f(x)$. Show that $g'(x) = 0$.

   (c) Use part (b) to show that $f(x) = (1+x)^r$.

75. (a) Show that

$$\arcsin x = \int_0^x \frac{dt}{\sqrt{1-t^2}} = x + \sum_{n=1}^{\infty} \frac{1 \cdot 3 \cdot 5 \cdots (2n-1)}{2 \cdot 4 \cdot 6 \cdots (2n)} \frac{x^{2n+1}}{2n+1}.$$

   (b) If $n \geq 1$ is an integer, then $\displaystyle\int_0^{\pi/2} \sin^{2n+1} x\, dx = \dfrac{2 \cdot 4 \cdot 6 \cdots (2n)}{3 \cdot 5 \cdot 7 \cdots (2n+1)}$. Use this fact to evaluate $\displaystyle\int_0^1 \frac{x^{2n+1}}{\sqrt{1-x^2}}\, dx.$

   (c) Use parts (a) and (b) to show that $\displaystyle\int_0^1 \frac{\arcsin x}{\sqrt{1-x^2}}\, dx = \sum_{k=0}^{\infty} \frac{1}{(2k+1)^2}.$

   (d) Use the substitution $u = \arcsin x$ to show that $\displaystyle\int_0^1 \frac{\arcsin x}{\sqrt{1-x^2}}\, dx = \frac{\pi^2}{8}.$

   (e) Use parts (c) and (d) to show that $\displaystyle\sum_{k=1}^{\infty} \frac{1}{k^2} = \frac{\pi^2}{6}.$

76. Let $f(x) = (1+x)^r$. In this exercise we prove that the binomial series converges to f.

   (a) Use part (e) of Exercise 19 in Section 9.2 to show that $R_n(x) = f(x) - P_n(x)$ is $R_n(x) = \dfrac{r \cdot (r-1) \cdot (r-2) \cdots (r-n)}{n!} \displaystyle\int_0^x \frac{(x-t)^n}{(1+t)^{n+1-r}}\, dt.$

   (b) Suppose that $0 \leq t \leq x < 1$. Show that $\dfrac{|x-t|}{1+t} \leq |x|.$

   (c) Use part (b) to show that if $0 \leq x < 1$, then $|R_n(x)| \leq \dfrac{|(r-1) \cdot (r-2) \cdots (r-n)|}{n!} x^n \left|(1+x)^r - 1\right|.$

   (d) Use part (c) to show that the binomial series converges to f if $0 \leq x < 1$.

   (e) Adapt the reasoning in parts (b)–(d) to show that the binomial series converges to f if $-1 < x < 0$.

77. (A proof that e is irrational.) Assume that $e = m/n$, where m and n are positive integers.

   (a) Explain why $m! \left| \dfrac{1}{e} - \displaystyle\sum_{k=0}^{m} \frac{(-1)^k}{k!} \right| \leq \dfrac{m!}{(m+1)!} = \dfrac{1}{m+1}.$

   (b) Explain why $m!/e$ is an integer.

   (c) Explain why $m! \displaystyle\sum_{k=0}^{m} \frac{(-1)^k}{k!}$ is an integer.

   [HINT: Start by explaining why $m!/k!$ is an integer if k is an integer and $0 \leq k \leq m$.]

   (d) Parts (a)–(c) imply that $N = m! \left| \dfrac{1}{e} - \displaystyle\sum_{k=0}^{m} \frac{(-1)^k}{k!} \right|$ is an integer that is less than or equal to $1/(m+1)$. Explain why it follows that $N = 0$.

   (e) Explain why the conclusion of part (d) is impossible and therefore e cannot be a rational number.

   [HINT: $\displaystyle\sum_{k=m+1}^{\infty} (-1)^k/k! \neq 0$.]

78. (Cauchy condensation theorem) Let $\sum_{n=1}^{\infty} a_n$ be a series of positive terms such that $a_{n+1} \leq a_n$ for all n.

   (a) Let $m \geq 1$ be an integer. Explain why

$$2^{m-1}a_{2^m} \leq a_{2^{m-1}+1} + a_{2^{m-1}+2} + \cdots + a_{2^m} \leq 2^{m-1}a_{2^{m-1}}.$$

   (b) Use part (a) to show that $\dfrac{1}{2} \displaystyle\sum_{k=1}^{m} 2^k a_{2^k} \leq \sum_{k=2}^{2^m} a_k.$

   (c) Use part (b) to show that if $\dfrac{1}{2} \displaystyle\sum_{k=1}^{\infty} 2^k a_{2^k}$ diverges, then $\displaystyle\sum_{k=1}^{\infty} a_k$ diverges.

   (d) Let m and n be integers such that $n \leq 2^m$. Use part (a) to show that $\displaystyle\sum_{k=2}^{n} a_k \leq \sum_{k=1}^{m} 2^{k-1}a_{2^{k-1}}.$

   (e) Use part (d) to show that if $\displaystyle\sum_{k=0}^{\infty} 2^k a_{2^k}$ converges, then $\displaystyle\sum_{k=2}^{\infty} a_k$ converges.

79. Use Exercise 78 to prove that the harmonic series diverges.

80. Use Exercise 78 and the properties of geometric series to prove that $\displaystyle\sum_{k=1}^{\infty} 1/k^p$ converges if $2^{1-p} < 1$ (i.e., $p > 1$) and diverges otherwise.

81. Use Exercise 78 and the result of the previous exercise to show that $\displaystyle\sum_{k=2}^{\infty} \frac{1}{k(\ln k)^p}$ converges if $p > 1$ and diverges otherwise.

82. Use the fact that $\displaystyle\int_0^{\infty} t^n e^{-t}\, dt = n!$ to show that $\displaystyle\int_0^{\infty} e^{-t} \sin(xt)\, dt = \dfrac{x}{1+x^2}$ if $|x| < 1$.

83. (a) Use the trigonometric identity $\tan x = \cot x - 2\cot(2x)$ to show that

$$\sum_{k=1}^{n} \frac{1}{2^k} \tan\left(\frac{x}{2^k}\right) = \frac{1}{2^n} \cot\left(\frac{x}{2^n}\right) - \cot x.$$

   (b) Use part (a) to show that

$$\sum_{k=1}^{\infty} \frac{1}{2^k} \tan\left(\frac{x}{2^k}\right) = \frac{1}{x} - \cot x.$$

84. Let $I_n = \int_0^{\pi/4} \tan^n x \, dx$ for any integer $n \geq 0$.

  (a) Show that $\{I_n\}$ is a nonincreasing sequence (i.e., $I_{n+1} \leq I_n$).

  (b) Show that $I_n + I_{n-2} = \dfrac{1}{n-1}$.

  (c) Use parts (a) and (b) to show that $\dfrac{1}{2(n+1)} \leq I_n \leq \dfrac{1}{2(n-1)}$. [HINT: Part (b) implies that $I_{n+2} + I_n = 1/(n+1)$.]

  (d) Show that, if $n \geq 2$, $I_n = \dfrac{1}{n-1} - \int_0^{\pi/4} \tan^{n-2} x \, dx$.

  (e) Use part (d) to show that $I_{2n} = (-1)^n \left(\dfrac{\pi}{4} + \sum_{k=1}^{n} \dfrac{(-1)^k}{2k-1} \right)$ for all integers $n \geq 1$.

  (f) Use parts (c) and (e) to show that $\displaystyle\sum_{k=1}^{\infty} \dfrac{(-1)^{k+1}}{2k-1} = \dfrac{\pi}{4}$.

  (g) Use part (d) to show that $I_{2n+1} = (-1)^n \left(\dfrac{1}{2} \ln 2 + \sum_{k=1}^{n} \dfrac{(-1)^k}{2k} \right)$ for all integers $n \geq 1$.

  (h) Use parts (c) and (g) to show that $\displaystyle\sum_{k=1}^{\infty} \dfrac{(-1)^k}{k} = -\ln 2$.

Fourier Series

Fourier series are the infinite analogues of **Fourier polynomials**, which we defined and explored briefly in Section 9.3. (In the same sense, Taylor series are the infinite version of Taylor polynomials.)

> **DEFINITION** Let $f(x)$ be a function defined on $[-\pi, \pi]$. The **Fourier series** for f is the series
>
> $$a_0 + a_1 \cos x + b_1 \sin x + a_2 \cos(2x) + b_2 \sin(2x) + \cdots$$
> $$+ a_n \cos(nx) + b_n \sin(nx) + \cdots$$
>
> with coefficients given by
>
> $$a_0 = \frac{1}{2\pi} \int_{-\pi}^{\pi} f(x)\, dx;$$
>
> $$a_k = \frac{1}{\pi} \int_{-\pi}^{\pi} f(x) \cos(kx)\, dx \qquad \text{if } k > 0;$$
>
> $$b_k = \frac{1}{\pi} \int_{-\pi}^{\pi} f(x) \sin(kx)\, dx \qquad \text{if } k > 0.$$

Naturally enough, Fourier series inherit many of their properties from Fourier polynomials.

PROBLEM 1 Let f and g be functions defined on $[-\pi, \pi]$; assume that f is odd and g is even.

(a) Show that the Fourier series of f has $a_k = 0$ for all $k \geq 0$.

(b) Show that the Fourier series of g has $b_k = 0$ for all $k \geq 1$.

PROBLEM 2 Show that the Fourier series for $f(x) = x$ has the form

$$\frac{2}{1} \sin x - \frac{2}{2} \sin(2x) + \frac{2}{3} \sin(3x) - \frac{2}{4} \sin(4x) + \frac{2}{5} \sin(5x) - \cdots .$$

What's the general "formula" for b_n?

PROBLEM 3 Show that the Fourier series for $f(x) = |x|$ has the form

$$\frac{\pi}{2} - \frac{4}{\pi} \cos x - \frac{4}{9\pi} \cos(3x) - \frac{4}{25\pi} \cos(5x) - \frac{4}{49\pi} \cos(7x) - \cdots .$$

(Note that $b_k = 0$ for all $k \geq 1$.)

Convergence and divergence of Fourier series Whether a Fourier series converges or diverges for a given value of x is, in general, a difficult question—much more so than for Taylor series. Although Taylor series converge for x in a particular *interval*, and diverge elsewhere, the convergence set of a Fourier series may be quite complicated. Under suitably strict conditions, however, a Fourier series *does* converge, and to the "right place"—to the function value $f(x)$ itself. The following theorem illustrates one set of (technical-looking) conditions under which this happens.

> **THEOREM 15** Let f be continuous on $[-\pi, \pi]$, with $f(-\pi) = f(\pi)$ and only finitely many local maximum and minimum points in the interval $[-\pi, \pi]$. Then the Fourier series of f converges to $f(x)$ for all x in $[-\pi, \pi]$.

PROBLEM 4 This problem is about the function $f(x) = |x|$. Observe that Theorem 15 applies to f.

(a) Apply Theorem 15 with $x = 0$ to show that

$$1 + \frac{1}{9} + \frac{1}{25} + \frac{1}{49} + \cdots = \frac{\pi^2}{8}.$$

(b) Apply Theorem 15 with $x = \pi/4$; what series identity results?

(c) Apply Theorem 15 with $x = \pi/2$; what series identity results?

PROBLEM 5 Experiment with Theorem 15, the function $f(x) = x^2$, and an appropriate value of x to show that

$$\sum_{k=1}^{\infty} \frac{1}{k^2} = 1 + \frac{1}{4} + \frac{1}{9} + \cdots = \frac{\pi^2}{6}.$$

12

CURVES AND VECTORS

12.1 THREE-DIMENSIONAL SPACE

Single-variable calculus is done mainly in the two-dimensional xy-plane. The Euclidean plane, also known as $\mathbb{R}^2$, is the natural home of many familiar calculus objects: the graph of $y = f(x)$, tangent lines to that graph at various points, and the various regions whose areas we might measure by integration.

To do *multivariable* calculus, we need more room. The graph of $z = f(x, y)$, where f is a function of *two* input variables, lives in *three*-dimensional xyz-space. With three dimensions to work in, we will "see" not only this graph but a variety of multivariable analogues of derivatives and integrals. This section explores three-dimensional Euclidean space, or $\mathbb{R}^3$, where we will spend most of our time.

In another sense, of course, we spend *all* of our time in $\mathbb{R}^3$. In everyday usage, "space" connotes three physical dimensions. The intuition we gain from living in three spatial dimensions is often useful in mentally picturing and manipulating the objects of multivariable calculus. Familiar as it is, however, three-dimensional space poses special problems for visualization. Two-dimensional pictures (on paper or on a computer screen) of three-dimensional objects are always more or less distorted or incomplete. Minimizing such problems is an active science (and an art) in its own right; doing so means carefully controlling viewpoint, perspective, shading, lighting, and other factors.

This book's main subject is multivariable calculus, not computer graphics (although we sometimes mention computer graphics), and so we draw pictures to illustrate ideas as simply as possible, not necessarily to look as lifelike as possible. It is worth remarking, however, that many of the basic tools and methods of computer graphics draw directly on the ideas we will develop in this chapter.

Cartesian coordinates in three dimensions

The idea of Cartesian coordinates is the same in both two and three dimensions, but the pictures look a little different. (See Figure 1.). Recall the formalities in the xy-plane: A Cartesian coordinate system consists of an origin, labeled O, and horizontal and vertical coordinate axes, labeled x and y, passing through O. On each axis we choose a positive direction (usually "east" and "north") and a unit of measurement (not necessarily the same on both axes).

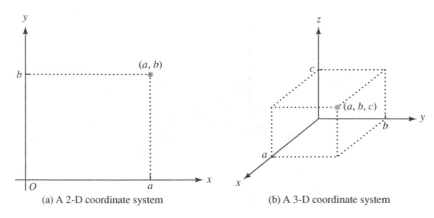

(a) A 2-D coordinate system   (b) A 3-D coordinate system

FIGURE 1
Coordinate systems

Given such a coordinate system, every point P in the plane corresponds to one and only one ordered pair (a, b) of real numbers, called the Cartesian coordinates of P. The pair (a, b) can be thought of as P's "Cartesian address." To reach P from the origin, move a units in the positive x-direction and b units in the positive y-direction. (If a or b is negative, go the other way.)

Coordinates in three-dimensional xyz-space work the same way—but with *three* coordinate axes labeled x, y, and z. Each axis is perpendicular to the other two. → To reach the point $P = (a, b, c)$ from the origin, go a units in the positive x-direction, b units in the positive y-direction, and c units in the positive z-direction. → As Figure 1 illustrates, the resulting point $P = (a, b, c)$ can also be thought of as a corner (the one opposite the origin) of a rectangular solid with dimensions $|a|$, $|b|$, and $|c|$.

There's "room" in $\mathbb{R}^3$ for three mutually perpendicular axes. $\mathbb{R}^2$ has room for only two.

If any of a, b, or c is negative, go the other way.

Quadrants and octants The two axes divide the xy-plane into four **quadrants**, defined by the pattern of positive or negative x- and y-coordinates. The analogous regions in xyz-space are called **octants**. The first octant, for instance, consists of all points (x, y, z) with all three coordinates positive. In the preceding picture, only the first octant is visible. The next picture gives another view. There are eight octants in all in xyz-space—one for each of the possible patterns of signs of the three coordinates: $(+, +, +), (+, +, -), (+, -, +),$ $(+, -, -), (-, +, +), (-, +, -), (-, -, +), (-, -, -)$.

Coordinate planes One can think of the first octant as a room with the origin at the lower left corner of the front wall. At this point, three walls meet, all at right angles. These "walls" are known as the **coordinate planes**: the yz-plane (the front wall), the xy-plane (the floor), and the xz-plane (the left wall). The coordinate planes correspond to simple equations in the variables x, y, and z. The yz-plane, for example, is the **graph** of the equation $x = 0$, that is, the set of all points (x, y, z) that satisfy the equation $x = 0$. Similarly, the xy- and xz-planes are graphs of the equations $z = 0$ and $y = 0$, respectively.

Many possible views The xy-plane, being "flat," is relatively easy to draw. Simulating three-dimensional space on a flat page or computer screen is much harder, and there is always some price to be paid in distortion. For example, in 3-D reality, the x-, y-, and z-axes are all perpendicular to each other, but no flat picture can really show this. The axes in Figure 2, for instance, do not make right angles on the page.

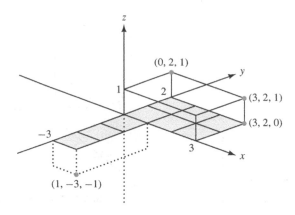

Another view of 3-D space

This view of xyz-space differs from that in Figure 1 in several ways:

- **Horizontal and vertical**   The xy-plane (in which the shaded "floor tiles" lie) is drawn to appear horizontal. The z-axis is vertical; the positive direction is up. This is a standard convention; we will follow it consistently.

- **Hidden lines**   The dashed lines in the picture lie "below" the xy-plane. They would be hidden from view if the xy-plane (the "floor") were opaque. How much of xyz-space is considered to be visible is a matter of choice. Sometimes only the first octant is shown.

- **Positive directions**   An arrow on each axis indicates the positive direction. The 3×2 block of shaded squares lies in the first quadrant of the xy-plane. The other shaded squares lie in the plane's fourth quadrant.

- **Plotting points: positive and negative coordinates**   Any point $P = (a, b, c)$ is plotted the same way: From the origin, move a, b, and c units in the positive x-, y-, and z-directions, respectively. (If a coordinate is negative, move in the opposite direction.)

- **Where's the viewer?**   Figure 2 is drawn as though the viewer were floating somewhere above the *fourth* quadrant of the xy-plane. In Figure 1(b), by contrast, the viewer seems to hover somewhere above the first quadrant. There is nothing sacred about either viewing angle; we will use various viewpoints as we go along. ◄

So do various computer plotting packages.

- **No perspective**   To a human viewer, rectangular boxes like those in Figures 1 and 2 would appear in *perspective*: The sides would taper toward a vanishing point. For the sake of simplicity, we ignore perspective effects in most of the pictures in this book.

An important moral is as follows: There is no single "best" picture of a 3-D object; choosing a good or convenient view may depend on properties of the object, what needs emphasis, or even the drawing technology at hand. ◄

Computer, calculator, pencil, sharp stick, ...

Distance and midpoints

Or from Q to P—it doesn't matter.

Let $P = (x_1, y_1)$ and $Q = (x_2, y_2)$ be any two points in the xy-plane. Recall that the distance from P to Q ◄ is given by the familiar Pythagorean formula

$$d(P, Q) = \sqrt{(x_2 - x_1)^2 + (y_2 - y_1)^2}$$

and that the midpoint M of the segment joining P to Q has these "averaged" coordinates:

$$M = \left(\frac{x_1 + x_2}{2}, \frac{y_1 + y_2}{2} \right).$$

The formulas in three dimensions are not much different.

> **DEFINITION**   The distance between $P = (x_1, y_1, z_1)$ and $Q = (x_2, y_2, z_2)$ is
>
> $$d(P, Q) = \sqrt{(x_2 - x_1)^2 + (y_2 - y_1)^2 + (z_2 - z_1)^2}.$$
>
> The midpoint of the segment joining P and Q has coordinates
>
> $$M = \left(\frac{x_1 + x_2}{2}, \frac{y_1 + y_2}{2}, \frac{z_1 + z_2}{2} \right).$$

Both definitions are simply three-dimensional versions of the corresponding formulas in the xy-plane. In both two and three dimensions, for example, distance is computed as the square root of the sum of the squared differences in coordinates. ➡

In either two or three dimensions, the distance formula reflects the Pythagorean rule. See the exercises for more details.

EXAMPLE 1   Consider the points $P = (0, 0, 0)$ and $Q = (2, 4, 6)$. Find the distance from P to Q and the midpoint M of the segment joining them. How far is M from P and from Q?

Solution   By the distance formula,

$$d(P, Q) = \sqrt{(2 - 0)^2 + (4 - 0)^2 + (6 - 0)^2} = \sqrt{56} \approx 7.483.$$

According to the formula, the midpoint is $M(1, 2, 3)$; each coordinate of M splits the difference between the corresponding coordinates of P and Q. To see why M deserves the name "midpoint," notice that

$$d(P, M) = \sqrt{(1 - 0)^2 + (2 - 0)^2 + (3 - 0)^2}$$

$$= \sqrt{(2 - 1)^2 + (4 - 2)^2 + (6 - 3)^2} = \sqrt{14} \approx 3.742.$$

Thus, M lies halfway between P and Q, as a midpoint should.   ■

Equations and their graphs

The graph of an equation in x and y is the set of all points (x, y) that satisfy the equation. The graph of $x^2 + y^2 = 1$, for instance, is a circle of radius 1 in the xy-plane, centered at the origin. The graph of the equation $x = 0$ is the y-axis. ➡

The same idea applies for three variables: The graph of an equation in x, y, and z is the set of points (x, y, z) in space that satisfy the equation. Figure 3 shows three simple examples, all in the first octant.

The graph of an equation may or may not be the graph of a function. The unit circle is not a function graph.

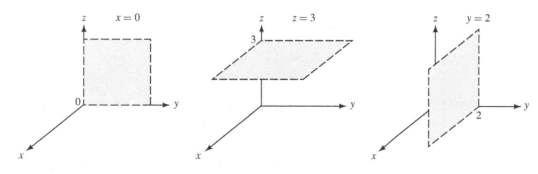

FIGURE 3
Three simple graphs

Solutions of the equation $x = 0$ are points of the form $(0, y, z)$, and so the graph is the yz-plane. Similarly, solutions of $z = 3$ are all points of the form $(x, y, 3)$, and thus the graph is a horizontal plane floating 3 units above the xy-plane. The graph of $y = 2$ is parallel to the xz-plane but moved 2 units in the positive y-direction.

Notice that, in each case, the graph of an equation in x, y, and z is a plane—a *two-dimensional* object. In contrast, the graph of one equation in x and y is usually a curve or a line—a *one-dimensional* object. The pattern is the same in both cases: The graph of an equation in n variables usually has dimension $n - 1$.

A few important types of graphs in xyz-space deserve special mention.

What goes wrong if $a = b = c = 0$?

Planes   **A linear equation** is one of the form $ax + by + cz = d$, where a, b, c, and d are constants and at least one of a, b, and c is nonzero. ◄ (All three equations plotted in Figure 3 are linear.) They illustrate an important general fact:

The graph of a linear equation is a plane in xyz-space.

To draw planes in xyz-space, we can use the fact that a plane is uniquely determined by three points.

EXAMPLE 2   Plot the linear equation $x + 2y + 3z = 3$ in the first octant.

Solution   Let's first find some points (x, y, z) that satisfy $x + 2y + 3z = 3$. There are infinitely many possibilities. Given *any* values for x and y, the equation determines a corresponding value for z. If, say, $x = 1$ and $y = 1$, then $x + 2y + 3z = 3$ can hold only if $z = 0$. Similarly, setting $y = 2$ and $z = 3$ forces $x = -10$. Among the simplest solutions are

$$P = (3, 0, 0); \qquad Q = \left(0, \frac{3}{2}, 0\right); \qquad R = (0, 0, 1).$$

These solutions are both easy to find (set any two coordinates to zero and solve for the third) and easy to plot (they lie on the coordinate axes). Figure 4 shows the resulting plane:

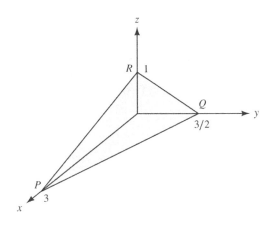

FIGURE 4
A plane in the first octant: $x + 2y + 3z = 3$

Not every line in the xy-plane intercepts both axes; not every plane in space intercepts all three axes. See this section's exercises for further details.

Observe some main features of the plane:

- **Intercepts**   A typical *line* in the xy-plane has x- and y-intercepts where the line intersects the coordinate axes. In a similar sense, a typical *plane* in xyz-space has x-, y-, and z-intercepts. The intercepts in Figure 4 are P, Q, and R. ◄

- **Traces** If we "slice" a surface in xyz-space with a plane, the intersection of the surface with the plane is called the **trace** of the surface in that plane. The plane p in Figure 4 meets each of the three coordinate planes in a straight line. Those three lines are therefore the traces of the surface $x + 2y + 3z = 3$ in the xy-plane, the xz-plane, and the yz-plane, respectively.

  It is easy to find equations for these traces. For example, a point (x, y, z) lies both in the plane p *and* in the xy-plane if and only if it satisfies both $x + 2y + 3z = 3$ and $z = 0$. Setting $z = 0$ in the first equation gives $x + 2y = 3$—as expected, the equation of a line in the xy-plane. This line is therefore the trace of p in the xy-plane. ➠   ■

Do you see this line in Figure 4?

Spheres In the plane, a circle of radius $r > 0$ and center $C = (a, b)$ is the set of points $P = (x, y)$ at distance r from (a, b). Translating this description into symbolic language produces the familiar formula for a circle in the plane:

$$d(P, C) = \sqrt{(x - a)^2 + (y - b)^2} = r, \quad \text{or} \quad (x - a)^2 + (y - b)^2 = r^2.$$

(Squaring both sides does no harm and simplifies the equation's appearance.)

The object in space that is analogous to a circle in the plane is a **sphere** of radius r. Like a circle, a sphere is "hollow," resembling an empty orange skin. Adding the interior (the edible part of the orange) produces a **ball**. Like a circle, a sphere is the set of points at some fixed distance—the radius—from a fixed center point. Given a radius $r > 0$ and a center point $C(a, b, c)$, the sphere of radius r centered at C is the set of points (x, y, z) such that

$$d(P, C) = \sqrt{(x - a)^2 + (y - b)^2 + (z - c)^2} = r,$$

or, equivalently,

$$(x - a)^2 + (y - b)^2 + (z - c)^2 = r^2.$$

The simplest example, the **unit sphere**, has center $(0, 0, 0)$ and radius 1. Its equation reduces to this simple form:

$$x^2 + y^2 + z^2 = 1.$$

Drawing circles in the xy-plane is easy, even by hand. Drawing spheres (or any "curved" objects, for that matter) convincingly by hand is admittedly harder, ➠ but rough sketches often suffice.

For instance, circles in space usually look "flattened."

One way to visualize any surface is to analyze its traces in various planes. Doing so for the sphere is relatively easy.

EXAMPLE 3 The unit sphere S has equation $x^2 + y^2 + z^2 = 1$. If we set $z = 0$ in this equation we get $x^2 + y^2 = 1$, which describes the *unit circle* in the xy-plane. This means, geometrically, that S intersects the xy-plane (where $z = 0$) in the unit circle $x^2 + y^2 = 1$. In other words, the unit circle is the trace of S in the xy-plane. ➠

In still other words, the unit circle is the unit sphere's "equator."

1. Find the traces of S in the yz-plane and in the xz-plane.
2. Find the traces of S in the planes $z = 1/2$, $z = 9/10$, $z = 1$, and $z = 2$.

Solution

1. Setting $x = 0$ in $x^2 + y^2 + z^2 = 1$ gives $y^2 + z^2 = 1$—the unit circle in the yz-plane. Similarly, setting $y = 0$ in $x^2 + y^2 + z^2 = 1$ gives $x^2 + z^2 = 1$—the unit circle in the xz-plane.

2. Setting $z = 1/2$ in $x^2 + y^2 + z^2 = 1$ gives $x^2 + y^2 = 3/4$—a circle of radius $\sqrt{3}/2$ centered at $(0, 0)$ in the plane $z = 1/2$. Similarly, setting $z = a$ in $x^2 + y^2 + z^2 = 1$ gives the equation

$$x^2 + y^2 = 1 - a^2.$$

If $a^2 < 1$, this describes a circle with radius $\sqrt{1 - a^2}$ and center $(0, 0)$, lying in the plane $z = a$. Notice that if $a = \pm 1$, the "circle" reduces to a single point, while if $a^2 > 1$ (e.g., if $a = 2$) the resulting equation has *no* solutions and thus the trace is empty. ■

Recognizing spheres   Recognizing *linear* equations is easy. Algebraic methods, such as **completing the square**, can help reveal the geometric form of more complicated equations.

EXAMPLE 4   Is the graph of $x^2 - 2x + y^2 - 4y + z^2 - 6z = 0$ a sphere? If so, which sphere?

Solution   First we complete the square in each variable separately:

$$x^2 - 2x + y^2 - 4y + z^2 - 6z = 0 \iff$$
$$(x^2 - 2x + 1) + (y^2 - 4y + 4) + (z^2 - 6z + 9) = 1 + 4 + 9 \iff$$
$$(x - 1)^2 + (y - 2)^2 + (z - 3)^2 = 14.$$

The last form shows that our equation describes a sphere of radius $\sqrt{14}$ centered at $(1, 2, 3)$. The equation also shows that the sphere passes through the origin. Figure 5 shows all:

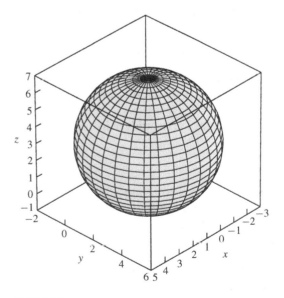

FIGURE 5
Graph of $x^2 - 2x + y^2 - 4y + z^2 - 6z = 0$ ■

Cylinders   What is the graph of the equation $y = x^2$? The answer depends on where we are working. In the xy-plane, the graph is the familiar parabola—the set of all points (x, x^2). As Figure 6 shows, the same equation looks different in xyz-space:

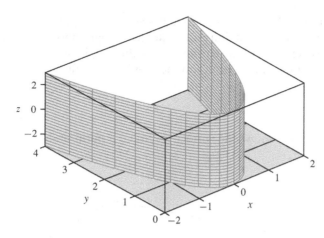

FIGURE 6
Graph of $y = x^2$ in xyz-space

Notice some properties of the graph:

- **A missing variable**   The graph is *unrestricted* in the z-direction—it contains all points that lie directly above or below the graph of $y = x^2$ in the xy-plane. ⇝ The graph has this property because z is "missing" in the equation $y = x^2$. This means that if (x, y) satisfies the equation, then so does *every* point (x, y, z) regardless of the value of z.

    In other words, the graph has "vertical walls."

- **What is a cylinder?**   Graphs like this one, in which at least one of the variables is unrestricted, are called **cylinders**. Any equation that omits one or more variables— $y = z$, say—has a cylindrical graph. Plotting cylinders is comparatively simple. If the equation involves only y and z, for instance, we first plot the equation in the yz-plane and then "extend" the graph in the x-direction.

In everyday speech "cylinder" usually means "circular tube," but as Figure 6 illustrates, the mathematical notion of a cylinder is more general. ⇝ Example 5 concerns another "cylindrical" graph.

Even the coordinate planes are "cylinders."

EXAMPLE 5   Discuss the graph in xyz-space of the equation $z = 2 + \sin y$. Interpret the result as a cylinder.

Solution   There's no variable x in the equation, and so the graph is unrestricted in the x-direction: It is a cylinder in x. Figure 7 gives a representative view: ⇝

The surface, like many graphs, continues forever; a picture shows only part of the graph.

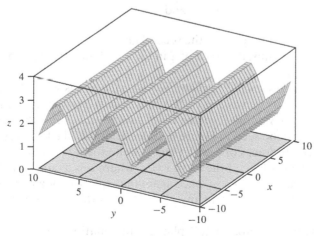

FIGURE 7
Graph of $z = 2 + \sin y$

The graph resembles the surface of an ideal ocean with regular waves moving parallel to the y-axis. Each wave is infinitely long, and all troughs and crests run parallel to the x-axis. Notice especially how the surface meets the yz-plane. The curve of intersection—the trace of the surface in the yz-plane—is the ordinary sine curve $z = 2 + \sin y$. ◄

The entire surface can be thought of as infinitely many identical copies of this curve, one for each value of x.

E X A M P L E 6   Figure 8 shows the surface S (called a **paraboloid**) defined by the equation $z = x^2 + y^2$.

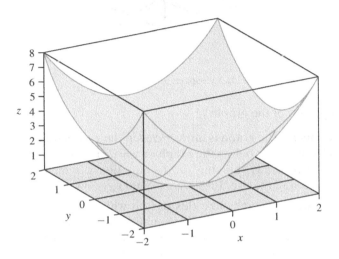

FIGURE 8
Graph of $z = x^2 + y^2$: a paraboloid

Discuss the traces of S in several planes Π.

Solution   The trace of S in any plane Π is the intersection of S and Π. Therefore, points on the trace of S in Π are those that satisfy *two* equations—those of S and of Π. Points on the trace of S in the xy-plane, for instance, satisfy both

$$z = 0 \quad \text{and} \quad z = x^2 + y^2.$$

Substituting the first equation into the second gives $0 = x^2 + y^2$; clearly, the only solution is $x = y = 0$. Thus, the trace of S in the xy-plane is a single point—the origin. (Find it in Figure 8.)

Now suppose that Π is the xz-plane, whose equation is $y = 0$. Substituting this equation into $z = x^2 + y^2$ gives $z = x^2$, the equation of a parabola in the xz-plane. Similarly, the trace of S in the yz-plane (with equation $x = 0$) is the parabola $z = y^2$. These traces, too, can be seen in Figure 8. Indeed, all the black curves superimposed on the surface shown are traces of S with planes of the form $x = a$ or $y = b$.

Now let Π be any plane of the form $x = a$, where a is a constant. Substituting $x = a$ into $z = x^2 + y^2$ gives $z = a^2 + y^2$—once again the equation of a parabola opening upward. Similarly, the trace of S in any plane $y = b$ is the parabola with equation $z = x^2 + b^2$. ◄

All these parabolas justify the name "paraboloid."

Finally, let Π be any "horizontal" plane, with equation $z = c$; then the trace equation is $c = x^2 + y^2$. If $c > 0$, this equation describes a circle with radius $\sqrt{c}$ and center $(0, 0)$; if $c < 0$, the equation $c = x^2 + y^2$ has no solutions and the trace is thus empty. The picture illustrates these facts, too. Notice, for example, that if $c < 0$, then the plane $z = c$ misses S entirely. ∎

BASIC EXERCISES

In Exercises 1–5, find an equation in x, y, and z for the graph in xyz-space of

1. a sphere of radius 2, centered at the origin.

2. a sphere of radius 1, centered at $(1, 1, 1)$.

3. a circular cylinder of radius 1, centered along the y-axis.

4. a circular cylinder of radius 2, centered along the z-axis.

5. a cylindrical surface that resembles an ocean with waves rolling in the x-direction.   [HINT: See Example 5.]

6. (a) Plot the equation $z = y^2$ in xyz-space. [HINT: Start in yz-space.]
   (b) Is the surface in part (a) a cylinder? If so, what is the unrestricted direction?

7. (a) Plot the equation $z = x^2$ in xyz-space.
   (b) Is the surface in part (a) a cylinder? If so, what is the unrestricted direction?

8. Plot the equation $x^2 + y^2 = 1$, first in xy-space and then in xyz-space. Discuss the relationship between the two graphs.

9. We said in this section that xyz-space contains eight different octants. List eight points, all with coordinates ± 1, one in each octant. Draw a picture showing all eight points.

10. Consider the plane p with equation $4x + 2y + z = 4$.
    (a) Find the x-, y-, and z-intercepts of p. Use them to draw a picture of p in the first octant.
    (b) Find the traces of p in each of the three coordinate planes. How do your answers appear in the picture in part (a)?

Let S be the surface with equation $z = x^2 + y$. In Exercises 11–16, describe the trace of S in each plane below. (Give an equation of the trace and describe its shape.)

11. $z = 0$

12. $z = 2$

13. $x - 0$

14. $x = 1$

15. $y = 0$

16. $y = 3$

17. Let S be the unit sphere $x^2 + y^2 + z^2 = 1$. Describe the trace of S in each plane below.
    (a) $z = 0$
    (b) $z = 1/2$
    (c) $z = 1$
    (d) $z = 2$

18. Let S be the sphere $x^2 + y^2 + z^2 = 8$. Describe the trace of S in each plane below.
    (a) $z = 0$
    (b) $x = 1$
    (c) $y = 2$
    (d) $z = 3$

Let S be the surface $x + 2y + 3z = 3$. In Exercises 19–22, describe the trace of S in the given plane.

19. $x = 0$

20. $y = 2$

21. $z = 1$

22. $z = 3$

23. Find an equation for the trace of the surface $y = x^2 - z^3$ in the yz-plane.

In Exercises 24–27, plot (by hand or using technology) the given equation in xyz-space and then describe the graph in words.

24. $y^2 + z^2 = -1$

25. $y^2 + z^2 = 0$

26. $y^2 - z^2 = 0$

27. $y^2 + z = -1$

FURTHER EXERCISES

28. The equation $z = 3$ omits two variables. Therefore, its graph in xyz-space should be a cylinder in both the x-direction and the y-direction. Is it? What is the trace of the graph in each of the coordinate planes?

29. The graph of $x^2 + y^2 - 6y + z^2 - 4z = 0$ is a sphere. Find the center and radius; then draw the sphere.

Exercises 30–33 concern the linear equation $Ax + By = C$ and its graph (a line) in the xy-plane. Here A, B, and C are constants, and we assume that A and B are not both zero.

30. What goes wrong if $A = B = 0$?

31. (a) Find the slope of the line $Ax + By = C$.
    (b) Which lines have undefined slope?

32. (a) Find the y-intercept of the line $Ax + By = C$.

(b) Which lines have no y-intercept?

33. (a) Find the x-intercept of the line $Ax + By = C$.

(b) Which lines have no x-intercept?

Exercises 34–36 consider the linear equation $Ax + By + Cz = D$ and its graph (a plane) in xyz-space. Here A, B, C, and D are all constants, and we assume that A, B, and C are not all zero.

34. We assumed that A, B, and C are not all zero. What goes wrong if $A = B = C = 0$?

35. (a) Find (if possible) an x-intercept of the plane $Ax + By + Cz = D$. (Set $y = 0$ and $z = 0$, and then solve for x.)

(b) Give an example of a plane with no x-intercept.

36. (a) Find (if possible) a z-intercept of the plane $Ax + By + Cz = D$.

(b) Give an example of a plane with no z-intercept.

We said in this section that, as a rule, a line in the xy-plane intercepts both coordinate axes, and a plane in xyz-space intercepts all three coordinate axes. Exercises 37–40 explore some exceptions.

37. Give an example of a line in the xy-plane that intercepts the x-axis but not the y-axis. Write an equation for your line in the form $ax + by = c$.

38. Consider the plane $x = 1$ in xyz-space. Find all possible intercepts with the three coordinate axes.

39. Consider the plane $x + 2y = 1$ in xyz-space. Find all possible intercepts with the three coordinate axes.

40. Give the equation of a plane in xyz-space that intersects the y-axis and the z-axis but not the x-axis.

41. Suppose that $P_1(x_1, y_1, z_1)$ and $P_2(x_2, y_2, z_2)$ both lie on the plane with equation $Ax + By + Cz = D$. Show that the midpoint of P_1 and P_2 also lies on this plane.

42. Is the graph of $y = 2$ a cylinder in xyz-space? Justify your answer.

43. Let $P = (1, 2, 3)$ and $Q = (4, 6, 8)$ be points in 3-space.

(a) Find the distance from P to Q.

(b) Write an equation for the sphere with center at P that passes through Q.

44. Find an equation for the smallest ball centered at the point $(4, -8, 1)$ that contains the origin.

45. The distance formula in xyz-space can be thought of as just another instance of the Pythagorean rule for right triangles. (The square of the hypotenuse is the sum of the squares of the sides.) This exercise illustrates why.

(a) Plot and label the points $O = (0, 0, 0)$, $P = (1, 0, 0)$, $Q = (1, 2, 0)$, and $R = (1, 2, 3)$ in an xyz-coordinate system. Observe that the triangles $\triangle OPQ$ and $\triangle OQR$ are both right triangles. Mark the sides OP, PQ, and QR with their lengths. (The lengths should be obvious from the picture.)

(b) Use the Pythagorean rule (not the distance formula) on the triangle $\triangle OPQ$ to find the length of OQ.

(c) Use the Pythagorean rule (not the distance formula) on the triangle $\triangle OQR$ to find the length of OR.

(d) For comparison, use the distance formula to compute the lengths of OQ and OR.

46. Any reasonable formula for distance should satisfy some commonsense requirements. For example, the distance $d(P, P)$ from any point P to itself should certainly be zero. So it is. If $P = (x, y, z)$ is any point, then the distance formula says $d(P, P) = \sqrt{(x-x)^2 + (y-y)^2 + (z-z)^2} = 0$. In the same spirit, use the distance formula to show that the following commonsense properties hold. Throughout, use the points $P = (x, y, z)$ and $Q = (a, b, c)$.

(a) If $P \neq Q$, then $d(P, Q) > 0$.

(b) $d(P, Q) = d(Q, P)$.

(c) If M is the midpoint of P and Q, then $d(P, M) = d(M, Q) = d(P, Q)/2$.

47. Let S be the surface defined by the equation $z = y^2/9 - x^2/16$. Describe the trace of S in the plane $y = b$, where b is a real number.

48. Let S be the surface defined by the equation $z = x^2/9 - y^2/16$. Describe the trace of S in the plane $z = c$, where c is a real number.

12.2 CURVES AND PARAMETRIC EQUATIONS

Introduction: all kinds of curves

Curves in the xy-plane come in an enormous variety. (Curves in xyz-space are equally plentiful; we will see some later in this section.) Figure 1 shows four basic examples. Let's take some closer looks:

- **Pieces of familiar graphs**   The curves C_1 and C_2 are pieces of familiar function graphs. Specifically, C_1 is that piece of the graph of $y = \sin x$ for which $-\pi \leq x \leq \pi$; C_2 is that

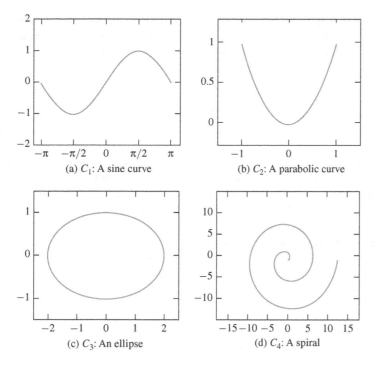

(a) C_1: A sine curve

(b) C_2: A parabolic curve

(c) C_3: An ellipse

(d) C_4: A spiral

FIGURE 1
Curves: basic examples

piece of the graph of $y = x^2$ for which $-1 \le x \le 1$. Using basic calculus properties of these functions and their derivatives, we can describe these curves in detail: where they rise and fall, their slopes at various points, their concavity, and so forth.

- **Graph vs. piece-of-a-graph** The difference between "graph" and "piece-of-a-graph" will sometimes matter to us. When it does matter, we will need to say carefully *which* piece we are interested in, usually by specifying (as we did above) a particular domain interval.

- **Not graphs of functions (but still important)** Curves C_3 and C_4 are *not* ordinary function graphs (or even pieces of such graphs) because some x-values on these curves correspond to more than one y-value. ➡ Nevertheless, curves like C_3 and C_4 are indispensable in real-life applications. A moving object, for instance, might well follow such a curve. Modeling physical motion—for centuries among the most important applications of calculus, and still so today—would be impossible without mathematical tools for handling general curves.

C_3 and C_4 fail the "vertical line test."

---

400 years of modeling motion.   Modeling physical motion has occupied mathematicians for centuries. Around 1600, the German astronomer Johannes Kepler asserted (among other things) that the planets follow *elliptical* orbits. In the late 1600s Isaac Newton used his new calculus to verify and extend Kepler's models of planetary motion.

   Modeling motion mathematically is still important. Choreographing the twists, turns, and extensions of an industrial robot arm, for instance, would be all but impossible without ideas and tools from multivariable calculus.

Before we can apply calculus ideas and methods to general curves, we will need to describe them in a concrete, mathematical manner. Parametric equations are the key. For the moment, we will restrict our attention to curves in the xy-plane.

Parametric equations

The idea   Imagine a point P wandering about in the xy-plane during a time interval $a \leq t \leq b$. The two coordinates of P, written $x = f(t)$ and $y = g(t)$ are both real-valued functions of t, defined for t in $[a, b]$.

Over the time interval $a \leq t \leq b$, the point $P = \big(f(t), g(t)\big)$ traces out some image figure, say C, in the xy-plane. As determined by f and g, the image figure C can have almost any shape: a line segment, a circular arc, a single point, a spiral, a sine curve, a jagged mess of line segments, or worse. Even if f and g are continuous functions, it is possible for the figure C to be surprisingly bizarre. ◀

C could be a filled-in square, for instance, or a spaghetti-like tangle.

Very often, however, f and g are sufficiently "well-behaved" that the image figure C turns out to be a "smooth curve" like C_1–C_4 above. (After some examples, we will say more precisely what "well-behaved" and "smooth curve" mean; here, we use these phrases informally.)

Parametric vocabulary   In sorting out similarities and differences between parametric equations and ordinary functions, it will help to agree on some standard terminology. Here is a quick summary:

Equations of the form

$$x = f(t) \qquad \text{and} \qquad y = g(t) \qquad \text{for } a \leq t \leq b$$

are called **parametric equations**. The functions f and g are **coordinate functions**. The variable t is called the **parameter**; ◀ the interval $a \leq t \leq b$ is the **parameter interval**.

In this setting, "parameter" refers to a variable. In other mathematical situations "parameters" may be constants.

The figure C traced out in the xy-plane by $P(t) = (f(t), g(t))$, for t in $[a, b]$, is called a **parametric curve**; the functions f and g and the parameter interval $[a, b]$ are said to **parametrize** the curve C.

A **parametric curve in space** is similar except that there are three coordinate functions: $x = f(t), y = g(t), z = h(t)$.

EXAMPLE 1   At each time t with $0 \leq t \leq 10$, the coordinates of P are given by the parametric equations

$$x = t - 2\sin t; \qquad y = 2 - 2\cos t.$$

What curve does P trace out over the time interval? Where is P at $t = 1$, and in which direction is P headed?

Solution   The simplest way to draw a curve is to calculate many points (x, y), plot each one, and "connect the dots." The first step is to calculate, for many inputs t, corresponding values of x and y. The following table shows some typical results rounded to two decimals:

Parametric plot points for $x = t - 2\sin t$, $y = 2 - 2\cos t$											
t	0	0.1	0.2	0.3	0.7	0.8	0.9	1.0	...	9.8	9.9	10
x	0	−0.10	−0.20	−0.29	−0.59	−0.63	−0.67	−0.68	...	10.53	10.82	11.09
y	0	0.01	0.04	0.09	0.47	0.61	0.76	0.92	...	3.86	3.78	3.68

Figure 2 shows the result, a curve C:

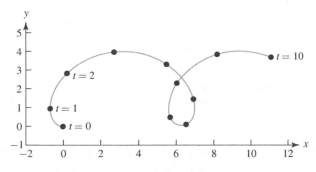

The parametric curve $x = t - 2\sin t$, $y = 2 - 2\cos t$, $0 \le t \le 10$

Notice:

- **Bullets** Points corresponding to *integer* values of t are shown bulleted. At $t = 1$, P has coordinates $(-0.68, 0.92)$; at this instant, P is heading almost due north.

- **Not a function graph** The curve C is *not* the graph of a function $y = f(x)$ because some x-values correspond to more than one y-value. ➡

- **No t-axis** The picture shows the x- and y-axes but not the t-axis: Particular t-values are indicated only by the bulleted points. On most curves, no such bullets are shown, and so the t-values can be found only from the coordinate functions.

- **Equal-time bullets** The bullets on the curve C appear at equal *time* intervals but not at equal *distances* from each other because P speeds up and slows down as it moves. ➡ We will soon see how to use derivatives of the coordinate functions to calculate the speed of P at any point along C.

- **Loops, vertical tangents, and other oddities.** Parametric curves can have many features that do not appear on ordinary function graphs. The curve above, for instance, has both loops and points with vertical tangent lines—even though both coordinates are given by familiar, well-behaved functions. We'll encounter other oddities (e.g., sharp corners) as we go along.

The input $x = 6$ is one example; note that the vertical line $x = 6$ cuts the curve C more than once.

When is P moving fastest? Slowest?

A sampler of plane parametric curves

Plane curves come in mind-boggling variety: Any choice of two equations $x = f(t)$ and $y = g(t)$ and a t-interval produces one. Surprisingly often the result is beautiful, useful, or interesting. The following examples hint at some of the possibilities and at connections between parametric curves and ordinary function graphs.

Ordinary function graphs Pieces of ordinary function graphs are easy to write in parametric form. The next example shows how.

EXAMPLE 2 Parametrize the curves C_1 and C_2 in Figure 1.

Solution One idea—letting x serve as the parameter t—works for both curves. For C_1:

$$x = t; \qquad y = \sin t; \qquad -2\pi \le t \le 2\pi.$$

For C_2:

$$x = t; \qquad y = t^2; \qquad -1 \le t \le 1.$$

Plotting these data produces the desired curves C_1 and C_2.

The same idea works for *any* function f defined on any interval $[a, b]$. Setting

$$x = t; \qquad y = f(t); \qquad a \le t \le b$$

gives a parametrization of the graph $y = f(x)$ from $x = a$ to $x = b$.

Line segments   Line segments are among the simplest and most useful "curves." In computer graphics, for instance, complicated curves are drawn by laying many ← line segments end to end.

Hundreds or thousands, sometimes.

EXAMPLE 3   (The line segment joining two points) Let $P = (a, b)$ and $Q = (c, d)$ be any two points in the xy-plane. Parametrize the segment from P to Q.

Other possibilities are outlined in the exercises.

Solution   Here is one possibility. ← Let

$$x(t) = a(1 - t) + ct; \qquad y(t) = b(1 - t) + dt; \qquad 0 \le t \le 1.$$

Is the resulting curve, C, really the segment we want? Notice first that

$$\big(x(0), y(0)\big) = (a, b) = P \qquad \text{and} \qquad \big(x(1), y(1)\big) = (c, d) = Q.$$

In other words, C starts and ends at the right places. It can also be shown that for all t in $[0, 1]$, $\big(x(t), y(t)\big)$ lies on the *line* through P and Q. See the exercises for details. Thus, C is indeed the line segment from P to Q.   ◼

Circles and circular arcs   Circles and circular arcs, like line segments, are important modeling and drawing tools. Fortunately, circles are easy to parametrize. The **unit circle**, with radius 1 and center $(0, 0)$, illustrates the main ideas.

EXAMPLE 4   Parametrize the unit circle $x^2 + y^2 = 1$.

*Recall: The cosine and sine functions are sometimes called **circular functions**.*

Solution   The simplest parametrization uses trigonometric ingredients. ← Let

$$x = \cos t; \qquad y = \sin t; \qquad 0 \le t \le 2\pi.$$

Figure 3 shows the result; some important t-values are marked with bullets.

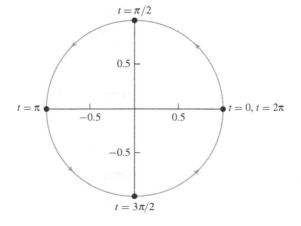

FIGURE 3
The parametric curve $x = \cos t$, $y = \sin t$, $0 \le t \le 2\pi$

This very important example deserves a close look.

- **Time or angle or arclength?**   On any parametric curve the parameter t can be understood as *time*—the time at which a moving point $P(t) = (x(t), y(t))$ arrives at a given point along the curve. Here, however, there are other useful ways to interpret the parameter t. One possibility is to think of t as the *angle* determined by the positive x-axis and the segment from the origin to $P(t)$. ➤ Another possibility is to interpret t as *arclength*—as the distance (measured along the circle) from the starting point $(1, 0)$ to the point $P(t)$. Which of these views is "best" depends on the situation.

  Angles are measured counterclockwise, in radians.

- **Varying the t-interval**   The full circle is traced once, counterclockwise, as t runs from $t = 0$ to $t = 2\pi$. (The curve starts and ends at the same point.) With larger or smaller t-intervals the circle would be traced more or less often. If, say, $0 \le t \le 4\pi$, then the circle is traversed *twice*. If, instead, we restrict t to the interval $[\pi/2, 3\pi/2]$, then only the "left" semicircle is covered.

- **Why the unit circle?**   Because $x = \cos t$ and $y = \sin t$, we get

$$x^2 + y^2 = (\cos t)^2 + (\sin t)^2 = 1$$

  for all t. This means that every point on our parametric curve satisfies the Cartesian equation $x^2 + y^2 = 1$ and therefore (as we planned) lies on the circle of radius 1 and center $(0, 0)$. ∎

Other circles and arcs   The idea in Example 4 extends to other circles. If (a, b) is any point in the plane and r is any positive number, then the parametrization

$$x = a + r \cos t; \qquad y = b + r \sin t; \qquad 0 \le t \le 2\pi$$

produces the circle with center (a, b) and radius r.

Eliminating the parameter   The technique just illustrated—trading *two* equations $x = f(t)$ and $y = g(t)$ for *one* equation involving only x and y—is called **eliminating the parameter**. Doing so can be easy or difficult as determined by the functions f and g. ➤ In the preceding calculation we used a standard trigonometric identity to eliminate t. In other cases, basic algebra may work.

For complicated functions f and g, eliminating the parameter may be impossible.

EXAMPLE 5   Consider the parametric curve with coordinate functions $x = f(t) = 1 + 2t$ and $y = g(t) = 3 + 4t$. Eliminate the parameter t. What does the form of the resulting equation in x and y say about the curve?

Solution   Each coordinate equation can be solved for t:

$$x = 1 + 2t \implies t = \frac{x-1}{2}; \qquad y = 3 + 4t \implies t = \frac{y-3}{4}.$$

Thus,

$$\frac{x-1}{2} = \frac{y-3}{4}, \qquad \text{and so} \qquad 4x - 4 = 2y - 6, \qquad \text{or} \qquad 2x + 1 = y.$$

The last equation describes a straight line—as we expect, because both coordinate functions are linear. ∎

Curves in space

Parametric curves in three-dimensional space are described exactly like curves in the plane, except that a third coordinate function is present, and the resulting curve wanders through space rather than through the plane. Adding a third dimension to the picture allows, in principle, very wild curves—imagine, say, a knotty, tangled mess of string. (We will stick to relatively tame space curves in this book.) On the other hand, essentially the same ideas and calculus methods apply in three dimensions as in two, so we'll content ourselves here with some basic examples.

The xy-plane is shaded.

E X A M P L E 6 Figure 4 shows two curves in space: ← a **helix** (or **spiral**) making two turns and an ordinary unit circle. (The xy-plane is shaded.)

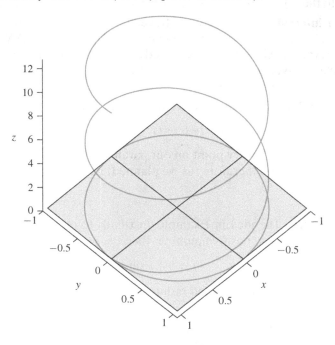

FIGURE 4
Two space curves: A helix and a circle

Find a parametrization for each curve; assume that both curves start at $(1, 0, 0)$ and turn counterclockwise.

Solution We have just seen how to parametrize the unit circle in the xy-plane. To place the curve in three dimensions we just give z a constant value:

$$x = \cos t; \qquad y = \sin t; \qquad z = 0; \qquad 0 \le t \le 2\pi.$$

The helix resembles the circle except that the helix rises as it turns. This suggests a parametrization similar to the one just given, but we will set $z = t$ to produce the desired "rising." To arrange *two* turns we double the parameter interval:

$$x = \cos t; \qquad y = \sin t; \qquad z = t; \qquad 0 \le t \le 4\pi.$$

The result is indeed what is shown in the picture—notice, in particular, that z takes values from 0 to $4\pi \approx 12.6$.

As always, there are many possible parametrizations. For instance, we could traverse the same helix in "half the time" with the parametrization

$$x = \cos(2t); \qquad y = \sin(2t); \qquad z = 2t; \qquad 0 \le t \le 2\pi,$$

while the "lazier" recipe

$$x = \cos(t/2); \qquad y = \sin(t/2); \qquad z = t/2; \qquad 0 \le t \le 8\pi$$

traverses the same curve in twice the original time. ∎

Plane curves to order

By combining such "elements" as line segments and circular arcs, we can draw various composite curves "to order."

EXAMPLE 7   A dog track is designed as in Figure 5, with three straight runs and three circular arcs.

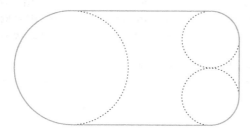

FIGURE 5

A simple racetrack

"Construct" the track mathematically by parametrizing each of its six elements running counterclockwise.

Solution   Imagine the track as lying in the xy-plane, as in Figure 6; the scales are chosen entirely for convenience: ➡

There's nothing special about these particular coordinates or units of size.

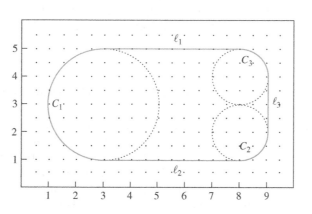

FIGURE 6

A racetrack in the xy-plane

We will parametrize each labeled curve separately.

- **The line segments**   We saw in Example 3 that setting

$$x(t) = a(1-t) + ct; \qquad y(t) = b(1-t) + dt; \qquad 0 \le t \le 1$$

parametrizes the segment from (a, b) to (c, d). The segment ℓ_1 runs from $(8, 5)$ to $(3, 5)$; substituting appropriate values for a, b, c, and d shows that

$$x(t) = 8(1-t) + 3t = 8 - 5t; \qquad y(t) = 5(1-t) + 5t = 5; \qquad 0 \le t \le 1$$

parametrizes ℓ_1. Similar reasoning shows that

$$x(t) = 3(1-t) + 8t = 3 + 5t; \qquad y(t) = 1(1-t) + 1t = 1; \qquad 0 \le t \le 1$$

parametrizes ℓ_2 and that

$$x(t) = 9(1-t) + 9t = 9; \qquad y(t) = 2(1-t) + 4t = 2 + 2t; \qquad 0 \le t \le 1$$

parametrizes ℓ_3.

- **The circular arcs**   Every circular arc has a parametrization of the general form

$$x = a + r\cos t; \qquad y = b + r\sin t; \qquad \alpha \le t \le \beta.$$

Choosing a, b, and r is easy here—each circle's center and radius are evident from Figure 6. Choosing appropriate parameter intervals $[\alpha, \beta]$ takes more care. The parameter interval $[0, 2\pi]$ produces a *full* circle; because we want only part of each circle, we need shorter parameter intervals.

The trick is to think of t in *angular* terms. Points on the arc C_1 correspond to angles with radian measure between $\pi/2$ and $3\pi/2$. Hence, the prescription

$$x(t) = 3 + 2\cos t; \qquad y(t) = 3 + 2\sin t; \qquad \pi/2 \le t \le 3\pi/2$$

parametrizes C_1. Similarly,

$$x(t) = 8 + \cos t; \qquad y(t) = 2 + \sin t; \qquad 3\pi/2 \le t \le 2\pi$$

parametrizes C_2, and

$$x(t) = 8 + \cos t; \qquad y(t) = 4 + \sin t; \qquad 0 \le t \le \pi/2$$

parametrizes C_3.

Notice, finally, that in parametrizing each track section we chose our t-intervals independently of each other. A slightly more "realistic" approach might be to use *successive* time intervals, such as $[0, 1], [1, 2], [2, 3], \ldots, [5, 6]$, to parametrize the six successive curve segments. This is easily done; we show how in Example 9. ∎

One road, many journeys

Many cars journey between Mission, South Dakota and Valentine, Nebraska, but there is only one road: U.S. Highway 83.

The same important difference holds between a curve in the plane, on the one hand, and any particular parametrization of that curve, on the other hand. A parametrization amounts, in effect, to prescribing a journey along the curve: At any time in some interval $a \le t \le b$, the coordinate functions $x = f(t)$ and $y = g(t)$ place a moving point (a car, say) somewhere along the curve (U.S. 83, for instance). Many journeys—in either direction, at various speeds, over various time intervals, and so on—are possible along that stretch of U.S. 83.

The recipe for a curve in space requires four ingredients.

Because every particular curve has many possible parametrizations, we will often emphasize parametrizations as much as the curves they determine. A parametrization depends on all three of its ingredients: (i) a coordinate function $x = x(t)$; (ii) a coordinate function $y = y(t)$; (iii) a parameter interval $a \le t \le b$. ←

We will return often to the theme of finding and comparing various parametrizations of a given curve or other object. Examples 8 and 9 illustrate the idea.

EXAMPLE 8   Give two different parametrizations of the parabola C_2 in Figure 1. How do the parametrizations differ?

Solution   Different pairs of parametric equations may produce the same geometric curve; labeling t-values can help show differences between the parametrizations. For example, both parametrizations

$$x = t; \qquad y = t^2; \qquad -1 \le t \le 1$$

and

$$x = t^3; \qquad y = t^6; \qquad -1 \le t \le 1$$

produce the same parabolic curve, as Figure 7 shows:

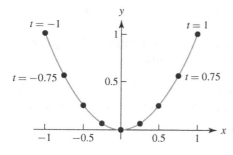

(a) The curve $x = t$, $y = t^2$, $-1 \le t \le 1$

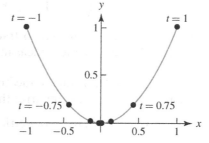

(b) The curve $x = t^3$, $y = t^6$, $-1 \le t \le 1$

FIGURE 7
Two parametrizations of the same curve

The bullets (they appear at 0.25-second time intervals) show, however, that the two parametrizations represent quite different "trips" along the same "road." Most important, the trips differ in speed. We will return to this idea in Section 12.6. ∎

Tricks of the trade

Various standard devices exist for trading one sort of parametrization for another. We mention some briefly here; others appear in the exercises.

Reversing direction U.S. Highway 83 runs *both* ways between Mission and Valentine. For similar reasons, every parametric curve can be "reversed." If C is parametrized by

$$x = f(t); \qquad y = g(t); \qquad a \le t \le b,$$

then the reversed curve (denoted by $-C$) can be parametrized by

$$x = f(a+b-t); \qquad y = g(a+b-t); \qquad a \le t \le b.$$

(As t runs from a to b, the quantity $a+b-t$ runs from b to a.) ➤

Use the formulas to convince yourself of this.

For the parabola C_2 on page 619, for instance, the reverse curve $-C_2$ can be parametrized as follows:

$$x = -t; \qquad y = (-t)^2 = t^2; \qquad -1 \le t \le 1.$$

Different parameter intervals U.S. 83 is the same route at all times of the day or night, regardless of how fast you drive it. For similar reasons, a parametric curve can be parametrized using *any* parameter interval.

For instance, if C is parametrized by

$$x = f(t); \qquad y = g(t); \qquad a \le t \le b,$$

then C can also be parametrized using the time interval $0 \le t \le 1$, using

$$x = f(a+(b-a)t); \qquad y = g(a+(b-a)t); \qquad 0 \le t \le 1.$$

Conversely, if C is parametrized by

$$x = f(t); \qquad y = g(t); \qquad 0 \le t \le 1,$$

then C can also be parametrized using the time interval $a \le t \le b$ via

$$x = f\left(\frac{t-a}{b-a}\right); \qquad y = g\left(\frac{t-a}{b-a}\right); \qquad a \le t \le b.$$

These formulas can be combined to trade any parameter interval $[a, b]$ for any other interval $[c, d]$. We illustrate these ideas in the next example.

EXAMPLE 9   Consider again the racetrack in Example 7. Use the parameter intervals $1 \le t \le 2$ and $2 \le t \le 3$ to parametrize the curve segments C_1 and ℓ_2, respectively.

Solution   In Example 7, we parametrized C_1 by

$$x = f(t) = 3 + 2\cos t; \qquad y = g(t) = 3 + 2\sin t$$

for the parameter interval $\pi/2 \le t \le 3\pi/2$. As indicated above, the equations

$$x = f\left(\frac{\pi}{2} + \pi t\right) = 3 + 2\cos\left(\frac{\pi}{2} + \pi t\right); \qquad y = g\left(\frac{\pi}{2} + \pi t\right) = 3 + 2\sin\left(\frac{\pi}{2} + \pi t\right)$$

produce the same curve for the new parameter interval $0 \le t \le 1$.

To switch to the new parameter interval $[1, 2]$, we can replace t above with $t - 1$. The resulting new parametrization for C_1 is

$$x = 3 + 2\cos\left(\frac{\pi}{2} + \pi(t-1)\right); \qquad y = 3 + 2\sin\left(\frac{\pi}{2} + \pi(t-1)\right); \qquad 1 \le t \le 2.$$

In Example 7, we parametrized ℓ_2 by

$$x = 3 + 5t; \qquad y = 1; \qquad 0 \le t \le 1.$$

If we prefer the parameter interval $2 \le t \le 3$, we simply replace t with $t - 2$ to get

$$x(t) = 3 + 5(t-2); \qquad y(t) = 1; \qquad 2 \le t \le 3.$$

BASIC EXERCISES

In Exercises 1–6, plot the parametric curve, mark the direction of travel, and label the points corresponding to $t = -1$, $t = 0$, and $t = 1$.

1. $x = t; y = \sqrt{1 - t^2}; -1 \le t \le 1$

2. $x = t; y = -\sqrt{1 - t^2}; -1 \le t \le 1$

3. $x = \sqrt{1 - t^2}; y = t; -1 \le t \le 1$

4. $x = -\sqrt{1 - t^2}; y = t; -1 \le t \le 1$

5. $x = \sin(\pi t); y = \cos(\pi t); -2 \le t \le 2$

6. $x = \cos(t); y = \sin(t); -2 \le t \le 2$

In Exercises 7–12, find a parametrization (there is more than one possibility!) for the curve. Check your answers using technology.

7. The line segment from $(0, 0)$ to $(1, 2)$.

8. The line segment from $(1, 2)$ to $(0, 0)$.

9. The entire unit circle, starting and ending at the east pole $(1, 0)$ but moving clockwise.

10. The left half of the unit circle, moving counterclockwise from the north pole to the south pole.

11. The entire unit circle, starting and ending at the east pole $(1, 0)$ but using the parameter interval $0 \le t \le 1$.

12. The part of the curve $y = x^2$ from $(0, 0)$ to $(3, 9)$.

In Exercises 13–18, the given parametric "curve" is actually a line segment. State the beginning point ($t = 0$) and ending point ($t = 1$) of the segment. Then state an equation in x and y for the line each segment determines.

13. $x = 2 + 3t; y = 1 + 2t; 0 \le t \le 1$

14. $x = 2 + 3(1 - t); y = 1 + 2(1 - t); 0 \le t \le 1$

15. $x = t; y = mt + b; 0 \le t \le 1$

16. $x = a + bt; y = c + dt; 0 \le t \le 1$

17. $x = x_0 + (x_1 - x_0)t; y = y_0 + (y_1 - y_0)t; 0 \le t \le 1$

18. $x = mt + b; y = t; 0 \le t \le 1$

19. Consider the curve C shown in Example 1; suppose that t tells time in seconds.

    (a) At which bulleted points would you expect P to be moving quickly? Slowly? Why?

    (b) *Estimate* the speed of P at $t = 3$. [HINT: Start by estimating how far P travels over the 1-second interval from $t = 2.5$ to $t = 3.5$.]

    (c) Estimate the speed of P at $t = 6$.

20. Plot the parametric curve $x = t^3$; $y = \sin(t^3)$; $-2 \le t \le 2$.

    (a) What familiar curve is produced? Why does the result happen?

    (b) Give another parametrization for the same curve.

21. Let (a, b) be any point in the plane and $r > 0$ any positive number. Consider the parametric equations $x = a + r \cos t$; $y = b + r \sin t$; $0 \le t \le 2\pi$.

    (a) Plot this parametric curve for $(a, b) = (2, 1)$ and $r = 2$. Describe your result in words.

    (b) Show by calculation that if x and y are as above, then $(x - a)^2 + (y - b)^2 = r^2$. Conclude that the curve defined above is the circle with center (a, b) and radius r.

    (c) Write parametric equations for the circle of radius $\sqrt{13}$ centered at $(2, 3)$.

    (d) What "curve" results from the equations above if $r = 0$?

22. Let a and b be any positive numbers, and let a parametric curve C be defined by $x = a \cos t$; $y = b \sin t$; $0 \le t \le 2\pi$. The resulting curve is an **ellipse**.

    (a) Plot the curve defined above for $a = 2$ and $b = 1$. Describe C in words. Where is the "center" of C? Why do you think the quantities $2a$ and $2b$ are called the **major** and **minor axes** of C?

    (b) What curve results if $0 \le t \le 4\pi$? Why?

    (c) Write parametric equations for an ellipse with major axis 10 and minor axis 6.

    (d) Write parametric equations for *another* ellipse with major axis 10 and minor axis 6.

    (e) Show that for all t, $x^2/a^2 + y^2/b^2 = 1$.

    (f) How does the "ellipse" look if $a = b$? How does its xy equation look?

    (g) How does an ellipse look if $a = 1000$ and $b = 1$?

    (h) How does an ellipse look if $a = 1$ and $b = 1000$?

FURTHER EXERCISES

23. Give two different parametrizations of the upper half of the ellipse C_3 on page 619. [HINT: One can be based on the graph of an ordinary function.]

24. If a parametric curve C has *linear* coordinate functions $x = f(t) = at + b$ and $y = g(t) = ct + d$, then C is a line (or part of a line).

    (a) Plot the parametric curve $x = 2t$; $y = 3t + 4$; $0 \le t \le 1$. Where does the curve start? Where does it end? What is its shape?

    (b) Eliminate the variable t in the two equations in part (a) to find a single equation in x and y for the line.

    (c) Suppose that a curve has coordinate functions $x = f(t) = at + b$ and $y = g(t) = ct + d$ for constants a, b, c, and d. (Assume that $a \ne 0$.) Show that $cx - ay = cb - ad$. Does this equation describe a line?

25. Look at the racetrack in Example 7. It resembles a belt drawn snugly around three circles. Consider another similar racetrack, this time in the shape of a belt drawn snugly around two circles, the first of radius 2 centered at $(3, 3)$ and the second of radius 2 centered at $(3, 7)$.

    (a) Draw the racetrack by hand; label its four elements. (Two are line segments and two are circular arcs.)

    (b) Parametrize each of the four arcs; all should run counterclockwise. (More than one method is possible.)

    (c) Use technology and your results from the previous part to draw the new racetrack.

26. Give a parametrization for the *left* half of the circle of radius 3 centered at the point $(1, 2)$.

27. Give a clockwise-oriented parametrization for the bottom half of a circle of radius 4 centered at $(2, -3)$.

28. Give two different parametric representations of the circular helix with radius 3 that makes two complete turns between $(3, 0, 0)$ and $(3, 0, 4)$.

29. Give a parametric representation of the circular helix (i.e., a spring) with radius R that makes N complete turns between $(R, 0, 0)$ and $(R, 0, H)$.

12.3 POLAR COORDINATES AND POLAR CURVES

Every point P in the xy-plane has a familiar and natural "address": its **rectangular** (or **Cartesian**) **coordinates**. The point P in Figure 1(a), for instance, has rectangular coordinates $(4, 3)$.

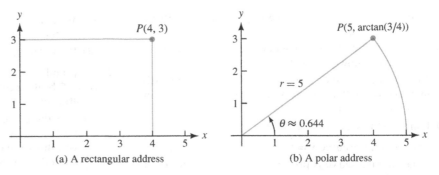

(a) A rectangular address   (b) A polar address

FIGURE 1
Polar and rectangular coordinates

The x- and y-coordinates of P measure the distances from P to the two perpendicular coordinate axes. To reach $P = (4, 3)$ from the origin, one moves 4 units right and 3 units up.

Polar coordinates offer another way of locating a point in the plane. In the **polar coordinate system**, a point P has coordinates r and θ; they tell, respectively, the *distance* from the origin O to P and the *angle* (in radians!) from the positive x-axis to the ray from O to P.

Convince yourself that $\theta = \arctan(3/4)$.

Figure 1(b) shows that the point P with rectangular coordinates $(4, 3)$ has polar coordinates $\left(5, \arctan(3/4)\right) \approx (5, 0.644)$. ◄

Polar coordinate systems

A **rectangular coordinate system** in the Euclidean plane starts with an origin O and two perpendicular coordinate axes. Usually the x-axis is horizontal and the y-axis is vertical; x-coordinates increase to the right, and y-coordinates increase upward.

A polar coordinate system starts with different ingredients: an origin O, called the **pole**, and a ray (i.e., a half-line) beginning at the origin, called the **polar axis**. The polar axis normally points to the right, along the positive x-axis. With these ingredients and a unit for measuring distance, we can assign polar coordinates (r, θ) to any point P:

> r *is the distance from* O *to* P; θ *is any angle from the polar axis to the segment* $\overline{OP}$.

Check carefully that each point's coordinates are correct.

Figure 2 shows several points with their polar coordinates, plotted on a polar grid. ◄

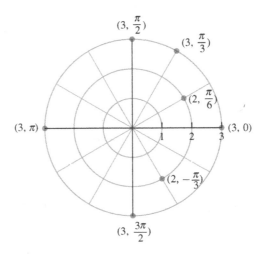

FIGURE 2
Points on a polar grid

Polar vs. rectangular grids A rectangular coordinate system leads naturally to a rectangular grid having vertical lines $x = a$ and horizontal lines $y = b$. In a polar system, holding the coordinates r and θ constant produces, respectively, concentric circles and radial lines. The result is a weblike polar grid. Figure 3 shows grids of both types:

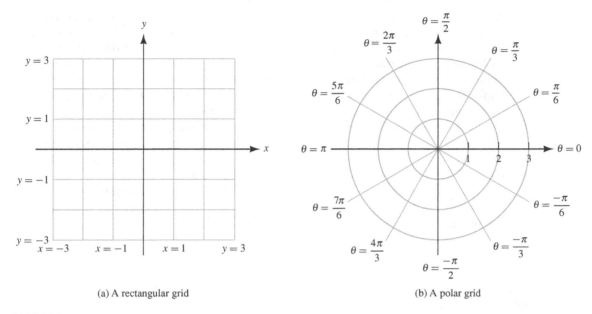

| (a) A rectangular grid | (b) A polar grid |

FIGURE 3
Two types of grids

On the scales. In a rectangular coordinate system, the two axes often have different scales of measurement. As a result, on a graph, a vertical inch and a horizontal inch may represent different distances. In particular, "circles" may look far from round.

A polar coordinate system, by contrast, has just one axis—the polar axis. As a result, distance does not depend on direction, and circles look round.

Polar coordinates: not unique A point in the plane—$P = (4, 3)$, for instance—has just one possible pair of rectangular coordinates. Different rectangular coordinate pairs (x_1, y_1) and (x_2, y_2) correspond to different points in the plane.

Polar coordinates, by contrast, are not unique. Every point in the plane has many possible pairs of polar coordinates. For example, all of the polar coordinate pairs

$$\left(2, \frac{\pi}{4}\right), \quad \left(2, \frac{9\pi}{4}\right), \quad \left(2, -\frac{7\pi}{4}\right), \quad \left(2, -\frac{15\pi}{4}\right), \quad \left(-2, \frac{5\pi}{4}\right), \quad \left(-2, -\frac{3\pi}{4}\right)$$

(and many others) represent the same point—the one with rectangular coordinates $(\sqrt{2}, \sqrt{2})$. Notice especially the last two pairs. A negative r-coordinate means that, to locate the point P, one moves r units in the direction *opposite* the θ-direction. The point $(-2, 5\pi/4)$, for instance, lies 2 units from the origin on the ray $\theta = \pi/4$. (This is the ray opposite $\theta = 5\pi/4$.) The origin O allows even more freedom: It is represented by *any* pair of the form $(0, \theta)$, regardless of θ.

This ambiguity of polar coordinates arises for a simple reason. All angles that differ by integer multiples of 2π determine the same direction. In practice, this ambiguity can be annoying but is seldom a serious problem. Two simple rules help.

- **Multiples of** 2π For any r and θ, the pairs (r, θ) and $(r, \theta + 2\pi)$ describe the same point.
- **Negative** r For any r and θ, the pairs (r, θ) and $(-r, \theta + \pi)$ describe the same point.

Polar coordinates on Earth.   The polar grid somewhat resembles an overhead view of Earth as if one were looking "down" at the North Pole. In cartographers' language, lines of **longitude** (or **meridians**) converge at the pole; the concentric circles are lines of **latitude** (or **parallels**). For hundreds of years, the **prime meridian** (polar axis, in calculus language), for which $\theta = 0$, has been taken to be the line of longitude that passes through the Greenwich Observatory just east of London, England. For the same reason, Greenwich Mean Time (the time of day along the prime meridian) is used worldwide as a reference point.

Why is Greenwich "prime" rather than, say, India or Arabia, where navigation and timekeeping flourished even in antiquity? There is no intrinsic reason; Greenwich just happened to be a center of attention when the terms were defined—an early (and quite literal) instance of Eurocentrism.

Polar coordinates in the *plane*, it should be said, aren't perfectly suited to measuring the (almost) spherical Earth. In practice, geographers use a related system called *spherical* coordinates. We will meet them in Chapter 14.

Polar graphs

The ordinary graph of an equation in x and y is the set of points (x, y) whose coordinates satisfy the equation. The graph of $x^2 + y^2 = 1$, for instance, is the circle of radius 1 about the origin. The point $(2, 3)$ does not lie on this graph because $2^2 + 3^2 \neq 1$.

The idea of a **polar graph** is similar but not quite identical. The graph of an equation in r and θ is the set of points whose polar coordinates r and θ satisfy the equation. For instance, the polar point $(3, 0)$ lies on the graph of $r = 2 + \cos\theta$ (because $3 = 2 + \cos 0$), but the polar point $(2, \pi)$ does not.

A warning   The fact that a point in the plane has more than one pair of polar coordinates means that polar plotting requires extra care. At first glance, for instance, the point P with polar coordinates $(-3, \pi)$ seems *not* to satisfy the polar equation $r = 2 + \cos\theta$. A closer look, however, shows that P can also be written with polar coordinates $(3, 0)$, which *do* satisfy the given equation. Here's the moral:

A point P lies on the graph of a polar equation if P has any pair of polar coordinates that satisfy the equation.

Drawing polar graphs

The simplest polar graphs come from functions, usually of the form $r = f(\theta)$. Given such a function and a specific θ-domain, it is a routine matter to tabulate points and then plot them. We illustrate by example.

EXAMPLE 1   Plot the equation $r = 2 + \cos\theta$ for $0 \le \theta \le 2\pi$.

Solution   Let $f(\theta) = 2 + \cos\theta$. We want the r-θ graph of f. First we tabulate some values.

Values of $r = f(\theta) = 2 + \cos\theta$												
θ	0	$\frac{\pi}{6}$	$\frac{\pi}{3}$	$\frac{\pi}{2}$	$\frac{2\pi}{3}$	$\frac{5\pi}{6}$	π	$\frac{7\pi}{6}$	$\frac{4\pi}{3}$	$\frac{3\pi}{2}$	$\frac{5\pi}{3}$	$\frac{11\pi}{6}$	2π
r	3	2.87	2.5	2	1.5	1.14	1	1.14	1.5	2	2.5	2.87	3

Next we plot the data (polar "graph paper" makes the job easier) and fill in the gaps smoothly. Figure 4 shows the result:

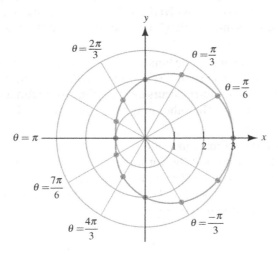

FIGURE 4
A polar graph: $r = 2 + \cos\theta$

Polar graphs: a sampler

Several polar graphs follow. We have already seen the simplest polar graphs of all—those of the equations $r = a$ and $\theta = b$—in Figure 3.

Cardioids and limaçons Graphs of the form $r = a \pm b\cos\theta$ and $r = a \pm b\sin\theta$, where a and b are positive numbers, are called **limaçons**; if $a = b$, the term **cardioid** ("heartlike") is used. (The graph in Example 1 is a limaçon.) The graphs in Figure 5 illustrate the variety of limaçons and show the effects of the constants a and b.

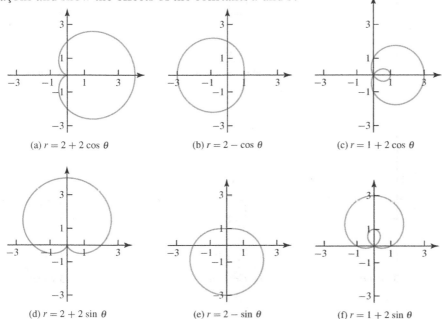

(a) $r = 2 + 2\cos\theta$ (b) $r = 2 - \cos\theta$ (c) $r = 1 + 2\cos\theta$

(d) $r = 2 + 2\sin\theta$ (e) $r = 2 - \sin\theta$ (f) $r = 1 + 2\sin\theta$

FIGURE 5
Six limaçons

Notice some features of these graphs.

- **What θ-range?**   The graphs were drawn by letting θ vary through the interval $[0, 2\pi]$. Because all functions involved are 2π-periodic, any other interval of length 2π would produce the same result. (We could have used the interval $-\pi \leq t \leq \pi$, for example.)

- **Symmetry**   Three of the preceding limaçons—those that involve the cosine function—are symmetric about the x-axis, that is the line $\theta = 0$. (The other three are symmetric about the y-axis.) This symmetry occurs because the cosine function is even: For every θ we have $\cos\theta = \cos(-\theta)$. The other graphs are symmetric about the y-axis because the sine function is odd.

- **Inner loops**   Each of the limaçons $r = 1 + 2\cos\theta$ and $r = 1 + 2\sin\theta$ has an inner loop. A close look at the graphs and the formulas reveals that these loops correspond to *negative* values of r. For $r = 1 + 2\cos\theta$, for instance, we have $r = 0$ when $\theta = 2\pi/3$ or $\theta = 4\pi/3$, and $r < 0$ for $2\pi/3 < \theta < 4\pi/3$. For these θ-values, therefore, the curve is drawn on the opposite side of the origin.

Roses   Equations of the form $r = a\cos(k\theta)$ and $r = a\sin(k\theta)$, where a is a constant and k is a positive integer, produce graphs called **roses**. The graphs in Figure 6 help explain the name.

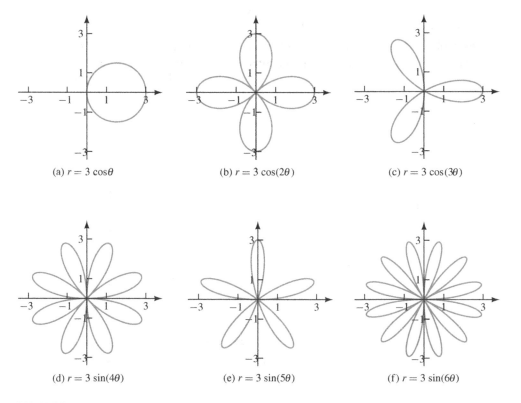

(a) $r = 3\cos\theta$ (b) $r = 3\cos(2\theta)$ (c) $r = 3\cos(3\theta)$

(d) $r = 3\sin(4\theta)$ (e) $r = 3\sin(5\theta)$ (f) $r = 3\sin(6\theta)$

FIGURE 6
A bouquet of roses

Notice some key features of the roses.

- **Symmetry**   Like limaçons (and for the same reason), all roses are symmetric about an axis—"cosine roses" about the *x*-axis and "sine roses" about the *y*-axis.
- **The rose's radius**   The coefficient *a* in $r = a\cos(k\theta)$ and $r = a\sin(k\theta)$ determines the rose's "radius."
- **How many petals?**   The coefficient *k* in $r = a\cos(k\theta)$ and $r = a\sin(k\theta)$ determines the number of "petals": *k* if *k* is odd, 2*k* if *k* is even. But here is a subtlety, best revealed by plotting some roses by hand:

> *If k is odd, then each petal is traversed twice for $0 \le \theta \le 2\pi$.*

In other words, for odd *k*, the rose $r = a\cos(k\theta)$ (or $r = a\sin(k\theta)$) has *k double* petals.

Trading polar and rectangular coordinates

How are the polar coordinates (r, θ) of a point *P* related to the rectangular coordinates (x, y) of the same point? How can either type of coordinates be found from the other? Figure 7 gives a useful view:

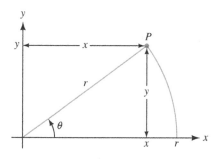

FIGURE 7
Relating polar and rectangular coordinates

The picture illustrates many relations among *x*, *y*, *r*, and θ. Here are some of the simplest:

$$x = r\cos\theta; \qquad y = r\sin\theta; \qquad r^2 = x^2 + y^2; \qquad \tan\theta = \frac{y}{x}.$$

(The last equation holds only if $x \ne 0$.)

These relations let us convert from one type of coordinates to the other. Equations in *x* and *y*, for instance, are easy to rewrite in terms of *r* and θ, as the next two examples illustrate.

E X A M P L E 2 Find a polar equation for the straight line $y = mx + b$, where $b \neq 0$.

Solution Substituting $x = r \cos\theta$ and $y = r \sin\theta$ into the equation for the line gives

$$y = mx + b \iff r \sin\theta = m(r\cos\theta) + b \iff r = \frac{b}{\sin\theta - m\cos\theta}.$$

(One moral is that rectangular coordinates are better suited to straight lines than are polar coordinates!)

Multiplying both sides of the first equation by r avoids square roots.

E X A M P L E 3 The graph of $r = 3\cos\theta$ looks like a circle. Is it a circle? Which circle?

Solution It *is* a circle. Changing to rectangular coordinates shows why:

$$r = 3\cos\theta \implies r^2 = 3r\cos\theta \implies x^2 + y^2 = 3x.$$

As expected, the last equation does define a circle. To decide *which* circle, complete the square:

$$x^2 + y^2 = 3x \iff x^2 - 3x + y^2 = 0 \iff \left(x - \frac{3}{2}\right)^2 + y^2 = \frac{9}{4}.$$

The circle therefore has radius 3/2 and center at $(3/2, 0)$—just as the picture suggests.

Polar curves and parametric equations

Suppose that a curve is described in polar coordinates by an equation of the form $r = f(\theta)$ for $\alpha \le \theta \le \beta$. Such a curve can also be written in parametric form with x and y given as separate functions of the parameter θ.

The key ingredients are the equations that relate polar and Cartesian coordinates:

$$x = r\cos\theta \quad \text{and} \quad y = r\sin\theta.$$

These equations show that each point on a polar curve $r = f(\theta)$ has a natural parametric form with θ as the parameter. Here is the result:

We often think of the parameter as time, but not always; here is one exception.

> **FACT (Writing polar curves parametrically)** The curve with polar equation $r = f(\theta)$, for $\alpha \le \theta \le \beta$, has parametric form
> $$x = r\cos\theta = f(\theta)\cos\theta; \quad y = r\sin\theta = f(\theta)\sin\theta; \quad \alpha \le \theta \le \beta.$$

Examples 4 and 5 illustrate the Fact (and a slight variation) in action.

EXAMPLE 4 Two curves are given in polar form as follows:

(i) $r = f(\theta) = 3;$ $0 \le \theta \le 2\pi;$

(ii) $r = f(\theta) = \sec\theta;$ $-\dfrac{\pi}{4} \le \theta \le \dfrac{\pi}{4}.$

Rewrite each curve in parametric form and identify its shape.

Solution Curve (i) is especially simple. Because $r = 3$ for all θ we have

$$x = r\cos\theta = 3\cos\theta; \qquad y = r\sin\theta = 3\sin\theta; \qquad 0 \le \theta \le 2\pi,$$

which is the familiar parametrization for a circle of radius 3 centered at the origin.

The parametric equations for curve (ii) reveal a little surprise:

$$x = \sec\theta\cos\theta = 1; \qquad y = \sec\theta\sin\theta = \tan\theta; \qquad -\frac{\pi}{4} \le \theta \le \frac{\pi}{4}.$$

Because the x-coordinate has constant value 1, curve (ii) is simply a vertical line segment with the y-coordinate running from $\tan(-\pi/4) = -1$ to $\tan(\pi/4) = 1$. ∎

EXAMPLE 5 A curve is described in polar coordinates by the equations

$$r = t; \qquad \theta = 3\cos t; \qquad 0 \le t \le 10.$$

Find parametric equations for x and y and plot the curve.

Solution Notice the slight difference from earlier situations: Here both r and θ, rather than x and y, are functions of a parameter t. Luckily, the new twist makes little difference. As before, we have $x = r\cos\theta$ and $y = r\sin\theta$; in terms of t we obtain

$$x = r\cos\theta = t\cos(3\cos t); \qquad y = r\sin\theta = t\sin(3\cos t); \qquad 0 \le t \le 10,$$

which we can plot parametrically as usual. Figure 8 shows the pretty result:

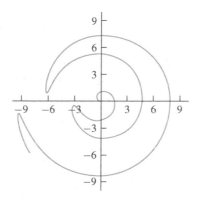

FIGURE 8
A polar-style parametric curve

We will soon develop tools to analyze such graphs in more detail, but we see even now how the radius $r = t$ increases with time while the angle $\theta = 3\cos t$ oscillates. ∎

In Exercises 1–4, a point is given in rectangular coordinates. Give three different pairs of polar coordinates for the point; r should be negative for at least one pair. [NOTE: Only familiar angles are involved, so give exact answers.]

1. $(\pi, 0)$ **3.** $(1, 1)$

2. $(0, \pi)$ **4.** $(-1, 1)$

In Exercises 5–8, a point is given in rectangular coordinates. Give three different pairs of polar coordinates for the point; r should be negative for at least one pair. [NOTE: Round answers to four decimal places.]

5. $(1, 2)$ **7.** $(1, 4)$

6. $(-1, 2)$ **8.** $(10, 1)$

In Exercises 9–12, a point is given in polar coordinates. Plot the point and label it with its rectangular coordinates. [NOTE: Give exact answers rather than decimal approximations.]

9. $(2, \pi/4)$ **11.** $(1, 13\pi/6)$

10. $(-2, 5\pi/4)$ **12.** $(42, 0)$

In Exercises 13–16, a point is given in polar coordinates. Plot the point and label it with its rectangular coordinates. [NOTE: Round answers to four decimal places.]

13. $(1, 1)$

14. $(-1, 2)$

15. $(2, \pi/4)$

16. $(3, \arctan 2)$

In Exercises 17–20, rewrite the equation in r and θ as an equivalent equation in x and y; then plot the graph.

17. $r = 2 \sec \theta$

18. $r = 4$

19. $\theta = \pi/3$

20. $r = 2 \sin \theta$

In Exercises 21–24, rewrite the equation in x and y as an equivalent equation in r and θ; then plot the graph.

21. $x^2 + y^2 = 9$

22. $y = 4$

23. $y = 2x$

24. $(x - 1)^2 + y^2 = 1$

25. We claimed in this section that, for any numbers r and θ, the pairs (r, θ), $(r, \theta + 2\pi)$, and $(-r, \theta + \pi)$ all describe the same point in the plane.

    (a) Show that the claim is true if $r = 1$ and $\theta = 0$.

    (b) Show that the claim is true if $r = -1$ and $\theta = \pi/4$.

    (c) The point with rectangular coordinates $(1, 0)$ can be written in polar coordinates as $(1, 2k\pi)$, where k is any integer, or as $(-1, (2k - 1)\pi)$, where k is any integer. In

the same sense, describe all the possible polar coordinates of the point with rectangular coordinates $(1, 1)$.

26. Consider the limaçon $r = 2 + \sin \theta$, $0 \le \theta \le 2\pi$.

    (a) Make a table of values like that in Example 1; let θ range from 0 to 2π in steps of $\pi/6$. (Round r-values to three decimals.)

    (b) Plot the points calculated in part (a) on a copy of the polar grid shown in Figure 3. Join the points with a smooth curve.

    (c) What is the axis of symmetry of this limaçon?

    (d) The r-values in the table in Example 1 are symmetric about $\theta = \pi$. What similar type of symmetry does your table from part (a) show?

27. Consider the cardioid $r = 1 + \cos \theta$, $0 \le \theta \le 2\pi$.

    (a) Make a table of values like that in Example 1; let θ range from 0 to 2π in steps of $\pi/6$. (Round r-values to three decimals.)

    (b) Plot the points calculated in part (a) on a copy of the polar grid shown in Figure 3. Join the points with a smooth curve.

    (c) What is the axis of symmetry of this cardioid?

    (d) How does the table of values in part (a) reflect the cardioid's symmetry?

28. Consider the limaçon $r = 1 - 2 \cos \theta$, $0 \le \theta \le 2\pi$.

    (a) Make a table of values like that in Example 1; let θ range from 0 to 2π in steps of $\pi/6$. (Round r-values to three decimals.)

    (b) Plot the points calculated in part (a) on a copy of the polar grid shown in Figure 3. Join the points with a smooth curve.

    (c) For what values of θ is $r = 0$? How do these values appear on the graph?

    (d) On what θ-interval is $r < 0$? How does this interval show up on the graph?

29. This problem explores the ideas of Example 5. Find Cartesian coordinate functions for each of the following parametrically described curves in the xy-plane. Then plot each curve to see its shape.

    (a) $r = t$, $\theta = \pi/4$, for $0 \le t \le 2$

    (b) $r = 1$, $\theta = t$, for $0 \le t \le 2\pi$

    (c) $r = t$, $\theta = t$, for $0 \le t \le 2\pi$

30. Consider the curve C given parametrically in polar coordinates by $r = t$ and $\theta = 3 \cos t$ for $0 \le t \le 10$. (See Figure 8.)

    (a) At what four t-values (for $0 \le t \le 10$) does C cross the x-axis?

    (b) At which points on the curve is θ greatest? Least? Which values of t correspond to these points?

31. Consider the curve C given parametrically by $r = t$ and $\theta = \sin t$ for $0 \le t \le 10\pi$.

(a) Plot C using technology. (Use ideas of Example 5.)

(b) For what values of t in the given interval does C cross the x-axis?

(c) At which points on C is θ greatest? Least? Which values of t correspond to these points?

(d) Explain why C never crosses the positive or negative y-axis.

In Exercises 32–34, write the polar curve in parametric form and then plot the result as a parametric curve.

32. The cardioid $r = 1 + \cos\theta$ for $0 \le \theta \le 2\pi$.

33. The circle $r = 2$ for $0 \le \theta \le 2\pi$.

34. The polar curve $r = \sin\theta$ for $0 \le \theta \le \pi$.

FURTHER EXERCISES

In Exercises 35–40, plot the given polar equation. (The cardioids and limaçons shown in Figure 5 should be helpful. Be sure to label your graphs with appropriate units.)

35. $r = 3 + 3\cos\theta$

36. $r = 3 - \cos\theta$

37. $r = 1 + \sqrt{3}\cos\theta$

38. $r = 4 + 4\sin\theta$

39. $r = 4 - 2\sin\theta$

40. $r = 2 - 4\sin\theta$

In Exercises 41–46, sketch the given polar rose. (The roses shown in Figure 6 should be helpful. Be sure to label your graphs with appropriate units.)

41. $r = 2\sin\theta$

42. $r = 2\sin(2\theta)$

43. $r = 2\sin(3\theta)$

44. $r = 2\cos(4\theta)$

45. $r = 2\cos(5\theta)$

46. $r = 2\cos(1001\theta)$ (A rough sketch is fine!)

In Exercises 47–54, plot the graph of the equation. (In each case, the graph is some sort of spiral.)

47. $r = \theta$ for $0 \le \theta \le 4\pi$ (an Archimedean spiral)

48. $r = 2\theta$ for $-4\pi \le \theta \le 0$ (another Archimedean spiral)

49. $r = \ln(\theta)$ for $1 \le \theta \le 4\pi$ (a logarithmic spiral)

50. $r = e^{\theta/2}$ for $-2\pi \le \theta \le 2\pi$ (an exponential spiral)

51. $r = \theta^2$ for $0 \le \theta \le 4\pi$ (a quadratic spiral)

52. $r = 1/\theta$ for $1/2 \le \theta \le 16$ (a hyperbolic spiral)

53. $r = \sqrt{\theta}$ for $1/4 \le \theta \le 9$ (a parabolic spiral)

54. $r = 1/\sqrt{\theta}$ for $1/4 \le \theta \le 16$ (a Lituus spiral)

55. The polar equation $r = a\cos\theta + b\sin\theta$, where a and b are real numbers, describes a circle. Find the radius and the center of this circle.

56. Suppose that $g(\theta) = f(\theta + \alpha)$, where $\alpha > 0$. Describe how the graph of the polar equation $r = g(\theta)$ is related to that of $r = f(\theta)$.

57. (a) Plot the curve $r = \cos^2\theta$, $0 \le \theta \le 2\pi$.

(b) Find an equation for this curve in Cartesian coordinates.

58. (a) Plot the curve $r = 1/(1 + \cos\theta)$, $0 \le \theta < \pi$.

(b) Find an equation for this curve in Cartesian coordinates.

59. Find an equation for the distance between the points with polar coordinates (r_1, θ_1) and (r_2, θ_2).

60. This exercise concerns limaçons of the form $r = 1 + a\cos\theta$, where a is any real constant. The following questions are open-ended. Answer them by experimenting with plots for various values of a: positive, negative, large, small, and so forth.

(a) For which positive values of a does the graph have an inner loop?

(b) What happens when $a = 1$?

(c) What happens as a tends to 0?

(d) How are the graphs for a and $-a$ (e.g., $r = 1 + 0.5\cos\theta$ and $r = 1 - 0.5\cos\theta$) related to each other?

(e) What happens as a tends to ∞?

12.4 VECTORS

A **vector** is a quantity that, like an arrow drawn in the xy-plane, has both magnitude and direction.

Vectors can "live" in any dimension—in the real line, in the xy-plane, in xyz-space, or in higher-dimensional spaces. This section is about vectors in the plane and in space: what they are, why they are important, how to describe them mathematically, and how to calculate with them. Vectors are so important in both pure and applied mathematics that a general mathematical area, **linear algebra**, is devoted to studying their theory, behavior, and generalizations. ➡

One possible generalization is to infinite-dimensional spaces.

Vectors in the plane

We will think of two-dimensional vectors either geometrically, as arrows in the xy-plane, or algebraically, as 2-tuples (a, b). (In three dimensions we use arrows in xyz-space and 3-tuples (a, b, c).) For the moment we concentrate on two-dimensional vectors. This is mainly for convenience—pictures are easier to draw in the plane than in space. Almost all of the main ideas, however, translate readily to vectors in three-dimensional space.

Concrete examples rather than abstract theory will drive our study of vectors. The next example, in particular, is worth special attention; it illustrates some of the ways we'll see vectors in action, and it introduces some important vocabulary. Formal definitions will come later.

EXAMPLE 1   Figure 1 shows the path of an ant walking on the xy-plane. At each marked point on the curve, an arrow—or vector—describes the ant's instantaneous velocity as it passes that point. What does the picture mean?

Solution   We will see later how the picture was drawn. Here we discuss what it means.

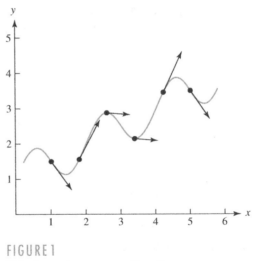

FIGURE 1
Velocity vectors on an ant's path

- **Velocity: a vector**   Velocity is among the simplest and most familiar vector quantities. We will return to it often in this book. At any instant, the velocity of a moving object involves both a **magnitude** (the object's speed at that instant) and a **direction** (the direction of movement at that instant).

  The arrows along the curve describe the ant's velocity in a convenient, compact form. Each arrow's direction is tangent to the curve (since the ant moves along the curve), and each arrow's length tells the ant's speed as the ant passes the point in question. The arrows show (among other things) that the ant moves from left to right along the curve and that it moves faster in the middle of the curve than at either end. At the leftmost marked point the arrow has length about 1, and so the ant's speed is about 1 distance unit per second.

- **Speed: a scalar**   Speed and velocity are sometimes equated in everyday language, but in mathematics and physics there is an important difference. Speed is an ordinary number—what a car's speedometer shows at any given instant. When vectors and numbers arise together, numbers are often called **scalars**. The name is appropriate because scalars describe the size, or *scale*, of vectors. Velocity, by contrast, is a vector quantity, incorporating both speed and direction. ◄

In a car, a "velocity-meter" would combine a speedometer and a compass.

- **Telling time** Both axes in the picture correspond to *spatial* directions; neither axis tells time. The velocity vectors, however, *are* related to time because they indicate *speed*. To draw such a picture, therefore, we had to know not only where the ant walked but also when it reached any given point—even though this information doesn't appear directly in the picture. (We will explain soon how we used this information.)

- **Movable arrows** The arrows at $x = 2$ and at $x = 4$ are *identical*: both have the same length and the same direction. Entomologically, this means that at these two points on its journey the ant had exactly the same velocity—even though the points themselves are different.

  The mathematical lesson is important:

  > *A vector is determined only by its length and direction—not by where in the plane the vector happens to be drawn.*

  The arrows at $x = 2$ and $x = 4$, in other words, are really two pictures of the same vector. We'll see that it is often useful to move vectors (without twisting or stretching!) from place to place to suit the purpose at hand.

What is a vector?

Vectors can be defined mathematically in more or less abstract ways on the basis of the setting and the purpose at hand. However, any reasonable definition of "vector" must capture the key idea of a quantity that has both magnitude and direction. Vectors in the xy-plane can be described either as arrows or as 2-tuples; both descriptions will be useful.

- **Vectors as arrows** A vector **v** can be described by an arrow in the xy-plane. Two arrows with the same length and the same direction describe the same vector—regardless of where the arrows are placed in the plane.

- **Vectors as 2-tuples** A vector **v** can be written as a 2-tuple of real numbers, as in $\mathbf{v} = (a, b)$. The numbers a and b are called the **scalar components** (or, alternatively, the **coordinates**) of the vector **v**. →

"Component" is used inconsistently in mathematics, referring sometimes to a scalar and sometimes to a vector. Whenever confusion seems possible we use either "scalar component" or "vector component."

Figure 2 illustrates these two different views of vectors and the connections between them:

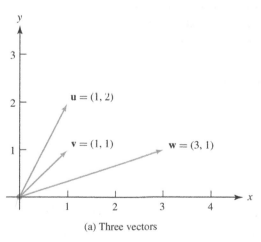

(a) Three vectors

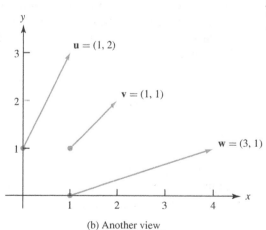

(b) Another view

FIGURE 2
Vectors as arrows and as 2-tuples

Some lessons from the pictures follow. Watch, too, for some important vocabulary and notation.

- **Different tail points, same vectors**   The two pictures show exactly the same vectors **u**, **v**, and **w**. The only difference concerns where in the plane the vectors are placed; in the pictures, the vectors' tail points are shown as dark dots.

- **Vectors based at the origin**   In the left-hand view all three vectors have their tails at the origin, and so each vector is of the form $\overrightarrow{OP}$, where O is the origin and P is the point at the tip. In this special case, the *components* of each vector are the same as the Cartesian *coordinates* of the tip point P. For this reason, a vector of the form $\mathbf{v} = \overrightarrow{OP}$ is often called the **position vector** of the point P.

  There is a technical difference between a point P with coordinates (x, y) and the vector from the origin O to P, with components (x, y). In practice, however, the difference seldom matters much, and so we will occasionally blur the distinction by thinking of (x, y) sometimes as a point and sometimes as a position vector. ◄

  This practice is sloppy but standard.

- **What the components say**   In the right-hand view, each vector's components are *not* the Cartesian coordinates of its tip point. Instead, they describe the **displacement** between the vector's head and its tail. The vector $\mathbf{w} = (3, 1)$, for instance, involves starting at the tail and then moving 3 units to the right and 1 unit up to locate the head.

- **The vector from P to Q**   If $P = (a, b)$ and $Q = (c, d)$ are any points in the plane, then the vector from P to Q, denoted by $\overrightarrow{PQ}$, has components $(c - a, d - b)$. In the right-hand view, for instance, the vector **w** starts at $(1, 0)$ and ends at $(4, 1)$, and thus its coordinates are indeed $(3, 1)$.

- **Finding lengths**   The **length** (or **magnitude**) of a vector is the distance between its head and its tail. The components make this easy to calculate using the distance formula in the xy-plane. If $\mathbf{v} = (a, b)$, then

$$\text{length of } \mathbf{v} = |\mathbf{v}| = \sqrt{a^2 + b^2}.$$

(The length of a vector is sometimes called its **norm**.) Notice that we use the same notation ($|\ |$) to denote both the length of a vector and the absolute value of a number. It should be clear from the context which is which. Using the same symbol makes sense because we are measuring size in both cases.

Bold vectors, faint scalars   Vectors and scalars are often seen together. To keep clear which is which, it is common to denote vectors by using either bold type, as in **v**, or an arrow on top, as in $\overrightarrow{PQ}$. (The latter is convenient when describing a vector with tail at P and head at Q or when writing by hand.) Scalars appear in ordinary type. For example, in the expression

$$a\,\mathbf{u} + b\,\mathbf{v} + c\,\mathbf{w},$$

a, b, and c denote scalars, while **u**, **v**, and **w** denote vectors.

Algebra with vectors

Vectors are useful largely because they allow algebraic operations like those with ordinary numbers. We begin with some easy samples: ◄

Soon we'll interpret these operations geometrically; the point here is how easy they are to perform.

$$(1, 2) + (2, 1) = (3, 3); \qquad -3(1, 2) = (-3, -6); \qquad 3(1, 2) - (3, 4) = (0, 2).$$

The first operation illustrates the **sum of two vectors**; the result is another vector. The second operation illustrates **scalar multiplication**; a scalar is multiplied by a vector to produce a new vector. The third operation involves both a scalar product and the **difference** of two vectors. Here are the formal rules:

> **DEFINITION (Operations on vectors)** Let $\mathbf{v} = (a, b)$ and $\mathbf{w} = (c, d)$ be plane vectors, and let r be a scalar. The **sum** of $\mathbf{v}$ and $\mathbf{w}$ is
>
> $$\mathbf{v} + \mathbf{w} = (a, b) + (c, d) = (a + c, b + d).$$
>
> **Scalar multiplication** of $\mathbf{v}$ by r is defined by
>
> $$r\,\mathbf{v} = r\,(a, b) = (ra, rb).$$

More enlightening than these simple formulas are their geometric meanings. Consider addition first, as illustrated in Figure 3:

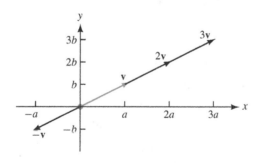

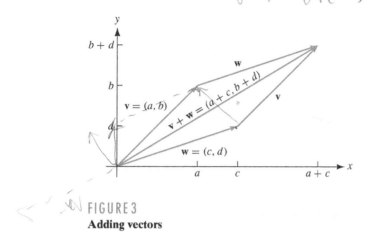

FIGURE 3
Adding vectors

The picture shows what it means geometrically to add two vectors $\mathbf{v} = (a, b)$ and $\mathbf{w} = (c, d)$. The sum $\mathbf{v} + \mathbf{w}$ has components $(a + c, b + d)$. Geometrically, the sum $\mathbf{v} + \mathbf{w}$ is the vector obtained by putting the tail of one vector at the head of the other. (The order doesn't matter.) The picture also illustrates the **parallelogram rule**: The sum $\mathbf{v} + \mathbf{w}$ is the diagonal of the parallelogram with sides $\mathbf{v}$ and $\mathbf{w}$.

Figure 4 shows scalar multiplication:

FIGURE 4
Scalar-multiplying vectors

(All vectors shown in Figure 4 have tails at the origin.) Figure 4 shows what it means geometrically to multiply the vector $\mathbf{v}$ by a scalar a. The result, written in the form $a\mathbf{v}$, is a vector with either the same direction as $\mathbf{v}$ (if $a > 0$) or the opposite direction (if $a < 0$).

Length, addition, and scalar multiplication Figures 3 and 4 show how vector operations affect the lengths of the vectors involved. For scalar multiplication, we see that the length

Why is the absolute value needed?

In a triangle, the length of each side is less than the sum of the other two lengths—hence the name.

of $a\mathbf{v}$ is $|a|$ times the length of $\mathbf{v}$. ← In math-speak, scalar multiplication by a **dilates** $\mathbf{v}$ by the factor $|a|$. In symbols: $|a\,\mathbf{v}| = |a|\,|\mathbf{v}|$.

Figure 3 also illustrates the famous **triangle inequality** ← for vectors. In words: *The length of a sum of vectors is less than or equal to the sum of the lengths.* In symbols:

$$|\mathbf{v} + \mathbf{w}| \le |\mathbf{v}| + |\mathbf{w}|\,.$$

The triangle inequality certainly looks believable, if not downright obvious. But an algebraic proof based purely on the definitions is slightly messy; we will defer giving one until we've developed a few more vector tools.

Figure 5 shows some "hybrids" of these vector operations:

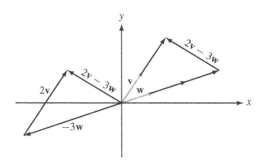

FIGURE 5
Algebra with vectors

Notice, especially, how the **difference** of two vectors appears:

> *If* $\mathbf{v}$ *and* $\mathbf{w}$ *have the same tail point, then the difference* $\mathbf{v} - \mathbf{w}$ *points from the tip of* $\mathbf{w}$ *to the tip of* $\mathbf{v}$.

(Figure 5 shows two copies of the difference vector $2\mathbf{v} - 3\mathbf{w}$.)

Properties of vector operations

The vector operations enjoy many of the pleasant properties we've come to expect of operations with ordinary numbers. We collect many such properties in one big theorem:

THEOREM 1 (Properties of vector operations)   Let $\mathbf{u}$, $\mathbf{v}$, and $\mathbf{w}$ be vectors and let r and s be scalars. The following algebraic properties hold:

(a) $\mathbf{u} + \mathbf{v} = \mathbf{v} + \mathbf{u}$ (commutativity);

(b) $\mathbf{u} + (\mathbf{v} + \mathbf{w}) = (\mathbf{u} + \mathbf{v}) + \mathbf{w}$ (associativity of addition);

(c) $r(s\mathbf{v}) = (rs)\mathbf{v}$ (associativity of scalar multiplication);

(d) $(r + s)\mathbf{v} = r\mathbf{v} + s\mathbf{v}$ (distributivity);

(e) $r(\mathbf{v} + \mathbf{w}) = r\mathbf{v} + r\mathbf{w}$ (distributivity again).

All parts of the theorem can be shown to hold by using similar properties of ordinary addition and multiplication.

Note carefully the different uses of similar symbols.

The zero vector   Among numbers, 0 has many special properties. The vector $\mathbf{O} = (0, 0)$, called the **zero vector**, is special in similar ways. ← It has length zero and, therefore, no meaningful direction. Algebraically, it behaves "as a zero should." If $\mathbf{v}$ is any vector and r

is any scalar, we have

$$\mathbf{O} + \mathbf{v} = \mathbf{v}, \qquad r\mathbf{O} = \mathbf{O}, \qquad \text{and} \qquad 0\mathbf{v} = \mathbf{O}.$$

Unit vectors   A vector $\mathbf{u} = (a, b)$ is called a **unit vector** if $\mathbf{u}$ has length 1, that is, if

$$|\mathbf{u}| = \sqrt{a^2 + b^2} = 1.$$

Any nonzero vector $\mathbf{v}$ can be multiplied (or divided) by an appropriate positive scalar to produce a unit vector with the same direction as $\mathbf{v}$. Suppose, for instance, that $\mathbf{v} = (3, 4)$. Then

$$|\mathbf{v}| = \sqrt{3^2 + 4^2} = 5;$$

thus, to scale $\mathbf{v}$ down to unit length we can divide by 5. ➡ This produces the unit vector $\mathbf{u} = (3/5, 4/5)$, which is called the **unit vector in the direction of v**. The same process works more generally:

Or, equivalently, multiply by 1/5.

FACT   Let $\mathbf{v} = (a, b)$ be any nonzero vector. Then the new vector

$$\mathbf{u} = \frac{\mathbf{v}}{\sqrt{a^2 + b^2}} = \left(\frac{a}{\sqrt{a^2 + b^2}}, \frac{b}{\sqrt{a^2 + b^2}} \right)$$

is a unit vector in the direction of $\mathbf{v}$.

Standard basis vectors   The two-dimensional vectors

$$\mathbf{i} = (1, 0) \qquad \text{and} \qquad \mathbf{j} = (0, 1)$$

are simple but useful. Both $\mathbf{i}$ and $\mathbf{j}$ have length 1, and they point in perpendicular directions—along the x- and y-axes, respectively. The vectors $\mathbf{i}$ and $\mathbf{j}$ are known as the **standard basis vectors** for $\mathbb{R}^2$. They deserve this important-sounding name for a good reason:

Every vector $\mathbf{v} = (a, b)$ in $\mathbb{R}^2$ can be written as a sum of scalar multiples of $\mathbf{i}$ and $\mathbf{j}$.

If, say, $\mathbf{v} = (2, 3)$, then

$$\mathbf{v} = (2, 3) = (2, 0) + (0, 3) = 2(1, 0) + 3(0, 1) = 2\mathbf{i} + 3\mathbf{j}.$$

A sum of the form $r\mathbf{v} + s\mathbf{w}$, where r and s are any scalar constants, is called a **linear combination** ➡ of $\mathbf{v}$ and $\mathbf{w}$. The calculation with $(2, 3)$ illustrates that every vector (a, b) in $\mathbb{R}^2$ can be written, if we wish, as a linear combination of $\mathbf{i}$ and $\mathbf{j}$. That is,

$$(a, b) = a\mathbf{i} + b\mathbf{j}.$$

Students who have taken linear algebra should recognize the words "basis" and "linear combination."

(Some authors use the **ij**-notation exclusively.)

What is a vector, exactly?   Defining the word "vector" in rigorous mathematical language is harder than one might think. A formal definition is usually first met seriously in a linear algebra course—and even there, the definition takes some getting used to. The definition describes vectors as *any* objects that can be added together and multiplied by real numbers, and these operations must satisfy some minimal restrictions. (Vector addition must be commutative, for instance.)

The good news is that vectors are easier to understand informally (as arrows, for instance, or as 2-tuples) than to define formally. An informal view of vectors is adequate for our use in multivariable calculus.

Vectors in space

Or even more than three

So far we have considered only vectors in the xy-plane—those that "fit" naturally into two dimensions. Sometimes, however, a third dimension ← is essential. Physical motion, for instance, may be inescapably three-dimensional, as when a fly buzzes aimlessly around the room or a maple seed spirals from a tree to the ground.

Vectors in space vs. vectors in the plane   In most respects vectors behave almost identically in two and three dimensions. In both settings, for instance, vectors can be thought of geometrically as arrows with both magnitude and direction. In the following paragraphs we collect some similarities and differences between two- and three-dimensional vectors.

- **Two-tuples and three-tuples**   A plane vector $\mathbf{v}$ corresponds to a 2-tuple (a, b). Similarly, a space vector $\mathbf{v}$ corresponds to a 3-tuple (a, b, c), and we sometimes simply write $\mathbf{v} = (a, b, c)$. The vector $\mathbf{v}$ can be thought of as the arrow (or **position vector**) from the origin $(0, 0, 0)$ to the point (a, b, c). The numbers a, b, and c are called the **coordinates** of $\mathbf{v}$.

- **Algebra with 3-vectors**   Three-dimensional vectors are added and scalar multiplied "term by term," exactly like two-dimensional vectors. If $\mathbf{v} = (v_1, v_2, v_3)$ and $\mathbf{w} = (w_1, w_2, w_3)$ are vectors, and a is any scalar, then

$$\mathbf{v} + \mathbf{w} = (v_1, v_2, v_3) + (w_1, w_2, w_3) = (v_1 + w_1, v_2 + w_2, v_3 + w_3);$$
$$a\mathbf{v} = (av_1, av_2, av_3).$$

- **Geometric meanings**   Vector addition and scalar multiplication have the same geometric meaning for vectors in space as for vectors in the plane. If $\mathbf{v}$ and $\mathbf{w}$ are space vectors and k is a scalar, then $\mathbf{v} + \mathbf{w}$ is the vector formed by laying $\mathbf{v}$ and $\mathbf{w}$ head to tail; the scalar multiple $k\mathbf{v}$ is formed by stretching $\mathbf{v}$ by the factor $|k|$. (If $k < 0$, we reverse the direction of $\mathbf{v}$.)

- **Length and the triangle inequality**   The **length** of a 3-vector is defined using the distance formula in xyz-space: If $\mathbf{v} = (v_1, v_2, v_3)$, then

$$|\mathbf{v}| = \text{length} = \sqrt{v_1^2 + v_2^2 + v_3^2}.$$

  The same algebraic properties of length hold in three dimensions as in two:

$$|a\mathbf{v}| = |a|\,|\mathbf{v}|; \qquad |\mathbf{v} + \mathbf{w}| \le |\mathbf{v}| + |\mathbf{w}|.$$

  The right-hand expression above is the **triangle inequality** again—this time for 3-vectors.

- **Standard basis vectors**   In three-dimensional space, there are *three* standard basis vectors:

$$\mathbf{i} = (1, 0, 0), \qquad \mathbf{j} = (0, 1, 0), \qquad \text{and} \qquad \mathbf{k} = (0, 0, 1).$$

  (Note that the symbols $\mathbf{i}$ and $\mathbf{j}$ denote slightly different objects in the 2-D and 3-D cases.) Each standard basis vector has length 1 and points in the positive direction along one of the coordinate axes. In space, as in the plane, every vector can be written as a linear combination of $\mathbf{i}$, $\mathbf{j}$, and $\mathbf{k}$:

$$(a, b, c) = a\,(1, 0, 0) + b\,(0, 1, 0) + c\,(0, 0, 1) = a\,\mathbf{i} + b\,\mathbf{j} + c\,\mathbf{k}.$$

BASIC EXERCISES

Let $\mathbf{u} = (1, 2)$, $\mathbf{v} = (2, 3)$, *and* $\mathbf{w} = (-2, 1)$. *In Exercises 1–7, find the vector:*

1. $\mathbf{u} + \mathbf{v}$

2. $\mathbf{u} - \mathbf{v}$

3. $2\mathbf{u} - 3\mathbf{v}$

4. $\mathbf{u} + \mathbf{v}/2 + \mathbf{w}/3$

5. the unit vector in the direction of $\mathbf{u}$

6. the unit vector in the direction of $\mathbf{v}$

7. the unit vector in the direction of $\mathbf{w}$

8. For each vector $\mathbf{v}$ below, find the unit vector $\mathbf{u}$ in the direction of $\mathbf{v}$. Then sketch all eight vectors on the same axes.

   (a) $\mathbf{v} = (1, 1)$ **(c)** $\mathbf{v} = (1, 2)$

   (b) $\mathbf{v} = (-1, 1)$ **(d)** $\mathbf{v} = (-1, -2)$

9. Let $\mathbf{v} = (3, 4)$. Find the vector of length 7 that points in the opposite direction as $\mathbf{v}$.

10. Let $\mathbf{v} = (1, 2, 3)$.

   (a) Find the length $|\mathbf{v}|$.

   (b) Find a unit vector $\mathbf{u}$ in the direction of $\mathbf{v}$. By hand, draw $\mathbf{u}$, $\mathbf{v}$, and $2\mathbf{u}$ on the same axes.

11. Repeat Exercise 10, but use $\mathbf{v} = (1, -2, 3)$.

In Exercises 12–15, $\mathbf{u}$ and $\mathbf{v}$ are the vectors in the xz-plane pictured below.

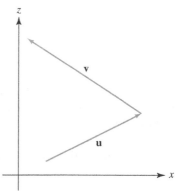

12. Draw the vector $\mathbf{w} = \mathbf{u} + \mathbf{v}$.

13. Draw the vector $\mathbf{a} = \mathbf{u} - \mathbf{v}$.

14. Draw the vector $\mathbf{b} = \mathbf{v} - \mathbf{u}$.

15. Draw the vector $\mathbf{c} = 2\mathbf{u} + \mathbf{v}$.

In Exercises 16–19, $\mathbf{u}$ and $\mathbf{v}$ are the vectors in the yz-plane pictured below.

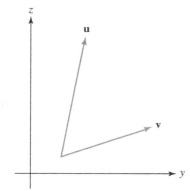

16. Draw the vector $\mathbf{w} = \mathbf{u} + \mathbf{v}$.

17. Draw the vector $\mathbf{a} = \mathbf{u} - \mathbf{v}$.

18. Draw the vector $\mathbf{b} = \mathbf{v} - \mathbf{u}$.

19. Draw the vector $\mathbf{c} = \mathbf{u} + 2\mathbf{v}$.

FURTHER EXERCISES

Consider the three vectors $\mathbf{u} = (a, b)$, $\mathbf{v} = (c, d)$, and $\mathbf{w} = (e, f)$, and let r be a scalar. By the definition of vector addition,

$$\mathbf{u} + \mathbf{v} = (a, b) + (c, d) = (a + c, b + d) = (c + a, d + b)$$
$$= (c, d) + (a, b) = \mathbf{v} + \mathbf{u}.$$

(The middle step holds because ordinary addition of numbers is commutative.) This argument shows that vector addition is commutative, as asserted in Theorem 1. In Exercises 20 and 21, use a similar argument to verify the identity.

20. $\mathbf{u} + (\mathbf{v} + \mathbf{w}) = (\mathbf{u} + \mathbf{v}) + \mathbf{w}$

21. $r(\mathbf{v} + \mathbf{w}) = r\mathbf{v} + r\mathbf{w}$

22. In this section we mentioned the parallelogram rule for vector addition: The sum $\mathbf{v} + \mathbf{w}$ is the diagonal of the parallelogram with adjacent sides $\mathbf{v}$ and $\mathbf{w}$.

   (a) Draw and label a picture to illustrate this fact for the vectors $\mathbf{v} = (2, 1)$ and $\mathbf{w} = (1, 3)$. (Base both vectors at the origin.) What are the components of the diagonal vector?

(b) The parallelogram you drew actually has two diagonals. You drew a northeast-pointing diagonal in the previous part. Now draw the northwest-pointing diagonal vector. What are its components? How is it related to **v** and **w**?

(c) Draw the southeast-pointing diagonal vector. What are its components? How is it related to **v** and **w**?

23. Let **v** and **w** be vectors. What basic fact from Euclidean geometry is expressed by the relation $|\mathbf{v} + \mathbf{w}| \leq |\mathbf{v}| + |\mathbf{w}|$?

12.5 VECTOR-VALUED FUNCTIONS, DERIVATIVES, AND INTEGRALS

The standard functions of single-variable calculus are *real-valued* functions of *one* real variable. The sine function, for instance, takes one real number, x, as input, and produces another real number, $\sin x$, as output.

With vectors available new types of functions become possible. A function might accept, say, real numbers as inputs and produce 2-vectors as outputs. Another function might accept 3-vectors as inputs and produce real numbers as outputs. The first function is called **vector-valued**; the second, **scalar-valued**. This variety of function types accounts for much of the remarkable power of multivariable calculus to model real-world phenomena.

Multivariable calculus handles functions that use all possible combinations of vectors and scalars as inputs and outputs. The good news is that most of the basic ideas of single-variable calculus—graph, derivative, linear approximation, integral, and so forth—extend in rather natural ways to functions with vector inputs or outputs (or both). But due care is needed, both to keep track of what type of object is under consideration at a given time and to find the "right" extension of a given single-variable idea to the multivariable setting.

Vector-valued functions

A **vector-valued function** is one that produces vectors as outputs. Consider, for example, the function defined by the rule

$$\mathbf{f}(t) = (\cos t, \sin t).$$

The notation $\mathbf{f} \colon \mathbb{R} \to \mathbb{R}^2$ makes sense for this function because $\mathbf{f}$ accepts a single number t as input and produces the 2-vector $(\cos t, \sin t)$ as output. Notice too that, because the function is vector-valued we use the boldface symbol $\mathbf{f}$ rather than an ordinary f to denote it. The two functions inside the parentheses are called **component functions**, or **coordinate functions**. (A function $\mathbf{g} \colon \mathbb{R} \to \mathbb{R}^3$ has 3-vectors as values, and so there are three component functions.)

Another notation The function $\mathbf{f}$ just considered can also be written in the form

$$\mathbf{f}(t) = \cos t\,\mathbf{i} + \sin t\,\mathbf{j},$$

We described the standard bases for $\mathbb{R}^2$ and $\mathbb{R}^3$ in Section 12.4.

where $\mathbf{i} = (1, 0)$ and $\mathbf{j} = (0, 1)$ are the **standard basis vectors** in $\mathbb{R}^2$. ← In a similar spirit, we can write either

$$\mathbf{g}(t) = (t, t^2, t^3) \qquad \text{or} \qquad \mathbf{g}(t) = t\mathbf{i} + t^2\mathbf{j} + t^3\mathbf{k}$$

to describe the same vector-valued function $\mathbf{g}$. Here $\mathbf{i} = (1, 0, 0)$, $\mathbf{j} = (0, 1, 0)$, and $\mathbf{k} = (0, 0, 1)$, and outputs of $\mathbf{g}$ are three-dimensional vectors. Both notations appear in textbooks; we will use each from time to time.

Parametric curves and vector-valued functions   In Section 1.2 we studied plane curves parametrized by two coordinate functions $x(t)$ and $y(t)$. (A space curve has three coordinate functions.) As the next example shows, such curves are close cousins to vector-valued functions.

EXAMPLE 1 Consider the vector-valued function $\mathbf{f}(t) = (\cos t, \sin t)$. How is $\mathbf{f}$ related to a parametric curve? Which curve?

Solution We have seen that setting

$$x = \cos t; \qquad y = \sin t; \qquad 0 \le t \le 2\pi$$

gives a parametrization for the unit circle $x^2 + y^2 = 1$, traversed once, counterclockwise, starting from the east pole. Thus, for any input t, the vector $\mathbf{f}(t) = (\cos t, \sin t)$ can be thought of as the position vector for a point on the unit circle. ➡ As t ranges through the domain $(-\infty, \infty)$, the tip of the vector $\mathbf{f}(t)$ traces the unit circle infinitely often, counterclockwise, once in each t-interval of length 2π.

The tail is pinned at the origin.

If, say, we want to trace only the right half of the unit circle, we restrict the parameter interval:

$$x = \cos t; \qquad y = \sin t; \qquad -\pi/2 \le t \le \pi/2.$$

In the language of vector-valued functions, this amounts to restricting the domain; we can write this economically as $\mathbf{f} : [-\pi/2, \pi/2] \to \mathbb{R}^2$. ◼

One idea, two views As the preceding example shows, the difference between a pair of parametric equations and a vector-valued function is slight; we will use both points of view and their corresponding notations more or less interchangeably.

When using vector-valued functions, it is sometimes useful to think of a curve as traced out by a collection of position vectors $\mathbf{f}(t)$, one for each input t; all tails are pinned at the origin. Figure 1 shows several of these position vectors for the function $\mathbf{f}(t) = (t, 3 + \sin t)$:

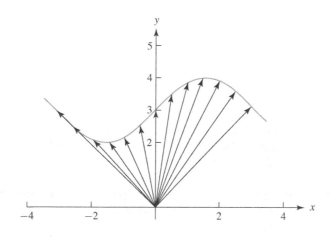

FIGURE 1
Position vectors tracing out a curve

Lines, line segments, and vector-valued functions

Vectors and vector-valued functions make quick, efficient work of describing lines and line segments.

Lines in the plane Any line ℓ in the xy-plane can be determined by two data: (i) a point $P_0 = (x_0, y_0)$ through which ℓ passes, and (ii) a **direction vector** $\mathbf{v} = (a, b)$ that points in the direction of ℓ. ➡ Figure 2 shows the geometric situation:

Recall that $-\mathbf{v}$ points in the opposite direction to $\mathbf{v}$.

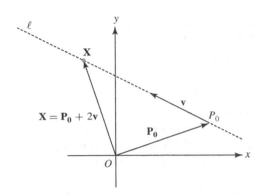

FIGURE 2

A point and a vector determine a line

As the picture illustrates, a point $X = (x, y)$ lies on ℓ if and only if the vector $\mathbf{X}$ is of the form $\mathbf{P_0} + t\mathbf{v}$, for some (positive or negative) scalar t. (For the point X shown, $t = 2$.) Therefore, points on the line are those that satisfy the vector equation

$$(x, y) = (x_0, y_0) + t(a, b)$$

for some real number t. This means that we can think of the line as the image of the vector-valued function $\mathbf{X} : \mathbb{R} \to \mathbb{R}^2$ defined for all real numbers t by $\mathbf{X}(t) = (x_0, y_0) + t(a, b)$. In the language of parametric equations, ℓ can be parametrized by

$$x(t) = x_0 + at; \qquad y(t) = y_0 + bt; \qquad -\infty < t < \infty.$$

As t ranges through all real numbers, $\mathbf{X}(t)$ traces out the entire line, which is infinite in both directions.

Line segments Suppose we want only part of a line, say the segment from $P = (x_0, y_0)$ to $Q = (x_1, y_1)$. Then we can use P as our fixed point and $\overrightarrow{PQ} = (x_1 - x_0, y_1 - y_0)$ as our direction vector. Then the line through P and Q has the form

$$\mathbf{X} = (x, y) = (x_0, y_0) + t(x_1 - x_0, y_1 - y_0).$$

(We could, instead, have used Q as our fixed point and $\overrightarrow{QP} = (x_0 - x_1, y_0 - y_1)$ as our direction vector.) Restricting t to the interval $0 \le t \le 1$ gives only the segment in question—$t = 0$ corresponds to P and $t = 1$ to Q.

Lines and segments in space A line ℓ in xyz-space is determined by a point $P_0 = (x_0, y_0, z_0)$ and a direction vector $\mathbf{v} = (a, b, c)$. A general point (x, y, z) lies on ℓ if and only if the vector $\mathbf{X} = (x, y, z)$ is of the form

$$\mathbf{X} = \mathbf{P_0} + t\mathbf{v},$$

where t is a real number and $\mathbf{P_0}$ is the position vector from the origin to (x_0, y_0, z_0). If we want only a line segment, we can restrict t to a subset of the real numbers. If we use, say, $0 \le t \le 1$, then the vector equation $\mathbf{X} = \mathbf{P_0} + t\mathbf{v}$ describes the segment that starts at P_0 and runs along ℓ for one length of the vector $\mathbf{v}$.

We can think of the same line as the image of the vector-valued function $\mathbf{X} : \mathbb{R} \to \mathbb{R}^3$ defined by

$$\mathbf{X}(t) = \mathbf{P_0} + t\mathbf{v} = (x_0, y_0, z_0) + t(a, b, c)$$

or, equivalently, as determined by three parametric equations:

$$x(t) = x_0 + at; \qquad y(t) = y_0 + bt; \qquad z(t) = z_0 + ct.$$

EXAMPLE 2 Give vector and parametric equations for the line through the point $(0.28, -0.96, 1.67)$ with direction vector $(0.96, 0.28, 0.33)$. ➥

We'll see in a moment why we chose these strange data.

Solution The line is determined either by the vector-valued function

$$\mathbf{X}(t) = (0.28, -0.96, 1.67) + t(.96, 0.28, 0.33)$$

or, equivalently, by the parametric equations

$$x = 0.28 + 0.96t; \qquad y = -0.96 + 0.28t; \qquad z = 1.67 + 0.33t,$$

whichever we please. For good measure, here is yet another equivalent description:

$$\mathbf{X}(t) = (0.28 + 0.96t)\mathbf{i} + (-0.96 + 0.28t)\mathbf{j} + (1.67 + 0.33t)\mathbf{k}.$$

Notice that time $t = 0$ corresponds to the point $(0.28, -0.96, 1.67)$—no matter the form. ∎

Derivatives of vector-valued functions

In single-variable calculus, derivatives describe rates of change. A vector-valued function $\mathbf{f}$ is, in one sense, simply a list of separate, scalar-valued functions. It is reasonable, therefore, to describe the rate of change of $\mathbf{f}$ by differentiating the component functions separately.

> **DEFINITION (Derivative of a vector-valued function)** Let the vector-valued function $\mathbf{f}: \mathbb{R} \to \mathbb{R}^2$ be defined by $\mathbf{f}(t) = (f_1(t), f_2(t))$. The derivative of $\mathbf{f}$ is the vector-valued function $\mathbf{f}': \mathbb{R} \to \mathbb{R}^2$ defined by
>
> $$\mathbf{f}'(t) = \frac{d}{dt}(f_1(t), f_2(t)) = (f_1'(t), f_2'(t)).$$
>
> If $\mathbf{g}: \mathbb{R} \to \mathbb{R}^3$ is defined by $\mathbf{g}(t) = (g_1(t), g_2(t), g_3(t))$, then $\mathbf{g}'(t) = (g_1'(t), g_2'(t), g_3'(t))$.

Calculating this new derivative amounts simply to finding several ordinary derivatives. For example,

$$\mathbf{f}(t) = (\cos t, \sin t) \implies \mathbf{f}'(t) = (-\sin t, \cos t);$$
$$\mathbf{g}(t) = (t, t^2, t^3) \implies \mathbf{g}'(t) = (1, 2t, 3t^2).$$

Interpreting the derivative The definition isn't surprising, but what does the derivative mean geometrically and in terms of rates? We will discuss these questions in detail below, but here is the short answer:

> **FACT** Let I be an interval and $\mathbf{f}: I \to \mathbb{R}^2$ a vector-valued function. Let $\mathbf{f}$ describe a curve C, and suppose that $\mathbf{f}'(t) \neq (0, 0)$ for all t in (a, b). For every t_0 in (a, b): (i) the vector $\mathbf{f}'(t_0)$ is tangent to C at the point $\mathbf{f}(t_0)$ and points in the direction of increasing t; (ii) the scalar $|\mathbf{f}'(t_0)|$ tells the instantaneous speed (in units of distance per unit of time) of $\mathbf{f}(t)$ at time t_0.

(Similar properties hold for a function $\mathbf{f}: I \to \mathbb{R}^3$.) The Fact says, in short, that if $\mathbf{f}(t)$ describes the position of a moving particle (a vector quantity) at time t, then the derivative $\mathbf{f}'(t)$ describes the particle's **velocity** (another vector quantity) at the same time. That the derivative links position and velocity is not a new idea: We saw the same phenomenon in single-variable calculus. That the same connection applies in higher dimensions is a crucial fact; it accounts for much of the modeling power of multivariable calculus.

We used the Fact to calculate a crawling ant's velocity vectors in Example 1, page 640. The following example illustrates the process for a less earthbound insect.

EXAMPLE 3   A fly buzzes through xyz-space. The fly's position $\mathbf{p}(t)$ at time t seconds is given by

$$\mathbf{p}(t) = \left(\cos t, \sin t, \frac{t}{3} \right).$$

The dotted points show the fly's position at $t = 1, 2, \ldots, 10$:

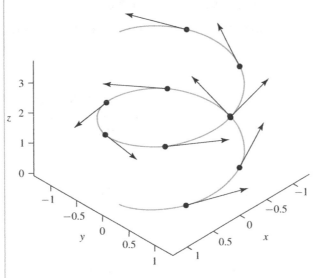

The black dots show the fly's position at successive seconds. The fly spirals upward.

FIGURE 3
Velocity vectors on a space curve

How was the tangent vector at each dotted point computed? What exactly was the fly doing at $t = 5$ seconds?

Solution   The tangent vectors were found by differentiation. From the formula for $\mathbf{p}(t)$ we see

$$\mathbf{p}'(t) = \left(-\sin t, \cos t, \frac{1}{3} \right).$$

All ten velocity vectors appear about the same length. Is this fact, coincidence, or a plotting quirk?

By the preceding Fact, at each time t the velocity vector $\mathbf{p}'(t)$ is tangent to the curve at the point $\mathbf{p}(t)$. ← At $t = 5$ seconds (which corresponds to the base of point of the darker vector), we have

$$\mathbf{p}(5) = (\cos 5, \sin 5, 5/3) \approx (0.28, -0.96, 1.67);$$
$$\mathbf{p}'(5) = (-\sin 5, \cos 5, 1/3) \approx (0.96, 0.28, 0.33).$$

The second vector, which tells velocity, has magnitude $|(0.96, 0.28, 0.33)| \approx 1.05$. Thus, at time 5 seconds the fly was at position $(0.28, -0.96, 1.67)$, moving in the direction of the vector $(0.96, 0.28, 0.33)$ at about 1.05 units per second.   ▪

EXAMPLE 4   We have seen that the vector-valued function

$$\mathbf{L}(t) = (x_0, y_0) + t\,(a, b) = (x_0 + at, y_0 + bt)$$

describes the line ℓ through (x_0, y_0) with direction vector (a, b). What does the derivative $\mathbf{L}'(t)$ tell us?

Solution An easy differentiation gives $\mathbf{L}'(t) = (a, b)$. Thus, in this case the derivative function has a *constant* vector value—which is not surprising given that every line has constant vector direction. We see, too, that the line's *slope* is b/a, the ratio of rise to run in the direction vector.

Our derivative calculation bears out both parts of the preceding Fact:

(i) The derivative (a, b) is indeed tangent to ℓ at every point, and (a, b) points in the direction of increasing t.

(ii) In each unit of time the position $\mathbf{L}(t)$ increases by (a, b); in particular, $\mathbf{L}(t)$ moves (at constant speed) through a distance of $|(a, b)| = \sqrt{a^2 + b^2}$.

Understanding vector derivatives

Our definition makes for easy calculation—we just differentiate the coordinate functions separately. But does the definition make good sense? Why does the vector function $\mathbf{f}'(t) = \left(f_1'(t), f_2'(t) \right)$ really deserve to be called a derivative? How do these new derivatives resemble—and differ from—one-variable derivatives, which involve difference quotients, limits, slopes, and tangent lines?

Vector derivatives as limits The following Fact shows that a vector derivative, just like an ordinary one, is a limit of difference quotients. Indeed, the vector derivative can be defined as such a limit.

> **FACT (The derivative as a limit)** Let $\mathbf{f}$ be a vector-valued function with derivative $\mathbf{f}'$ as defined above, and let t_0 be in the domain of $\mathbf{f}$. Then
> $$\mathbf{f}'(t_0) = \lim_{h \to 0} \frac{\mathbf{f}(t_0 + h) - \mathbf{f}(t_0)}{h}.$$

We "unpack" this Fact in the following paragraphs.

- **Interpreting the difference quotient** The ratio
$$\frac{\mathbf{f}(t_0 + h) - \mathbf{f}(t_0)}{h}$$
resembles the difference quotient of one-variable calculus—except that here the numerator is a vector and the denominator a scalar. (This ensures that the quotient makes sense; it is a vector quantity.) The numerator, being a difference of vectors, can be drawn as an arrow joining two nearby points on the curve, as in Figure 4.

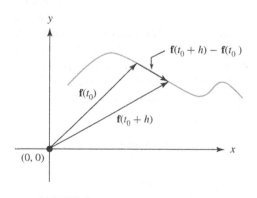

FIGURE 4
The difference of nearby vectors

In particular, for small positive h the vector $\mathbf{f}(t_0 + h) - \mathbf{f}(t_0)$ points approximately in the direction of the curve. Because the quotient vector

The scalar is $1/h$.

$$\frac{\mathbf{f}(t_0 + h) - \mathbf{f}(t_0)}{h}$$

is a scalar multiple ← of the numerator, the quotient vector also points approximately in the direction of the curve. (In other words, the quotient vector is approximately tangent to the curve at $\mathbf{f}(t)$.)

We can apply similar reasoning to the *magnitude* of the difference quotient vector. Figure 4 illustrates that $|\mathbf{f}(t_0 + h) - \mathbf{f}(t_0)|$ (the length of the numerator) is approximately the length of the small curve segment from t_0 to $t_0 + h$, that is, the distance traveled by a point moving along the curve from t_0 to $t_0 + h$. Therefore, dividing by h (the "time" elapsed) is approximately the average speed of $\mathbf{f}(t)$ from time t_0 to time $t_0 + h$.

- **Interpreting the limit: instantaneous velocity**   The preceding discussion explains why, as h tends to zero, the difference quotient vector tends to a limit vector that has two key properties:

  - **Direction**   If $\mathbf{f}'(t_0) \neq (0, 0)$, then the vector $\mathbf{f}'(t_0)$ is tangent to the curve C at $\mathbf{f}(t_0)$

As opposed to the opposite direction.

  and points in the direction of increasing t. ←
  - **Speed**   The magnitude $|\mathbf{f}'(t_0)|$ tells the instantaneous speed (at the instant $t = t_0$, in units of distance per unit of time t) with which the point $\mathbf{f}(t)$ moves along C.

  These two properties justify calling the derivative $\mathbf{f}'(t_0)$ the **velocity vector** of $\mathbf{f}(t)$ at $t = t_0$.

- **The symbolic view**   Writing $\mathbf{f}(t) = \big(f_1(t),\, f_2(t) \big)$ shows symbolically why the Fact holds:

$$\lim_{h \to 0} \frac{\mathbf{f}(t+h) - \mathbf{f}(t)}{h} = \lim_{h \to 0} \frac{\big(f_1(t+h),\, f_2(t+h) \big) - \big(f_1(t),\, f_2(t) \big)}{h}$$

$$= \lim_{h \to 0} \frac{\big(f_1(t+h) - f_1(t),\, f_2(t+h) - f_2(t) \big)}{h}$$

$$= \lim_{h \to 0} \left(\frac{f_1(t+h) - f_1(t)}{h},\, \frac{f_2(t+h) - f_2(t)}{h} \right)$$

$$= \left(\lim_{h \to 0} \frac{f_1(t+h) - f_1(t)}{h},\, \lim_{h \to 0} \frac{f_2(t+h) - f_2(t)}{h} \right)$$

$$= \big(f_1'(t),\, f_2'(t) \big).$$

Derivatives, slopes, and tangent lines   Elementary calculus teaches that, for a scalar-valued function $y = f(x)$, the derivative $f'(x_0)$ gives the slope of the line that is tangent to the graph of f at $x = x_0$. A similar relation holds between derivatives and tangent lines for vector-valued functions $\mathbf{f}(t)$: The derivative vector $\mathbf{f}'(t_0)$ is tangent at $t = t_0$ (i.e., at the point $\mathbf{f}(t_0)$) to the curve with position vector $\mathbf{f}(t)$. We can use this information to find tangent lines and, in the two-dimensional case, slopes of these lines. (In multivariable calculus one seldom bothers with slopes because (i) the idea of slope makes sense only in two dimensions, and (ii) tangent vectors are often easier to calculate and give more

But it's fine to remember slope now and then for old time's sake.

information than slopes.) ←

EXAMPLE 5   Use vector derivatives to find a vector equation for the line tangent to the unit circle at the point $(1/\sqrt{2}, 1/\sqrt{2})$. What is the slope of the circle at this point?

Solution   We can use the position function $\mathbf{f}(t) = (\cos t, \sin t)$ to describe the unit circle; the point in question corresponds to $t = \pi/4$. Now $\mathbf{f}'(t) = (-\sin t, \cos t)$, and so

$\mathbf{f}'(\pi/4) = (-\sqrt{2}/2, \sqrt{2}/2)$. Thus, the tangent line passes through $(1/\sqrt{2}, 1/\sqrt{2})$ with direction vector $\mathbf{v} = (-1/\sqrt{2}, 1/\sqrt{2})$. Knowing a point and a direction vector, we can describe the line in vector notation:

$$\mathbf{X}(t) = \left(1/\sqrt{2}, 1/\sqrt{2}\right) + t\left(-1/\sqrt{2}, 1/\sqrt{2}\right).$$

A simple diagram shows that the line and the circle are indeed tangent at the target point. The circle's slope at this point is simply the ratio of rise to run in the tangent vector: ➔

Draw the simple diagram by hand to see what's happening.

$$\text{slope} = \frac{1/\sqrt{2}}{-1/\sqrt{2}} = -1.$$

EXAMPLE 6   Find an equation for the tangent line at $t = 5$ to the fly's path shown in Example 3. ➔

A distracted fly might move along this line if it stopped "turning" at $t = 5$.

Solution   We found several equations for this very line in Example 2.

Tangent lines to polar curves   Any curve given by a polar equation $r = f(\theta)$ can, instead, be thought of as a parametric curve with parameter θ and coordinate functions

$$x = f(\theta)\cos\theta \qquad \text{and} \qquad y = f(\theta)\sin\theta.$$

(See the Fact on page 636.) If we now combine these formulas into one vector-valued function (let's call it $\mathbf{g}$) in the equation

$$\mathbf{g}(\theta) = (f(\theta)\cos\theta, \, f(\theta)\sin\theta),$$

then we can use vector derivatives to find tangent lines to polar curves.

EXAMPLE 7   Figure 5 shows the cardioid $r = 1 + \cos\theta$; points with horizontal and vertical tangent lines are marked with dots.

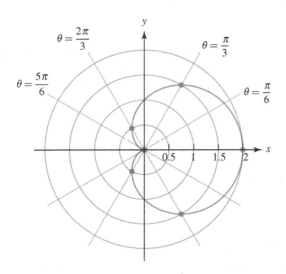

FIGURE 5
Horizontal and vertical points on a cardioid

Find exact coordinates for the marked points. What happens at the origin (where $\theta = \pi$)?

Solution We first write the polar curve as a vector-valued function of θ:

$$\mathbf{g}(\theta) = (x(\theta), y(\theta)) = ((1 + \cos\theta)\cos\theta, \ (1 + \cos\theta)\sin\theta),$$

We omit the symbolic details, but work them out for yourself.

where $0 \le \theta \le 2\pi$. Differentiation (and some algebra) gives the derivative ◄

$$\mathbf{g}'(\theta) = (x'(\theta), y'(\theta)) = (-\sin\theta(2\cos\theta + 1), \ (2\cos\theta - 1)(\cos\theta + 1)).$$

The cardioid can have a *vertical* tangent line only where the derivative vector is vertical—that is, where $x'(\theta) = 0$ but $y'(\theta) \ne 0$. Similarly, the tangent line can be *horizontal* only where $y'(\theta) = 0$ but $x'(\theta) \ne 0$.

A careful look at the derivative formulas shows that the $x'(\theta)$ has four roots: $\theta = 0$, $\theta = \pi$, $\theta = 2\pi/3$, and $\theta = 4\pi/3$. Similar calculations reveal three roots of $y'(\theta)$, at $\theta = \pi$, $\theta = \pi/3$, and $\theta = 5\pi/3$. We see, too, that $\theta = \pi$ is a root of *both* $x'(\theta)$ and $y'(\theta)$.

Now Figure 5 shows just what the preceding analysis suggests: The cardioid is vertical at three of the four roots of $x'(\theta)$ and horizontal at two of the three roots of $y'(\theta)$.

A little glitch occurs at the common root $\theta = \pi$, which corresponds to the cardioid's "cusp" at the origin. Here both $x'(\pi) = 0$ and $y'(\pi) = 0$ are zero, and so no clear

With slightly different ground rules, one could argue for a horizontal tangent line at the origin.

candidate for a tangent line emerges. ◄   ∎

Algebraic rules for vector derivatives

Scalar- and vector-valued functions can be combined in various ways to form new functions. If $\mathbf{f}$ and $\mathbf{g}$ are vector-valued functions, and a is a scalar-valued function, then we can form such combinations as $\mathbf{f}(t) + \mathbf{g}(t)$ and $a(t)\mathbf{g}(t)$. Derivatives of such combinations can be found using rules that closely resemble the ordinary sum and product rules.

> **THEOREM 2 (Sum and product rules)** Let $\mathbf{f}$ and $\mathbf{g}$ be differentiable vector-valued functions, and let a be a differentiable scalar-valued function. Then the combined functions $\mathbf{f} + \mathbf{g}$ and $a\mathbf{g}$ are differentiable with derivatives as follows:
>
> - **sum rule** $(\mathbf{f}(t) + \mathbf{g}(t))' = \mathbf{f}'(t) + \mathbf{g}'(t)$
> - **scalar multiple rule** $(a(t)\,\mathbf{f}(t))' = a'(t)\,\mathbf{f}(t) + a(t)\,\mathbf{f}'(t)$

Proof The theorem is proved by writing $\mathbf{f}$ and $\mathbf{g}$ in components and using the corresponding derivative rule for scalar-valued functions. To prove the sum rule in dimension two, for example, we first write $\mathbf{f} = (f_1, f_2)$ and $\mathbf{g} = (g_1, g_2)$. ◄ Then,

We omit the input variable t to reduce clutter.

$$
\begin{aligned}
(\mathbf{f} + \mathbf{g})' = (f_1 + g_1, f_2 + g_2)' &= ((f_1 + g_1)', (f_2 + g_2)') && \text{definition of derivative} \\
&= (f_1' + g_1', f_2' + g_2') && \text{ordinary sum rule} \\
&= (f_1', f_2') + (g_1', g_2') && \text{vector addition} \\
&= \mathbf{f}' + \mathbf{g}',
\end{aligned}
$$

as desired.   ∎

EXAMPLE 8 Differentiate $\mathbf{g}(t) = t(\cos t, \sin t)$.

Solution The scalar multiple rule applies with $a(t) = t$ and $\mathbf{f}(t) = (\cos t, \sin t)$. It says that

$$\mathbf{g}'(t) = (t(\cos t, \sin t))' = 1(\cos t, \sin t) + t(-\sin t, \cos t).$$

This is not very exciting—after all, we could have multiplied through by t before differentiating—but it works.

Antiderivatives and integrals

Derivatives of vector-valued functions are found component by component. For the same reason, vector antiderivatives are found the same way. (In this section we will just calculate some vector antiderivatives. In the next section we'll use them to model physical motion.)

Put tersely in symbols: If $\mathbf{f}(\mathbf{t}) = (f_1(t), f_2(t))$, then

$$\int \mathbf{f}(t)\,dt = \int (f_1(t), f_2(t))\,dt = \left(\int f_1(t)\,dt, \int f_2(t)\,dt \right).$$

The following example shows the idea in action.

EXAMPLE 9 Let $\mathbf{f}(t) = (1,2) + t(3,4)$. Show that

$$\int \mathbf{f}(t)\,dt = t(1,2) + \frac{t^2}{2}(3,4) + (C_1, C_2),$$

where C_1 and C_2 are arbitrary constants.

Solution If we first write $\mathbf{f}(t) = (1+3t, 2+4t)$, then antidifferentiating separately in each component gives

$$\int \mathbf{f}(t)\,dt = \int (1+3t, 2+4t)\,dt = \left(t + 3\frac{t^2}{2} + C_1, 2t + 2t^2 + C_2 \right),$$

where C_1 and C_2 are arbitrary constants. Factoring like powers of t out of both components gives another way to write the answer:

$$\left(t + 3\frac{t^2}{2} + C_1, 2t + 2t^2 + C_2 \right) = t(1,2) + \frac{t^2}{2}(3,4) + (C_1, C_2).$$

(Both forms of the answer are correct, but the second form probably shows more clearly where the answer came from.)

Using derivatives and antiderivatives

Vector derivatives and antiderivatives, suitably employed, are basic tools for modeling phenomena of motion. In the next section we look quite carefully at modeling physical motion in the presence of gravity. Here, we introduce basic ideas.

Speed and direction We have seen that if C is a curve in the xy-plane, described by a position vector function $\mathbf{p}(t)$, and we think of t as time, then, for any $t = t_0$, the vector derivative

$$\mathbf{p}'(t_0) = (x'(t_0), y'(t_0))$$

conveys two useful types of information: (i) the magnitude $|\mathbf{p}'(t_0)|$ tells the instantaneous *speed* at which $\mathbf{p}(t)$ moves along C; (ii) if $\mathbf{p}'(t_0) \neq (0,0)$, then $\mathbf{p}'(t_0)$ is tangent to C at $(x(t_0), y(t_0))$.

EXAMPLE 10 The position of a moving point along a curve C is given parametrically by

$$\mathbf{p}(t) = (2\cos t, \sin t); \quad 0 \le t \le 2\pi.$$

What is the point's speed at time $t = \pi/3$? Find a *linear* curve $\mathbf{l}(t)$ that passes through $\mathbf{p}(\pi/3)$ with direction vector $\mathbf{p}'(\pi/3)$. Plot both curves together.

Solution   It is easy to check that

$$\mathbf{p}\left(\frac{\pi}{3}\right) = \left(1, \frac{\sqrt{3}}{2}\right), \quad \text{and} \quad \mathbf{p}'\left(\frac{\pi}{3}\right) = \left(-\sqrt{3}, \frac{1}{2}\right).$$

The point's speed at this moment is the magnitude of the velocity: $|\mathbf{p}'(\pi/3)| = \sqrt{13}/2 \approx 1.803$, and the linear curve

$$\mathbf{l}(t) = \mathbf{p}\left(\frac{\pi}{3}\right) + \left(t - \frac{\pi}{3}\right)\mathbf{p}'\left(\frac{\pi}{3}\right) = \left(1, \frac{\sqrt{3}}{2}\right) + \left(t - \frac{\pi}{3}\right)\left(-\sqrt{3}, \frac{1}{2}\right)$$

"matches" $\mathbf{p}(t)$ in the desired sense. (In particular, $\mathbf{l}(\pi/3) = \mathbf{p}(\pi/3)$.) Figure 6 shows $\mathbf{p}$ and $\mathbf{l}$ together.

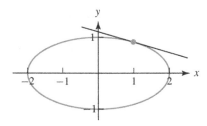

FIGURE 6
A curve and a tangent line

The line appears, as advertised, to be tangent to the curve (an ellipse) at the point $\mathbf{p}(\pi/3)$. ∎

Finding arclength   Consider a curve C with parametrization

$$\mathbf{p}(t) = \big(x(t), y(t)\big); \qquad a \le t \le b.$$

How *long* is C? The answer is easy to find for the simplest curves, such as line segments and circles. For the general case, speed is the key to an answer.

We said above that the speed at which C is traversed is given, at time t, by

$$|\mathbf{p}'(t)| = \sqrt{x'(t)^2 + y'(t)^2}.$$

This formula shows that, for most curves, the speed is a nonconstant function of t. In the (rare) exceptional case, we speak of a **constant-speed parametrization**.

Recall from single-variable calculus that to find the total distance a moving object travels over the t-interval $a \le t \le b$ (also called the **arclength**), we integrate the speed function. The same integration strategy turns out to work here—and it works both in the plane and in space:

FACT (Arclength by integration)   Let C be a curve parametrized by continuously differentiable coordinate functions $x(t)$ and $y(t)$ for $a \le t \le b$. Then

$$\text{arclength of } C = \int_a^b \sqrt{x'(t)^2 + y'(t)^2}\, dt.$$

EXAMPLE 11   How far did the fly in Example 3 fly over the 10-second period illustrated in Figure 3?

Solution   We calculated the fly's *velocity* in Example 3. Finding the fly's *speed* function is now easy:

$$\mathbf{p}'(t) = \left(-\sin t, \cos t, \frac{1}{3}\right) \implies |\mathbf{p}'(t)| = \sqrt{1+1/9} = \frac{\sqrt{10}}{3}.$$

Hence, the fly moves at *constant* speed, and its total distance—the arclength of its path—is simply

$$\int_0^{10} \frac{\sqrt{10}}{3}\,dt = 10\frac{\sqrt{10}}{3} \approx 10.54 \qquad \text{units of distance.}$$

EXAMPLE 12   Find the length of the upper half of the unit circle; use the parametrization $\mathbf{p}(t) = (\cos t, \sin t), 0 \le t \le \pi$.

Solution   Since $x'(t) = -\sin t$ and $y'(t) = \cos t$, we see that $\sqrt{x'(t)^2 + y'(t)^2} = 1$—in this case, therefore, the speed is always 1.

$$\text{arclength} = \int_0^\pi 1\,dt = \pi.$$

What could be simpler?

EXAMPLE 13   Repeat the preceding calculation, but this time with another parametrization: $\mathbf{p}(t) = (\cos(t^2), \sin(t^2)), 0 \le t \le \sqrt{\pi}$.

Solution   This time the speed is not constant: $x'(t) = -2t\sin(t^2)$ and $y'(t) = 2t\cos(t^2)$. But it doesn't make much difference:

$$\sqrt{x'(t)^2 + y'(t)^2} = \sqrt{4t^2\sin^2(t^2) + 4t^2\cos^2(t^2)} = 2t,$$

and so

$$\text{arclength} = \int_0^{\sqrt{\pi}} 2t\,dt = \pi.$$

Different parametrizations but only one length   The calculations in Examples 12 and 13 produced easy answers. More important, they produced the same answer—as we'd expect, given that only one curve is involved. An important principle lurks here:

The length of a smooth curve is a property of the curve, not of a particular parametrization. Different parametrizations give the same length.

Easy or hard?   Examples 12 and 13 might suggest that finding arclengths exactly is a piece of cake. Alas, it isn't: In practice, the pesky square root in the integrand often foils symbolic integration. When that occurs, however, we can use numerical integration methods.

EXAMPLE 14   How long—exactly—is the curve $\mathbf{p}(t) = (t, \sin t)$ for $0 \le t \le \pi$? (The curve is one arch of a sine function.) How long is it approximately?

Solution   The arclength integral is easy to write down: ➡   *Check it!*

$$\text{arclength} = \int_0^\pi \sqrt{1 + \cos^2 t}\,dt.$$

The integral, unfortunately, does not yield in symbolic form even to *Maple* or *Mathematica* because the integrand has no convenient antiderivative. But numerical techniques, such as the midpoint rule with 20 subdivisions, are usually available for definite integrals. ← We get

Your calculator or computer can help you check such calculations.

$$M_{20} = 3.820197791,$$

which means that our curve is approximately 3.82 units long. ∎

BASIC EXERCISES

1. Let ℓ be the line through the point $P = (1, 2)$ in the direction of the vector $\mathbf{v} = (2, 3)$.

   (a) The line ℓ is the image of the vector-valued function $\mathbf{L}(t) = (1, 2) + t(2, 3)$ for all real t. Draw a picture to explain why; label the points on ℓ that correspond to $t = 0$, $t = 1$, $t = 2$, and $t = -1$.

   (b) What is the image of the function $\mathbf{L}(t) = (1, 2) + t(2, 3)$ if the domain is restricted to $t \geq 0$?

   (c) What is the image of the function $\mathbf{L}(t) = (1, 2) + t(2, 3)$ if the domain is restricted to $-1 \leq t \leq 1$?

2. Repeat Exercise 1, but let ℓ be the line through the point (a, b) in the direction of the vector (c, d). (Assume that c and d are not both zero.)

3. Suppose that the motion of a particle is described by the parametric equations $x(t) = 1 - 2t + t^3$, $y(t) = 2t^4$. What is the particle's speed at time $t = 1$?

4. Let $P = (1, 3)$ and $Q = (2, 6)$ be points in the plane. Find a vector-valued function $\mathbf{R}(t) = \mathbf{R_0} + t\mathbf{v}$ such that $\mathbf{R}(t)$ describes the line through P and Q.

5. Let $P = (-2, 1)$ and $Q = (3, 2)$ be points in the plane. Find a vector-valued function $\mathbf{S}(t) = \mathbf{S_0} + t\mathbf{u}$ such that $\mathbf{S}(t)$ describes the line through P and Q.

6. Let $\mathbf{P}$ be the vector-valued function $\mathbf{P}(t) = (\sin t, \cos t)$.

   (a) What curve is traced by $\mathbf{P}$? Draw a picture of this curve.

   (b) Draw the vector $\mathbf{P}'(\pi/3)$ at the point corresponding to $t = \pi/3$ on the picture you produced in part (a).

   (c) Find a vector equation for the line tangent to the curve at $t = \pi/3$.

7. Let $\mathbf{P}$ be the vector-valued function $\mathbf{P}(t) = (2\cos t, \sin t)$.

   (a) What curve is traced by $\mathbf{P}$? Draw a picture of this curve.

   (b) Draw the vector $\mathbf{P}'(\pi/6)$ at the point corresponding to $t = \pi/6$ on the picture you produced in part (a).

   (c) Find a vector equation for the line tangent to the curve at $t = \pi/6$.

8. Let C be the curve described by the parametric equations $x(t) = t$, $y(t) = t^3$.

   (a) Sketch a graph of C.

   (b) Find the vector-valued function $\mathbf{f}(t)$ associated with this parameterization of the curve C.

(c) Find the velocity vector and the speed of $\mathbf{f}(t)$ at $t = 1$.

9. Find an equation of the line tangent to the curve $\mathbf{f}(t) = (t^2, 2t, t^3)$ at $t = 1$. What is the speed of $\mathbf{f}(t)$ at $t = 1$?

10. Find the velocity of $\mathbf{g}(t) = (t^2 - 1, t + 1, t^3)$ at $t = 2$. What is the speed of $\mathbf{g}(t)$ at $t = 2$?

11. Let $\mathbf{P}(t) = t\mathbf{i} + t^2\mathbf{j}$.

    (a) Draw the vectors $\mathbf{P}(-2)$, $\mathbf{P}(-1)$, $\mathbf{P}(0)$, $\mathbf{P}(1)$, and $\mathbf{P}(2)$.

    (b) Add a sketch of the curve described by $\mathbf{P}$ to the figure you prepared in part (a).

    (c) Add the vectors $\mathbf{P}'(-2)$, $\mathbf{P}'(-1)$, $\mathbf{P}'(1)$, and $\mathbf{P}'(2)$ to the figure you prepared in part (a).

12. Let $\mathbf{Q}(t) = t\mathbf{i} + t^{-1}\mathbf{j}$.

    (a) Draw the vectors $\mathbf{Q}(1)$, $\mathbf{Q}(2)$, $\mathbf{Q}(3)$, $\mathbf{Q}(4)$, and $\mathbf{Q}(5)$.

    (b) Add a sketch of the curve described by $\mathbf{Q}$ to the figure you prepared in part (a).

    (c) Add the vectors $\mathbf{Q}'(1)$, $\mathbf{Q}'(2)$, $\mathbf{Q}'(3)$, $\mathbf{Q}'(4)$, and $\mathbf{Q}'(5)$ to the figure you prepared in part (a).

In Exercises 13–17, plot the given curve and estimate its arclength by eye. Set up the appropriate integral. If possible, evaluate it exactly by antidifferentiation; otherwise, estimate the answer using a midpoint approximating sum with 20 subdivisions.

13. $\mathbf{r}(t) = (3 + t, 2 + 3t)$; $0 \leq t \leq 1$.

14. $\mathbf{r}(t) = (\cos(2t), \sin(2t))$; $0 \leq t \leq \pi$

15. $\mathbf{r}(t) = (\sin(3t), \cos(3t))$; $0 \leq t \leq 2\pi/3$

16. $\mathbf{r}(t) = (3\sin t, \cos t)$; $0 \leq t \leq 2\pi$

17. $\mathbf{r}(t) = (t\cos t, t\sin t)$; $0 \leq t \leq 4\pi$

In Exercises 18–21, find a linear curve $\mathbf{l}(t)$ that passes through $\mathbf{r}(t_0)$ with direction vector $\mathbf{r}'(t_0)$. Check your work by plotting. (Do these exercises in the spirit of Example 10.) [HINT: When plotting, be sure to use the same units on the x- and y-axes.]

18. $\mathbf{r}(t) = (t^2, t^3)$; $t_0 = 1$

19. $\mathbf{r}(t) = (\cos t, \sin t)$; $t_0 = \pi/4$

20. $\mathbf{r}(t) = (t, \sin t)$; $t_0 = \pi/4$

21. $\mathbf{r}(t) = (t\sin t, t\cos t)$; $t_0 = \pi$

22. Let $P = (0, 1, 2)$ and $Q = (3, 1, 6)$ be points in space.
  (a) Find a vector-valued function $\mathbf{R}(t)$ such that $\mathbf{R}(0) = P$ and $\mathbf{R}(t)$ describes the line through P and Q and $|\mathbf{R}'(t)| = 5$.
  (b) Find a vector-valued function $\mathbf{S}(t)$ such that $\mathbf{S}(0) = Q$ and $\mathbf{S}(t)$ describes the line through P and Q and $|\mathbf{S}'(t)| = 10$.

23. Suppose that $\mathbf{R}(t) = \mathbf{P}_0 + t\mathbf{v}$ and $\mathbf{S}(t) = \mathbf{P}_0 + t^2\mathbf{v}$, where $\mathbf{P}_0$ and $\mathbf{v} \neq \mathbf{0}$ are vectors. Do $\mathbf{R}$ and $\mathbf{S}$ describe the same curve? Justify your answer.

24. Suppose that $\mathbf{R}(t) = \mathbf{P}_0 + t\mathbf{v}$ and $\mathbf{S}(t) = \mathbf{P}_0 + t^3\mathbf{v}$, where $\mathbf{P}_0$ and $\mathbf{v}$ are vectors. Do $\mathbf{R}$ and $\mathbf{S}$ describe the same curve? Justify your answer.

25. Suppose that $\mathbf{P}(0) = \mathbf{u}$ and $\mathbf{P}'(t) = \mathbf{v}$, where $\mathbf{u}$ and $\mathbf{v}$ are constant vectors. Describe the curve traced by $\mathbf{P}(t)$.

26. Suppose that $\mathbf{P}(0) = \mathbf{u}$ and $\mathbf{P}'(t) = t\mathbf{v}$, where $\mathbf{u}$ and $\mathbf{v}$ are constant vectors. Describe the curve traced by $\mathbf{P}(t)$.

27. Let H be the helix described by the vector-valued function $\mathbf{f}(t) = \cos t\,\mathbf{i} + \sin t\,\mathbf{j} + t\,\mathbf{k}$.
  (a) Show that $\mathbf{f}(t)$ moves along H at a constant speed.
  (b) Find a vector-valued function $\mathbf{g}(t)$ that moves along H at constant speed 1.

28. Consider the vector-valued function $\mathbf{p}(t) = (\cos t, \sin t, t)$.
  (a) Plot the curve defined by $\mathbf{p}(t)$; let $0 \leq t \leq 4\pi$.
  (b) Find a vector equation for the tangent line ℓ at $t = \pi$ to the curve defined by $\mathbf{p}(t)$.
  (c) On one set of axes, plot both the curve and the tangent line from part (a).
  (d) Show that $\mathbf{p}(t) = (\cos t, \sin t, t)$ has constant speed. Find the arclength from $t = 0$ to $t = 4\pi$.

29. Repeat Exercise 28 using the vector-valued function $\mathbf{p}(t) = (\cos(t/2), \sin(t/2), t)$.

30. Find the length of the curve $\mathbf{x}(t) = (t, \sin(4t), \cos(4t))$ between $t = 0$ and $t = \pi$.

31. Find the length of the logarithmic spiral $r = e^\theta$ between $\theta = 0$ and $\theta = \pi$.

32. Which helix is longer, one of radius 5 centimeters and height 4 centimeters that makes three complete turns or one of radius 3 centimeters and height 4 centimeters that makes five complete turns? Justify your answer.

12.6 MODELING MOTION

On March 13, 1986, the Giotto spacecraft approached within 600 kilometers of Halley's comet. This fly-by had been planned since Giotto's launch on July 2, 1985. An unplanned event occurred, too—Giotto was severely damaged by space dust. (For a color photograph of the nucleus of Halley's comet, taken by Giotto, see

`http://nssdc.gsfc.nasa.gov/planetary/giotto.html` on the World Wide Web.)

Halley's comet had last been seen near Earth in 1910. On its third previous pass, in 1682, the comet had been observed by the English astronomer Edmund Halley (1656–1742); a contemporary and supporter (intellectual and financial) of Isaac Newton's. Halley's theory of orbits correctly predicted that the comet of 1682 would return to Earth's vicinity on a 76-year cycle.

Predicting the orbit of a comet and arranging a rendezvous with a spacecraft would be impossible without calculus-based tools and techniques. So, for that matter, would much humbler tasks, such as estimating that the 206-mile drive from Chattanooga to Tuscaloosa takes 3 hours and 49 minutes—unless one takes the longer route through Talladega, which adds about 57 minutes to the trip.

Comets and spacecraft are difficult and expensive to observe directly, and so it is essential to model their motion mathematically. To model a physical phenomenon mathematically means first to describe it in mathematical language, using mathematical objects such as vectors, functions, derivatives, and integrals. Once this is done, mathematical consequences can be deduced and their physical meaning interpreted.

Modeling physical motion was historically among the most important applications of calculus. To a large extent, in fact, the invention and development of calculus in the 17th and 18th centuries were spurred by the desire to understand and predict motion, whether the orbits of planets or the trajectories of mortar shells. Modeling motion remains just as important today, whether for predicting when we will get to Grandma's or for programming a robot arm.

How many variables? One-variable calculus nicely models one-dimensional motion, such as that of a car on a straight east–west road like Interstate 94 through North Dakota. But the curves and mountains of Montana are not far away. When we get there, an extra variable or two will prove very handy.

The physical world is three-dimensional, and so it might seem that we would *always* need three variables to describe motion realistically. Many types of motion, however, are essentially one- or two-dimensional. A westbound car on Interstate 94 through North Dakota moves, for practical purposes, on a (one-dimensional) straight line. If the driver turns south at Mandan and takes the scenic loop through Carson, Elgin, Mott, and Richardton, rejoining I-94 at Dickinson, the car's motion is essentially two-dimensional. Nothing much happens in the third dimension, altitude, until the mountainous stretch ahead in Montana from Billings to Butte. (Even there we might choose to ignore altitude for the sake of convenience or approximation.)

Position, velocity, and acceleration

"Particle" is the quaint-sounding term for any object that can be thought of as a point.

In one dimension Imagine a particle ← moving along the x-axis, with its position at time t seconds given by a **position function** $p(t)$. The particle has a corresponding **velocity function** $v(t)$ and an **acceleration function** $a(t)$. As elementary calculus teaches, these functions are related by derivatives:

$$v(t) = p'(t); \qquad a(t) = v'(t); \qquad a(t) = p''(t).$$

We'll always assume this unless something is said to the contrary.

(We assume here, of course, that all the needed derivatives exist.) ← We used these facts in elementary calculus to model various phenomena of motion. A typical example follows: Starting from an acceleration function, we *antidifferentiate* to work "back" to find the velocity and position functions.

> EXAMPLE 1 A particle has constant acceleration $a(t) = -3$ units per second per second. At time $t = 0$, the particle has velocity $v(0) = 12$ units per second and position $p(0) = 0$. Find formulas for $v(t)$ and $p(t)$. Describe the particle's journey over the next 8 seconds.
>
> Solution From the derivative relations above, we have
>
> $$v'(t) = a(t) = -3 \implies v(t) = \int -3\,dt = -3t + C,$$
>
> where C is some constant. Since $v(0) = 12$, we must have $C = 12$, and so $v(t) = -3t + 12$. Arguing similarly for position gives
>
> $$p'(t) = v(t) = -3t + 12 \implies p(t) = \int (-3t + 12)\,dt = \frac{-3t^2}{2} + 12t + D,$$
>
> where D is a constant. The condition $p(0) = 0$ implies that $D = 0$. Putting the pieces together gives
>
> $$v(t) = -3t + 12 \qquad \text{and} \qquad p(t) = \frac{-3t^2}{2} + 12t.$$

Or their graphs.

The formulas ← show that $t = 4$ is especially important: $v(4) = 0$, $p(4) = 24$, and $v(t)$ changes sign (from positive to negative) at $t = 4$. Now we can describe the particle's itinerary: It moves to the right until $t = 4$, reaching $x = 24$; then it moves to the left, returning to the origin, with velocity -12 units per second, at $t = 8$. ■

In higher dimensions The vector calculus tools and techniques we have developed permit us to model many types of two- and three-dimensional motion. (For simplicity we will stick mainly to plane motion, but the ideas are essentially the same in three variables.) The key idea is that *vector-valued* acceleration, velocity, and position functions are all derivatives and antiderivatives of each other just as in one dimension:

$$\mathbf{v}(t) = \mathbf{p}'(t); \qquad \mathbf{a}(t) = \mathbf{v}'(t); \qquad \mathbf{a}(t) = \mathbf{p}''(t).$$

There are differences, however, between these equations and their one-dimensional counterparts: The position, velocity, and acceleration are now all *vector-valued* functions of t, and the derivatives must now be interpreted in the sense appropriate for such functions. →

Recall: Boldface letters denote vectors.

Starting with acceleration In modeling physical motion, it is often most natural to begin with information about acceleration and to work toward formulas for velocity and position. Acceleration crops up naturally in practice because of its close connection to force. **Newton's second law** puts it like this:

> *A force acting on an object produces an acceleration that is directly proportional to the force and inversely proportional to the object's mass.*

Forces (gravitational force, air drag, sliding friction, a rocket's thrust, etc.) can often be measured directly and—thanks to Newton's law—converted into information about acceleration. For these reasons, acceleration rather than velocity or position is often the "given" in physical modeling problems.

Zero acceleration In the simplest possible case, an object's acceleration is the zero vector. In this case it is easy to find velocity and position by antidifferentiation. For velocity,

$$\mathbf{a}(t) = (0,0) \Longrightarrow \mathbf{v}(t) = \int (0,0)\,dt = \left(\int 0\,dt, \int 0\,dt \right) = (C_1, C_2),$$

where C_1 and C_2 are constants. → For position,

Notice the antiderivative of a vector-valued function.

$$\mathbf{v}(t) = (C_1, C_2) \Longrightarrow \mathbf{p}(t) = \int (C_1, C_2)\,dt = \left(\int C_1\,dt, \int C_2\,dt \right)$$
$$= (C_1 t + C_3, \ C_2 t + C_4),$$

where C_3 and C_4 are arbitrary constants.

In specific cases, the four constants are evaluated using additional information such as the velocity and position of the particle at time $t = 0$. Notice, however, that regardless of the numerical values of these constants, the calculation illustrates an interesting physical fact: In the absence of outside forces, an object has *zero* acceleration, *constant* velocity (and speed), and a *linear* position function.

EXAMPLE 2 At time $t = 0$, a particle in the plane has velocity vector $(1,2)$ and position vector $(3,4)$. (These stipulations are called **initial conditions**.) No external force acts, and so the acceleration is $\mathbf{a}(t) = (0,0)$ at all times. Describe the particle's movement. Where is the particle at time $t = 100$?

Solution As we just calculated, the particle has constant velocity function $\mathbf{v}(t) = (C_1, C_2)$ and linear position function $\mathbf{p}(t) = t(C_1, C_2) + (C_3, C_4)$. The initial conditions give $C_1 = 1$, $C_2 = 2$, $C_3 = 3$, and $C_4 = 4$, and so

$$\mathbf{v}(t) = (1,2), \qquad \text{and} \qquad \mathbf{p}(t) = (3,4) + t(1,2).$$

Thus, the particle moves away from $(3,4)$ with constant speed $\sqrt{5}$ in the direction of the vector $(1,2)$. At $t = 100$ the particle's position is $\mathbf{p}(100) = (103, 204)$. ∎

Gravity and acceleration   Everyone knows that Earth's gravity pulls objects toward the center of the Earth, producing a "downward" acceleration. It is also true (but not so obvious until around 1600, when Galileo tossed various items off the Leaning Tower of Pisa) that the acceleration due to gravity has the *same* magnitude for all objects, regardless of their mass. ◄

Heavier objects fall harder but not faster.

The practical result is that gravity induces the same acceleration on all objects near the surface of the Earth. Considered as a vector, the acceleration due to gravity is an arrow pointing toward the center of the Earth; it has the same magnitude for all objects. (This magnitude is usually denoted by g; it is called the **acceleration due to gravity**.) In the metric system, $g \approx 9.8$ meters per second per second. (In English units, $g \approx 32$ feet per second per second.)

Displacement and definite integrals   In Example 2 we antidifferentiated the velocity function $\mathbf{v}$ to get the position function $\mathbf{p}$; then we found $\mathbf{p}(100)$. A slightly different strategy involves the *definite* integral of the velocity function over the time interval in question:

> FACT   Let $\mathbf{v}(t)$ be a particle's velocity and $\mathbf{p}(t)$ its position for $a \le t \le b$. Then
>
> $$\int_a^b \mathbf{v}(t)\, dt = \mathbf{p}(t)\Big]_a^b = \mathbf{p}(b) - \mathbf{p}(a).$$

The vector quantity $\mathbf{p}(b) - \mathbf{p}(a)$ is called the **displacement** of the particle over the interval $a \le t \le b$; it tells how the particle's position changes over the time interval.

EXAMPLE 3   In the situation of Example 2, use a definite integral to find the particle's displacement over the interval $0 \le t \le 100$. What is the particle's final position?

Solution   Note first that if $\mathbf{v}(t) = (1, 2)$, then $t\,(1, 2) + (C_1, C_2)$ is an antiderivative for *any* constants C_1 and C_2. Since we are working with a definite integral, we can safely set both constants to zero. Thus, the displacement vector is

$$\int_0^{100} (1, 2)\, dt = t(1, 2)\Big]_0^{100} = (100, 200).$$

Adding the displacement to the starting position gives the particle's final position: $\mathbf{p}(100) = (3, 4) + (100, 200) = (103, 204)$. ◄

We found the same answer in Example 2.

Constant acceleration   In the next simplest case, an object's acceleration is a nonzero constant vector. This case is especially important; it arises, for instance, when a constant force (such as gravity, under appropriate conditions) acts on an object, producing a constant acceleration. Suppose, then, that an object has constant acceleration vector $\mathbf{a}(t) = (a_1, a_2)$. Then

$$\mathbf{v}(t) = \int (a_1, a_2)\, dt = t(a_1, a_2) + (C_1, C_2),$$

where C_1 and C_2 are arbitrary constants. Since $\mathbf{v}(0) = (C_1, C_2)$, the constants represent the object's initial velocity.

To find position we antidifferentiate again: ➡

$$\mathbf{p}(t) = \int \big(t(a_1, a_2) + (C_1, C_2)\big)dt = \frac{t^2}{2}(a_1, a_2) + t(C_1, C_2) + (C_3, C_4),$$

Convince yourself that these antiderivatives are correct.

where C_3 and C_4 are again arbitrary constants. Since $\mathbf{p}(0) = (C_3, C_4)$, these constants represent the object's initial position.

We see, in particular, that *constant* acceleration begets *linear* velocity and *quadratic* position functions. ➡

Unless some of the constants happen to be zero.

EXAMPLE 4   An object starts from rest at the origin and has constant acceleration $\mathbf{a} = (1, 2)$. What happens? What are the object's position and velocity at $t = 100$ seconds?

Solution   The calculations in Example 3, together with the initial conditions, mean that the object's velocity and position functions are

$$\mathbf{v}(t) = t(1, 2) \qquad \text{and} \qquad \mathbf{p}(t) = \frac{t^2}{2}(1, 2).$$

Thus, the particle moves along the curve defined by $\mathbf{p}(t)$. (This "curve" is actually a straight line—but that doesn't matter to the calculation.)

The velocity at $t = 100$ $\mathbf{v}(100) = (100, 200)$; the speed is $|\mathbf{v}(100)| = 100\sqrt{5}$, and the position is $\mathbf{p}(100) = (5000, 10000)$. ■

Free fall

A moving object that is affected *only* by gravity is said to be in **free fall**. In the real world, falling objects are always affected to some degree by other forces such as air drag, passing breezes, and stray pigeons. These forces may or may not be negligible; in any event, modeling free fall is useful as a first step toward understanding more complex combinations of forces.

Free fall is conveniently modeled in the xy-plane; we will let the negative y-direction represent "downward" (that is, toward the Earth's center); the x- and y-units will be meters. Then the acceleration due to gravity is of the form $\mathbf{a}(t) = (0, -g)$, where $g \approx 9.8$ meters per second per second is the (constant) acceleration due to gravity. With calculations like those above we can model free-fall phenomena.

Trajectories   What shape of path does a shotput or baseball follow (ignoring wind resistance and other stray forces) after being thrown or hit? Newton deduced the answer—a parabola—from the fact that gravity's acceleration is constant. So can we.

EXAMPLE 5   A free-falling projectile leaves the origin at time $t = 0$, with initial speed 100 meters per second, at an angle α above the horizontal. What path does the projectile follow? Where does it land? (Let the x-axis represent ground level.)

Solution   The given conditions say the following: ➡

$$\mathbf{a}(t) = (0, -g), \qquad \mathbf{v}(0) = 100(\cos\alpha, \sin\alpha), \qquad \text{and} \qquad \mathbf{p}(0) = (0, 0).$$

Calculating as in Example 4 gives

$$\mathbf{v}(t) = t(0, -g) + 100(\cos\alpha, \sin\alpha); \qquad \mathbf{p}(t) = \frac{t^2}{2}(0, -g) + 100t(\cos\alpha, \sin\alpha),$$

Do you see why the initial velocity vector $\mathbf{v}(0)$ has the form claimed? Draw your own figure.

and we see that the curve $\mathbf{p}(t)$ has parametric equations

$$x(t) = 100t \cos \alpha; \qquad y(t) = \frac{-gt^2}{2} + 100t \sin \alpha.$$

Figure 1 shows a sample of such curves for various initial angles α:

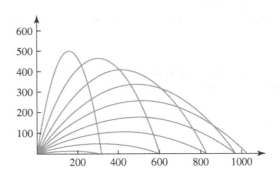

FIGURE 1

Trajectories for various initial angles

This section's exercises explore the symbolic reasons.

The picture suggests Newton's insight: All trajectories are parabolas. ◄ The projectile lands at the time t when $y(t) = 0$. Solving this equation for t (we want the positive solution) gives

$$\frac{-gt^2}{2} + 100t \sin \alpha = 0 \quad \Longrightarrow \quad t = \frac{200 \sin \alpha}{g}.$$

For this t, we have

$$x(t) = 100t \cos \alpha = 100 \frac{200 \sin \alpha}{g} \cos \alpha;$$

this is the projectile's **range**. If, say, $\alpha = \pi/4$, then the range is $10000/g \approx 1020$ meters.

(These numbers are not completely unrealistic, by the way. One simple mortar used in World War II had initial velocity 82 meters per second and range 650 meters.) ∎

BASIC EXERCISES

1. In Example 1, we claimed: ... *the particle moves to the right until $t = 4$, reaching $x = 24$; then it moves to the left, returning to the origin, with velocity -12 units per second, at $t = 8$.* Plot the function $\mathbf{p}(t)$ for $0 \le t \le 8$. Then explain how the graph "agrees with" all parts of the italicized statement.

Exercises 2–5 use the same notation as Example 1. In each exercise, find equations for $\mathbf{p}(t)$ and $\mathbf{v}(t)$. What happens over the interval $t = 0$ to $t = 10$? Find and interpret $\mathbf{v}(10)$ and $\mathbf{p}(10)$.

2. Assume that $\mathbf{a}(t) = 0$, $\mathbf{v}(0) = 0$, and $\mathbf{p}(0) = 1$.

3. Assume that $\mathbf{a}(t) = 0$, $\mathbf{v}(0) = 1$, and $\mathbf{p}(0) = 0$.

4. Assume that $\mathbf{a}(t) = 1$, $\mathbf{v}(0) = 0$, and $\mathbf{p}(0) = 0$.

5. Assume that $\mathbf{a}(t) = t$, $\mathbf{v}(0) = 0$, and $\mathbf{p}(0) = 0$.

6. This exercise is about Example 4.

  (a) By hand, sketch the "curve" defined by $\mathbf{p}(t)$ for $0 \le t \le$ 100. What very simple shape does the "curve" seem to have? [HINT: The curve is parametrized by $x(t) = p_1(t)$; $y(t) = p_2(t)$; $0 \le t \le 100$.]

  (b) Eliminate the variable t in the parametric equations for the curve $\mathbf{p}(t)$. What is the resulting equation? Does it agree with what you found in the previous part?

  (c) Find the arclength of the curve $\mathbf{p}(t)$ from $t = 0$ to $t = 100$.

7. An object starts from rest at the origin and has constant acceleration vector $\mathbf{a} = (1, -1)$.

  (a) By hand, sketch the "curve" defined by $\mathbf{p}(t)$ for $0 \le t \le$ 100. What very simple shape does the "curve" seem to have?

  (b) Eliminate the variable t in the parametric equations for the curve $\mathbf{p}(t)$. What is the resulting equation? Does it agree with what you found in the previous part?

  (c) Find the arclength of the curve $\mathbf{p}(t)$ from $t = 0$ to $t = 100$.

8. At time $t = 0$ seconds a particle is at the origin and has velocity vector $(4, 4)$. It undergoes constant acceleration $\mathbf{a}(t) = (0, -1)$.

  (a) Find formulas for the velocity function $\mathbf{v}(t)$ and the position function $\mathbf{p}(t)$.

  (b) Plot the path taken by the particle $\mathbf{p}(t)$ for $0 \le t \le 10$. What familiar shape does the path seem to have?

  (c) Eliminate the variable t in the parametric equations for the curve $\mathbf{p}(t)$. What is the resulting equation? Does it agree with what you found in the previous part?

  (d) Find the arclength of the curve $\mathbf{p}(t)$ from $t = 0$ to $t = 10$. [HINT: Set up the integral by hand; either solve it exactly using a table of integrals or use technology to approximate the integral numerically.]

9. A projectile is at $(0, 0)$ at time $t = 0$ seconds. Its initial speed is 100 meters per second, its initial angle is $\pi/3$, and it travels under free fall conditions.

  (a) Find equations for the velocity and position functions.

  (b) When does the projectile touch down?

  (c) Plot the projectile's trajectory from takeoff to landing.

  (d) At what time is the projectile at maximum height? How high is this? [HINT: Find the time at which the velocity vector is horizontal.]

  (e) Find the speed and the velocity at the moment the projectile is highest.

10. Suppose that the velocity of an object at time t seconds is $(5t^2 + 3t - 4, 1 - t)$ meters/second. At time $t = 2$ the object is at the point $(1, 0)$.

  (a) Find the object's acceleration vector at time $t = 1$.

  (b) Find the position of the object at time $t = 0$.

  (c) Express the distance traveled by the object between time $t = 0$ and time $t = 2$ as an integral. (Don't evaluate this integral.)

11. Suppose that the motion of a particle is described by the parametric equations $x = t^3 - 3t$, $y = t^2 - 2t$.

  (a) Does the particle ever come to a stop? If so, when and where?

  (b) Is the particle ever moving straight up or straight down? If so, when and where?

  (c) Is the particle ever moving horizontally left or right? If so, when and where?

12. A particle moves in the xy-plane with constant acceleration vector $\mathbf{a}(t) = (0, -1)$. At time $t = 0$ the particle is at the point $\mathbf{p}(0) = (0, 0)$ and has velocity $\mathbf{v}(0) = (1, 0)$.

  (a) Give a formula for the particle's velocity function $\mathbf{v}(t)$.

  (b) Give a formula for the particle's position function $\mathbf{p}(t)$.

  (c) Write down an integral whose value is the distance traveled by the particle between $t = 0$ and $t = 5$. (Don't evaluate this integral.)

FURTHER EXERCISES

13. Example 5 shows that a projectile with initial speed 100 meters per second and initial angle α follows a parabolic path given parametrically by

$$x(t) = 100t \cos \alpha; \qquad y(t) = \frac{-gt^2}{2} + 100t \sin \alpha.$$

  (a) Eliminate t to find an equation in x and y for the parabola. [HINT: Since $x = 100t \cos \alpha$, $t = x/(100 \cos \alpha)$. Substitute this into the equation for y.]

  (b) Show that the maximum range is obtained if $\alpha = \pi/4$. What *is* the maximum range?

14. A projectile is at $(0, 0)$ at time $t = 0$ seconds. Its initial speed is s_0 meters per second, its initial angle is α, and it travels under free fall conditions.

  (a) Find equations for the velocity and position functions.

  (b) At what time is the projectile highest? [HINT: The answer depends on both s_0 and α.]

  (c) Find the speed and the velocity at the moment when the projectile is highest.

  (d) Find the speed and the velocity at the moment when the projectile lands.

15. Suppose that the position of a particle at time t is $\mathbf{p}(t) = (3 \cos t, 3 \sin t, t)$. If the particle flies off on a tangent at $t = \pi/2$, where is the particle at $t = \pi$?

16. An astronaut is flying in a spacecraft along the path described by $\mathbf{r}(t) = (t^2 - t, 2 + t, -3/t)$, where t is given in hours. The engines are shut off when the spacecraft reaches the point $(6, 5, -1)$. Where is the astronaut 2 hours later?

17. An object moving with constant velocity passes through the point $(1, 1, 1)$ and then passes through the point $(2, -1, 3)$ 5 seconds later.

  (a) Find the object's velocity vector.

  (b) Find the object's acceleration vector.

18. Suppose that the velocity of an object at time t seconds is $(4e^{-t}, 3t^2, 5 \cos(\pi t))$ meters/second. At time $t = 1$ the object is at the point $(1, 2, 3)$.

  (a) Find the object's speed at time $t = 0$.

  (b) Find the object's acceleration vector at time $t = 1$.

  (c) Find the position of the object at time $t = 2$.

19. Suppose that a particle moves clockwise once around a circle of radius 3 centered at the point $(4, 5)$ in the xy-plane, that it moves with constant speed 2, and that it starts at the position $(1, 5)$. Give a position function that models the motion of the particle.

12.7 THE DOT PRODUCT

We have already met two algebraic operations on vectors, each of which produces new vectors from old ingredients. If **v** and **w** are vectors and r is a scalar, then both the sum **v** + **w** and the scalar multiple r**v** are vectors. In this section we meet a new operation, the **dot product**, also known, sometimes, as the **scalar product**. The latter name stems from an important property: The dot product **v** · **w** of two vectors is a *scalar*. And a very interesting scalar at that—few other mathematical tools reveal so much information at so little cost in calculation.

Defining the dot product

The algebraic definition is essentially the same in two and three variables:

> **DEFINITION (The dot product)** Let $\mathbf{v} = (v_1, v_2)$ and $\mathbf{w} = (w_1, w_2)$ be vectors. The dot product of **v** and **w** is the scalar defined by
> $$\mathbf{v} \cdot \mathbf{w} = v_1 w_1 + v_2 w_2.$$
> If $\mathbf{v} = (v_1, v_2, v_3)$ and $\mathbf{w} = (w_1, w_2, w_3)$, then
> $$\mathbf{v} \cdot \mathbf{w} = v_1 w_1 + v_2 w_2 + v_3 w_3.$$

The definition is simple. What the dot product *means*, algebraically and geometrically, is our main question. We will study it first in two variables and later in three.

EXAMPLE 1 Consider the vectors $\mathbf{i} = (1, 0)$, $\mathbf{j} = (0, 1)$, $\mathbf{v} = (2, 3)$, and $\mathbf{w} = (-3, 2)$. Find some dot products.

Solution Calculations are easy:

$$\mathbf{i} \cdot \mathbf{j} = (1, 0) \cdot (0, 1) = 0 + 0 = 0;$$
$$\mathbf{i} \cdot \mathbf{i} = (1, 0) \cdot (1, 0) = 1;$$
$$\mathbf{i} \cdot \mathbf{v} = (1, 0) \cdot (2, 3) = 2;$$
$$\mathbf{j} \cdot \mathbf{v} = (0, 1) \cdot (2, 3) = 3;$$
$$\mathbf{v} \cdot \mathbf{v} = (2, 3) \cdot (2, 3) = 2^2 + 3^2 = 13;$$
$$\mathbf{v} \cdot \mathbf{w} = (2, 3) \cdot (-3, 2) = -6 + 6 = 0.$$

What is remarkable, as we'll see, is that such simple calculations can tell us so much. ■

Example 1 suggests some patterns. When, for instance, is a dot product zero? Let's look at some more general examples.

EXAMPLE 2 Let $\mathbf{v} = (a, b)$ and $\mathbf{w} = (-b, a)$. Find all possible dot products. What do the answers mean geometrically?

Solution The dot products are easy to find:

$$\mathbf{v} \cdot \mathbf{v} = (a, b) \cdot (a, b) = a^2 + b^2 = |\mathbf{v}|^2;$$
$$\mathbf{w} \cdot \mathbf{w} = (-b, a) \cdot (-b, a) = b^2 + a^2 = |\mathbf{w}|^2;$$
$$\mathbf{v} \cdot \mathbf{w} = (a, b) \cdot (-b, a) = -ab + ba = 0.$$

The first two results illustrate a useful connection between the dot product and *lengths* of vectors. The third result, **v** · **w** = 0, occurs because, as Figure 1 shows, **v** and **w** are perpendicular:

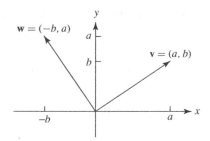

FIGURE 1
Perpendicular vectors

We summarize these observations formally in Theorems 3 and 4.

 The picture suggests an important fact about perpendicular vectors and the dot product that we will state formally in a moment. ➥

▪ *But can you guess it now?*

Theorem 3 collects basic properties of the dot product.

THEOREM 3 (Algebraic properties of the dot product)   Let **u**, **v**, and **w** be vectors, and let r be any scalar. Then

(a) $\mathbf{u} \cdot \mathbf{v} = \mathbf{v} \cdot \mathbf{u}$;

(b) $(r\mathbf{u}) \cdot \mathbf{v} = \mathbf{u} \cdot (r\mathbf{v}) = r\,(\mathbf{u} \cdot \mathbf{v})$;

(c) $\mathbf{u} \cdot (\mathbf{v} + \mathbf{w}) = \mathbf{u} \cdot \mathbf{v} + \mathbf{u} \cdot \mathbf{w}$;

(d) $\mathbf{u} \cdot \mathbf{u} = |\mathbf{u}|^2$.

All parts of the theorem are readily checked using the definition of the dot product; the exercises have more details.

Geometry of the dot product

What does the dot product mean geometrically? **Unit vectors** turn out to be key to an answer. Figure 2 shows unit vectors in 12 directions; all angles shown are multiples of $\pi/6$:

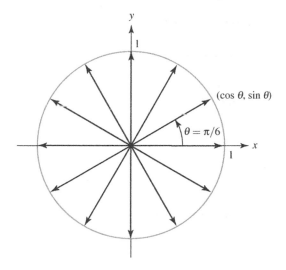

FIGURE 2
Unit vectors in 12 directions

For any angle θ, the vector $\mathbf{u}_\theta$ that makes angle θ with the x-axis has coordinates $(\cos\theta, \sin\theta)$. Every vector of this form, moreover, has length 1. The dot product shows why:

$$|\mathbf{u}_\theta|^2 = (\cos\theta, \sin\theta) \cdot (\cos\theta, \sin\theta) = \cos^2\theta + \sin^2\theta = 1.$$

We will call the vector $\mathbf{u}_\theta = (\cos\theta, \sin\theta)$ the **unit vector in the θ-direction**. For example,

$$\mathbf{u}_{\pi/6} = \left(\cos\left(\frac{\pi}{6}\right), \sin\left(\frac{\pi}{6}\right)\right) = \left(\frac{\sqrt{3}}{2}, \frac{1}{2}\right).$$

Polar form of a vector Every nonzero vector $\mathbf{v} = (a, b)$ can be written as a scalar multiple of some $\mathbf{u}_\theta$. The idea is the same as that for polar coordinates: If a Cartesian point (a, b) has polar coordinates (r, θ), then

$$\mathbf{v} = (a, b) = (r\cos\theta, r\sin\theta) = r(\cos\theta, \sin\theta) = r\mathbf{u}_\theta,$$

where $r = \sqrt{a^2 + b^2}$ and $\tan\theta = b/a$.

Recall: $\theta = \arctan(y/x)$.

EXAMPLE 3 Write the vector $\mathbf{v} = (4, 3)$ in the form $r\mathbf{u}_\theta$. What are r and θ?

Solution The vector $(4, 3)$ has length 5, and so the unit vector in the same direction is $\mathbf{u} = (4/5, 3/5)$. The point $(x, y) = (4, 3)$ has polar coordinates $r = 5$ and $\theta = \arctan(3/4) \approx 0.644$. ← This means that $\mathbf{v}$ makes angle $\theta \approx 0.644$ with the x-axis. ∎

We complete our preliminary tour of the dot product with a final important example.

EXAMPLE 4 Let α and β be angles; consider the unit vectors

$$\mathbf{u}_\alpha = (\cos\alpha, \sin\alpha) \qquad \text{and} \qquad \mathbf{u}_\beta = (\cos\beta, \sin\beta)$$

in the directions of α and β, respectively. Find $\mathbf{u}_\alpha \cdot \mathbf{u}_\beta$. What does this answer mean geometrically?

Solution The calculation is easy enough:

$$\mathbf{u}_\alpha \cdot \mathbf{u}_\beta = (\cos\alpha, \sin\alpha) \cdot (\cos\beta, \sin\beta) = \cos\alpha \, \cos\beta + \sin\alpha \, \sin\beta.$$

But what does it *mean*? The answer comes from recognizing the right-hand side above as part of a standard trigonometric identity:

$$\cos(\alpha - \beta) = \cos\alpha \, \cos\beta + \sin\alpha \, \sin\beta.$$

The result is important:

> **FACT** For unit vectors $\mathbf{u}_\alpha$ and $\mathbf{u}_\beta$, the dot product is the cosine of the angle between $\mathbf{u}_\alpha$ and $\mathbf{u}_\beta$.

Figure 3 illustrates the situation:

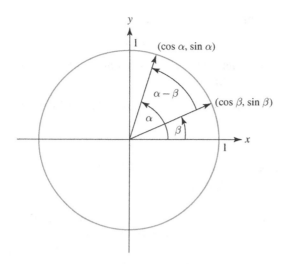

The angle between two vectors

Two special cases are most important: (i) If $\alpha = \beta$ then $\alpha - \beta = 0$, and so $\mathbf{u}_\alpha \cdot \mathbf{u}_\beta = \cos 0 = 1$. (ii) If $\alpha - \beta = \pi/2$ (so $\mathbf{u}_\alpha$ and $\mathbf{u}_\beta$ are perpendicular), then $\mathbf{u}_\alpha \cdot \mathbf{u}_\beta = \cos(\pi/2) = 0$. ◼

Now we can state—and prove—the main geometric property of the dot product. ➔

We omit many proofs, but this one is instructive.

> **THEOREM 4 (The dot product geometrically)**   Let $\mathbf{v}$ and $\mathbf{w}$ be any two vectors, and let θ be the (smaller) angle between $\mathbf{v}$ and $\mathbf{w}$. Then
>
> $$\mathbf{v} \cdot \mathbf{w} = |\mathbf{v}|\,|\mathbf{w}|\,\cos\theta.$$

Proof   Suppose that $\mathbf{v}$ and $\mathbf{w}$ make angles α and β, respectively, with the x-axis and that $\theta = \alpha - \beta$. Writing $\mathbf{v}$ and $\mathbf{w}$ in polar form gives

$$\mathbf{v} = |\mathbf{v}|\mathbf{u}_\alpha \qquad \text{and} \qquad \mathbf{w} = |\mathbf{w}|\mathbf{u}_\beta,$$

where $\mathbf{u}_\alpha$ and $\mathbf{u}_\beta$ are the unit vectors in the α- and β-directions, respectively. From Theorem 3(b) and Example 4 we have

$$\mathbf{v} \cdot \mathbf{w} = (|\mathbf{v}|\mathbf{u}_\alpha) \cdot (|\mathbf{w}|\mathbf{u}_\beta) = |\mathbf{v}|\,|\mathbf{w}|\,(\mathbf{u}_\alpha \cdot \mathbf{u}_\beta) = |\mathbf{v}|\,|\mathbf{w}|\,\cos\theta$$

as desired. ◼

Theorem 4 deserves some closer looks.

- **Right angles**   Theorem 4 makes it easy to tell whether two nonzero vectors meet at right angles. ➔ In this case, $\theta = \pi/2$, and so $\cos\theta = 0$. In general:

If either vector is zero, there is no meaningful angle between them.

> **FACT**   Two nonzero vectors $\mathbf{v}$ and $\mathbf{w}$ are perpendicular if and only if $\mathbf{v} \cdot \mathbf{w} = 0$.

(In math-speak any vectors $\mathbf{v}$ and $\mathbf{w}$ that satisfy the condition $\mathbf{v} \cdot \mathbf{w} = 0$ are called **orthogonal**. This term is slightly more general than "perpendicular"—it makes sense even when vectors and angles have no convenient geometric interpretation.)

- **Identical vectors**   At the opposite extreme from being perpendicular, two vectors might be identical. In this case, the angle between them is zero, and Theorem 4 says (as did Theorem 3) that

$$\mathbf{v} \cdot \mathbf{v} = |\mathbf{v}||\mathbf{v}| \cos 0 = |\mathbf{v}|^2.$$

- **The sign of the dot product** Theorem 4 implies that the *sign* of the dot product $\mathbf{v} \cdot \mathbf{w}$ depends entirely on $\cos \theta$. More precisely:

  The angle between $\mathbf{v}$ and $\mathbf{w}$ is **acute** if $\mathbf{v} \cdot \mathbf{w} > 0$ and **obtuse** if $\mathbf{v} \cdot \mathbf{w} < 0$.

Using the dot product

The dot product helps us calculate many quantities that depend on the lengths of vectors and angles between them. We illustrate several by example.

Projecting one vector onto another The idea of **projecting** one vector onto another is useful in various settings. Projections arise in physics, for instance, in writing a force vector as the sum of two or three force vectors in perpendicular directions. Figure 4 illustrates two instances of projecting one vector $\mathbf{v}$ onto another vector $\mathbf{w}$:

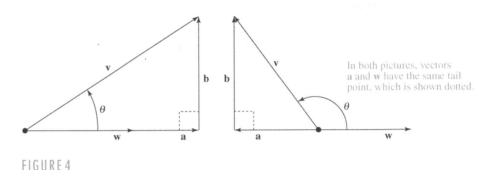

In both pictures, vectors a and w have the same tail point, which is shown dotted.

FIGURE 4
Projecting one vector onto another

The vector labeled **a** in each diagram is called the **vector projection** of $\mathbf{v}$ on $\mathbf{w}$; another common name for **a** is the **vector component** of $\mathbf{v}$ in the direction of $\mathbf{w}$. Notice that **a** and $\mathbf{w}$ point in either the same or in opposite directions, depending on whether the angle θ between $\mathbf{v}$ and $\mathbf{w}$ is acute or obtuse. ←

What happens if $\theta = \pi/2$?

The **scalar projection** of $\mathbf{v}$ on $\mathbf{w}$ (also known as the **scalar component** of $\mathbf{v}$ in the direction of $\mathbf{w}$) is the magnitude $|\mathbf{a}|$ in the left-hand diagram and the opposite, $-|\mathbf{a}|$, in the right-hand diagram (in which the vector projection points opposite to $\mathbf{w}$).

Notice that (in both diagrams) the vector **a** is parallel to $\mathbf{w}$, while **b** is perpendicular to $\mathbf{w}$, and so the magnitudes satisfy the Pythagorean rule $|\mathbf{a}|^2 + |\mathbf{b}|^2 = |\mathbf{v}|^2$.

EXAMPLE 5 Use the dot product to find expressions for $|\mathbf{a}|$, **a**, and **b**, all in terms of $\mathbf{v}$ and $\mathbf{w}$.

Solution Figure 4 and basic trigonometry give $|\mathbf{a}| - |\mathbf{v}| \cos \theta$. By Theorem 4 we have $\mathbf{v} \cdot \mathbf{w} = |\mathbf{v}||\mathbf{w}| \cos \theta$. Combining these equations gives a useful formula:

$$\text{scalar projection of } \mathbf{v} \text{ on } \mathbf{w} = |\mathbf{v}| \cos \theta = \frac{\mathbf{v} \cdot \mathbf{w}}{|\mathbf{w}|}.$$

To find the *vector* **a**, we multiply the scalar projection just found by the unit vector parallel to $\mathbf{w}$:

$$\text{vector projection} = \frac{\mathbf{v} \cdot \mathbf{w}}{|\mathbf{w}|} \frac{\mathbf{w}}{|\mathbf{w}|} = \frac{\mathbf{v} \cdot \mathbf{w}}{|\mathbf{w}|^2} \mathbf{w} = \frac{\mathbf{v} \cdot \mathbf{w}}{\mathbf{w} \cdot \mathbf{w}} \mathbf{w}.$$

Figure 4 also shows that $\mathbf{a} + \mathbf{b} = \mathbf{v}$, which implies that

$$\mathbf{b} = \mathbf{v} - \mathbf{a} = \mathbf{v} - \frac{\mathbf{v} \cdot \mathbf{w}}{|\mathbf{w}|^2}\,\mathbf{w}.$$

These formulas may look a little scary, but in concrete cases everything reduces to numbers.

E X A M P L E 6   Let $\mathbf{v} = (2, 3)$ and $\mathbf{w} = (1, 1)$. Find perpendicular vectors $\mathbf{a}$ and $\mathbf{b}$ (as shown in Figure 4) such that $\mathbf{a} + \mathbf{b} = \mathbf{v}$.

Solution   Working through the formulas in Example 5 with these particular vectors gives

$$\mathbf{a} = \frac{\mathbf{v} \cdot \mathbf{w}}{|\mathbf{w}|^2}\,\mathbf{w} = \frac{(2,3) \cdot (1,1)}{2}(1,1) = \left(\frac{5}{2}, \frac{5}{2}\right).$$

Since $\mathbf{a} + \mathbf{b} = \mathbf{v}$, we must have

$$\mathbf{b} = \mathbf{v} - \mathbf{a} = \left(-\frac{1}{2}, \frac{1}{2}\right).$$

Notice, too, that $\mathbf{a}$ and $\mathbf{b}$ are indeed perpendicular. ➡

Use the dot product to convince yourself.

Distance from a point to a line   Vector projections lead to an easy way of calculating the shortest distance from any point to any line. ➡ Figure 5 shows the idea.

The shortest distance is measured perpendicular to the line.

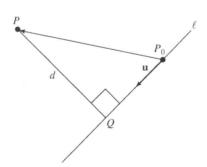

FIGURE 5
Measuring distance from a point to a line

The line ℓ is determined, as shown, by a point P_0 and a direction vector $\mathbf{u}$. Here P is a point not on ℓ. We want the distance d from P to a point Q on ℓ with the perpendicularity property shown. Now a close look at Figure 5 shows that the vector $\overrightarrow{P_0Q}$ is the projection of the vector $\overrightarrow{P_0P}$ on $\mathbf{u}$. The distance d can therefore be found using the Pythagorean rule. In symbols:

$$d^2 = |\overrightarrow{P_0P}|^2 - |\overrightarrow{P_0Q}|^2 = |\overrightarrow{P_0P}|^2 - \frac{(\overrightarrow{P_0P} \cdot \mathbf{u})^2}{\mathbf{u} \cdot \mathbf{u}}.$$

EXAMPLE 7 How far is the point $P = (2, 3)$ from the line through $P_0 = (1, 1)$ with direction vector $\mathbf{u} = (-2, 3)$? Which point Q on the line is closest to P?

Solution Here we have $\overrightarrow{P_0P} = (2, 3) - (1, 1) = (1, 2)$ and $\mathbf{u} = (-2, 3)$, and the preceding formula gives

$$d^2 = |\overrightarrow{P_0P}|^2 - \frac{(\overrightarrow{P_0P} \cdot \mathbf{u})^2}{\mathbf{u} \cdot \mathbf{u}} = 5 - \frac{16}{13},$$

which yields $d = 7/\sqrt{13} \approx 1.94$. To find the closest point (let's call it Q), we calculate that the projection of $\overrightarrow{P_0P}$ on $\mathbf{u}$ is the vector $(-8/13, 12/13)$. This implies that

$$Q = (1, 1) + \left(-\frac{8}{13}, \frac{12}{13}\right) = \left(\frac{5}{13}, \frac{25}{13}\right).$$

The dot product in three variables

The dot product of 3-vectors $\mathbf{v} = (v_1, v_2, v_3)$ and $\mathbf{w} = (w_1, w_2, w_3)$ is

$$\mathbf{v} \cdot \mathbf{w} = v_1w_1 + v_2w_2 + v_3w_3.$$

Given the similarity to the two-dimensional formula, it is not surprising that the same algebraic properties hold for dot products in three variables:

$$\mathbf{v} \cdot \mathbf{v} = |\mathbf{v}|^2; \qquad \mathbf{u} \cdot (\mathbf{v} + \mathbf{w}) = \mathbf{u} \cdot \mathbf{v} + \mathbf{u} \cdot \mathbf{w}; \qquad a\mathbf{v} \cdot \mathbf{w} = \mathbf{v} \cdot a\mathbf{w} = a(\mathbf{v} \cdot \mathbf{w}).$$

As in the two-dimensional case, these formulas are readily proved by writing everything out in coordinates. More remarkable is the fact that the same *geometric* properties of the dot product hold also in three dimensions:

THEOREM 5 Let $\mathbf{v}$ and $\mathbf{w}$ be 3-vectors, and let θ be the angle between $\mathbf{v}$ and $\mathbf{w}$. Then

$$\mathbf{v} \cdot \mathbf{w} = |\mathbf{v}|\,|\mathbf{w}|\,\cos\theta.$$

In particular, $\mathbf{v}$ and $\mathbf{w}$ are perpendicular if and only if $\mathbf{v} \cdot \mathbf{w} = 0$.

The law of cosines generalizes the Pythagorean rule. A proof is outlined in the exercises.

Proof We use the **law of cosines:** ⬅

If a, b, and c are the sides of any triangle and θ is the angle between sides a and b, then $c^2 = a^2 + b^2 - 2ab\cos\theta$.

Draw your own picture of this triangle.

To prove the theorem, we apply the law of cosines to the triangle whose sides are the vectors $\mathbf{v}$, $\mathbf{w}$, and $\mathbf{w} - \mathbf{v}$. ⬅ The triangle's sides have lengths $a = |\mathbf{v}|$, $b = |\mathbf{w}|$, and $c = |\mathbf{w} - \mathbf{v}|$, and the law of cosines gives

$$|\mathbf{w} - \mathbf{v}|^2 = |\mathbf{v}|^2 + |\mathbf{w}|^2 - 2|\mathbf{v}||\mathbf{w}|\cos\theta.$$

If we now substitute

$$|\mathbf{w} - \mathbf{v}|^2 = (\mathbf{w} - \mathbf{v}) \cdot (\mathbf{w} - \mathbf{v}) = \mathbf{w} \cdot \mathbf{w} - 2\mathbf{w} \cdot \mathbf{v} + \mathbf{v} \cdot \mathbf{v}$$
$$= |\mathbf{v}|^2 + |\mathbf{w}|^2 - 2\mathbf{w} \cdot \mathbf{v}$$

and let the symbol dust settle, the theorem follows immediately.

Unit vectors, the dot product, and direction cosines A space vector of length one is called a **unit vector**. If $\mathbf{v} = (a, b, c)$ is *any* nonzero vector, then

$$\frac{\mathbf{v}}{|\mathbf{v}|} = \frac{\mathbf{v}}{\sqrt{a^2 + b^2 + c^2}}$$

is a unit vector parallel to $\mathbf{v}$.

The dot product can help us interpret the coordinates of a unit vector geometrically. Let $\mathbf{u} = (u_1, u_2, u_3)$ be any unit vector. Then $\mathbf{u} \cdot \mathbf{i} = u_1 = \cos(\theta_1)$, where θ_1 is the angle between $\mathbf{u}$ and $\mathbf{i}$. Similarly,

$$\mathbf{u} \cdot \mathbf{j} = u_2 = \cos(\theta_2) \qquad \text{and} \qquad \mathbf{u} \cdot \mathbf{k} = u_3 = \cos(\theta_3),$$

where θ_2 and θ_3 are, respectively, the angles between $\mathbf{u}$ and $\mathbf{j}$ and between $\mathbf{u}$ and $\mathbf{k}$. For this reason the coordinates of a unit vector $\mathbf{u}$ are sometimes called the **direction cosines** of $\mathbf{u}$.

EXAMPLE 8 What angles θ_1, θ_2, and θ_3 does the vector $\mathbf{v} = (1, 2, 3)$ make with the three positive coordinate axes?

Solution Dividing $\mathbf{v}$ by $|\mathbf{v}| = \sqrt{14}$ produces the unit vector

$$\mathbf{u} = \left(\frac{1}{\sqrt{14}}, \frac{2}{\sqrt{14}}, \frac{3}{\sqrt{14}} \right),$$

which has the same direction as $\mathbf{v}$. The coordinates of $\mathbf{u}$ are cosines of the desired angles, and so

$$\theta_1 = \arccos\left(\frac{1}{\sqrt{14}}\right) \approx 1.30; \quad \theta_2 = \arccos\left(\frac{2}{\sqrt{14}}\right) \approx 1.01; \quad \theta_3 = \arccos\left(\frac{3}{\sqrt{14}}\right) \approx 0.64$$

with all answers in radians. ∎

Projections and components The idea of projecting one vector onto another makes sense in any dimension, and the formulas derived earlier continue to hold. ➤ If $\mathbf{v}$ and $\mathbf{w}$ are any vectors, then

See Example 5 and the preceding discussion for the derivation in dimension two.

$$\frac{\mathbf{v} \cdot \mathbf{w}}{|\mathbf{w}|} \qquad \text{and} \qquad \frac{\mathbf{v} \cdot \mathbf{w}}{|\mathbf{w}|^2}\mathbf{w}$$

are, respectively, the **scalar projection** and the **vector projection** of $\mathbf{v}$ onto $\mathbf{w}$. (As in two dimensions, these quantities are also known as the vector and scalar **components** of $\mathbf{v}$ in the direction of $\mathbf{w}$.)

EXAMPLE 9 Find the vector and scalar projections of $\mathbf{v} = (a, b, c)$ in each of the directions $\mathbf{i}$, $\mathbf{j}$, and $\mathbf{k}$.

Solution The projection formulas just given are especially simple if $\mathbf{w}$ is any of the standard basis vectors $\mathbf{i}$, $\mathbf{j}$, and $\mathbf{k}$. If $\mathbf{w} = \mathbf{i}$, for instance, then we have

$$\frac{\mathbf{v} \cdot \mathbf{i}}{|\mathbf{i}|} = (a, b, c) \cdot (1, 0, 0) = a \qquad \text{and} \qquad \frac{\mathbf{v} \cdot \mathbf{i}}{|\mathbf{i}|^2}\mathbf{i} = (a, 0, 0) = a\mathbf{i}.$$

Similarly, the scalar projections of $\mathbf{v}$ on $\mathbf{j}$ and $\mathbf{k}$ are b and c, respectively, and the respective vector projections are $b\mathbf{j}$ and $c\mathbf{k}$. Figure 6 shows how the parts fit together: Observe what the picture shows:

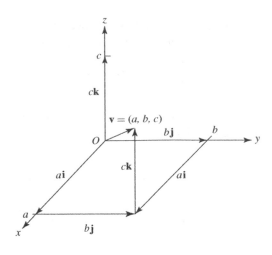

FIGURE 6
Vector and scalar projections

- **Three vectors** The vector $\mathbf{v} = (a, b, c)$ is a simple linear combination of the standard basis $\mathbf{i}$, $\mathbf{j}$, and $\mathbf{k}$:

$$\mathbf{v} = a\mathbf{i} + b\mathbf{j} + c\mathbf{k}.$$

The summands on the right are, respectively, the vector components of $\mathbf{v}$ in the $\mathbf{i}$-, $\mathbf{j}$-, and $\mathbf{k}$-directions.

- **Three scalars** The scalar projections of $\mathbf{v} = (a, b, c)$ on $\mathbf{i}$, $\mathbf{j}$, and $\mathbf{k}$ are a, b, and c—in this case, simply the *coordinates* of $\mathbf{v}$. ∎

Derivatives and the dot product

If $\mathbf{f}$ and $\mathbf{g}$ are vector-valued functions, then their dot product $\mathbf{f} \cdot \mathbf{g}$ is a new *scalar*-valued function. The derivative of the product function can be found by a rule that closely resembles the familiar product rule of elementary calculus:

> **THEOREM 6 (Dot product rule for derivatives)** Let $\mathbf{f}$ and $\mathbf{g}$ be differentiable vector-valued functions. Then the dot product function $\mathbf{f} \cdot \mathbf{g}$ is is differentiable, and
>
> $$\big(\mathbf{f}(t) \cdot \mathbf{g}(t)\big)' = \mathbf{f}'(t) \cdot \mathbf{g}(t) + \mathbf{f}(t) \cdot \mathbf{g}'(t).$$

We show the 2-D case.

Proof The theorem is proved by writing out $\mathbf{f} = (f_1, f_2)$ and $\mathbf{g} = (g_1, g_2)$ in components and using derivative rules from elementary calculus: ◂

$$
\begin{aligned}
(\mathbf{f} \cdot \mathbf{g})' &= (f_1 g_1 + f_2 g_2)' = (f_1 g_1)' + (f_2 g_2)' && \text{sum rule}\\
&= f_1' g_1 + f_1 g_1' + f_2' g_2 + f_2 g_2' && \text{ordinary product rule}\\
&= f_1' g_1 + f_2' g_2 + f_1 g_1' + f_2 g_2' && \text{rearranging summands}\\
&= \mathbf{f}' \cdot \mathbf{g} + \mathbf{f} \cdot \mathbf{g}',
\end{aligned}
$$

as desired. ∎

Vector-valued functions of constant magnitude As a striking application of the dot product rule, consider a vector-valued function $\mathbf{f}$ for which the magnitude $|\mathbf{f}(t)|$ is a constant, say k; then $\mathbf{f}(t) \cdot \mathbf{f}(t) = k^2$. The right side is constant, and so its derivative is zero. The left side can be differentiated using the dot product rule. The result is

$$\mathbf{f}'(t) \cdot \mathbf{f}(t) + \mathbf{f}(t) \cdot \mathbf{f}'(t) = 2\mathbf{f}(t) \cdot \mathbf{f}'(t) = 0.$$

The last equality is interesting in its own right:

> **FACT** Let $\mathbf{f}$ be a differentiable vector-valued function with constant magnitude $|\mathbf{f}(t)|$. Then the vectors $\mathbf{f}(t)$ and $\mathbf{f}'(t)$ are perpendicular for all t.

Figure 7 illustrates the Fact for the function $\mathbf{f}(t) = (\cos t, \sin t)$. As predicted, the velocity vector $\mathbf{f}'(t)$ at each point on the circle is perpendicular to the corresponding position vector $\mathbf{f}(t)$.

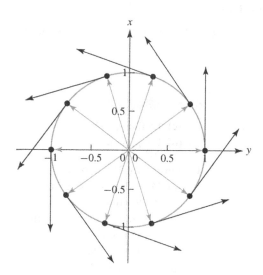

FIGURE 7

Position and velocity vectors on the unit circle

Work and the dot product

Work (in the physicist's sense) occurs when a force, say $\mathbf{F}$, acts along a displacement, say $\mathbf{d}$. (Both $\mathbf{F}$ and $\mathbf{d}$ are vector quantities.) In the simplest possible case, such as lifting a weight vertically, a *constant* force is exerted in the same direction as the displacement. In this case, the work done is the ordinary product of the magnitude of the force and the distance through which it acts. In symbols,

$$\text{work} = |\mathbf{F}|\,|\mathbf{d}|.$$

Often, however, the force $\mathbf{F}$ acts at an angle θ to the displacement $\mathbf{d}$. This occurs, for instance, when the force of gravity causes an object to slide down an inclined plane: gravity acts downward, but the displacement is oblique.

Sailing with the dot product. Sailors have relied for thousands of years on the phenomenon of forces acting obliquely to displacement. Thanks to its keel and rudder, a boat under sail normally travels at some angle to the force of the wind on its sail. Otherwise, boats could sail only straight downwind—and world history would have been much different.

The dot product $\mathbf{F} \cdot \mathbf{d}$ is just what's needed when we calculate the work done by an oblique force $\mathbf{F}$ along a displacement $\mathbf{d}$. In this case, work is done only by the vector component of the force parallel to the displacement. This vector component, as we have seen, has magnitude $|\mathbf{F}| \cos\theta$, and it acts through distance $|\mathbf{d}|$. Thus, the work done is precisely

$$\mathbf{F} \cdot \mathbf{d} = |\mathbf{d}||\mathbf{F}| \cos\theta.$$

EXAMPLE 10   An inclined plane makes an angle of $\pi/3$ radians with the downward vertical. How much work, in foot-pounds, is done by gravity as a 10-pound object slides 10 feet down the plane?

Solution   The force $\mathbf{F}$ acts in the straight downward direction with magnitude 10 pounds. The displacement $\mathbf{d}$ has magnitude 10 feet and makes an angle of $\pi/3$ with $\mathbf{F}$. The work done, therefore, is

$$\mathbf{F} \cdot \mathbf{d} = 10 \times 10 \times \cos(\pi/3) = 50 \text{ foot-pounds.} \qquad \blacksquare$$

BASIC EXERCISES

In Exercises 1–6, find the two coordinates of the unit vector in the θ-direction. (Do these by hand—only famous values of θ are involved.)

1. $\theta = 0$

2. $\theta = \pi/4$

3. $\theta = -\pi/4$

4. $\theta = 2\pi/3$

5. $\theta = 1234\pi$

6. $\theta = 1234\pi/3$

In Exercises 7–10, find a number $r > 0$ and an angle θ such that $\mathbf{v} = r\mathbf{u}_\theta$. (Do these by hand—only famous values of θ are involved.)

7. $\mathbf{v} = (2, 0)$

8. $\mathbf{v} = (2, 2)$

9. $\mathbf{v} = (\sqrt{3}, 1)$

10. $\mathbf{v} = (1, -\sqrt{3})$

11. This exercise is about Theorem 3. Using the notation $\mathbf{u} = (u_1, u_2)$, $\mathbf{v} = (v_1, v_2)$, and $\mathbf{w} = (w_1, w_2)$, part (a) is easy to show: $\mathbf{u} \cdot \mathbf{v} = u_1 v_1 + u_2 v_2 = v_1 u_1 + v_2 u_2 = \mathbf{v} \cdot \mathbf{u}$. (The middle step works because ordinary multiplication is commutative.) Do the parts below in the same spirit.

    (a) Show part (b) of the theorem.

    (b) Show part (c) of the theorem.

12. Let $\mathbf{v} = (3, 2)$ and $\mathbf{w} = (1, 0)$.

    (a) Find the vector $\mathbf{a}$ that is the projection of $\mathbf{v}$ in the $\mathbf{w}$-direction.

    (b) Find the vector $\mathbf{b}$ that is the projection of $\mathbf{v}$ perpendicular to the $\mathbf{w}$-direction.

    (c) Evaluate $\mathbf{a} \cdot \mathbf{b}$.

    (d) Evaluate $\mathbf{a} + \mathbf{b}$.

13. Let $\mathbf{v} = (3, 2)$ and $\mathbf{w} = (0, 1)$.

    (a) Find the vector $\mathbf{a}$ that is the projection of $\mathbf{v}$ in the $\mathbf{w}$-direction.

    (b) Find the vector $\mathbf{b}$ that is the projection of $\mathbf{v}$ perpendicular to the $\mathbf{w}$-direction.

    (c) Evaluate $\mathbf{a} \cdot \mathbf{b}$.

    (d) Evaluate $\mathbf{a} + \mathbf{b}$.

14. Let $\mathbf{v} = (3, 2)$ and $\mathbf{w} = (1, 1)$.

    (a) Find the vector $\mathbf{a}$ that is the projection of $\mathbf{v}$ in the $\mathbf{w}$-direction.

    (b) Find the vector $\mathbf{b}$ that is the projection of $\mathbf{v}$ perpendicular to the $\mathbf{w}$-direction.

    (c) Evaluate $\mathbf{a} \cdot \mathbf{b}$.

    (d) Evaluate $\mathbf{a} + \mathbf{b}$.

15. Suppose that a 10-pound object is moved by gravity down a 10-foot inclined plane tilted at angle $\pi/4$ to the vertical.

    (a) How much work is done by gravity?

    (b) How far, vertically, does the object descend?

    (c) How much work would have been done if the ramp were twice as long but had the same vertical drop?

16. Suppose that a 20-pound object is moved by gravity down a 5-foot inclined plane tilted at angle $\pi/6$ to the vertical.

    (a) How much work is done by gravity?

    (b) How far, vertically, does the object descend?

    (c) How much work would have been done if the ramp were twice as long?

In Exercises 17–20, show that the position vector $\mathbf{f}(t)$ *and the velocity vector* $\mathbf{f}'(t)$ *are perpendicular for all values of* t.

17. $\mathbf{f}(t) = (\cos t, \sin t); \quad 0 \le t \le 2\pi$

18. $\mathbf{f}(t) = (t, \sqrt{4 - t^2}); \quad -2 < t < 2$

19. $\mathbf{f}(t) = (\sin(2t), \cos(2t)); \quad 0 \le t \le \pi$

20. $\mathbf{f}(t) = (\sin(t^2), \cos(t^2)); \quad 0 \le t \le \sqrt{2\pi}$

21. What do the four curves in Exercises 17–20 have in common geometrically?

22. Let $\mathbf{u} = (1, -2)$ and $\mathbf{v} = (-3, 5)$. Find the angle between the vectors $\mathbf{u}$ and $\mathbf{v}$. [HINT: Write your answer as the value of an inverse trigonometric function.]

23. Find a vector of length 3 that is perpendicular to $\mathbf{u} = (1, 2)$. How many such vectors are there?

24. Find a vector of length 4 that is perpendicular to $\mathbf{u} = (1, 2, 3)$. How many such vectors are there?

25. Are the vectors $\mathbf{u} = (1, -2, 3)$ and $\mathbf{v} = (-3, 0, 1)$ perpendicular? Justify your answer.

26. Find the vector projection of $\mathbf{u} = (2, -3, 4)$ in the direction $\mathbf{v} = (1, 1, \sqrt{2})$.

27. Let $\mathbf{v} = (1, 2, 3)$.
   (a) Find the length $|\mathbf{v}|$.
   (b) Find the cosine of the angle that $\mathbf{v}$ makes with each of the standard basis vectors $\mathbf{i}$, $\mathbf{j}$, and $\mathbf{k}$.

28. Repeat Exercises 27 using $\mathbf{v} = (1, -2, 3)$.

In Exercises 29–32, find the distance from

29. the point $P = (2, 3)$ to the line ℓ through $(0, 0)$ with direction vector $\mathbf{i}$.

30. the point $P = (2, 3)$ to the line ℓ through $(1, 1)$ with direction vector $(-1, 1)$.

31. the point $P = (2, 3)$ to the line ℓ through $(1, 1)$ and $(0, 2)$.

32. the origin to the line with Cartesian equation $y = mx + b$. (Assume that $b \ne 0$.)

FURTHER EXERCISES

33. (a) Suppose that the vectors $\mathbf{u}$ and $\mathbf{v}$ have the same length. Show that the vectors $\mathbf{u} + \mathbf{v}$ and $\mathbf{u} - \mathbf{v}$ are orthogonal.
   (b) Show that an angle inscribed in a semicircle is a right angle. (Such an angle has its vertex on the circle, and its sides pass through the ends of a diameter of the circle.) [HINT: Let $\mathbf{x}$ be the vector from the center of the circle to the vertex of the angle, and let $\mathbf{y}$ be the vector from the center of the circle to the point on the circle intersected by one of the sides of the angle.]

34. Show that the line segment connecting the midpoints of two sides of a triangle is parallel to, and half as long as, the third side.

35. Suppose that $\mathbf{x}(t)$ and $\mathbf{x}'(t)$ are always perpendicular. Show that $|\mathbf{x}(t)|$ is constant. [HINT: Show that $|\mathbf{x}(t)|^2$ is constant.]

36. A rectangle has dimensions 2 by 3. Find the angles between its sides and a diagonal.

37. Show that the diagonals of a parallelogram are perpendicular if and only if all the sides of the parallelogram are equal.

38. Show that an equilateral triangle also has all angles equal.

39. Suppose that $\mathbf{r}(t)$ is a differentiable vector-valued function and that $\mathbf{r}(t) \ne \mathbf{0}$ for all t. Show that

$$\frac{d}{dt}\left(\frac{\mathbf{r}}{|\mathbf{r}|}\right) = \frac{\mathbf{r}'}{|\mathbf{r}|} - \frac{\mathbf{r} \cdot \mathbf{r}'}{|\mathbf{r}|^3}\mathbf{r}.$$

40. Suppose that the path of a particle in space is described by a differentiable curve and that the particle's speed is always nonzero. Show that the particle's velocity and acceleration vectors are perpendicular whenever the particle's speed is a local maximum or a local minimum.

41. Find the angle between the diagonal of a cube and any one of its edges.

42. The curves $\mathbf{x}(t) = (1 + t, t^2, t^3)$ and $\mathbf{y}(s) = (\sin s, \cos s, s - \pi/2)$ intersect at the point $(1, 0, 0)$. Find the angle between the curves at this point (i.e., find the angle between their tangent vectors).

43. Suppose that P, Q, and R are points in space. Explain how the dot product can be used to determine whether these points are collinear. [HINT: Consider the vectors $\overrightarrow{PQ}$ and $\overrightarrow{PR}$.]

44. Is the angle between the vectors $(1, -2, 3)$ and $(-3, 2, 1)$ less than $\pi/2$? Justify your answer.

45. This exercise is about applying the Fact on page 677 to the vector-valued function $\mathbf{f}(t) = (\cos t, \sin t)$.
   (a) Find $\mathbf{f}(t)$ and $\mathbf{f}'(t)$ for $t = 0$, $t = \pi$, $t = \pm\pi/2$, and $t = \pi/4$.
   (b) Plot all the vectors you found in the previous part along with the unit circle. (Plot the vectors $\mathbf{f}(t)$ with tails at the origin; plot each vector $\mathbf{f}'(t)$ with its tail at the appropriate point on the unit circle.) Is your picture consistent with the Fact?

46. Repeat Exercise 45 using $\mathbf{f}(t) = (\sin t, \cos t)$.

47. Let $\mathbf{f}(t) = (t, \sin t)$.
   (a) Evaluate $\mathbf{f}(t)$ and $\mathbf{f}'(t)$ for $t = 0$, $t = \pi/4$, $t = \pi/2$, and $t = \pi$.
   (b) Plot the vectors you found in part (a).
   (c) The vectors $\mathbf{f}(t)$ and $\mathbf{f}'(t)$ are not perpendicular for every value of t. Why doesn't this contradict the Fact on page 677?

48. Let $\mathbf{v} = (1, 2, 3)$ and $\mathbf{w} = (2, 3, -4)$.

    (a) Find the cosine of the angle θ between $\mathbf{v}$ and $\mathbf{w}$. Is θ obtuse or acute (i.e., more or less than a right angle)? How do you know?

    (b) Find the vector projection of $\mathbf{v}$ in the direction of $\mathbf{w}$.

    (c) Find a nonzero vector $\mathbf{x} = (x_1, x_2, x_3)$ that is perpendicular to *both* $\mathbf{v}$ and $\mathbf{w}$. [HINT: To be perpendicular to $\mathbf{v}$, the components of $\mathbf{x}$ must satisfy $x_1 + 2x_2 + 3x_3 = 0$. Use this idea to set up two equations in the three unknowns x_1, x_2, and x_3. Then solve the two equations simultaneously. There are infinitely many solutions.]

49. Repeat Exercise 48 using $\mathbf{v} = (1, 2, 0)$ and $\mathbf{w} = (2, 3, 0)$. [NOTE: The hint given in part (c) of Exercise 48 doesn't apply directly.]

50. Let $\mathbf{v} = (v_1, v_2, v_3)$ and $\mathbf{w} = (w_1, w_2, w_3)$ be two vectors. Use $\mathbf{v}$ and $\mathbf{w}$ to define a third vector $\mathbf{u}$ by the formula

$$\mathbf{u} = (v_2 w_3 - v_3 w_2, \ v_3 w_1 - v_1 w_3, \ v_1 w_2 - v_2 w_1).$$

    (a) Show that $\mathbf{u}$ is perpendicular to $\mathbf{v}$ and to $\mathbf{w}$.

    (b) Show that

$$|\mathbf{u}|^2 = |\mathbf{v}|^2 |\mathbf{w}|^2 - (\mathbf{v} \cdot \mathbf{w})^2 = |\mathbf{v}|^2 |\mathbf{w}|^2 |\sin \theta|^2,$$

    where θ is the (positive) angle between $\mathbf{v}$ and $\mathbf{w}$. [HINT: It's a slightly long-winded calculation, but it works out in the end.]

    [NOTE: The vector $\mathbf{u}$ in this exercise is called the **cross product** of $\mathbf{v}$ and $\mathbf{w}$. We study the cross product in more detail in Section 12.9.]

In Exercises 51–55, let $\mathbf{v} = (v_1, v_2, v_3)$ and $\mathbf{w} = (w_1, w_2, w_3)$ be vectors, and let a be any scalar.

51. Show that $|a\mathbf{v}| = |a||\mathbf{v}|$.

52. Show that $\mathbf{v} \cdot \mathbf{v} = |\mathbf{v}|^2$.

53. Show that $|\mathbf{v} + \mathbf{w}| \le |\mathbf{v}| + |\mathbf{w}|$. [HINT: Show first that $|\mathbf{v} + \mathbf{w}|^2 = (\mathbf{v} + \mathbf{w}) \cdot (\mathbf{v} + \mathbf{w}) \le (|\mathbf{v}| + |\mathbf{w}|)^2.$]

54. Show that $\mathbf{u} \cdot (\mathbf{v} + \mathbf{w}) = \mathbf{u} \cdot \mathbf{v} + \mathbf{u} \cdot \mathbf{w}$.

55. Show that $a(\mathbf{v} \cdot \mathbf{w}) = (a\mathbf{v}) \cdot \mathbf{w}$.

56. Show the **law of cosines**: If a triangle has sides a, b, and c, with angle θ between sides a and b, then $c^2 = a^2 + b^2 - 2ab \cos \theta$. [HINT: The sides of the triangle can be represented as vectors $\mathbf{a}$, $\mathbf{b}$, $\mathbf{c}$ such that $\mathbf{a} = \mathbf{b} + \mathbf{c}$, $|\mathbf{a}| = a$, $|\mathbf{b}| = b$, and $|\mathbf{c}| = c$.]

57. Let $\mathbf{v}$ and $\mathbf{w}$ be any nonzero vectors. Let $\mathbf{a}$ be the projection

of $\mathbf{v}$ in the direction of $\mathbf{w}$, and let $\mathbf{b} = \mathbf{v} - \mathbf{a}$. Draw a picture to show the situation. Then show, using the dot product, that $\mathbf{b}$ is perpendicular to $\mathbf{w}$.

In Exercises 58 and 59, $\mathbf{u}$ and $\mathbf{v}$ are the vectors in the xz-plane pictured below, $|\mathbf{u}| = 3$, $|\mathbf{v}| = 4$, and $\phi = \pi/3$.

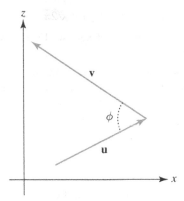

58. Evaluate $\mathbf{u} \cdot \mathbf{v}$.

59. Evaluate $|\mathbf{u} + \mathbf{v}|$.

In Exercises 60 and 61, $\mathbf{u}$ and $\mathbf{v}$ are the vectors in the yz-plane pictured below, $|\mathbf{u}| = 4$, $|\mathbf{v}| = 3$, and $\phi = \pi/3$.

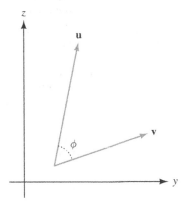

60. Evaluate $\mathbf{u} \cdot \mathbf{v}$.

61. Evaluate $|\mathbf{u} - \mathbf{v}|$.

62. Suppose that $\mathbf{x}(t) = (\cos t, \sin t, t)$, and $\mathbf{y}(t) = (2, t^2, -4t)$. Evaluate $\dfrac{d}{dt}\big(\mathbf{x}(t) \cdot \mathbf{y}(t)\big)$.

12.8 LINES AND PLANES IN THREE DIMENSIONS

In Section 12.5, for example

Lines and planes have already appeared informally in several settings. We have used derivatives to calculate tangent lines to curves ← and have seen that the graph of a linear equation $ax + by + cz = d$ is a *plane* in xyz-space. In this section we focus specifically on lines and planes in space. Our newly acquired vector tools and notation will simplify and streamline our work.

Lines

A line in *any* dimension is completely determined by (i) a **point** P_0 through which it passes and (ii) a **direction vector v**. In practice, one or both of these data may need to be ferreted out from other sorts of information. A direction vector, for instance, is sometimes deduced from knowledge of two points on the line.

Lines and segments in $\mathbb{R}^3$ can be described in various ways. We collect several below, with examples.

Vector equations for lines Let ℓ be the line through $P_0 = (x_0, y_0, z_0)$ in the direction of the vector $\mathbf{v} = (a, b, c)$. A point $X = (x, y, z)$ lies on ℓ if and only if the vector joining X to P_0 is a scalar multiple, say $t\mathbf{v}$, of the direction vector. This is another way of saying that the line is the image of the vector-valued function $\mathbf{X}(t)$ defined for all real numbers t by

$$\mathbf{X}(t) = \mathbf{P_0} + t\mathbf{v},$$

where $\mathbf{P_0} = (x_0, y_0, z_0)$ is the position vector of the point P_0. ➧ This equation is called a **vector equation**, or a **vector parametrization**, for ℓ.

The difference between the point P_0 and the position vector $\mathbf{P_0}$—note the boldface type for the latter—is mainly in our point of view.

We said "a" rather than "the" in the preceding sentence because more than one vector equation can determine the same line. This should not be too surprising: we have seen already that *every* curve can be parametrized in many different ways.

> **EXAMPLE 1** Explain why the two vector equations
>
> $$\mathbf{X}(t) = (0, 0, 0) + t(1, 2, 3) \qquad \text{and} \qquad \mathbf{Y}(t) = (1, 2, 3) + t(-2, -4, -6)$$
>
> describe the same line. Is there *any* difference between the two equations? ➧

There's nothing sacred about the letter t. Soon we'll use other letters.

> **Solution** Let ℓ_1 and ℓ_2 be the lines determined by $\mathbf{X}(t)$ and $\mathbf{Y}(t)$. To show that ℓ_1 and ℓ_2 are the same, it is enough to find any two points in common. ➧ In fact, it is easy to see that $(0, 0, 0)$ lies on both ℓ_1 and ℓ_2 (set $t = 0$ in the first equation and $t = 1/2$ in the second). Similarly, $(1, 2, 3)$ lies on both lines (set $t = 1$ in the first equation and $t = 0$ in the second).

Only one line passes through two given points.

> Though both vector-valued functions define the same line, they differ in other ways. One difference is in the respective velocities:
>
> $$\mathbf{X}'(t) = (1, 2, 3) \qquad \text{but} \qquad \mathbf{Y}'(t) = (-2, -4, -6).$$
>
> In effect, the second equation parametrizes the line at twice the speed of the first and in the opposite direction.   ∎

Parametric equations for lines Let ℓ (again) be the line through $P_0 = (x_0, y_0, z_0)$ with direction vector $\mathbf{v} = (a, b, c)$. Writing out the vector equation $\mathbf{X}(t) = \mathbf{P_0} + t\mathbf{v}$ in its three coordinates gives **parametric equations** ➧ for ℓ:

The parameter is t.

$$x = x_0 + at; \qquad y = y_0 + bt; \qquad z = z_0 + ct.$$

> **EXAMPLE 2** At what point does the line ℓ through $(1, 2, 3)$ and $(3, 5, 7)$ intersect the xy-plane? Where does ℓ intersect the plane Π with equation $z = x + y - 4$?

> **Solution** To describe ℓ, we need a point and a direction vector. Given the information at hand, we may as well use $(1, 2, 3)$ as our point and the difference vector $(3, 5, 7) - (1, 2, 3) = (2, 3, 4)$ as our direction vector. These data determine parametric equations for ℓ:
>
> $$x = 1 + 2t; \qquad y = 2 + 3t; \qquad z = 3 + 4t.$$
>
> Now ℓ pierces the xy-plane where $z = 3 + 4t = 0$, that is, at $t = -3/4$. For this t the

parametric equations give

$$x = 1 - 2 \cdot \frac{3}{4} = -\frac{1}{2} \quad \text{and} \quad y = 2 - 3 \cdot \frac{3}{4} = -\frac{1}{4},$$

and so ℓ pierces the xy-plane at $(-1/2, -1/4, 0)$.

Check these details for yourself.

To find where ℓ intersects Π, we can substitute $x = 1 + 2t$, $y = 2 + 3t$, and $z = 3 + 4t$ into the plane equation $z = x + y - 4$. The resulting equation has just one variable, t, and its solution is $t = 4$. ← The intersection occurs at $(9, 14, 19)$.   ▪

Symmetric scalar equations   Given the parametric form

$$x = x_0 + at; \qquad y = y_0 + bt; \qquad z = z_0 + ct$$

of a line ℓ through (x_0, y_0, z_0) with direction vector (a, b, c), we can solve all three equations for t to get

$$t = \frac{x - x_0}{a} = \frac{y - y_0}{b} = \frac{z - z_0}{c}.$$

We've seen other instances of "eliminating the variable" t.

The value of t is now immaterial, so we drop t. ← The remaining equations,

$$\frac{x - x_0}{a} = \frac{y - y_0}{b} = \frac{z - z_0}{c},$$

are known as **symmetric scalar equations** for ℓ. This form exhibits some minor behavioral quirks:

- **No t in sight**   No parameter t appears; instead, ℓ is defined as the solution set of *two* equations in the *three* variables x, y, and z. (These numbers might have been guessed by readers with some linear algebra background: *One* linear equation determines a plane in $\mathbb{R}^3$, and so *two* linear equations determine the intersection of two planes—a line in $\mathbb{R}^3$.)

- **Vanishing denominators**   If any of a, b, and c happens to be zero, then the symmetric form needs a little help, but no real harm is done. If, say, $a = 0$, then the parametric equations become

$$x = x_0; \qquad y = y_0 + bt; \qquad z = z_0 + ct;$$

eliminating t from the last *two* equations gives

$$x = x_0; \qquad \frac{y - y_0}{b} = \frac{z - z_0}{c}.$$

As before, the result is *two* linear equations in x, y, and z.

- **Not unique**   As with the vector and parametric forms of a line, and for exactly the same reasons, different sets of symmetric scalar equations can be given for the same line. In math-speak, the symmetric scalar equations are "not unique."

EXAMPLE 3   Find and compare symmetric scalar equations for the following lines:

$$\ell_1: \text{ through } (0, 0, 0) \text{ in the direction of } (1, 2, 3)$$
$$\ell_2: \text{ with vector equation } \mathbf{X}(t) = (1, 2, 3) + t(-2, -4, -6).$$

Solution   Line ℓ_1 perfectly fits the pattern above; its symmetric scalar equations are

$$\frac{x - 0}{1} = \frac{y - 0}{2} = \frac{z - 0}{3}.$$

Line ℓ_2 passes through $(1, 2, 3)$, with direction vector $(-2, -4, -6)$, and so its symmetric scalar equations are

$$\frac{x - 1}{-2} = \frac{y - 2}{-4} = \frac{z - 3}{-6}.$$

Thus, ℓ_1 and ℓ_2 have different-looking equations. Nevertheless (as we saw in Example 1), ℓ_1 and ℓ_2 are actually the same line. ▪

Parallel, skew, and intersecting lines Two lines in the xy-plane are called parallel if their slopes are equal. This definition works fine in single-variable calculus, but it works less well in multivariable calculus. It isn't even clear, for example, what the idea of "slope" should mean in three dimensions.

Vectors to the rescue! Using them we can define and quickly detect parallel lines in *any* dimension:

> Two lines are **parallel** if their direction vectors are scalar multiples of each other.

Although parallel lines are defined the same way in two and three dimensions, the implications of being parallel or not are different in the two cases. For example, every pair of nonparallel lines in the xy-plane intersects at exactly one point. The situation in three-dimensional space is quite different: there is plenty of "room" for two or more nonparallel lines to miss each other. (Two nonparallel lines that miss each other are called **skew lines**.)

In practice, lines in space are much more likely to miss each other than to meet, as basic geometric intuition suggests. Basic equation-counting suggests the same thing: One line in xyz-space corresponds to *two* equations in three variables, and so two lines determine *four* equations in three variables. When equations outnumber variables, solutions are unlikely but not impossible; the next example illustrates both possibilities.

E X A M P L E 4 Three lines are given in vector form as follows: ➡

$$\ell_1 : \mathbf{X}(r) = (1, 1, 1) + r(1, 2, 3);$$
$$\ell_2 : \mathbf{Y}(s) = (-3, 2, 4) + s(2, 1, 1);$$
$$\ell_3 : \mathbf{Z}(t) = (0, 0, 0) + t(1, 3, 1).$$

Why use three different variables r, s, and t? Read on.

Do any of the lines intersect? If so, where?

Solution The lines ℓ_1 and ℓ_2 meet if $\mathbf{X}(r) = \mathbf{Y}(s)$ for some values r and s. (Using different variable names lets us avoid assuming, unnecessarily, that the lines meet at equal values of the parameter.)

Writing this condition in vector form and simplifying slightly gives

$$(1, 1, 1) + r(1, 2, 3) = (-3, 2, 4) + s(2, 1, 1) \iff (4, -1, -3) = -r(1, 2, 3) + s(2, 1, 1).$$

The right-hand vector equation produces three scalar equations in two unknowns:

$$4 = -r + 2s; \qquad -1 = -2r + s; \qquad -3 = -3r + s.$$

Routine algebra shows that $r = 2$ and $s = 3$ is a solution. ➡

Convince yourself.

This means that ℓ_1 and ℓ_2 meet at the common point $(1, 1, 1) + 2(1, 2, 3) = (-3, 2, 4) + 3(2, 1, 1) = (3, 5, 7)$. (But they meet at different parameter values!)

The same method applied to ℓ_1 and ℓ_3 leads to the vector equations

$$(1, 1, 1) + r(1, 2, 3) = (0, 0, 0) + t(1, 3, 1) \iff (1, 1, 1) = -r(1, 2, 3) + t(1, 3, 1)$$

and thus to three more equations in two unknowns:

$$1 = -r + t; \qquad 1 = -2r + 3t; \qquad 1 = -3r + t.$$

Convince yourself of this, too. This time, routine algebra shows that no solutions exist. ◄

Planes

Straight lines in the xy-plane are the simplest "curves." They're useful, therefore, for describing and approximating more complicated curves. We take this point of view often in elementary calculus when we approximate a smooth curve by its tangent line at a point. Indeed, it is an important fact of single-variable calculus that *every* smooth curve can be closely approximated near every point by a straight line. ◄

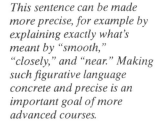

This sentence can be made more precise, for example by explaining exactly what's meant by "smooth," "closely," and "near." Making such figurative language concrete and precise is an important goal of more advanced courses.

Planes in xyz-space—tangent planes, especially—play similar roles in multivariable calculus. Planes, being "flat," are the simplest surfaces in $\mathbb{R}^3$. They're useful, therefore, for modeling and approximating more complicated surfaces and functions. First, however, we need simple, convenient ways to represent planes, using such familiar ingredients as algebraic formulas and parametric equations. Like lines, planes can be described in various ways; we explore some of them.

Planes described by points and normal vectors   One way to determine a plane Π in $\mathbb{R}^3$ is to specify (i) a point P_0 through which Π passes and (ii) a **normal vector n**, which is perpendicular to Π. ◄ Figure 1 shows the idea:

"Normal" is a rough synonym for "perpendicular."

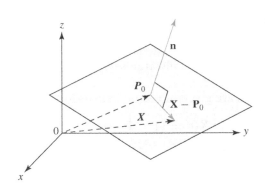

FIGURE 1

Determining a plane by a point and a normal vector

As the picture suggests, the plane consists of all points for which the vector $\mathbf{X} = (x, y, z)$ has the property that $\mathbf{X} - \mathbf{P_0}$ is perpendicular to $\mathbf{n}$. The plane's defining equation, therefore, naturally involves the dot product, which detects perpendicularity.

> **DEFINITION**   Suppose that the plane Π passes through $P_0 = (x_0, y_0, z_0)$ and has normal vector $\mathbf{n} = (a, b, c)$. Then
>
> $$(\mathbf{X} - \mathbf{P_0}) \cdot \mathbf{n} = 0.$$
>
> is a **vector equation** for Π. Rewriting the equation in scalar form gives
>
> $$a(x - x_0) + b(y - y_0) + c(z - z_0) = 0,$$
>
> a **scalar equation** for Π.

Observe:

- **One linear equation**   The scalar equation for Π can be rewritten in the form $ax + by + cz = d$ (where d is the constant $ax_0 + by_0 + cz_0$). In Section 12.1 we called such equations **linear equations** and observed that every linear equation in x, y, and z determines a plane in xyz-space.

- **Not unique**   These equations are *not* uniquely determined by a plane Π. Any point P_0 on Π does as well as any other, and we can multiply the normal vector **n** by any nonzero scalar without changing its direction. It is easy to see, for instance, that the equations $x + y + z = 1$ and $2x + 2y + 2z = 2$ define the same plane. ➡

- **Reading information from the equation**   The preceding discussion includes a handy property of a linear equation $ax + by + cz = d$:

  The coefficients of x, y, and z are the components of a normal vector (a, b, c) to the plane described by the equation.

We will apply this property in the next example.

It's also easy to see how to change one equation into the other.

EXAMPLE 5   Two planes Π_1 and Π_2 are defined by linear equations:

$$\Pi_1: \quad x + 2y + 3z = 6; \qquad \Pi_2: \quad y = 3.$$

For each plane, find a normal vector and a point on the plane; use them to write a vector equation for each plane. Describe Π_1 and Π_2 geometrically.

Solution   The italicized remark above lets us simply read off suitable normal vectors: $\mathbf{n} = (1, 2, 3)$ is normal to Π_1 and $\mathbf{n} = (0, 1, 0) = \mathbf{j}$ is normal to Π_2. Finding a point on each plane is also easy. Indeed, a linear equation in three unknowns has *infinitely many* solutions, and we need just one. A little guessing reveals that $(1, 1, 1)$ satisfies equation Π_1, while $(0, 3, 0)$ lies on Π_2. Putting these data together gives the desired vector equations:

$$\Pi_1: \quad \big(\mathbf{X} - (1, 1, 1)\big) \cdot (1, 2, 3) = 0; \qquad \Pi_2: \quad \big(\mathbf{X} - (0, 3, 0)\big) \cdot (0, 1, 0) = 0.$$

Figure 2 shows the first-octant part of Π_1:

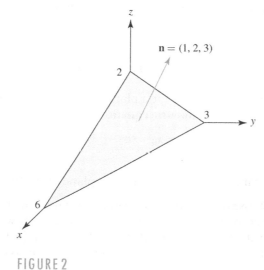

FIGURE 2
The plane $x + 2y + 3z = 6$

The plane Π_2 should be easy to visualize, and we leave it to you.   ■

We describe such subsets by restricting the parameter interval.

Parametrizing planes and parts of planes We have described lines in space both by equations in x, y, and z and in parametric form. One advantage of the parametric form is in describing just *part* of a line, such as a ray or a line segment. ◄

Planes (and parts of planes) can also be described parametrically—but we will need two parameters, not one. Figure 3 suggests the idea:

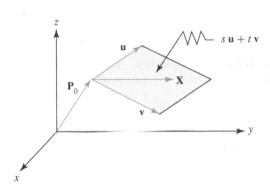

FIGURE 3
Parametrizing a plane patch

In the picture, the shaded piece (or "patch") of a plane is determined by (i) a point P_0 (indicated by the position vector $\mathbf{P_0}$); and (ii) two **spanning vectors** $\mathbf{u}$ and $\mathbf{v}$. A point on the plane has a position vector of the form

$$\mathbf{X}(s,t) = \mathbf{P_0} + s\mathbf{u} + t\mathbf{v},$$

where s and t are real numbers. (One such point X is shown in the picture.) In other words, points in the plane differ from $\mathbf{P_0}$ by a **linear combination** of $\mathbf{u}$ and $\mathbf{v}$. As s and t range through all real numbers, the vector $\mathbf{P_0} + s\mathbf{u} + t\mathbf{v}$ traces out the entire plane spanned by $\mathbf{u}$ and $\mathbf{v}$. Restricting the values of s and t, on the other hand, gives various subsets of the plane. The shaded patch corresponds to restricting s and t to the intervals $0 \le s \le 1$ and $0 \le t \le 1$. ◄

What difference would it make if we took $0 \le s \le 2$ and $0 \le t \le 2$?

The equation

$$\mathbf{X}(s,t) = \mathbf{P_0} + s\mathbf{u} + t\mathbf{v}$$

is called a **vector parametric equation** for a plane; looking at the components separately gives the corresponding **scalar parametric equations**:

$$x = x_0 + su_1 + tv_1; \qquad y = y_0 + su_2 + tv_2; \qquad z = z_0 + su_3 + tv_3.$$

EXAMPLE 6 Write parametric equations for the plane Π_1 shown in Example 5.

Solution We will choose a point and two spanning vectors on Π_1. (There are many ways to do this—our choices are not sacred.) As the picture shows, the plane passes through the points $(6, 0, 0)$, $(0, 3, 0)$, and $(0, 0, 2)$. We will use $(6, 0, 0)$ as our fixed point and the vectors joining $(6, 0, 0)$ to the other two known points as our spanning vectors. These spanning vectors are, respectively,

$$\mathbf{u} = (-6, 3, 0) \qquad \text{and} \qquad \mathbf{v} = (-6, 0, 2).$$

Thus, our plane can be written in vector parametric form as

$$\mathbf{X}(s, t) = (6, 0, 0) + s(-6, 3, 0) + t(-6, 0, 2).$$

In scalar form, we get

$$x = 6 - 6s - 6t; \qquad y = 3s; \qquad z = 2t.$$

By restricting the values of s and t, we can parametrize as much or as little of the plane as we wish. ■

Planes vs. lines The function $\mathbf{X}(s, t)$ above is a vector-valued function of *two* variables s and t. Notice the close resemblance between the vector parametrizations of lines and planes:

$$\text{line: } \mathbf{X}(t) = \mathbf{P_0} + t\mathbf{v}; \qquad \text{plane: } \mathbf{X}(s, t) = \mathbf{P_0} + s\mathbf{u} + t\mathbf{v}.$$

The main difference is in the number of parameters. In effect, lines and planes are both "flat" objects, but it takes *one* vector to span a line and *two* vectors to span a plane.

BASIC EXERCISES

In Exercises 1–4, find a vector parametric equation for the line in $\mathbb{R}^3$ and then plot the line over a convenient parameter interval.

1. The y-axis.

2. The line through $(1, 2, 3)$ and $(2, 3, 4)$.

3. The line with symmetric scalar equations

$$\frac{x-1}{2} = \frac{y-2}{3} = \frac{z-3}{4}.$$

4. The line tangent to the curve $x = \cos t$, $y = \sin t$, $z = t$ at the point $(1, 0, 0)$ (i.e., where $t = 0$).

In Exercises 5–8, decide whether the given lines intersect. If so, find the point of intersection; if not, explain why not.

5. The lines with parametric equations $\mathbf{X}(t) = (1, 2, 3) + t(1, 1, 1)$ and $\mathbf{Y}(s) = (5, 3, 5) + s(-3, 6, 3)$.

6. The x-axis and the line with symmetric scalar equations

$$\frac{x-1}{2} = \frac{y-2}{3} = \frac{z-3}{4}.$$

7. The lines ℓ_2 and ℓ_3 of Example 4.

8. The z-axis and the line through $(1, 2, 3)$ and $(2, 4, 3)$.

In Exercises 9–14, find vector and scalar equations for the plane.

9. The plane through $(1, 2, 3)$ that is perpendicular to $\mathbf{n} = (3, 4, 5)$.

10. The xz-plane. [Hint: Find a suitable normal vector first.]

11. The plane through $(1, 2, 3)$ with normal vector parallel to the line through $(0, 1, 2)$ and $(3, 3, 3)$.

12. The plane through $(0, 0, 0)$ that is perpendicular to the line with symmetric scalar equations

$$\frac{x-1}{2} = \frac{y-2}{3} = \frac{z-3}{4}.$$

13. The graph of the function $L(x, y) = 2x + 3y + 5$.

14. The set of points (x, y, z) for which $x = y$.

15. In Example 6 we found the scalar parametric equations $x = 6 - 6s - 6t$; $y = 3s$; $z = 2t$ for a certain plane. Eliminate the variables s and t in the three equations above to get one equation in x, y, and z.

16. The plane $x + 2y + 3z = 6$ (shown in Example 5) passes through $(1, 1, 1)$. Using this point as $\mathbf{P_0}$, write vector and scalar parametric equations for the plane. (Use the method of Example 6.)

In Exercises 17–21, write parametric equations for the plane (or plane piece).

17. The xy-plane.

18. The plane through $(1, 2, 3)$ spanned by $\mathbf{i}$ and $\mathbf{j}$.

19. The plane through $(1, 2, 3)$ spanned by $\mathbf{i}$ and $\mathbf{i} + \mathbf{j}$.

20. The first quadrant of the xy-plane.

21. The rectangle $0 \le x \le 1$, $0 \le y \le 2$ in the xy-plane.

22. Find an equation of the line through the point $(3, 2, 1)$ that is perpendicular to the plane $x + y = 0$.

23. Does the line with vector equation $\mathbf{X}(t) = (1, 1, 2) + t(2, 3, 4)$ intersect the plane $x + 2y - 2z = 0$? If so, where? If not, why not?

24. Let ℓ be the line $(2, 1, 0) + t(1, 1, 1)$ and Π be the plane $x - 3y + 2z = 4$. Do ℓ and Π intersect? Justify your answer.

25. Find the line of intersection of the two planes $x + 2y + 3z = 6$ and $x + y + z = 3$. Write the answer in parametric form. [HINT: First find two points that are on both planes.]

26. Let $\mathbf{v}$ be a vector and $\mathbf{u}$ be a unit vector.

    (a) Show that the vector $\mathbf{w} = \mathbf{v} - (\mathbf{v} \cdot \mathbf{u})\mathbf{u}$ is perpendicular to $\mathbf{u}$.

    (b) Let $\mathbf{x} = (1, 2, 3)$. Use part (a) to write the vector $(4, 5, 6)$ as the sum of a vector parallel to $\mathbf{x}$ and a vector perpendicular to $\mathbf{x}$.

27. Does the line $(2, -3, 1) + t(1, 2, -3)$ intersect the line $x = 1 - 2s$, $y = 2 + 3s$, $z = 4 + 6s$? Justify your answer.

28. Find an equation for the line through $(1, 1, 3)$ that is parallel to the line $x = 2 - t$, $y = -t$, $z = 3 + 3t$.

29. Are the lines $(0, -2, 0) + t(1, -2, -3)$ and $(0, 1, 0) + s(-1, -2, 1)$ perpendicular? Justify your answer.

30. Find the scalar equation of the plane through $(1, -2, 5)$ perpendicular to $\mathbf{n} = (3, -4, 1)$.

31. The line through the origin perpendicular to a plane intersects the plane at the point $(2, -1, 1)$. Find an equation of the plane.

32. A plane through $(2, -2, 5)$ is perpendicular to the line through $(1, 1, 1)$ and $(-2, 3, 1)$. Find a scalar equation of the plane.

33. Find an equation of the plane through $(1, 2, 3)$ that is parallel to the plane described by the equation $-x + 3y - 4z = 2$.

34. Let ℓ be the line $(2, 1, 1) + t(-1, 3, 2)$ and Π be the plane $x - 3y - 2z = 11$.

    (a) Find the point where ℓ and Π intersect.

    (b) Is ℓ perpendicular to Π? Justify your answer.

35. Does the line $(3, 1, -2) + t(1, -1, 3)$ intersect the plane $2x - y - z = 5$? Justify your answer.

36. Show that if the line $\mathbf{x_0} + t\mathbf{v}$ and the plane $\mathbf{n} \cdot (\mathbf{x} - \mathbf{x_1}) = 0$ do not intersect, then $\mathbf{n} \cdot \mathbf{v} = 0$.

37. Let P_0 and P_1 be points in space, $\mathbf{P}_0$ and $\mathbf{P}_1$ the corresponding vectors, $\mathbf{n}$ a unit vector, and Π the plane $\mathbf{n} \cdot (\mathbf{X} - \mathbf{P}_0) = 0$. Show that the (perpendicular) distance between the point P_1 and the plane Π is $d = |\mathbf{n} \cdot \mathbf{P}_0 - \mathbf{n} \cdot \mathbf{P}_1|$.

38. Find an equation for the plane containing the lines $(2, 0, -1) + s(3, -3, -1)$ and $(3, -1, 1) + t(1, -1, 2)$.

39. Find an equation of the plane that contains the line $(2, 3, 0) + t(-1, 2, 4)$ and is perpendicular to the plane $x + 2y - z = 3$.

40. Show that the distance D from the point $P_0 = (x_0, y_0, z_0)$ to the plane $ax + by + cz = d$ is

$$D = \frac{|ax_0 + by_0 + cz_0 - d|}{\sqrt{a^2 + b^2 + c^2}}.$$

41. Let Π_1 denote the plane $ax + by + cz = d_1$ and let Π_2 denote the plane $ax + by + cz = d_2$.

    (a) Explain why the planes Π_1 and Π_2 are parallel.

    (b) Find the distance between the planes Π_1 and Π_2.

42. Let ℓ_1 be the line $\mathbf{x} = (1, 1, 2) + t(3, -1, 4)$ and ℓ_2 be the line $\dfrac{x-1}{6} = \dfrac{y}{-2} = \dfrac{z-3}{8}$.

    (a) Are the lines ℓ_1 and ℓ_2 parallel? Explain.

    (b) The point $P = (1, 1, 2)$ is on ℓ_1, and the point $Q = (1, 0, 3)$ is on ℓ_2. Find $|\overrightarrow{PQ}|$.

    (c) Find the scalar projection $\overrightarrow{PQ}$ in the direction of $\mathbf{v} = (3, -1, 4)$.

    (d) Find the distance between ℓ_1 and ℓ_2.

43. Find the distance between the point $(8, 7, 9)$ and the line $\dfrac{x-1}{6} = \dfrac{y-2}{5} = \dfrac{z+3}{4}$.

44. The position of a particle in $\mathbb{R}^3$ at time t is $\mathbf{x}(t) = (t^2, 3t, t^3 - 12t + 5)$.

    (a) Find the velocity of the particle at time t.

    (b) Find all times at which the particle is traveling parallel to the xy-plane.

    (c) Find all times at which the particle's velocity vector is perpendicular to the plane $2x + 6y + 5z = 7$. If there are no such times, carefully explain how you know this.

12.9 THE CROSS PRODUCT

This section introduces a new operation, the **cross product** on vectors in $\mathbb{R}^3$. If $\mathbf{v}$ and $\mathbf{w}$ are any vectors in space, then their cross product, denoted by $\mathbf{v} \times \mathbf{w}$, is another *vector*; by comparison, $\mathbf{v} \cdot \mathbf{w}$ is a *scalar*.

Before getting into details we acknowledge some unusual properties of the cross product:

- $\mathbf{v} \times \mathbf{w}$ is defined only for vectors in $\mathbb{R}^3$ (by contrast, $\mathbf{v} \cdot \mathbf{w}$ is defined for vectors in any dimension);
- the cross product is not commutative—on the contrary, $\mathbf{v} \times \mathbf{w} = -(\mathbf{w} \times \mathbf{v})$;
- the recipe for finding $\mathbf{v} \times \mathbf{w}$ from the components of $\mathbf{v}$ and $\mathbf{w}$ is slightly complicated. ↠

This is not a fatal objection. Good recipes are often complicated.

Despite these apparent drawbacks, the cross product is a useful basic tool—useful enough to be "known" to mathematical software ranging from *Maple* and *Mathematica* to modest graphing calculators.

The idea and definition

Let $\mathbf{v}$ and $\mathbf{w}$ be any two vectors. Can we find a third vector that is perpendicular to both $\mathbf{v}$ and $\mathbf{w}$?

If we work entirely in $\mathbb{R}^2$, the answer is usually no: A nonzero *plane* vector cannot be perpendicular to two different directions. ↠ In $\mathbb{R}^3$, by contrast, there is more "room." Two nonparallel vectors $\mathbf{v}$ and $\mathbf{w}$ determine a plane Π in $\mathbb{R}^3$, and any vector perpendicular to Π is automatically perpendicular to both $\mathbf{v}$ and $\mathbf{w}$. In fact, there are infinitely many such space vectors, because if $\mathbf{n}$ is perpendicular to both $\mathbf{v}$ and $\mathbf{w}$ then so is $k\mathbf{n}$ for any scalar k.

If $\mathbf{v}$ and $\mathbf{w}$ happen to be parallel, an answer can be found.

The cross product chooses—from among the infinitely many possibilities—one particular vector $\mathbf{v} \times \mathbf{w}$ that is perpendicular to both $\mathbf{v}$ and $\mathbf{w}$. As we will see, this choice is cleverly made to ensure that both the length and the direction of $\mathbf{v} \times \mathbf{w}$ give valuable information.

DEFINITION (Cross product) Let $\mathbf{v} = (v_1, v_2, v_3)$ and $\mathbf{w} = (w_1, w_2, w_3)$ be 3-vectors. Their **cross product** is the 3-vector

$$\mathbf{v} \times \mathbf{w} = (v_2 w_3 - v_3 w_2, \ v_3 w_1 - v_1 w_3, \ v_1 w_2 - v_2 w_1).$$

This slightly cumbersome product turns out to have many interesting algebraic and geometric properties. Here are several:

- **Perpendicularity** It's easy to check by direct calculation that $\mathbf{v} \times \mathbf{w}$ is indeed perpendicular to both $\mathbf{v}$ and $\mathbf{w}$. For instance,

$$\mathbf{v} \cdot (\mathbf{v} \times \mathbf{w}) = v_1(v_2 w_3 - v_3 w_2) + v_2(v_3 w_1 - v_1 w_3) + v_3(v_1 w_2 - v_2 w_1)$$
$$= v_1 v_2 w_3 - v_1 v_3 w_2 + v_2 v_3 w_1 - v_2 v_1 w_3 + v_3 v_1 w_2 - v_3 v_2 w_1;$$

the last quantity is zero because all terms cancel in pairs. ↠

Check for yourself that $\mathbf{w} \cdot (\mathbf{v} \times \mathbf{w}) = 0$, too.

- **Anti-commutativity** The dot product is commutative: $\mathbf{v} \cdot \mathbf{w} = \mathbf{w} \cdot \mathbf{v}$. The cross product, by contrast, is *anti-commutative*:

$$\mathbf{v} \times \mathbf{w} = -(\mathbf{w} \times \mathbf{v}).$$

The minus signs in the definition explain this peculiarity. Their presence means that reversing the roles of $\mathbf{v}$ and $\mathbf{w}$ changes the sign of each component, and hence reverses the product vector itself. Notice what this means about the cross product of a vector with *itself*:

$$\mathbf{v} \times \mathbf{v} = -(\mathbf{v} \times \mathbf{v}) = \mathbf{0} = (0, 0, 0).$$

(The second equality holds because only the zero vector is its own opposite.)

In each case, the product is perpendicular to both factors.

- **Standard basis vectors**   Cross products of the standard basis vectors with each other are especially simple: ◄

$$\mathbf{i} \times \mathbf{j} = \mathbf{k}; \qquad \mathbf{j} \times \mathbf{k} = \mathbf{i}; \qquad \mathbf{k} \times \mathbf{i} = \mathbf{j}.$$

As was just seen, reversing the order of factors introduces a minus sign:

$$\mathbf{j} \times \mathbf{i} = -\mathbf{k}; \qquad \mathbf{k} \times \mathbf{j} = -\mathbf{i}; \qquad \mathbf{i} \times \mathbf{k} = -\mathbf{j}.$$

One way to remember these rules is to think of the basis vectors as a repeating cycle: **i j k i j k** .... Then, reading from left to right, the cross product of any two adjacent vectors is the next one.

Basic properties of matrices and determinants are outlined in Appendix A.

- **Remembering the formula**   One useful device for remembering the definition is to think of the cross product as a 3×3 **determinant** ◄

$$\begin{vmatrix} \mathbf{i} & \mathbf{j} & \mathbf{k} \\ v_1 & v_2 & v_3 \\ w_1 & w_2 & w_3 \end{vmatrix},$$

expanded along the first row:

$$\mathbf{v} \times \mathbf{w} = (v_2 w_3 - v_3 w_2)\mathbf{i} - (v_1 w_3 - v_3 w_1)\mathbf{j} + (v_1 w_2 - v_2 w_1)\mathbf{k}.$$

Another memory device begins with this "double" array:

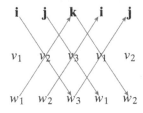

See for yourself that this device gives the cross product.

Then the cross product is the sum of six *threefold diagonal products* with southeast-pointing diagonals counted positive and northeast-pointing diagonals negative. ◄ (Still another method of remembering the cross product formula appears in the exercises.)

> **E X A M P L E 1**   Find a vector **u** that is perpendicular to both $\mathbf{v} = (1, 2, 3)$ and $\mathbf{w} = (2, 3, 4)$.
>
> **Solution**   The formula gives $\mathbf{u} = \mathbf{v} \times \mathbf{w} = (-1, 2, -1)$, and we see easily that $\mathbf{u} \cdot \mathbf{v} = 0 = \mathbf{u} \cdot \mathbf{w}$.   ■

Algebra with the cross product

The cross product enjoys many of the algebraic properties one hopes for from any well-mannered mathematical operation. ◄ The next theorem collects several such properties.

But not all—the cross product isn't commutative.

> **THEOREM 7 (Cross product algebra)**   Let **u**, **v**, and **w** be vectors in $\mathbb{R}^3$ and a a scalar. Then
>
> - (Scalars factor out) $a\mathbf{v} \times \mathbf{w} = \mathbf{v} \times a\mathbf{w} = a(\mathbf{v} \times \mathbf{w})$
> - (Anti-commutativity) $\mathbf{v} \times \mathbf{w} = -(\mathbf{w} \times \mathbf{v})$
> - (Distributivity) $(\mathbf{u} + \mathbf{v}) \times \mathbf{w} = \mathbf{u} \times \mathbf{w} + \mathbf{v} \times \mathbf{w}$

The theorem is readily proved by careful but slightly tedious algebra with the vectors' components. We give one example and leave others as exercises.

EXAMPLE 2 Prove the first part of Theorem 7.

Solution Let $\mathbf{v} = (v_1, v_2, v_3)$ and $\mathbf{w} = (w_1, w_2, w_3)$. Then

$$a\mathbf{v} \times \mathbf{w} = (av_1, av_2, av_3) \times (w_1, w_2, w_3)$$
$$= (av_2 w_3 - av_3 w_2, av_3 w_1 - av_1 w_3, av_1 w_2 - av_2 w_1)$$
$$= a(v_2 w_3 - v_3 w_2, v_3 w_1 - v_1 w_3, v_1 w_2 - v_2 w_1) = a(\mathbf{v} \times \mathbf{w}).$$

The proof that $\mathbf{v} \times a\mathbf{w} = a(\mathbf{v} \times \mathbf{w})$ is almost identical. ■

Parallel vectors The dot product detects *perpendicularity*: nonzero vectors $\mathbf{v}$ and $\mathbf{w}$ are perpendicular if and only if $\mathbf{v} \cdot \mathbf{w} = 0$. In a similar way, the cross product detects *parallelism*:

> *Vectors $\mathbf{v}$ and $\mathbf{w}$ are scalar multiples if and only if $\mathbf{v} \times \mathbf{w} = \mathbf{0}$.*

To see why, suppose that $\mathbf{w} = a\mathbf{v}$, for some scalar a. Then

$$\mathbf{v} \times \mathbf{w} = \mathbf{v} \times a\mathbf{v} = a(\mathbf{v} \times \mathbf{v}) = \mathbf{0}.$$

We leave as an exercise the proof that $\mathbf{v} \times \mathbf{w} = \mathbf{0}$ implies that $\mathbf{v}$ and $\mathbf{w}$ are parallel. →

For convenience we take the zero vector to be parallel to every vector.

Finding equations for planes The cross product often helps us find equations for planes.

EXAMPLE 3 Find scalar and vector equations for the plane Π through the three points $A = (1, 1, 1)$, $B = (1, 2, 3)$, and $C = (4, 5, 6)$. Does Π pass through the origin?

Solution We will need a point (A will do) on Π and a normal vector. The vectors

$$\overrightarrow{AB} = (1, 2, 3) - (1, 1, 1) = (0, 1, 2) \quad \text{and} \quad \overrightarrow{AC} = (4, 5, 6) - (1, 1, 1) = (3, 4, 5)$$

lie *on* Π, and so their cross product

$$\overrightarrow{AB} \times \overrightarrow{AC} = (-3, 6, -3)$$

is a suitable normal vector. Thus, Π has the vector equation

$$(\mathbf{X} - (1, 1, 1)) \cdot (-3, 6, -3) = 0.$$

The scalar version is $-3x + 6y - 3z = 0$, or, equivalently, $x - 2y + z = 0$, and it is now easy to see that the origin is one solution. ■

Geometry of the cross product

Like the dot product, the cross product has important geometric interpretations:

FACT Let $\mathbf{v}$ and $\mathbf{w}$ be vectors in $\mathbb{R}^3$, and let θ be the angle between them. Then $\mathbf{v} \times \mathbf{w}$ has these geometric properties:

- **Length** $|\mathbf{v} \times \mathbf{w}| = |\mathbf{v}||\mathbf{w}| \sin \theta$
- **Direction** If $\mathbf{v}$ and $\mathbf{w}$ are not parallel, then $\mathbf{v} \times \mathbf{w}$ is perpendicular to both $\mathbf{v}$ and $\mathbf{w}$ and points in the direction determined by the **right-hand rule**: If the fingers of the right hand sweep from $\mathbf{v}$ to $\mathbf{w}$, then the thumb points in the direction of $\mathbf{v} \times \mathbf{w}$.

Both parts of the Fact deserve some comment:

- **Why it holds** A proof based on straightforward calculation is outlined in the exercises; it is better worked through than read.

- **More on the angle** θ Because the angle between **v** and **w** never exceeds π radians, $\sin\theta \geq 0$. In particular, $\sin\theta = 0$ if θ is either 0 or π—in either case, **v** and **w** must be parallel. ◂

We saw this earlier by algebraic means.

- **Interpreting the magnitude as area**   Figure 1 shows why the quantity $|\mathbf{v}||\mathbf{w}|\sin\theta$ (the magnitude of $\mathbf{v} \times \mathbf{w}$) is the *area* of the parallelogram spanned by **v** and **w**:

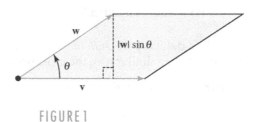

FIGURE 1
The area spanned by v and w

The parallelogram's base is $|\mathbf{v}|$ and its height is $|\mathbf{w}|\sin\theta$. Notice, in particular, that the second factor is nonnegative, and zero *only* if **v** and **w** are parallel.

*Sweep the fingers of your right hand from **v** to **w**. Notice that your right thumb points* upward.

Try this too.

- **More on the right-hand rule**   Applying the right-hand rule to Figure 1 shows that $\mathbf{v} \times \mathbf{w}$ points straight up—toward you, the reader. Try it! ◂ If **v** and **w** are switched, then the area they span does not change but the direction of $\mathbf{v} \times \mathbf{w}$ is reversed. ◂

  Figure 2 illustrates another way to think of the right-hand rule.

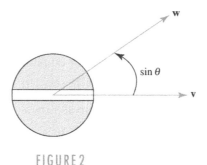

FIGURE 2
The turn of the screw

Imagine driving an ordinary wood screw into the plane of **v** and **w**. Turning the screw head from **v** to **w** causes the screw to move either into or out of the plane—the direction of motion is that of $\mathbf{v} \times \mathbf{w}$. In the situation of Figure 2, the indicated turn *loosens* the screw, causing it to rise out of the page. ◂

Standard bolts and screws from the hardware store are sometimes called "right-threaded." Left-threaded bolts and screws are sometimes used for special purposes.

E X A M P L E 4   Find the area of the triangle with corners at $A = (1, 1, 1)$, $B = (1, 2, 3)$, and $C = (4, 5, 6)$.

Solution   The triangle is half the parallelogram spanned by the vectors $\overrightarrow{AB} = (0, 1, 2)$ and $\overrightarrow{AC} = (3, 4, 5)$. The parallelogram has area

$$|\overrightarrow{AB} \times \overrightarrow{AC}| = |(-3, 6, -3)| = \sqrt{54},$$

and so the triangle has area $\sqrt{54}/2 \approx 3.67$. ∎

The cross product as torque: a physical interpretation

Physicists interpret the cross product $\mathbf{v} \times \mathbf{w}$ as **torque**, a vector quantity that describes the tendency of a force to twist a body about an axis. From this point of view, we can think of the vector $\mathbf{v}$ as a rigid bar pinned at its tail and the vector $\mathbf{w}$ as a force exerted at the *head* of $\mathbf{v}$. For physical reasons the twisting tendency at the tail of $\mathbf{v}$ is the product lf, where l is the length of $\mathbf{v}$ and f is the scalar component of the force $\mathbf{w}$ perpendicular to $\mathbf{v}$. ➡ In other words, the torque at the tail of $\mathbf{v}$ has magnitude

This is essentially the principle of the lever, which has been known since Archimedes.

$$|\mathbf{v}||\mathbf{w}| \sin \theta = |\mathbf{v} \times \mathbf{w}|,$$

and the direction of the torque vector is given by the right-hand rule. Observe:

- **Parallel vectors** If the force $\mathbf{w}$ is applied parallel to $\mathbf{v}$, then no twisting occurs—as the equation $\mathbf{v} \times \mathbf{w} = \mathbf{0}$ predicts.
- **Perpendicular vectors** The torque is greatest if $\mathbf{w}$ is applied perpendicular to $\mathbf{v}$.
- **The torque's direction** The torque's direction is consistent with right-hand threading of a bolt or screw.

EXAMPLE 5 Figure 3 shows a large wrench, 0.6 meters long as measured along the vector $\mathbf{v}$. The vector $\mathbf{w}$ represents a force with magnitude 100 newtons, exerted in the direction of $\mathbf{w}$. ➡

A force of 100 newtons corresponds to about 22 pounds.

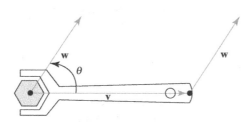

FIGURE 3
Measuring torque

Find in terms of θ the magnitude and direction of the resulting torque exerted at the center of the bolt (the left-hand black dot).

Solution The preceding discussion implies that the torque has magnitude

$$|\mathbf{v}||\mathbf{w}| \sin \theta = 0.6 \cdot 100 \cdot \sin \theta = 60 \sin \theta \text{ newton-meters;}$$

the torque's direction is determined by the right-hand rule.

In Figure 3 the angle θ is about 0.9 radians, which implies that the torque vector points upward (out of the page) and so has magnitude $60 \sin(0.9) \approx 47$ newton-meters. ∎

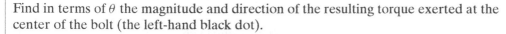

BASIC EXERCISES

In Exercises 1–8, find (by hand) the cross product $\mathbf{v} \times \mathbf{w}$ of each pair of vectors below.

1. $\mathbf{v} = (1, 2, 3); \mathbf{w} = (4, 5, 6)$

2. $\mathbf{v} = (4, 5, 6); \mathbf{w} = (2, 3, 4)$

3. $\mathbf{v} = (1, 2, 3); \mathbf{w} = (40, 50, 60)$

4. $\mathbf{v} = (10, 20, 30); \mathbf{w} = (4, 5, 6)$

5. $\mathbf{v} = \mathbf{i} + \mathbf{j}; \mathbf{w} = \mathbf{j} + \mathbf{k}$

6. $\mathbf{v} = (v_1, v_2, v_3); \mathbf{w} = \mathbf{i}$

7. $\mathbf{v} = (v_1, v_2, v_3); \mathbf{w} = \mathbf{j}$

8. $\mathbf{v} = (v_1, v_2, v_3); \mathbf{w} = \mathbf{k}$

In Exercises 9–11, find a scalar equation for the given plane. [HINT: Use the cross product as needed to find a normal vector.]

9. The plane through $(0, 0, 0)$, $(1, 2, 3)$, and $(-1, 1, 2)$.

10. The plane through $(1, 2, 3)$ with spanning vectors $\mathbf{i}$ and $\mathbf{i} + \mathbf{j}$.

11. The plane with parametric equations $x = 1 + 2s + 3t$, $y = -1 + 3s - 2t$, $z = s + t$.

12. Find an equation for the plane that contains the points $(1, 2, 3)$, $(1, 1, 1)$, and $(1, 0, 2)$.

13. Let $\mathbf{u} = (1, 2, 3)$ and $\mathbf{v} = (4, 5, 6)$. Find a unit vector that is perpendicular to both $\mathbf{u}$ and $\mathbf{v}$.

14. Find a unit vector orthogonal to both $\mathbf{i} + \mathbf{k}$ and $2\mathbf{j} - 3\mathbf{k}$.

15. Let $\mathbf{u} = (1, 2, 3)$ and $\mathbf{v} = (1, 3, 5)$. Find a unit vector that is perpendicular to both $\mathbf{u}$ and $\mathbf{v}$.

16. Let $\mathbf{u} = (1, 2, 3)$ and $\mathbf{v} = (2, 3, 4)$. Find a unit vector that is orthogonal to both $\mathbf{u}$ and $\mathbf{v}$.

17. This problem is about the situation of Example 5. In each part, find the magnitude and direction of the torque vector under the given assumptions:

   (a) The wrench is 0.6 meters long, the force $\mathbf{w}$ has magnitude 200 newtons, and $\theta = 0.9$.

   (b) The wrench is 0.6 meters long, the force $\mathbf{w}$ has magnitude 200 newtons, and $\theta = 0$.

   (c) The wrench is 0.6 meters long, the force $\mathbf{w}$ has magnitude 200 newtons, and $\theta = \pi$.

Exercises 18–20 are about the situation of Example 5.

18. For which angle θ does the torque have largest magnitude? (Give two possible answers.)

19. For which angles θ between 0 and 2π does the torque vector point downward?

20. Suppose the wrench is 0.6 meters long and the force $\mathbf{w}$ has magnitude 100 newtons. Find an angle θ for which the torque vector points upward and has magnitude 10 newton-meters.

FURTHER EXERCISES

21. Let $\mathbf{v} = (v_1, v_2, v_3)$ and $\mathbf{w} = (w_1, w_2, w_3)$. Show from the definition that $\mathbf{v} \times \mathbf{w}$ is perpendicular to $\mathbf{w}$.

22. Let $\mathbf{v} = (v_1, v_2, v_3)$ and $\mathbf{w} = (w_1, w_2, w_3)$. Show from the definition that $\mathbf{v} \times \mathbf{w} = -(\mathbf{w} \times \mathbf{v})$.

23. Is the cross product associative? In other words, is $\mathbf{u} \times (\mathbf{v} \times \mathbf{w}) = (\mathbf{u} \times \mathbf{v}) \times \mathbf{w}$ for all vectors $\mathbf{u}$, $\mathbf{v}$, and $\mathbf{w}$? Either show that the cross product is associative or give an example to show that it is not.

24. Let $\mathbf{v} = (v_1, v_2, v_3)$ and $\mathbf{w} = (w_1, w_2, w_3)$ be nonzero vectors. Show that if $\mathbf{v} \times \mathbf{w} = (0, 0, 0)$, then $\mathbf{v}$ and $\mathbf{w}$ are parallel, that is each is a scalar multiple of the other. [HINTS: Since $\mathbf{v} \neq (0, 0, 0)$, it is OK to assume that some component, say v_1, is not zero. Let $t = w_1/v_1$. Now use the assumption that $\mathbf{v} \times \mathbf{w} = (0, 0, 0)$ to show that $\mathbf{w} = t\mathbf{v}$.]

25. Here is another way to make sense of the algebraic formula for the cross product. We will assume three basic cross products:

$$\mathbf{i} \times \mathbf{j} = \mathbf{k}; \quad \mathbf{j} \times \mathbf{k} = \mathbf{i}; \quad \mathbf{k} \times \mathbf{i} = \mathbf{j}.$$

We will also assume the algebraic properties of Theorem 7, page 690. From these assumptions, we will deduce the general formula for $\mathbf{v} \times \mathbf{w}$.

   Let $\mathbf{v} = (v_1, v_2, v_3)$ and $\mathbf{w} = (w_1, w_2, w_3)$. Cite a brief reason for each computational step below:

$$\begin{aligned}
\mathbf{v} \times \mathbf{w} &= (v_1\mathbf{i} + v_2\mathbf{j} + v_3\mathbf{k}) \times (w_1\mathbf{i} + w_2\mathbf{j} + w_3\mathbf{k}) \\
&= (v_1w_1)\mathbf{i} \times \mathbf{i} + (v_2w_1)\mathbf{j} \times \mathbf{i} + (v_3w_1)\mathbf{k} \times \mathbf{i} + (v_1w_2)\mathbf{i} \times \mathbf{j} \\
&\quad + (v_2w_2)\mathbf{j} \times \mathbf{j} + (v_3w_2)\mathbf{k} \times \mathbf{j} + (v_1w_3)\mathbf{i} \times \mathbf{k} \\
&\quad + (v_2w_3)\mathbf{j} \times \mathbf{k} + (v_3w_3)\mathbf{k} \times \mathbf{k} \\
&= (v_2w_1)\mathbf{j} \times \mathbf{i} + (v_3w_1)\mathbf{k} \times \mathbf{i} + (v_1w_2)\mathbf{i} \times \mathbf{j} \\
&\quad + (v_3w_2)\mathbf{k} \times \mathbf{j} + (v_1w_3)\mathbf{i} \times \mathbf{k} + (v_2w_3)\mathbf{j} \times \mathbf{k} \\
&= (v_2w_3 - v_3w_2)\mathbf{i} + (v_3w_1 - v_1w_3)\mathbf{j} + (v_1w_2 - v_2w_1)\mathbf{k}.
\end{aligned}$$

26. The Fact on page 691 says that $|\mathbf{v} \times \mathbf{w}| = |\mathbf{v}|\,|\mathbf{w}| \sin\theta$, where θ is the angle between $\mathbf{v}$ and $\mathbf{w}$. This exercise outlines a proof; the underlying idea is to relate the cross product to the dot product.

   (a) Show (by direct calculation) that $|\mathbf{v} \times \mathbf{w}|^2 = |\mathbf{v}|^2 |\mathbf{w}|^2 - (\mathbf{v} \cdot \mathbf{w})^2$.

   (b) Use the result of (a) to show that $|\mathbf{v} \times \mathbf{w}| = |\mathbf{v}|\,|\mathbf{w}| \sin\theta$, as the Fact claims.

   [HINTS: In (b), write $(\mathbf{v} \cdot \mathbf{w})^2 = |\mathbf{v}|^2|\mathbf{w}|^2 \cos^2\theta$. Then take the square root of both sides of (a).]

27. Let $\mathbf{u}$, $\mathbf{v}$, and $\mathbf{w}$ be vectors in $\mathbb{R}^3$. The quantity $\mathbf{u} \cdot (\mathbf{v} \times \mathbf{w})$ is called the **triple scalar product** of $\mathbf{u}$, $\mathbf{v}$, and $\mathbf{w}$ in that order. We will investigate the following useful property of the triple product:

The absolute value $|\mathbf{u} \cdot (\mathbf{v} \times \mathbf{w})|$ is the volume of the parallelepiped in $\mathbb{R}^3$ (i.e., the three-dimensional solid) spanned by $\mathbf{u}$, $\mathbf{v}$, and $\mathbf{w}$.

Here is a possible picture:

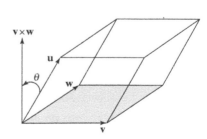

FIGURE 4
The solid spanned by three vectors

   (a) Calculate the triple scalar product of $\mathbf{i}$, $\mathbf{j}$, and $\mathbf{k}$. Is the result consistent with the italicized fact above?

(b) Calculate the triple scalar product of $a\mathbf{i}, b\mathbf{j}$, and $c\mathbf{k}$, where a, b, and c are positive constants. Is the result consistent with the italicized fact above?

(c) Suppose that $\mathbf{u}$ is in the plane spanned by $\mathbf{v}$ and $\mathbf{w}$. What is the value of $\mathbf{u} \cdot (\mathbf{v} \times \mathbf{w})$? What does this mean about the parallelepiped?

(d) Find the area of the base of the solid (shown shaded).

(e) Explain why $|\mathbf{u}|\,|\cos\theta|$ is the height (perpendicular to the shaded plane) of the solid.

(f) Explain why $|\mathbf{u} \cdot (\mathbf{v} \times \mathbf{w})|$ is the claimed volume. [HINT: Volume = base times height.]

28. Let $\mathbf{u}$ and $\mathbf{v}$ be vectors in $\mathbb{R}^3$. Give a geometric explanation of the identity $(\mathbf{u} \times \mathbf{v}) \cdot \mathbf{v} = 0$.

29. Let L be the line through the point $(1, 1, 2)$ that is perpendicular to the plane Π described by the equation $3x + 5y + 8z = 13$. Find the point where L and Π intersect.

30. Find an equation for the line that passes through the point $(2, 3, 1)$ and is parallel to the line of intersection of the planes with equations $2x - 2y - 3z = 2$ and $3x - 2y + z = 1$. [HINT: The line of intersection is perpendicular to normal lines for either plane.]

31. Let ℓ be the line through the points $(1, 2, 3)$ and $(4, 5, 6)$. Write an equation for the plane that is perpendicular to ℓ and passes through the point $(7, 8, 9)$.

32. Find the area of the parallelogram in space formed by the vectors $\mathbf{u} = (1, 5, 1)$ and $\mathbf{v} = (-2, 1, 3)$.

33. Show that $\mathbf{u} \cdot (\mathbf{v} \times \mathbf{w}) = (\mathbf{u} \times \mathbf{v}) \cdot \mathbf{w}$.

34. Suppose that $\mathbf{X}_0 + s\mathbf{u}$ and $\mathbf{X}_1 + t\mathbf{v}$ are two skew (i.e., nonintersecting, nonparallel) lines in space. Find an expression for the distance between the lines. [HINT: The vector $\mathbf{u} \times \mathbf{v}$ is perpendicular to both lines.]

35. Show that $\dfrac{d}{dt}\big(\mathbf{x}(t) \times \mathbf{y}(t)\big) = \mathbf{x}'(t) \times \mathbf{y}(t) + \mathbf{x}(t) \times \mathbf{y}'(t)$.

36. Suppose that $\mathbf{x}(t) = (\cos t, \sin t, t)$ and $\mathbf{y}(t) = (2, t^2, -4t)$. Evaluate $\dfrac{d}{dt}\big(\mathbf{x}(t) \times \mathbf{y}(t)\big)$.

37. Show that if $\mathbf{u}$, $\mathbf{v}$, and $\mathbf{w}$ are distinct nonzero vectors, then $\mathbf{u} \times \mathbf{v} = \mathbf{u} \times \mathbf{w}$ if and only if $\mathbf{u}$ is parallel to $\mathbf{v} - \mathbf{w}$.

38. Show that $(\mathbf{v} + \mathbf{w}) \times (\mathbf{v} - \mathbf{w}) = 2\mathbf{w} \times \mathbf{v}$.

39. Show that if three vectors form a triangle, then the lengths of their pairwise cross products are all equal.

40. Explain why the vector $\mathbf{u} \times (\mathbf{v} \times \mathbf{w})$ must lie in the plane spanned by $\mathbf{v}$ and $\mathbf{w}$.

41. Let L be a line with direction vector $\mathbf{v}$ and P a point not on L. Let $\mathbf{w}$ be a vector from some point on L to P. Show that the distance from P to L is given by $|\mathbf{w} \times \mathbf{v}|\,/|\mathbf{v}|$.

42. Suppose that the plane spanned by the nonzero vectors $\mathbf{a}$ and $\mathbf{b}$ and the plane spanned by the nonzero vectors $\mathbf{c}$ and $\mathbf{d}$ are not parallel. Explain why the vector $(\mathbf{a} \times \mathbf{b}) \times (\mathbf{c} \times \mathbf{d})$ is parallel to the line of intersection of the planes.

43. Suppose that $\mathbf{u}$ and $\mathbf{v}$ are the vectors in the xz-plane pictured below, $|\mathbf{u}| = 3$, $|\mathbf{v}| = 4$, and $\phi = \pi/3$. Evaluate $\mathbf{u} \times \mathbf{v}$.

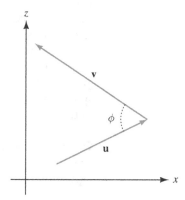

44. Suppose that $\mathbf{u}$ and $\mathbf{v}$ are the vectors in the yz-plane pictured below, $|\mathbf{u}| = 4$, $|\mathbf{v}| = 3$, and $\phi = \pi/3$. Evaluate $\mathbf{u} \times \mathbf{v}$.

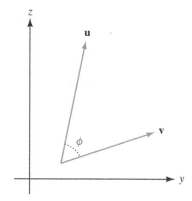

45. (a) Explain how the cross product can be used to determine whether three points in space are collinear.

(b) Describe a method for determining whether four points lie in the same plane.

46. Let Π_1 and Π_2 be planes in $\mathbb{R}^3$ that both contain the origin. Furthermore, suppose that the vector $\mathbf{n}_1 = (1, 2, 1)$ is orthogonal to the plane Π_1 and that the vector $\mathbf{n}_2 = (-3, -5, 0)$ is orthogonal to the plane Π_2. Find an equation for the line of intersection of Π_1 and Π_2.

47. Show that any two lines in space that do not intersect must lie in parallel planes.

This chapter introduced some of the main objects and tools needed to extend the basic ideas of calculus to the multivariable setting.

Three-dimensional space The xy-plane is the natural place to study an ordinary calculus function $y = f(x)$; we use one dimension for inputs and one for outputs. Because multivariable calculus treats functions of *several* input variables, at least three dimensions are needed to accommodate graphs and other visual representations. Having introduced xyz-space and some of the challenges posed by projecting three-dimensional view onto a flat page or screen, we explored basic properties of graphs and equations in the new setting, emphasizing both differences from, and similarities to, the two-dimensional setting. The graph of a function $z = f(x, y)$, for example, is usually a *two*-dimensional surface in xyz-space; by contrast, the graph of a function $y = f(x)$ in the plane is a *one*-dimensional curve in the xy-plane.

Curves and parametric equations Most of the curves seen in single-variable calculus are graphs of functions $y = f(x)$. A broader class of curves can be given by **parametric equations** of the form

$$x = f(t) \qquad \text{and} \qquad y = g(t),$$

where f and g are ordinary functions of the **parameter** t, and t ranges through some **parameter interval** $a \le t \le b$. Allowing x and y to be defined independently produces a huge variety of curves in the plane and in space and begins to suggest the advantages of working with several variables at once. With a little ingenuity we can produce almost any imaginable curve "to order." Any given curve, moreover, has many possible **parametrizations**. We gave several possibilities for such basic examples as line segments and arcs of circles.

Polar coordinates and polar curves A point in the xy-plane can be located either through its Cartesian "address" (x, y) (from the origin, move x units right and y units up) or through its **polar** address (r, θ) (from the origin, move r units of distance at an angle θ to the **polar axis**). After reviewing basic properties of **polar coordinates**, we took the geometric view, discussing **polar graphs** and their relations to equations in the polar variables r and θ. Certain regions and functions, we found, are better and more simply described in polar than in rectangular coordinates; we will exploit these advantages in later chapters.

Vectors Vectors, represented either as "tuples" or as arrows in the plane or in space, are used to represent quantities that have both **magnitude** and **direction**. (**Scalars**, by contrast, represent magnitudes but permit only two directions—positive and negative.) We introduced basics of vector notation, vector algebra, and geometric representation of vectors and their combinations.

Vector-valued functions, derivatives, and integrals A vector-valued function $\mathbf{f}(t)$ accepts a single real variable as input but produces a **vector**, often written as a 2-tuple, as output. Such functions permit many of the usual operations and calculations and have graphs, derivatives, and integrals much like those for ordinary functions, but with somewhat different forms and meanings. The derivative $\mathbf{f}'(t_0)$ of a vector-valued function $\mathbf{f}$, for instance, is a *vector* that is tangent to the graph of $\mathbf{f}$ at the point $\mathbf{f}(t_0)$. By contrast, the derivative $f'(x_0)$ of an ordinary function f is a *scalar* that measure slope of the f-graph at the point $(x_0, f(x_0))$.

Modeling motion Derivatives and integrals are useful for modeling motion even in the one-variable case, where (under suitable assumptions and with appropriate units of measurement) position, velocity, and acceleration are all derivatives and antiderivatives of each other. The same ideas apply, but more flexibly and powerfully, to vector-valued functions. If $\mathbf{p}(t)$ denotes the **position** vector at time t of an object moving in the plane or in space, then the derivatives $\mathbf{p}'(t) = \mathbf{v}(t)$ and $\mathbf{p}''(t) = \mathbf{v}'(t) = \mathbf{a}(t)$ describe, respectively, the (vector) **velocity** and **acceleration** of the object at time t. Combining these familiar ideas with some basic physical principles lets us model a wide variety of motions in the plane and in space.

The dot product and the cross product   Given two vectors $\mathbf{v} = (v_1, v_2)$ and $\mathbf{w} = (w_1, w_2)$ in the plane (or $\mathbf{v} = (v_1, v_2, v_3)$ and $\mathbf{w} = (w_1, w_2, w_3)$ in space), we can form their **dot product**, a scalar, by a simple calculation:

$$\mathbf{v} \cdot \mathbf{w} = v_1 w_1 + v_2 w_2 \quad \text{or} \quad \mathbf{v} \cdot \mathbf{w} = v_1 w_1 + v_2 w_2 + v_3 w_3,$$

as determined by the dimension. The result, although easy to compute, conveys a remarkable amount of information. Vectors are **perpendicular**, for example, only if their dot product is zero. The dot product can also be used to calculate both the lengths and the angle between any two nonzero vectors in the plane or in space; it also leads to a useful formula for the perpendicular **projection** of one vector onto another.

The **cross product**, by contrast, is defined only for vectors in three-dimensional space. For two such vectors $\mathbf{v} = (v_1, v_2, v_3)$ and $\mathbf{w} = (w_1, w_2, w_3)$, the cross product—another *vector*—has the pretty but slightly complicated formula

$$\mathbf{v} \times \mathbf{w} = (v_2 w_3 - v_3 w_2, \ v_3 w_1 - v_1 w_3, \ v_1 w_2 - v_2 w_1).$$

Like the dot product, the cross product has useful geometric meaning: $\mathbf{v} \times \mathbf{w}$ is a vector with three main properties: (i) it is perpendicular both to $\mathbf{v}$ and to $\mathbf{w}$, (ii) its direction is determined by the **right-hand rule**, and (iii) the length $|\mathbf{v} \times \mathbf{w}|$ is the area of the parallelogram spanned by $\mathbf{v}$ and $\mathbf{w}$. (If $\mathbf{v}$ and $\mathbf{w}$ are collinear, then this area is zero—and $\mathbf{v} \times \mathbf{w}$ is the zero vector.)

Lines and planes   Vectors and vector operations, including the dot and cross products, offer convenient algebraic and geometric descriptions of lines and planes in two and three dimensions. For instance, a **plane** in $\mathbb{R}^3$ passing through the point (x_0, y_0, z_0) can be described in **scalar form** as the graph of a linear equation $a(x - x_0) + b(y - y_0) + c(z - z_0) = 0$, where $a, b,$ and c are constants. The same plane can be described in **vector form** by defining the **normal vector** $\mathbf{n} = (a, b, c)$ and a (vector) point $\mathbf{X_0} = (x_0, y_0, z_0)$. Then the plane consists of all points (vectors) $\mathbf{X}$ for which $(\mathbf{X} - \mathbf{X_0}) \cdot \mathbf{n} = 0$. The vector form is more compact than the scalar, and it shows more clearly the sense in which a plane is perpendicular to a fixed vector through any point on the plane.

REVIEW **E**XERCISES

1. Write an equation for the plane passing through the point $(1, 2, 3)$, with normal vector $(2, 3, 4)$.

2. Parametrize the line segment starting at $(3, 4)$ and ending at $(1, 1)$.

3. Are the points $P = (1, 2, 3)$, $Q = (-3, -3, -3)$, and $R = (9, 12, 15)$ collinear? Justify your answer.

4. Let $\mathbf{v} = (1, 2, 3)$ and let $\mathbf{w} = (2, 3, 4)$. Find $\cos\theta$, where θ is the angle between $\mathbf{v}$ and $\mathbf{w}$.

5. Give two different curves joining $(1, 0)$ to $(0, 1)$ in the xy-plane and give a parametrization for each.

6. Let Π be the the plane passing through the point $(1, 2, 3)$, with normal vector $(4, 5, 6)$.

    (a) Write an equation for Π.

    (b) Does the point $(7, 8, 9)$ lie on Π? Justify your answer.

7. Let C be the left half of the circle with radius 3 that is centered at $(1, 2)$. Give a parametrization for C that is oriented counterclockwise.

8. Are the points $(1, 2, 3)$, $(2, 4, 6)$, and $(3, 5, 7)$ collinear? Justify your answer.

9. Let $\mathbf{v} = (1, 2, 3)$ and $\mathbf{w} = (2, 3, 4)$. Find a unit vector that is perpendicular to both $\mathbf{v}$ and $\mathbf{w}$.

10. A particle moves in the xy-plane with constant acceleration vector $\mathbf{a}(t) = (0, -1)$. At time $t = 0$ the particle has position $(0, 0)$ and velocity $(1, 0)$.

    (a) Give formulas for the particle's velocity function $\mathbf{v}(t)$ and position function $\mathbf{p}(t)$.

    (b) Write down (but don't evaluate) an integral that gives the length of the particle's path from $t = 0$ to $t = 10$.

11. **(a)** Write an equation for the line passing through the point $(1, 2, 3)$ with direction vector $(2, 3, 4)$.

    (b) Does the point $(51, 77, 103)$ lie on the line in part (a)? Justify your answer.

12. Let C be the circle with radius 5 that is centered at $(3, 4)$. Give a parametrization for C that is oriented clockwise.

13. Let Π be the plane containing the three points $(0, 0, 0)$, $(1, 2, 3)$, and $(2, 3, 4)$.

    (a) Find an equation for Π.

    (b) Does the point $(51, 77, 103)$ lie on this plane? Justify your answer.

14. Find a unit vector that is perpendicular to $(1, 2, 3)$.

15. A particle moves in the xy-plane with acceleration vector $\mathbf{a}(t) = (2, 6t)$. At time $t = 0$ the particle has position $(0, 0)$ and velocity $(0, 0)$.

    (a) Give formulas for the particle's velocity function $\mathbf{v}(t)$ and position function $\mathbf{p}(t)$.

    (b) Write down (but don't evaluate) an integral that gives the length of the particle's path from $t = 0$ to $t = 10$.

16. **(a)** Write an equation for the line passing through the point $(2, 3, 4)$ with direction vector $(1, 2, 3)$.

    (b) Does the point $(9, 19, 29)$ lie on the line in part (a)? Justify your answer.

17. Give a counterclockwise-oriented parametrization of the circle with radius 13 centered at $(12, 5)$.

18. Let Π be the plane containing the three points $(0, 0, 0)$, $(2, 3, 4)$, and $(3, 4, 5)$.

    (a) Find an equation for Π.

    (b) Does the point (π, π, π) lie on Π? Justify your answer.

19. Find a unit vector that is perpendicular to the plane with equation $x + 2y + 3z = 6$.

20. A particle moves in the xy-plane with acceleration vector $\mathbf{a}(t) = (6t, -2)$. At time $t = 0$ the particle has position $(0, 0)$ and velocity $(0, 0)$.

    (a) Give formulas for the particle's velocity function $\mathbf{v}(t)$ and position function $\mathbf{p}(t)$.

    (b) Write down (but don't evaluate) an integral that gives the length of the particle's path from $t = 0$ to $t = 10$.

21. Find a scalar equation for the plane through the point $(1, 2, 3)$ that is perpendicular to $\mathbf{i} + \mathbf{j} + \mathbf{k}$.

22. Find a scalar equation for the plane through the points $(0, 0, 0)$, $(1, 2, 3)$, $(0, 1, 0)$.

23. How far is the point $(1, 2, 3)$ from the plane Π defined by the equation $3x + 5y + 7z = 0$?

24. Let ℓ be the line through $(1, 2, 3)$ in the direction of $\mathbf{i} + \mathbf{j} + \mathbf{k}$.

    (a) Write scalar parametric equations for ℓ.

    (b) Write symmetric scalar equations for ℓ.

    (c) Does ℓ intersect the line $x/2 = y/3 = z/4$? If so, where?

In Exercises 25–29, $\mathbf{a} = (1, 2, 3)$ *and* $\mathbf{b} = (1, 1, 1)$.

25. Compute the component of $\mathbf{b}$ in the direction of $\mathbf{a}$.

26. The vector $\mathbf{a}$ makes angles α, β, and γ with the three positive coordinate axes. Find $\cos\alpha$, $\cos\beta$, and $\cos\gamma$.

27. Compute the area of the parallelogram generated by $\mathbf{a}$ and $\mathbf{b}$.

28. Write a scalar equation for the plane Π through the origin containing $\mathbf{a}$ and $\mathbf{b}$.

29. Find the distance from $(-1, 2, 3)$ to the plane Π defined in Exercise 28.

30. At what point (x, y, z)—if any—do the lines $\mathbf{r_1} = (1, 1, 4) + t(2, -1, 0)$ and $\mathbf{r_2} = (-3, 2, 6) + u(1, 0, -1)$ intersect?

31. A certain plane Π has scalar equation $x + 2y + 3z = 4$. Find a vector equation for Π.

In Exercises 32–34, let C be the space curve described by $\mathbf{f}(t) = (3\cos t, 3\sin t, 2t)$.

32. Write a vector equation for the line tangent to C at $t = \pi$.

33. Compute the length of C from $t = 0$ to $t = \pi$.

34. Think of C as describing the path of a moving object. Find the velocity, speed, and acceleration at $t = \pi$. Indicate clearly whether your answers are vectors or scalars.

In Exercises 35–38, $\mathbf{v} = (1, 2, 3)$, $\mathbf{w} = (-1, 2, 1)$, $P = (4, 5, 6)$, $Q = (1, 1, 1)$, *and* $R = (3, 4, 5)$.

35. Write an equation for the line through P in the direction of $\mathbf{v}$.

36. **(a)** Write an equation for the line through P and Q.

    (b) Does R lie on the line in part (a)? Justify your answer.

37. Write an equation for the plane through P with normal vector $\mathbf{v} + \mathbf{w}$.

38. Find a unit vector that is perpendicular to both $\mathbf{v}$ and $\mathbf{w}$.

39. Write equations for any two *lines* in 3-D space that do not intersect.

40. Write equations for any two *planes* in 3-D space that do not intersect.

41. **(a)** If $\mathbf{A}$ and $\mathbf{B}$ are any two vectors, what geometric information does $|\mathbf{A} \times \mathbf{B}|$ convey? (Draw a sketch if you like.)

    (b) What does it mean geometrically if $\mathbf{A} \times \mathbf{B} = (0, 0, 0)$?

In Exercises 42 and 43, $P = (1, 2, 3)$ *and* $Q = (4, 6, 8)$.

42. Find the distance from P to Q.

43. Write an equation in x, y, and z for the sphere with center at P that passes through Q.

In Exercises 44–47, $P = (2, 1, 3)$, $Q = (4, 5, 6)$, $R = (2, 1)$, $S = (4, 5)$, $T = (0, -3)$, and C is the circle with center at R that passes through S.

44. Write an equation in x, y, and z for the sphere with center at P that passes through Q.

45. Write an equation in x and y for the circle C. [HINT: First find the distance from R to S.]

46. Does the circle C pass through T? Justify your answer.

47. Give parametric equations for the circle C traversed once counterclockwise. (Be sure to state the parameter interval.)

48. Parametrize the line segment from $(1, 2)$ to (e, π). (State your answer as a vector-valued function of t; be sure to state the parameter interval.)

49. A particle moves in the plane; its position at time t is $\mathbf{f}(t) = (t^2 + 1, 3t + 5)$.

    (a) Find the particle's position, velocity, speed, and acceleration at $t = 2$.

    (b) Give a vector equation for the line that is tangent to the particle's path at $t = 2$.

50. Give two different parametrizations of the upper half of the unit circle—one counterclockwise and one clockwise. (Be sure to indicate which is which.)

In Exercises 51–55, $\mathbf{v} = (1, 2)$ and $\mathbf{w} = (3, 1)$.

51. Find $\cos\theta$, where θ is the angle between $\mathbf{v}$ and $\mathbf{w}$.

52. Find a unit vector in the same direction as $\mathbf{v}$.

53. On one set of axes, draw the vectors $\mathbf{v}$, $\mathbf{w}$, and $\dfrac{\mathbf{v} \cdot \mathbf{w}}{|\mathbf{w}|^2}\mathbf{w}$; make it obvious how the third vector is related to the first two.

54. Write a vector equation for the line through the point $(4, 5)$ in the direction of the vector $\mathbf{v}$.

55. Find the distance from the origin to the line in Exercise 54.

56. Let C be the curve parametrized by $x = t^2$, $y = e^t$, $-1 \le t \le 1$. Write a vector equation for the line that is tangent to C at the point $(0, 1)$ (i.e., at the point for which $t = 0$).

In Exercises 57–63, $P = (1, 2, 3)$, $Q = (2, 3, 4)$, $R = (5, 3, 5)$, and $S = (2, 9, 8)$. Also, Π_1, Π_2, and Π_3 are the planes

$$\Pi_1\colon x + y + z = 1; \quad \Pi_2\colon x + 2y + 3z = 3; \quad \Pi_3\colon x + y - z = 1.$$

57. Are P, Q, and R collinear? Justify your answer.

58. Find an equation for the plane through P, Q, and S.

59. Are P, Q, R, and S coplanar? Why or why not?

60. Does the line through P and Q intersect the line through R and S? If so, where? If not, why not?

61. Give an equation for the line of intersection between Π_1 and Π_2.

62. Find a plane that is perpendicular to both Π_2 and Π_3.

63. Is there a plane that is perpendicular to all three of Π_1, Π_2, and Π_3? If so, find one. If not, briefly explain why not.

64. Consider the curve $\mathbf{f}(t) = \left(t - \sin(t), 1 - \cos(t)\right)$ in the xy-plane; use the t-range $0 \le t \le 6\pi$. This curve is called a cycloid. It represents the path followed by a point on the rim of a wheel of radius 1 as the wheel rolls (without slipping) three turns to the right

    (a) Find the velocity and acceleration functions. Describe the point's position, velocity, speed, and acceleration at $t = 1$.

    (b) Find an equation for the tangent line at $t = \pi/2$. Plot both the curve and the tangent line on the same axes. Use equal-sized units on both axes.

    (c) At what times is the speed zero? How do these points appear on your graph?

    (d) Find the length of one arch of the cycloid. [HINT: The half-angle formula $2\sin^2(t/2) = 1 - \cos(t)$ will be handy.]

65. Write a vector equation for the line through the point $(4, 5)$ in the direction of the vector $\mathbf{v} = (3, 2)$. Then find the distance from the origin to the line in question.

66. The two equations $x + 2y + 3z = 3$ and $3x + 2y + z = 1$ describe two different planes. These two planes intersect in a line. Write a vector equation for this line. [HINT: Any two points determine a line.]

In Exercises 67–69, $\mathbf{X} = (x, y, z)$ and ℓ_1, ℓ_2, and ℓ_3 are the lines

    $\ell_1\colon$ given by $\mathbf{X}(t) = (1, 1, 2) + t(1, 2, 3)$;

    $\ell_2\colon$ given by $\mathbf{X}(s) = (9, 8, 11) + s(2, 1, 1)$;

    $\ell_3\colon$ given by $\mathbf{X}(r) = (0, 0, 0) + r(2, 4, 6)$.

67. Do the lines ℓ_1 and ℓ_2 intersect? If so, at what point? If not, why not?

68. Do the lines ℓ_2 and ℓ_3 intersect? If so, at what point? If not, why not?

69. Does the line ℓ_1 intersect the plane $x + y + z = 10$? If so, at what point? If not, why not?

70. Write a scalar equation for the yz-plane.

71. Write a scalar equation for the plane through $(1, 2, 3)$, perpendicular to the line through $(1, 2, 3)$ and $(2, 3, 4)$.

72. Write a scalar equation for any plane that passes through the origin and the point $(1, 2, 3)$.

73. Write a scalar equation for the plane through $(1, 2, 3)$ that is perpendicular to the line through $(1, 2, 3)$ and $(4, 6, 8)$.

74. Write a scalar equation for the plane through $(-1, 0, 2)$, $(1, -2, 0)$, and $(2, 7, 3)$.

75. Write a scalar equation for the plane through the origin spanned by $\mathbf{i} + \mathbf{j}$ and $\mathbf{i} - \mathbf{j}$.

76. Write a scalar equation for the plane that contains both the line with vector equation $\mathbf{X}(s) = (-5, 5, 5) + s(-3, 2, 3)$ and the line with vector equation $\mathbf{X}(t) = (-5, -2, -1) + t(2, 1, 0)$.

77. The equations $2x + 3y + 4z = 20$ and $3x + 2y + z = 10$ describe two different planes. These two planes intersect in a line. Write a vector equation for this line.

In Exercises 78–80, give a parametrization of the curve.

78. The right half of a circle having radius 2 that is centered at $(2, 3)$, and traversed counterclockwise.

79. One full cycle of the curve $y = \sin x$, from $x = 0$ to $x = 2\pi$, traversed from left to right.

80. The curve in Exercise 79, but rotated $\pi/4$ radians counterclockwise with center of rotation at the origin.

81. Suppose that a moving object has constant acceleration $\mathbf{a}(t) = (1, -2)$, that it starts from the origin in the xy-plane at time $t = 0$, and that it has initial velocity vector $\mathbf{v}(0) = (0, 10)$.

   (a) Find formulas for the velocity function $\mathbf{v}(t)$ and the position function $\mathbf{p}(t)$.

   (b) By eliminating the variable t, find an equation in x and y for the path taken by the object.

   (c) Find the time at which the object "lands" (i.e., returns to height $y = 0$). Find the acceleration, velocity, speed, and position at this time.

82. Let a, b, and c be positive numbers and consider the triangle with vertices $(0, 0)$, $(0, a)$, and (b, c). Use the cross product to find the area of the triangle. [HINT: Think of the vertices as the 3-D points $(0, 0, 0)$, $(0, a, 0)$, and $(b, c, 0)$.]

83. A particle moves in the xy-plane. Assume that (i) the particle's acceleration at time t seconds is $\mathbf{a}(t) = (e^{-t}, -1)$; (ii) the particle is at the origin at time $t = 0$; (iii) the particle has velocity $\mathbf{v}(0) = (0, 0)$ at time $t = 0$.

   (a) Find formulas for the velocity function $\mathbf{v}(t)$ and the position function $\mathbf{p}(t)$.

   (b) Find the particle's position, acceleration, velocity, and speed at time $t = 5$; be sure to give appropriate units.

   (c) Suppose that another particle has the same position and velocity as the first particle at $t = 5$ but that the second particle's acceleration is always the zero vector. Find a formula for the second particle's position function $\mathbf{p}(t)$.

   (d) On one set of axes, plot the curves traveled by both particles described above from $t = 0$ to $t = 10$.

Beyond Free Fall

In the real world, most falling objects do not undergo what physicists call free fall (i.e., falling under the *sole* influence of gravity). And a good thing, too—otherwise we'd all be dead, brained by falling raindrops. Luckily, most falling objects are influenced by some "resistive" force, such as air drag, that acts in the direction opposite to the direction of falling. (For objects falling in media other than air, such as a ball bearing in oil, the viscosity of the medium has similar resistive effects.) This project explores the life-saving effects of air drag.

To model the situation, let the positive y-direction represent "up," with the units on both axes in meters. We will use the usual notations $\mathbf{a}(t)$, $\mathbf{v}(t)$, and $\mathbf{p}(t)$ to denote the acceleration, velocity, and position vectors, respectively, for an object falling through the xy-plane. These are vector-valued functions; we'll sometimes write them using components in the form

$$\mathbf{a}(t) = \big(a_1(t), a_2(t)\big), \qquad \mathbf{v}(t) = \big(v_1(t), v_2(t)\big), \qquad \text{and} \qquad \mathbf{p}(t) = \big(p_1(t), p_2(t)\big).$$

For a small, light object falling at moderate speed—such as a raindrop in air—the resistive force of air drag is approximately proportional to the velocity. This means that the acceleration function $\mathbf{a}(t)$ has the form

$$\mathbf{a}(t) = \big(a_1(t), a_2(t)\big) = (0, -g) - k\big(v_1(t), v_2(t)\big),$$

where $k > 0$ is some positive constant, and $g \approx 9.8$ m/s^2 is (as always) the magnitude of the acceleration due to gravity. The vector equation above is equivalent to the two scalar equations

$$a_1(t) = -kv_1(t) \qquad \text{and} \qquad a_2(t) = -g - kv_2(t).$$

Throughout this interlude we assume for simplicity that the falling object starts at the origin (so $p_1(0) = 0$ and $p_2(0) = 0$) and that its initial velocity is horizontal; thus, $v_2(0) = 0$. (This might occur, e.g., if the raindrop were blown horizontally by a sudden gust of wind.)

PROBLEM 1 The acceleration is the derivative of the velocity. This means that the two scalar equations just above are equivalent to the two differential equations (DEs)

$$v_1'(t) = -kv_1(t) \qquad \text{and} \qquad v_2'(t) = -g - kv_2(t).$$

Show that for any constants C_1 and C_2, the functions

$$v_1(t) = C_1 e^{-kt} \qquad \text{and} \qquad v_2(t) = \frac{C_2 e^{-kt} - g}{k}$$

are solutions of the DEs above. (Note: You can simply *check* these claims by differentiation. But students who have studied separable differential equations should also try to *derive* the solutions by separation of variables.)

PROBLEM 2 Explain why the constant C_1 represents the initial horizontal velocity. Also, we are assuming that $v_2(0) = 0$; use this to show that

$$v_2(t) = -\frac{g}{k}\big(1 - e^{-kt}\big).$$

Are v_1 and v_2 positive or negative for $t > 0$?

PROBLEM 3 Integrate the formulas for v_1 and v_2 to show that

$$p_1(t) = -\frac{C_1}{k}e^{-kt} + D_1 \quad \text{and} \quad p_2(t) = -\frac{g}{k}\left(t + \frac{e^{-kt}}{k}\right) + D_2,$$

where D_1 and D_2 are constants of integration. Then use the fact that $\mathbf{p}(0) = (0, 0)$ to rewrite the position functions as

$$p_1(t) = \frac{C_1}{k}(1 - e^{-kt}) \quad \text{and} \quad p_2(t) = \frac{g}{k^2}(1 - e^{-kt} - kt).$$

PROBLEM 4 Find the limits

$$\lim_{t \to \infty} v_1(t); \quad \lim_{t \to \infty} v_2(t); \quad \lim_{t \to \infty} p_1(t); \quad \lim_{t \to \infty} p_2(t).$$

(The answers involve the constants.) What do these numbers mean about the physical situation?

PROBLEM 5 For an average-sized raindrop, the value $k = 1$ is reasonable. Let $s(t)$ be the object's speed at time t. The limit $\lim_{t \to \infty} s(t)$ is called the object's **terminal speed**. Find the raindrop's terminal speed in meters per second. (The result explains, among other things, why we don't die in rainstorms.)

PROBLEM 6 Plot the position function for each initial velocity condition below. (Assume always that $k = 1$ and use $g = 9.8$.) Always use the plotting window $[0, 100] \times [-100, 0]$. If possible, plot all the position functions on the same axes.

(a) $\mathbf{v}(0) = (0, 0)$

(b) $\mathbf{v}(0) = (10, 0)$

(c) $\mathbf{v}(0) = (40, 0)$

(d) $\mathbf{v}(0) = (100, 0)$

DERIVATIVES

13.1 FUNCTIONS OF SEVERAL VARIABLES

The derivative and the integral are the most important concepts of calculus—in one and several variables. This chapter is about derivatives of functions of several variables. We use the plural "derivatives" intentionally because there is more than one way to generalize the basic idea of derivative from functions of one variable setting to the multivariable setting. Indeed, we have already seen derivatives in one of their multivariate forms: In Chapter 12 we found derivatives of vector-valued functions; we interpreted them either geometrically, as tangent vectors to curves, or physically, in the language of velocity and speed.

Most of the functions of Chapter 12 produced vectors as *outputs*, but they were all functions of only one *input* variable. ➡ Thus, in a sense, the functions of Chapter 12 are merely lists of ordinary, single-variable calculus functions. In this chapter, by contrast, we study derivatives of functions of more than one input variable.

We often used t as our (single) input variable.

Functions of one variable are the basic objects of single-variable calculus. Functions of two (or more) variables play a similar role in multivariable calculus. In this section we meet such functions in their own right and consider some of their rudimentary properties.

Functions of one or more variables

The squaring function, defined for all real numbers x by $f(x) = x^2$, is typical of the functions of beginning calculus: f accepts one number, x, as input and assigns another number, x^2, as output. If, say, $x = 2$, then $f(2) = 4$.

Consider, by contrast, the function g defined by $g(x, y) = x^2 + y^2$. Unlike f, the function g accepts a *pair* (x, y) of real numbers as inputs. The output is a third real number, $x^2 + y^2$. If, say, $x = 2$ and $y = 1$, then $g(2, 1) = 4 + 1 = 5$.

Naturally enough, f is called a function of one variable and g a function of two variables. ➡ The difference has to do with domains: The domain of f is the one-dimensional real number line; the domain of g is the two-dimensional xy-plane.

"One" and "two" count the input variables to f and g.

The function f corresponds to the equation $y = x^2$; x is the **independent variable**, and y is the **dependent variable**. The function g corresponds to the **equation** $z = x^2 + y^2$; now both x and y are **independent variables** and z is the **dependent variable**.

We will use f and g below to illustrate various similarities and differences between functions of one and two variables. Notice, however, that there is nothing sacred about *two* variables—we could (and will) discuss such functions as

$$h(x, y, z) = x^2 + y^2 + z^2 \quad \text{and} \quad k(x, y, z, w) = x^2 + y^2 + z^2 + w^2,$$

which accept three or more variables. For now we will keep things simple by sticking mainly to functions of two variables.

Multivariable functions: why bother? Single-variable calculus is challenging enough. Why complicate things by adding more variables?

There are purely mathematical answers, too.

It's a fair question. One good practical answer ◄ is that functions of several variables are essential for describing and predicting phenomena we care about—both natural and human-made. In economics, for instance, a manufacturer's profit depends on many "input" variables: labor costs, distance to markets, tax rates, and so on. In physics, a satellite's motion through space depends on a variety of forces. In biology, populations rise and fall with variations in climate, food supply, predation, and other factors. The weather varies with both longitude and latitude. Our world, in short, is multidimensional: modeling it successfully requires multivariable tools.

The National Weather Service plots multivariable functions every day—they may appear on the back page of your newspaper. For instance, Figure 1 shows noon surface temperatures on a relatively warm winter day:

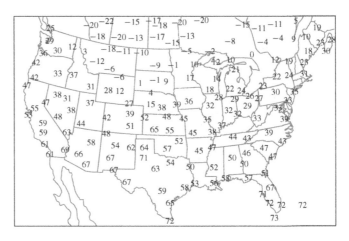

FIGURE 1
Surface temperatures on a January day

The map shows, quite literally, how the temperature function varies across its domain: continental North America.

Functional vocabulary and notation   The basic words and notations for functions of several variables are similar to those for functions of one variable. Roughly speaking, a function is a "machine" that accepts inputs and assigns outputs. A bit more formally:

• **Domain**   The **domain** of a function is the set of permissible inputs. The function $g(x, y) = x^2 + y^2$, for instance, accepts any 2-tuple (x, y) as input, and so its domain is the set $\mathbb{R}^2$.

• **Range**   The **range** of a function is the set of outputs. For g above, the range is the set $[0, \infty)$ of nonnegative real numbers. (The sum of squares can't be negative.)

• **Rule**   The **rule** of a function is the method for assigning an output, or "value," to each given input. For g, the rule is given by the algebraic formula $g(x, y) = x^2 + y^2$.

Not every function has a simple symbolic rule. We will often meet functions given by tables, by graphs, or in other ways.

- **"Arrow" notations**   Above we used the notation $g(x, y) = x^2 + y^2$ to describe a certain function of two variables. Other notations involving arrows are sometimes convenient. The notation

$$g : \mathbb{R}^2 \to \mathbb{R}$$

says that g is a function that accepts *two* real numbers as input and produces *one* real number as output. If we want to specify the rule by which g sends inputs to outputs, we can write

$$g : (x, y) \mapsto x^2 + y^2.$$

These notations remind us that a function *begins* with an input (a 2-tuple in this case) and *ends* with an output (a single number in this case).

- **Vector variables**   It is sometimes convenient to use vector notation in describing functions of several variables. For instance, if we write $\mathbf{X} = (x, y)$, then $\mathbf{X} \cdot \mathbf{X} = (x, y) \cdot (x, y) = x^2 + y^2$, and thus the function $g(x, y) = x^2 + y^2$ could also be written in the form $g(\mathbf{X}) = \mathbf{X} \cdot \mathbf{X}$. In three variables, if we write $\mathbf{X} = (x, y, z)$ and $\mathbf{A} = (1, 2, 3)$, then the notations

$$h(x, y, z) = x + 2y + 3z + 4 \qquad \text{and} \qquad h(\mathbf{X}) = \mathbf{A} \cdot \mathbf{X} + 4$$

convey the same information.

Graphs in one and several variables

Let's compare graphs of the functions $f(x) = x^2$ and $g(x, y) = x^2 + y^2$ discussed earlier. The graph of f is the set of all points (x, y) for which $y = f(x) = x^2$. For example, $f(2) = 4$, and so the point $(2, 4)$ lies on the graph. Geometrically, this graph is a curve—a parabola, in this case—in the the xy-plane. Like a straight line, the curve $y = x^2$ is a one-dimensional object ➡ that lives in a two-dimensional space—the xy-plane. Figure 2 shows part of the familiar graph:

To an ant walking along a parabola, it "looks" like a (one-dimensional) straight line.

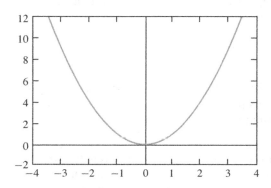

FIGURE 2
Graph of $f(x) = x^2$: a parabola

The graph of g is the set of all points (x, y, z) for which $z = g(x, y) = x^2 + y^2$. For example, since $g(2, 1) = 5$, the point $(2, 1, 5)$ lies on the g-graph. Figure 3 shows some of the graph of g:

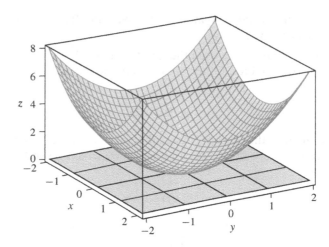

FIGURE 3
Graph of $g(x, y) = x^2 + y^2$: a paraboloid

We discussed this same paraboloid in Section 12.1.

To an ant walking along the g-graph, it "looks" like a (two-dimensional) flat plane—just as Earth's surface looks flat from a human vantage point.

The graph, a **paraboloid**, ← is quite different from the **parabola** in Figure 2. Notice especially the dimensions involved: The graph of g is a two-dimensional surface ← hovering in three-dimensional xyz-space. As a rule, the "dimension" of any graph is the number of *input* variables the function accepts. (Some very ill-behaved functions violate this rule.)

In other respects, the graphs of f and g are quite similar to each other. For both functions, the height of the graph above a given domain point (a typical input is x_0 for f and (x_0, y_0) for g) tells the corresponding output value (typical outputs are $y_0 = f(x_0)$ and $z_0 = g(x_0, y_0)$, respectively).

And to the interest.

Multivariable graphs: beware . . . Graphs are at least as important in multivariable calculus as in elementary calculus, but multivariable graphs are usually more complicated and so need extra care in handling. Choosing a "good" viewing window, for example, takes some care even for functions of one variable, and the problem can be stickier still for functions of two variables. The fact that multivariable graphs "live" naturally in three-dimensional (or even higher-dimensional) space—not on a flat page or computer screen— only adds to the problem. ←

Different views of multivariable functions

Given the difficulty of visualizing multivariable graphs, it is helpful to "see" functions from as many points of view as possible. Level curves, contour maps, and numerical tables are especially useful for functions of two variables.

Level curves and contour maps Let $f(x, y)$ be a function of two variables, and let c be a number in the range of f. Then $f(x, y) = c$ is an equation in x and y; its graph is (usually) a curve in the xy-plane. ← Such curves have a special name:

Occasionally this curve is just a point.

> **DEFINITION** Let $f(x, y)$ be a function and c a constant. The set of all (x, y) for which $f(x, y) = c$ is called a **level curve** (or **level set**) of f. A collection of level curves drawn together is called a **contour map** of f.

Figure 4 shows several level curves; we study them in detail in Example 1. In the meantime, observe:

- **Why "level"?** The word "level" makes good sense here because $f(x, y)$ has the same value, namely c, at each point (x, y) along the level curve. In other words, the graph of f is "level" above the level curve $f = c$.

- **Which level curves to draw?** Each number c in the range of f has its own level curve. Because most functions have infinitely many output values, we can't draw *all* of the level curves. In practice, therefore, we draw some convenient selection of curves usually corresponding to evenly spaced values of the output c. The spacing *between* curves reflects how fast the function increases or decreases.

- **Level curves or level sets?** The phrase "level curve" makes good sense for functions of two variables, which are usually constant on lines or curves in the xy-plane. In other dimensions, the phrase "level set" may be better because level sets may be points, surfaces, or higher-dimensional sets.

- **Labels** Level curves are sometimes labeled with their corresponding output values. Typical labels have the form $z = c$, $f = c$, or just c. Sometimes, as in Figure 4, color or shading is used to indicate levels.

E X A M P L E 1 Figure 4 shows a sample of level curves for the function $g(x, y) = x^2 + y^2$. Label each level curve with the appropriate output value. The graph of g is a paraboloid. How is this shape reflected in the level curves shown in Figure 4?

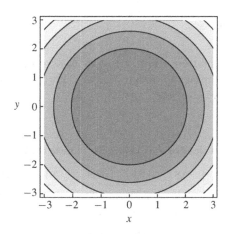

FIGURE 4
Level curves of $g(x, y) = x^2 + y^2$

Solution All level curves are circles about the origin. ➔ Each circle is the graph of an equation $x^2 + y^2 = c$ for some c. The circles, from smallest to largest, correspond to the levels $z = 1, 4, 7, 10, 13, 16, 19$. (The shading corresponds to levels—darker shades are "lower.")

   Notice that level curves of g get closer and closer together as we move outward from the origin. This reflects the fact that the g-graph is a paraboloid—it gets steeper and steeper as we move away from the origin. ■

Some circles are not fully visible.

EXAMPLE 2 Drawing level curves on a temperature map makes the map easier to read and interpret. (Newspapers usually do this.) Figure 5 shows another version of the temperature map in Figure 1; the boundaries of shaded regions correspond to the level curves of the temperature function; they are called **isotherms**.

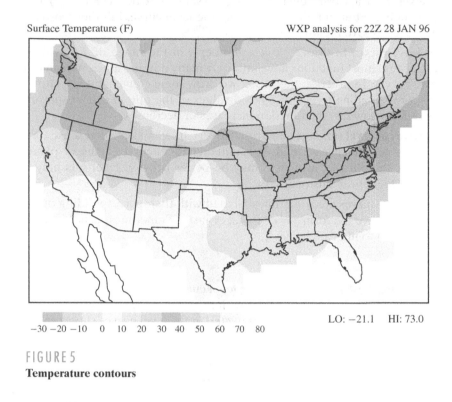

Surface Temperature (F)   WXP analysis for 22Z 28 JAN 96

LO: −21.1   HI: 73.0

−30 −20 −10   0   10   20   30   40   50   60   70   80

FIGURE 5
Temperature contours

Functions as tables Information about a function is sometimes given numerically, in tabular form, rather than graphically or symbolically. Here, for instance, are a few output values of the function $g(x, y) = x^2 + y^2$: ◄

For some functions we may have only a table of values and no symbolic formula.

y \ x	**0.8**	**0.9**	**1.0**	**1.1**	**1.2**
2.2	5.48	5.65	5.84	6.05	6.28
2.1	5.05	5.22	5.41	5.62	5.85
2.0	4.64	4.81	5.00	5.21	5.44
1.9	4.25	4.42	4.61	4.82	5.05
1.8	3.88	4.05	4.24	4.45	4.68

Values of $g(x, y) = x^2 + y^2$ near $(1, 2)$

The table alone reveals nothing about how g behaves anywhere else—for that we would need a graph, a formula, or a larger table.

The table contains a lot of information about the function g for inputs near $(1, 2)$. ◄ For instance, the pattern of increase in the table means that, for inputs (x, y) near $(1, 2)$, the function's output values increase as either x or y increases. We see, too, that the table entries increase *faster and faster* as we move to the right or upward. Notice that the graph and the contour map convey the same information in their own (geometric) ways.

Linear functions

A linear function of one variable is one that can be written in the form $L(x) = a + bx$, where a and b are constants. A linear function in several variables has a similar algebraic form:

> **DEFINITION** A **linear function** in two variables is one that can be written in the form
> $$L(x, y) = a + bx + cy,$$
> where a, b, and c are constants.

Linear functions in three (or more) variables are similar. A linear function of three variables has the form $L(x, y, z) = a + bx + cy + dz$, where a, b, c, and d are constants. In vector notation, with $\mathbf{X} = (x, y, z)$ and $\mathbf{A} = (a, b, c)$, a linear function has the simple formula $L(\mathbf{X}) = \mathbf{A} \cdot \mathbf{X} + d$, where d is a constant.

Why "linear"? Why do linear functions deserve that name? In the xy-plane, there is no mystery: The graph of a linear function $y = a + bx$ is a straight line. In xyz-space, however, the graph of a linear function $z = a + bx + cy$ is not a line but a plane. (In $xyzw$-space, the graph of a linear function $w = a + bx + cy + dz$ is even less like a line. It's a three-dimensional solid called a **hyperplane**.) Nevertheless, we will call any function linear that involves only constants and the first power of each variable, regardless of the number of variables. In the same spirit, we will call a function quadratic—no matter how many variables are involved—if the variables appear in nonnegative integer powers, none of which exceeds two. (The function $f(x, y) = x^2 + xy + y^2$ is quadratic in this sense.)

Readers who have studied linear algebra ⇢ will recall that the word "linear" (as in "linear transformation") is used in still another way in that subject. (The functions we call "linear" in this book are sometimes referred to, in linear algebra jargon, as "affine" functions.) Like other very useful English words, "linear" has spawned a whole family of related but not identical meanings.

Yet another use of "linear"!

Planes and linear functions We saw in Chapter 1 that every plane Π in $\mathbb{R}^3$ has a scalar equation of the form $ax + by + cz = d$ with a, b, c, and d all constants. We saw, too, that the vector (a, b, c) is normal to Π at any point of Π. These facts can help us understand and visualize planes either as graphs of linear functions or in terms of points and normal vectors.

Let $L(x, y) = Ax + By + C$ be a linear function. ⇢ The graph of L is the set of points (x, y, z) that satisfy the equation

We use capital letters here because we used a, b, c, and d a little differently above.

$$z = Ax + By + C, \quad \text{or, equivalently,} \quad Ax + By - z = -C.$$

From the last equation we can simply read off the normal vector; the conclusion follows:

> **FACT** Let $L(x, y) = Ax + By + C$ define a linear function $L : \mathbb{R}^2 \to \mathbb{R}$. The graph of L is the plane with scalar equation $Ax + By - z = -C$. This plane passes through the point $(0, 0, C)$ and has normal vector $(A, B, -1)$.

EXAMPLE 3 Describe the graphs of the linear functions $L(x, y) = 3$ and $M(x, y) = -\dfrac{x}{3} - \dfrac{2y}{3} + 2$.

Solution The preceding Fact says that the graph of L, a plane, has normal vector $(A, B, -1) = (0, 0, -1) = -\mathbf{k}$. This means that the graph is perpendicular to the z-axis. This is easy to visualize because the plane has equation $z = 3$ and so is parallel to the xy-plane.

The graph of M is the plane with equation

$$z = -\frac{x}{3} - \frac{2y}{3} + 2, \qquad \text{or, equivalently,} \qquad x + 2y + 3z = 6.$$

Multiplying a normal vector by a nonzero constant doesn't affect perpendicularity.

By the Fact, the vector $(-1/3, -2/3, -1)$ is normal to the plane; therefore, $-3(-1/3, -2/3, -1) = (1, 2, 3)$ is also normal to the plane. ← This plane and the normal vector $(1, 2, 3)$ are shown in Example 5, page 685. ◼

Why linear functions matter Linear functions are simple, useful, and easy to work with. Most important for us, linear functions are prototypes or models for all differentiable functions. Indeed, any differentiable function, in any number of variables, might be called "almost linear," or "locally linear," in much the same sense that an ordinary calculus function $y = f(x)$ looks like a straight line if we zoom in repeatedly on a typical point on its graph. We will return often to this theme.

Vector-valued functions of several variables; matrix notations So far in this section we have considered only *scalar-valued* functions of several variables—functions that accept several variables as inputs and produce scalars as outputs. But *vector-valued* functions of several variables are also possible and useful. For instance, consider the function $\mathbf{f} \colon \mathbb{R}^2 \to \mathbb{R}^2$ defined by

$$\mathbf{f} \colon (x, y) \mapsto (x + 2y + 3, 4x + 5y + 6).$$

Notice that $\mathbf{f}$ accepts a 2-vector as input and produces another 2-vector as output. Using vectors and matrices, we could, instead, write the formula for $\mathbf{f}(x, y)$ in the form

$$\mathbf{f}(\mathbf{X}) = \begin{bmatrix} 1 & 2 \\ 4 & 5 \end{bmatrix} \begin{bmatrix} x \\ y \end{bmatrix} + \begin{bmatrix} 3 \\ 6 \end{bmatrix}.$$

See Appendix J for a brief introduction to (or review of) matrices and matrix multiplication.

Matrix multiplication can simplify and unclutter treatment of some multivariable ideas; we will use it later in this book. ← For the moment, the main point is simply to acknowledge the variety of functions that multivariable calculus can treat.

Vector-valued functions of several variables have important practical applications—especially in modeling physical phenomena such as forces that vary over space. We will return to vector-valued functions and their uses in later chapters. For the moment, our main purpose will be to study the components from which functions like $\mathbf{f}$ are built—scalar-valued functions of several variables.

Multivariate limits, continuity, and differentiability: a first glimpse at theory

The theory of calculus—in one *and* several variables—relies on certain tacit assumptions about the functions under study. For example, the mean value theorem of one-variable calculus requires that the function in question be *continuous* on a closed interval $[a, b]$ and *differentiable* on the open interval (a, b). Not every function has these properties. For example, the absolute value function $f(x) = |x|$ is continuous on every interval but not differentiable at $x = 0$. On the other hand, the standard elementary functions, such as polynomial, trigonometric, and exponential functions, are continuous and differentiable wherever defined, and so we can often use them without undue fuss.

Similar technical requirements apply in multivariable calculus. We will often assume (sometimes without explicit mention) that the functions we use are "well-behaved" in the sense of being continuous or differentiable, having limits, and the like. These technical properties turn out to be harder to define and study rigorously in several variables than in one variable; doing so is important in more advanced courses. As a rule, we will treat such matters informally, with occasional glances at the theoretical side of the subject.

Consider, for example, the question of **continuity** for a function $f(x, y)$. A rigorous definition of continuity involves a limit—and limits themselves require rigorous definition for functions of several variables. ➡ Intuitively, however, a continuous function of two variables is one whose graph (a surface) has no "holes" or "tears"; a **differentiable** function is one whose graph is "smooth," without kinks or corners. (We will consider multivariate continuity and differentiability in a little more detail in Section 13.8.)

We'll discuss multivariate limits briefly later in this chapter.

Well- and ill-behaved functions   As one might expect, multivariable functions, such as $g(x, y) = x^2 + y^2$, that are built from standard elementary function ingredients are differentiable and continuous wherever they are defined. Graphs of such functions (the one in Figure 3, page 706, is a good example) are smooth, unbroken, and otherwise relatively tame. But—as in single-variable calculus—not every function given by a simple formula behaves quite so well.

EXAMPLE 4   A function f is defined by
$$f(x, y) = \begin{cases} \dfrac{xy}{x^2 + y^2} & \text{if } (x, y) \neq (0, 0); \\ 0 & \text{if } (x, y) = (0, 0). \end{cases}$$

Figure 6 shows part of the graph (and part of the xy-plane for reference):

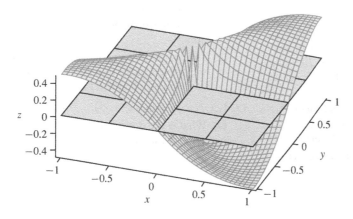

FIGURE 6
An ill-behaved function

What do the graph and formula suggest about the function's behavior?

Solution   The surface appears to be torn or folded near the origin, that is, for inputs near $(x, y) = (0, 0)$, but (from what we can see) it is well-behaved elsewhere. The algebraic formula suggests the same thing: The quotient expression is undefined at $(x, y) = (0, 0)$, ➡ but *only* there. We might (correctly) guess, then, that f is **discontinuous** at $(0, 0)$, but continuous and differentiable everywhere else.

That's why we defined $f(0, 0)$ separately.

Without precise definitions at hand, these properties can only be understood informally. We will return to the matter briefly at the end of this chapter. ■

BASIC EXERCISES

In Exercises 1–6, find the domain and the range of the function.

1. $g(x, y) = x^2 + y^2$

2. $h(x, y) = x^2 + y^2 + 3$

3. $j(x, y) = 1/(x^2 + y^2)$

4. $k(x, y) = e^{-(x^2+y^2)}$

5. $m(x, y, z) = \sqrt{1 - x^2 - y^2 - z^2}$

6. $f(x, y) = \cos(xy^2z^3)$

7. For any nonzero constant a, the graph of $z = ax^2 + ay^2$ is a **circular paraboloid**. This exercise explores the reasons for this terminology.

 (a) Find the traces of the surface $z = x^2 + y^2$ in the plane $z = c$ for $c = 0$, $c = 1$, and $c = 2$.

 (b) Explain why, for each real number c such that $ac > 0$, the trace of the surface $z = ax^2 + ay^2$ in the plane $z = c$ is a circle.

 (c) Describe the trace of the surface $z = ax^2 + ay^2$ in the plane $x = c$. Find equations for the traces for $c = 0, c = 1$, and $c = 2$.

 (d) Describe the trace of the surface $z = ax^2 + ay^2$ in the plane $y = c$. Find equations for the traces for $c = 0, c = 1$, and $c = 2$.

8. Repeat the previous exercise, but for the **hyperbolic paraboloid** $z = x^2 - y^2$. (In this case the cross sections $z = c$ are hyperbolas, not circles.)

9. Let $f(x, y) = x^2 + y^2$ and let $g(x, y) = x^2 + y^2 + 1$.

 (a) In the rectangle $[-3, 3] \times [-3, 3]$, draw and label the level curves of f that correspond to $z = 0$, $z = 2$, $z = 4$, $z = 6$, and $z = 8$. What is the shape of each level curve?

 (b) In the rectangle $[-3, 3] \times [-3, 3]$, draw and label the level curves of g that correspond to $z = 1$, $z = 3$, $z = 5$, $z = 7$, and $z = 9$. What is the shape of each level curve?

 (c) How are the results of parts (a) and (b) similar? How are they different?

 (d) Use technology to plot the graphs $z = f(x, y)$ and $z = g(x, y)$, for (x, y) in $[-3, 3] \times [-3, 3]$. Describe briefly, in words, how the two graphs are related to each other.

10. Let $f(x, y) = x^2 + y^2$ and let $g(x, y) = \sqrt{x^2 + y^2}$.

 (a) In the rectangle $[-4, 4] \times [-4, 4]$, draw and label the level curves of f that correspond to $z = 0$, $z = 1$, $z = 4$, $z = 9$, and $z = 16$. What is the shape of each level curve?

 (b) In the rectangle $[-4, 4] \times [-4, 4]$, draw and label the level curves of g that correspond to $z = 0$, $z = 1$, $z = 2$, $z = 3$, and $z = 4$. What is the shape of each level curve?

 (c) Use technology to plot the graphs $z = f(x, y)$ and $z = g(x, y)$, for (x, y) in $[-4, 4] \times [-4, 4]$.

 (d) How are the level curves of f and g similar? How are they different?

11. Let $f(x, y) = y - x^2$ and let $g(x, y) = x - y^2$.

 (a) In the rectangle $[-3, 3] \times [-3, 3]$, draw and label the level curves of f that correspond to $z = -3, z = -2, \ldots$, $z = 2, z = 3$. What is the shape of each level curve?

 (b) In the rectangle $[-3, 3] \times [-3, 3]$, draw and label the level curves of g that correspond to $z = -3, z = -2, \ldots$, $z = 2, z = 3$. What is the shape of each level curve?

 (c) How are the results of parts (a) and (b) similar? How are they different?

 (d) Use technology to plot the graphs $z = f(x, y)$ and $z = g(x, y)$, for (x, y) in $[-3, 3] \times [-3, 3]$. Describe briefly, in words, how the two graphs are related to each other.

12. Let f and g be the functions $f(x, y) = 2 + x^2$ and $g(x, y) = 2 + y^2$.

 (a) In the rectangle $[-3, 3] \times [-3, 3]$, draw and label the level curves of f that correspond to $z = 0$, $z = 2$, $z = 4$, $z = 6$, and $z = 8$. What is the shape of each level curve?

 (b) In the rectangle $[-3, 3] \times [-3, 3]$, draw and label the level curves of g that correspond to $z = 0$, $z = 2$, $z = 4$, $z = 6$, and $z = 8$. What is the shape of each level curve?

 (c) How are the results of parts (a) and (b) similar? How are they different?

 (d) Use technology to plot the graphs $z = f(x, y)$ and $z = g(x, y)$ for (x, y) in $[-3, 3] \times [-3, 3]$. Describe briefly, in words, how the two graphs are related to each other.

 (e) The graphs of f and g are both "cylinders" (in the sense we defined in Section 12.1). How do the contour maps of f and g reflect this fact?

13. Let $f(x, y) = \ln(4x^2 + y^2)$.

 (a) What is the domain of f?

 (b) What is the range of f?

 (c) Describe the level curves of f.

14. Let $g(x, y) = e^{xy}$.

 (a) What is the domain of g?

 (b) What is the range of g?

 (c) Describe the level curves of g.

15. Let f and g be the linear functions $f(x, y) = 2x - 3y$ and $g(x, y) = -2x + 3y$.

 (a) In the rectangle $[-3, 3] \times [-3, 3]$, draw and label the level curves of f that correspond to $z = -5$, $z = -3$, $z = -1, z = 1, z = 3$, and $z = 5$. What is the shape of each level curve?

 (b) In the rectangle $[-3, 3] \times [-3, 3]$, draw and label the level curves of g that correspond to $z = -5$, $z = -3$, $z = -1, z = 1, z = 3$, and $z = 5$. What is the shape of each level curve?

(c) How are the results of parts (a) and (b) similar? How are they different?

(d) What special properties do the level curves of a *linear* function have?

(e) Use technology to plot the graphs $z = f(x, y)$ and $z = g(x, y)$ for (x, y) in $[-3, 3] \times [-3, 3]$. Describe briefly, in words, how the two graphs are related to each other.

16. The following table shows some values of a linear function $L(x, y)$. (No explicit symbolic formula for L is given.)

y \ x	−3	−2	−1	0	1	2	3
3	−15	−12	−9	−6	−3	0	3
2	−13	−10	−7	−4	−1	2	5
1	−11	−8	−5	−2	1	4	7
0	−9	−6	−3	0	3	6	9
−1	−7	−4	−1	2	5	8	11
−2	−5	−2	1	4	7	10	13
−3	−3	0	3	6	9	12	15

(a) All level curves of L are straight lines. Using this fact, draw (all in the rectangle $[-3, 3] \times [-3, 3]$) and label the level curves $z = -12$, $z = -8$, $z = -4$, $z = 0$, $z = 4$, $z = 8$, $z = 12$.

(b) Find an equation in x and y for the level line $z = 0$.

(c) Because L is a linear function, its formula has the form $L(x, y) = a + bx + cy$ for some constants a, b, and c. Find numerical values for a, b, and c. [HINT: The table says that $L(0, 0) = 0$. Therefore, $L(0, 0) = a + b \cdot 0 + c \cdot 0 = 0$, so $a = 0$. Use similar reasoning to find values for b and c.]

(d) Use technology to plot $L(x, y)$ over the rectangle $[-3, 3] \times [-3, 3]$. Is the shape of the graph consistent with the level curves you plotted in part (a)?

17. Let f be the linear function $f(x, y) = 2x + 3y + 4$.

(a) Describe the graph of f.

(b) Find an equation of the line through the point $(0, 0, 4)$ that is perpendicular to the graph of f.

(c) Let c be a constant. Show that the level curve of f that corresponds to $z = c$ is $y = -2x/3 + (c - 4)/3$.

(d) Describe the contour map of f.

18. Imagine a map of the United States in the usual position. The positive x-direction is east, and the positive y-direction is north. Suppose that the units of x and y are miles and that Los Angeles, California, has coordinates $(0, 0)$. (Several approximations are involved here of course: The Earth's surface is not really flat, and the real Los Angeles occupies more than a single point.) Let $T(x, y)$ be the temperature, in degrees Celsius, at the location (x, y) at noon, Central Standard Time, on January 1, 2003.

(a) What does it mean in weather language to say that $T(0, 0) = 15$?

(b) What do the level curves of T mean in weather language? As a rule, would you expect, level curves of T to run north and south or east and west? Why?

(c) International Falls, Minnesota, is about 1400 miles east and 1100 miles north of Los Angeles. The noon temperature in International Falls on January 1, 2003, was $-15°$C. What does this imply about $T(x, y)$?

(d) Suppose that International Falls was the coldest spot in the country at the time in question. How would you expect the level curves to look near International Falls?

19. Describe the level surfaces of $f(x, y, z) = x^2 + y^2 + z^2$.

20. Describe the level surfaces of $g(x, y, z) = x + 2y + 3z$.

FURTHER EXERCISES

21. For any point (x, y) in the xy-plane, let $f(x, y)$ be the distance from (x, y) to the origin. Then f has the formula $f(x, y) = \sqrt{x^2 + y^2}$.

(a) Find the domain and range of f.

(b) Plot f. Describe the graph in words.

(c) Draw the level curve of f that passes through $(3, 4)$.

(d) All level curves of f have the same shape. What is it?

22. For any point (x, y) in the xy-plane, let $f(x, y)$ be the distance from (x, y) to the line $x = 1$.

(a) Find a formula for $f(x, y)$.

(b) Plot f. Describe the graph in words.

(c) Draw the level curve that passes through $(3, 4)$.

(d) All level curves of f have the same shape. What is it?

23. Suppose that the level curves of a function $g(x, y)$ are horizontal lines. What does this imply about g?

24. Suppose that the level curves of the function f are parallel straight lines. Must f be a plane? Justify your answer.

*In Exercises 25–28, indicate whether the statement **must** be true, **might** be true, or **cannot** be true. Justify your answers.*

25. The level curve of the function $f(x, y)$ corresponding to $z = 7$ consists of the lines $y = -2x$ and $y = 5x$.

26. The level curve of the linear function $g(x, y)$ corresponding to $z = 7$ is the line $y = 1 - 2x$, and the level curve corresponding to $z = 1$ is the line $y = 5x + 3$.

27. The level curve of the linear function $h(x, y)$ that corresponds to $z = 0$ is the parabola $y = 1 - x^2$.

28. The level curve of the function $g(x, y)$ that corresponds to $z = -3$ is the ellipse $x^2/25 + y^2/9 = 1$, and the level curve that corresponds to $z = 2$ is the circle $x^2 + y^2 = 16$.

In Exercises 29–32, describe the level sets $f(\mathbf{X}) = 0$ and $f(\mathbf{X}) = 3$, where $\mathbf{X} = (x, y)$. (The level sets are subsets of $\mathbb{R}^2$.)

29. $f(\mathbf{X}) = \mathbf{X} \cdot \mathbf{i}$

30. $f(\mathbf{X}) = \mathbf{X} \cdot \mathbf{X}$

31. $f(\mathbf{X}) = \mathbf{X} \cdot (\mathbf{i} + \mathbf{j})$

32. $f(\mathbf{X}) = \mathbf{X} \cdot (2, 3)$

In Exercises 33–36, describe the level sets $f(\mathbf{X}) = 0$ and $f(\mathbf{X}) = 3$, where $\mathbf{X} = (x, y, z)$. (The level sets are subsets of $\mathbb{R}^3$.)

33. $f(\mathbf{X}) = \mathbf{X} \cdot \mathbf{i}$

34. $f(\mathbf{X}) = \mathbf{X} \cdot \mathbf{X}$

35. $f(\mathbf{X}) = \mathbf{X} \cdot (\mathbf{i} + \mathbf{j} + \mathbf{k})$

36. $f(\mathbf{X}) = \mathbf{X} \cdot (1, 2, 3) + 4$

13.2 PARTIAL DERIVATIVES

Derivatives in one variable — interpretations

Let f be a function of one variable. Recall some familiar properties of the derivative function f': ←

We assume, to avoid distractions, that f and f' are continuous functions.

- **Slope**   For any fixed input $x = x_0$, the derivative $f'(x_0)$ tells the slope of the f-graph at the point $(x_0, f(x_0))$. The sign of $f'(x_0)$, in particular, tells whether f is increasing or decreasing ← at $x = x_0$.

Rising or falling, in everyday speech

- **Rate of change**   The derivative f' can also be interpreted as the *rate function* associated to f as follows: For any input x_0, the derivative $f'(x_0)$ tells the instantaneous rate of change of $f(x)$ with respect to x. If, say, $f(x)$ gives the position of a moving object at time x, then $f'(x)$ gives the corresponding velocity at time x. ←

In a specific example we'd need to specify appropriate units for everything.

- **Limit**   The derivative $f'(x_0)$ is defined as a *limit* of difference quotients:

$$f'(x_0) = \lim_{h \to 0} \frac{f(x_0 + h) - f(x_0)}{h}.$$

For any $h > 0$, the difference quotient can be thought of either as the average rate of change of f over the interval $[x_0, x_0 + h]$ or as the slope of a secant line on the f-graph over the same interval. ← Taking the limit as $h \to 0$ corresponds to finding the instantaneous rate of change of f or, equivalently, the slope of the tangent line to the f-graph at $x = x_0$.

Similar interpretations hold if $h < 0$, but $h = 0$ is taboo.

- **Linear approximation**   At a point $(x_0, f(x_0))$ on the curve $y = f(x)$, the tangent line has slope $f'(x_0)$. This tangent line is the graph of the linear function L with equation

$$y = L(x) = f(x_0) + f'(x_0)(x - x_0).$$

See the exercises for more on linear approximation.

The function L is called the **linear approximation to f at x_0**. ← The name makes sense because the graphs of f and L are close together near x_0. In symbols,

$$L(x) \approx f(x) \qquad \text{when} \qquad x \approx x_0.$$

Derivatives in several variables

Derivatives are just as important in multivariable calculus as in one-variable calculus, but the idea is—not surprisingly—more complicated for functions of several variables.

Take slope, for instance. The graph of a one-variable function $y = f(x)$ is a curve in the xy-plane. The slope of the graph at (x_0, y_0)—a single number—completely describes the graph's "direction" at (x_0, y_0). ← The graph of a two-variable function $z = f(x, y)$, by contrast, is a *surface*, which has no single "slope" at a point. A surface's steepness at a point depends on the direction (uphill, downhill, along the "contour," etc.) that one follows from

Here $y_0 = f(x_0)$.

the point. ➧ To put it another way, the graph of a one-variable function can be approximated near a given point by a one-dimensional tangent *line*. The graph of a two-variable function, as we will see, can be approximated near a fixed point by a two-dimensional tangent *plane*.

Every hiker knows this. How steep a mountain "feels" depends on the direction of the trail.

Here is the moral: To suit multivariable calculus, our notion of derivative must go beyond the simple idea of slope. We will see later (and often) that the idea of linear approximation—not slope—turns out to be the key to extending the idea of derivative to functions of more than one variable.

In this section we start to extend the derivative idea to functions of several variables. **Partial derivatives** are the simplest multivariable analogues of ordinary derivatives. ➧

"Partial" suggests (correctly) that there's more to the derivative story.

Partial derivatives: the idea The basic idea of a partial derivative is to differentiate with respect to *one* variable while holding all the others constant. An easy example will illustrate the idea and introduce some useful terminology and notation. ➧

The fine print will come later.

E X A M P L E 1   Let $f(x, y) = x^2 - 3xy + 6$. Find $f_x(x, y)$ and $f_y(x, y)$, the partial derivatives of f with respect to x and y, respectively. Find the numerical values $f_x(2, 1)$ and $f_y(2, 1)$.

S o l u t i o n   To find f_x, we differentiate $f(x, y) = x^2 - 3xy + 6$ with respect to x, treating y as a constant: ➧

$$f_x(x, y) = 2x - 3y.$$

Check this calculation and the next carefully. Do you agree?

To find f_y, we treat x as a constant:

$$f_y(x, y) = -3x.$$

Setting $x = 2$ and $y = 1$ in these formulas gives

$$f_x(2, 1) = 2 \cdot 2 - 3 \cdot 1 = 1; \qquad f_y(2, 1) = -3 \cdot 2 = -6. \qquad ■$$

The calculations were easy, but what do the results mean? What do they say about how f behaves near $(x, y) = (2, 1)$? Can we interpret the results graphically and numerically? Understanding multivariable derivatives fully is a long-term proposition, but here are some starters.

Holding variables constant To find the partial derivative f_x of a function f with respect to x, we treat all the *other* variables as constants. This produces a function of just one variable, x, which we differentiate in the "usual" way. For example, suppose we fix $y = 3$ in $f(x, y) = x^2 - 3xy + 6$. Then, $f(x, y) = f(x, 3) = x^2 - 9x + 6$, and

$$\frac{d}{dx}[f(x, 3)] = f_x(x, 3) = 2x - 9.$$

This agrees, as it should, with the general formula $f_x(x, y) = 2x - 3y$ found above.

Directional rates of change Partial derivatives (like ordinary derivatives) can be interpreted as rates or slopes—but with an important proviso about directions. For a function $f(x, y)$ and a point (x_0, y_0) in its domain, the partial derivative $f_x(x_0, y_0)$ tells the rate of change of $f(x, y)$ with respect to x, ➧ that is, how fast $f(x, y)$ increases as the input (x, y) moves away from (x_0, y_0) in the positive x-direction. The other partial derivative, $f_y(x_0, y_0)$, tells how fast f increases near (x_0, y_0) as y increases. The next example illustrates what this means numerically.

We fixed y, and so x is the only variable that can "move."

E X A M P L E 2   The function $f(x, y) = x^2 - 3xy + 6$ has partial derivatives $f_x(x, y) = 2x - 3y$ and $f_y(x, y) = -3x$. ➧ What does this say about rates of change of f at $(x, y) = (2, 1)$? At $(x, y) = (1, 2)$?

We saw this above.

S o l u t i o n   We will start at $(2, 1)$. The formulas give $f_x(2, 1) = 1$ and $f_y(2, 1) = -6$. These numbers represent rates of change of f with respect to x and y, respectively,

The boxed row and column meet at (2, 1), our target point.

at (2, 1). A table of f-values "centered" at (2, 1) shows what this means: ←

Values of $f(x, y) = x^2 - 3xy + 6$ near (2, 1)							
y \ x	**1.97**	**1.98**	**1.99**	**2.00**	**2.01**	**2.02**	**2.03**
1.03	3.7936	3.8022	3.8110	3.8200	3.8292	3.8386	3.8482
1.02	3.8527	3.8616	3.8707	3.8800	3.8895	3.8992	3.9091
1.01	3.9118	3.9210	3.9304	3.9400	3.9498	3.9598	3.9700
1.00	3.9709	3.9804	3.9901	4.0000	4.0101	4.0204	4.0309
0.99	4.0300	4.0398	4.0498	4.0600	4.0704	4.0810	4.0918
0.98	4.0891	4.0992	4.1095	4.1200	4.1307	4.1416	4.1527
0.97	4.1482	4.1586	4.1692	4.1800	4.1910	4.2022	4.2136

Reading *up* the boxed column (at each step y *increases* by 0.01) shows successive corresponding values of f *decreasing* by about 0.06. Thus, $-.06/.01 = -6$ is the rate of change of f with respect to y at (2, 1)—equivalently, $f_y(2, 1) = -6$. Similarly, the fact that $f_x(2, 1) = 1$ suggests that values of $f(x, 1)$ should increase at about the same rate as x if $x \approx 2$. Reading across the boxed *row* confirms this expectation. ←

Convince yourself of this. Use the formulas to check these values.

Similar reasoning applies at (1, 2). The values $f_x(1, 2) = -4$ and $f_y(1, 2) = -3$ ← mean that f *decreases* (at different rates) with respect to both x and y near (1, 2). Numerical f-values bear this out:

Values of $f(x, y) = x^2 - 3xy + 6$ near (1, 2)					
y \ x	**0.98**	**0.99**	**1.00**	**1.01**	**1.02**
2.02	1.0216	0.9807	0.9400	0.8995	0.8592
2.01	1.0510	1.0104	0.9700	0.9298	0.8898
2.00	1.0804	1.0401	1.0000	0.9601	0.9204
1.99	1.1098	1.0698	1.0300	0.9904	0.9510
1.98	1.1392	1.0995	1.0600	1.0207	0.9816

Reading either across or up shows f-values decreasing at rates of about -4 and -3, respectively. ■

Formal definitions

Partial derivatives are defined formally as limits, much like ordinary derivatives:

DEFINITION Let $f(x, y)$ be a function of two variables. The partial derivative with respect to x of f at (x_0, y_0), denoted $f_x(x_0, y_0)$, is defined by

$$f_x(x_0, y_0) = \lim_{h \to 0} \frac{f(x_0 + h, y_0) - f(x_0, y_0)}{h}$$

if the limit exists. The partial derivative with respect to y at (x_0, y_0), denoted $f_y(x_0, y_0)$, is defined by

$$f_y(x_0, y_0) = \lim_{h \to 0} \frac{f(x_0, y_0 + h) - f(x_0, y_0)}{h}$$

if the limit exists.

The definition says, in effect, that $f_x(x_0, y_0)$ is the ordinary derivative at $x = x_0$ of an ordinary function of one variable—namely, the function given by the rule $x \mapsto f(x, y_0)$. Here are some further comments and observations:

- **Do they exist?**   If either limit does not exist, then neither does the corresponding partial derivative. It is possible, for instance, that $f_x(x_0, y_0)$ exists but $f_y(x_0, y_0)$ does not. ➡ For most functions we'll see in this book, however, both partial derivatives *do* exist.

  The exercises explore this possibility a little further.

- **About domains**   In order for the limits in the definition above to exist, $f(x, y)$ must be defined at (x_0, y_0) and at nearby points (x, y). In practice, this condition seldom causes trouble. In particular, there is no problem if (x_0, y_0) lies in the interior of the domain of f, that is, if f is defined both at and near (x_0, y_0). (For the record, trouble is likeliest if (x_0, y_0) lies on the edge of the domain of f.)

- **Other notations**   As with ordinary derivatives, various notations are used to denote partial derivatives. If, say, $z = f(x, y) = x \sin y$, then all of the expressions

$$f_x, \quad \frac{\partial f}{\partial x}, \quad \frac{\partial z}{\partial x}, \quad \text{and} \quad \frac{\partial}{\partial x}(x \sin y)$$

mean the same thing, as do

$$f_x(x_0, y_0), \quad \frac{\partial f}{\partial x}(x_0, y_0), \quad \frac{\partial z}{\partial x}\bigg|_{(x_0, y_0)}, \quad \text{and} \quad \frac{\partial}{\partial x}(x \sin y)\bigg|_{(x_0, y_0)}.$$

Partial derivatives with respect to y use similar notations. Notice especially the "curly-d" symbol ∂; it's read aloud as "partial" or as "del."

Partial derivatives and contour maps

We saw in Example 2 how to estimate partial derivatives from a table of function values. Contour maps can be used for the same purpose. Thinking of $f_x(x_0, y_0)$ and $f_y(x_0, y_0)$ as rates suggests how to estimate partial derivatives: Use the contour map to measure how fast $z = f(x, y)$ rises or falls near (x_0, y_0) as either x or y increases. ➡

On a topographic map oriented the "usual" way, with north pointing "up," the partial derivatives f_x and f_y describe the steepness of the terrain in, respectively, the eastward and northward directions.

EXAMPLE 3   Figure 1 shows a contour map of the function $f(x, y) = x^2 - 3xy + 6$, centered at $(2, 1)$. Use contours to estimate the partial derivatives $f_x(2, 1)$ and $f_y(2, 1)$.

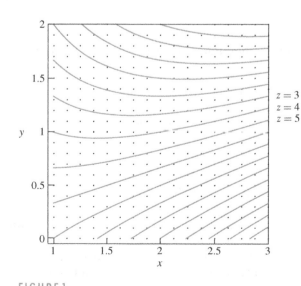

z = 3
z = 4
z = 5

A few contours are labeled. The dots help us locate specific points.

FIGURE 1
Level curves of $f(x, y) = x^2 - 3xy + 6$

Take a close look at the contour map to convince yourself of these claims.

Solution The level curves represent successive integer values of $z = f(x, y)$. In general, values of $f(x, y)$ increase as (x, y) moves toward the lower right. ←

First we will estimate $f_x(2, 1)$. A close look at the picture suggests that $f(2.1, 1) \approx 4.1$. Increasing x by 0.1 increases f by 0.1; this suggests that $f_x(2, 1) \approx 0.1/0.1 = 1$. Similarly $f(2, 1.1) \approx 3.4$, and so increasing y by 0.1 increases f by −0.6; thus, $f_y(2, 1) \approx -0.6/0.1 = -6$. ∎

Partial derivatives, slicing, and traces

We discussed traces in Section 12.1.

Fixing $y = y_0$, as we do when finding a partial derivative $f_x(x_0, y_0)$, can be thought of geometrically as slicing the surface $z = f(x, y)$ with the plane $y = y_0$. (The curve of intersection between this plane and the surface is the **trace** of the plane in the surface.) ← Then the partial derivative $f_x(x_0, y_0)$ is the *slope* of the trace at $x = x_0$. We illustrate the idea with an example.

EXAMPLE 4 Consider the function $f(x, y) = 3 + \cos(x)\sin(2y)$; its graph is shown in Figure 2.

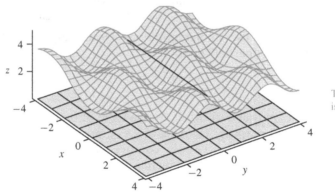

The horizontal line on the surface is the trace with the plane $y = 0$.

FIGURE 2
The surface $z = 3 + \cos(x)\sin(2y)$

Calculate the partial derivatives. What are their numerical values at the origin? What do the answers mean about traces?

Solution The symbolic calculations are easy:

$$\frac{\partial f}{\partial x}(x, y) = -\sin(x)\sin(2y) \quad \text{and} \quad \frac{\partial f}{\partial y}(x, y) = 2\cos(x)\cos(2y).$$

At the origin, therefore,

$$\frac{\partial f}{\partial x}(0, 0) = 0 \quad \text{and} \quad \frac{\partial f}{\partial y}(0, 0) = 2.$$

Both answers can be interpreted as slopes. Have a close look: Slicing the surface with the plane $y = 0$ produces the curve $z = 3$, which is "flat" at $x = 0$—as the partial derivative $f_x(0, 0) = 0$ suggests. Similarly, slicing the surface with the plane $x = 0$ produces the curve $z = 3 + \sin(2y)$, which has slope 2 at $y = 0$; correspondingly, $f_y(0, 0) = 2$. ∎

Partial derivatives and stationary points

In several variables, as in one, points where derivatives vanish are of special interest.

In one variable For a one-variable function $f(x)$, a domain point x_0 at which $f'(x_0) = 0$ is called a **stationary point**. Stationary points are natural places to look for maximum and minimum values of f, but it is possible that f may have neither a maximum nor a minimum point at x_0. For example, $f(x) = x^3$ is stationary at $x = 0$, but $f(x)$ takes neither a maximum nor a minimum value at $x = 0$. On the other hand, if a function f does assume a local maximum or local minimum value at x_0, and $f'(x_0)$ exists, then, necessarily, $f'(x_0) = 0$.

In several variables The same terminology (and some of the same reasoning) applies to functions of several variables. A domain point (x_0, y_0) is called a **stationary point** for a function $f(x, y)$ if *both* partial derivatives vanish there:

$$f_x(x_0, y_0) = 0; \qquad f_y(x_0, y_0) = 0.$$

(For a function of three variables, all three partial derivatives must vanish.)

As in the one-variable case, a stationary point in several variables need not correspond either to a local maximum or to a local minimum of the function at hand. The phrases "local maximum" and "local minimum" mean that $f(x_0, y_0)$ is either larger or smaller than $f(x, y)$ for all nearby values of (x, y)—exactly as for functions of one variable. Indeed, the extra "room" in several variables permits $f(x, y)$ to assume many different types of behavior near a stationary point. On the other hand:

> **THEOREM 1 (Extreme points and partial derivatives)** Suppose that $f(x, y)$ has a local maximum or a local minimum at (x_0, y_0). If both partial derivatives exist, then
>
> $$f_x(x_0, y_0) = 0 \quad \text{and} \quad f_y(x_0, y_0) = 0.$$

The result probably sounds plausible, but why is it true? Consider the case of a local maximum. Geometrically, the surface $z = f(x, y)$ has a "peak" above the domain point (x_0, y_0). If we slice the surface with any vertical plane, say the plane $y = y_0$, then the resulting curve—the trace of the surface $z = f(x, y)$ in the plane $y = y_0$—has the equation $z = f(x, y_0)$, and *this* curve must peak at $x = x_0$. From one-variable calculus we know, therefore, that if the function $x \mapsto f(x, y_0)$ is differentiable, then its derivative must be zero at x_0. In other words,

$$\frac{dz}{dx}(x_0) = f_x(x_0, y_0) = 0.$$

For similar reasons, $f_y(x_0, y_0) = 0$.

The following example hints at the variety of possible behaviors for a multivariable function near a stationary point.

EXAMPLE 5 Let $f(x, y) = x^2 + y^2$ and $g(x, y) = xy$. Find all the stationary points of f and g. What happens at each one?

Solution Finding the partial derivatives is easy:

$$f_x(x, y) = 2x; \qquad f_y(x, y) = 2y; \qquad g_x(x, y) = y; \qquad g_y(x, y) = x.$$

Both f and g, therefore, are stationary only at the origin $(0, 0)$. For f, the origin is a minimum point because $f(x, y) = x^2 + y^2 \geq 0$ for all (x, y). For g, however, the origin is *neither* a maximum nor a minimum point because $g(x, y)$ assumes both positive and negative values near $(0, 0)$. (For example, $g(0.1, -0.1) < 0$, but $g(0.1, 0.1) > 0$.) Contour maps of f and g illustrate their very different behavior near the stationary point. (In the following pictures the level curves are the edges of the shaded regions. Note the "key" to the right of each contour map.) Figure 3 shows f:

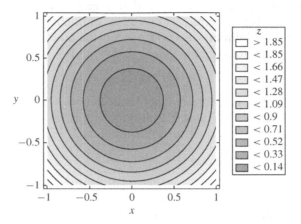

FIGURE 3
Level curves of $f(x, y) = x^2 + y^2$ are circles

Level curves of f are circles centered on the "basin" at $(0, 0)$. Thus, the picture suggests ◄ that f has a local *minimum* at the stationary point $(x, y) = (0, 0)$. The suggestion is correct: $f(0, 0) = 0$ and $f(x, y) \geq 0$ for all (x, y), and so f assumes a local (and even global) minimum value at $(0, 0)$.

But take a careful look. Notice, in particular, that darker regions are "lower."

Now look at g near its stationary point:

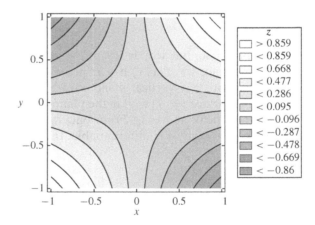

FIGURE 4
Level curves of $g(x, y) = x^2 - y^2$ are hyperbolas

Think about this carefully. Can you "see" the saddle in the contour map?

We didn't do so to encourage you to study the contour map.

Level curves of g show what is called a **saddle point** at $(0, 0)$. (If the surface were literally a saddle, the horse would be walking either "northeast" or "southwest.") ◄ The surface rises above the first and third quadrants and falls below the second and fourth quadrants. Plotting the graph as a surface in xyz-space would reveal similar features. ◄

Maxima and minima—more to the story   The full story of finding maximum and minimum values of functions of several variables is more complicated than these simple examples convey. We will return to it in Section 13.6 after developing some more sophisticated calculus tools.

BASIC EXERCISES

In Exercises 1–10, find the partial derivative with respect to each variable.

1. $f(x, y) = x^2 - y^2$

2. $f(x, y) = x^2 y^2$

3. $f(x, y) = \dfrac{x^2}{y^2}$

4. $f(x, y) = \cos(xy)$

5. $f(x, y) = \cos x \cos y$

6. $f(x, y) = \dfrac{\cos x}{\cos y}$

7. $f(x, y, z) = xy^2 z^3$

8. $f(x, y, z) = \cos(xyz)$

9. $f(x, y, z) = \dfrac{x + y}{1 + xyz}$

10. $f(x, y, z,) = \dfrac{1}{\sqrt{x^2 + y^2 + z^2}}$

In Exercises 11–16, find the indicated partial derivative.

11. $\dfrac{\partial}{\partial T}\left(\dfrac{2\pi r}{T}\right)$

12. $\dfrac{\partial F}{\partial m_1}$ if $F = \dfrac{Gm_1 m_2}{r^2}$

13. $\dfrac{\partial}{\partial y}\sin(3x^4 y + 2x^3 y^5)$

14. $\dfrac{\partial}{\partial x}\left(xe^{\sqrt{xy}}\right)$

15. $\dfrac{\partial}{\partial y}\left(\dfrac{1}{\sqrt{x^2 + y^2 + z^2}}\right)$

16. $\dfrac{\partial}{\partial x}\ln(x^2 + y^2)$

In Exercises 17–20, use the values of the function $g(x, y)$ shown in the table to estimate the partial derivative. (No explicit symbolic formula for g is given.)

$g(x, y)$ for Exercises 17–20

y \ x	−0.0100	0.0000	0.0100	0.0200	...	0.9900	1.0000	1.0100	1.0200
1.02	2.0603	2.0604	2.0603	2.0600	...	1.0803	1.0604	1.0403	1.0200
1.01	2.0300	2.0301	2.0300	2.0297	...	1.0500	1.0301	1.0100	0.9897
1.00	1.9999	2.0000	1.9999	1.9996	...	1.0199	1.0000	0.9799	0.9596
0.99	1.9700	1.9701	1.9700	1.9697	...	0.9900	0.9701	0.9500	0.9297
⋮	⋮	⋮	⋮	⋮	⋮	⋮	⋮	⋮	⋮
0.02	0.0203	0.0204	0.0203	0.0200	...	−0.9597	−0.9796	−0.9997	−1.0200
0.01	0.0100	0.0101	0.0100	0.0097	...	−0.9700	−0.9899	−1.0100	−1.0303
0.00	−0.0001	0.0000	−0.0001	−0.0004	...	−0.9801	−1.0000	−1.0201	−1.0404
−0.01	−0.0100	−0.0099	−0.0100	−0.0103	...	−0.9900	−1.0099	−1.0300	−1.0503

17. $g_x(1, 1)$ and $g_y(1, 1)$

18. $g_x(0, 0)$ and $g_y(0, 0)$

19. $g_x(1, 0)$ and $g_y(1, 0)$

20. $g_x(0, 1)$ and $g_y(0, 1)$

Suppose that the function $f : \mathbb{R}^2 \to \mathbb{R}$ has the values shown in this table:

$f(x, y)$ for Exercises 21–24

y \ x	1.5	2.0	2.5	3.0
3.0	4	6	9	6
2.5	6	9	7	5
2.0	4	8	6	4
1.5	3	5	5	7

In Exercises 21–24, estimate the partial derivatives.

21. $f_x(2.5, 2.5)$ and $f_y(2.5, 2.5)$.

22. $f_x(2.0, 2.0)$ and $f_y(2.0, 2.0)$.

23. $f_x(3.0, 1.5)$ and $f_y(3.0, 1.5)$.

24. $f_x(2.0, 1.5)$ and $f_y(2.0, 1.5)$.

25. Let $f(x, y) = 2x - 3y$.

(a) Draw a contour map of f in the rectangle $[-3, 3] \times [-3, 3]$. Show the level curves that correspond to $z = -5$, $z = -4, z = -3, \ldots, z = 4$, and $z = 5$.

(b) Use your contour map to find $f_x(0, 0)$ and $f_y(0, 0)$.

(c) The formula for f implies that both f_x and f_y are constant functions. How does the contour map of f reflect this fact?

(d) The formula for f implies that, for any (x, y), $f_x(x, y) = 2$ and $f_y(x, y) = -3$. How does the contour map reflect the fact that $f_x(x, y)$ is positive, but $f_y(x, y)$ is negative?

26. Let $f(x, y) = 2y - x$.

(a) Draw a contour map of f in the rectangle $[-3, 3] \times [-3, 3]$. Show the level curves that correspond to $z = -5$, $z = -4, z = -3, \ldots, z = 4$, and $z = 5$.

(b) Use your contour map to find $f_x(0, 0)$ and $f_y(0, 0)$.

(c) The formula for f implies that $f_x(x, y) = -1$ and $f_y(x, y) = 2$ for all (x, y). How does the contour map reflect these facts? In particular, how does the contour map show that $f_x(x, y)$ is negative but $f_y(x, y)$ is positive?

27. Let $f(x, y) = \sin x$.

(a) Draw a contour map of f in the square $-\pi \le x \le \pi$, $-\pi \le y \le \pi$. Show the level curves that correspond to $z = \pm 1, z = \pm 0.75, z = \pm 0.5, z = \pm 0.25$, and $z = 0$.

(b) Use the level curve diagram to estimate $f_x(0, 0)$ and $f_y(0, 0)$.

(c) Use the level curve diagram to estimate $f_x(\pi/2, 0)$ and $f_y(\pi/2, 0)$.

(d) The formula shows that $f_y(x, y) = 0$ for all (x, y). How does the contour map reflect this fact?

(e) The formula shows that $f_x(x, y)$ is independent of y. How does the contour map of f reflect this fact?

28. Let $f(x, y) = \cos y$.

(a) Draw a contour map of f in the square $-\pi \le x \le \pi$, $-\pi \le y \le \pi$. Show the level curves that correspond to $z = \pm 1, z = \pm 0.75, z = \pm 0.5, z = \pm 0.25$, and $z = 0$.

(b) Use the level curve diagram to estimate $f_x(0, 0)$ and $f_y(0, 0)$.

(c) Use the level curve diagram to estimate $f_x(\pi/2, 0)$ and $f_y(\pi/2, 0)$.

(d) The formula for f shows that $f_x(x, y) = 0$ for all (x, y). How does the contour map of f reflect this fact?

(e) The formula for f shows that $f_y(x, y)$ is independent of x. How does the contour map of f reflect this fact?

29. Let $f(x, y) = x^2 - 3xy + 6$.

(a) Use the level curve diagram shown in Figure 1 to determine whether $f_x(1.5, 1.75)$ is positive or negative. Justify your answer.

(b) Use the level curve diagram shown in Figure 1 to estimate $f_x(2.5, 0.25)$ and $f_y(2.5, 0.25)$.

30. Let $g(x, y) = x^2 - y^2$. Use the contour map shown in Figure 4 to determine the sign (positive or negative) of $f_x(-0.75, -0.5)$, $f_y(-0.75, -0.5)$, $f_x(0.5, 0.75)$, and $f_y(0.5, 0.75)$.

31. Let $f(x, y) = x^2 - 3xy + 6$. (See Examples 1 and 2.)

(a) Explain why the trace of f in the plane $y = 1$ is the curve $z = x^2 - 3x + 6$.

(b) Use part (a) to find a formula for $f_x(x, 1)$.

(c) Use the formula you found in part (b) to evaluate $f_x(2, 1)$, $f_x(0, 1)$, and $f_x(-1, 1)$.

(d) Find an equation for the trace of f in the plane $x = 2$.

(e) Use part (d) to find a formula for $f_y(2, y)$.

(f) Use the formula you found in part (e) to evaluate $f_y(2, 1)$, $f_y(2, 0)$, and $f_y(2, -2)$.

32. Let $f(x, y) = 3 + \cos(x) \sin(2y)$. (A graph of f is shown in Figure 2.)

(a) Find an equation for the trace of f in the plane $x = \pi$.

(b) Plot the equation you found in part (a) over the interval $0 \le y \le \pi$.

(c) Use part (b) to explain why $f_y(\pi, \pi/4) = 0$.

(d) Find an equation for the trace of f in the plane $y = -\pi/4$.

(e) Plot the equation you found in part (d) over the interval $0 \le x \le \pi$.

(f) Explain how the graph in part (e) could be used to estimate $f_x(2, -\pi/4)$.

33. Let $f(x, y) = 3 + \cos(x) \sin(2y)$. (See Example 4.)

(a) Show that f has a stationary point at $(0, \pi/4)$.

(b) Is $(0, \pi/4)$ a local maximum, a local minimum, or a saddle point of f? Justify your answer.

34. Let $f(x, y) = x^2 - 3xy + 6$.

(a) f has only one stationary point. Find it. [HINT: See Example 1.]

(b) Is the stationary point of f a local maximum, a local minimum, or a saddle point? Justify your answer.

FURTHER EXERCISES

35. Suppose that f is the linear function $f(x, y) = ax + by + c$, where a, b, and c are constants.

(a) Let k be a constant. Show that the trace of f in the plane $y = k$ is a line with slope a.

(b) Let k be a constant. Show that the trace of f in the plane $x = k$ is a line with slope b.

(c) Use parts (a) and (b) to explain why $f_x(x, y) = a$ and $f_y(x, y) = b$.

(d) Use part (c) to show that $f(x, y) = f(0, 0) + f_x(0, 0)x + f_y(0, 0)y$.

36. Suppose that f is the linear function $f(x, y) = ax + by + c$, where $a, b,$ and c are constants. Show that $f(x, y) = f(1, 2) + f_x(1, 2)(x - 1) + f_y(1, 2)(y - 2)$.

37. Suppose that $f(x, y)$ is a linear function. Furthermore, suppose that $f(0, 0) = 1$, $f_x(0, 0) = -1$, and $f_y(0, 0) = 2$.

(a) Explain why $f_x(x, y) = -1$.

(b) Evaluate $f(2, 3)$.

38. Suppose that $g(x, y)$ is a linear function. Furthermore, suppose that $g(1, 0) = -1$, $g_x(1, 0) = 2$, and $g_y(1, 0) = 0$.

(a) Evaluate $g_y(3, 4)$.

(b) Evaluate $g(2, 3)$.

39. Suppose that the trace of the surface $z = f(x, y)$ in the plane $x = 3$ is the curve $z = 3 - y^3$. Suppose also that the trace of the surface $z = f(x, y)$ in the plane $y = -2$ is the curve $z = 2 + x^2$. Evaluate $f_x(3, -2)$.

40. Suppose that the trace of the surface $z = g(x, y)$ in the plane $x = -1$ is the curve $z = 1 + 3y^2$. Suppose also that the trace of the surface $z = g(x, y)$ in the plane $y = 2$ is the curve $z = 2x^3 + 4x + 5$. Evaluate $g_y(-1, 2)$.

Suppose that the level curves of the surface $z = f(x, y)$ are straight lines parallel to the x-axis. In Exercises 41–44, determine whether there is a function f with this additional property. If so, give an example of such a function f. If not, explain why no such function f exists.

41. $f_x(1, -1) = 2$

42. $f_y(-1, 1) = 2$

43. $f(1, 2) = 1$ and $f(3, 2) = -1$

44. f is a linear function

Suppose that the level curves of the surface $z = g(x, y)$ are straight lines parallel to the line $y = 2x$. In Exercises 45–48, determine whether there is a function g with this additional property. If so, give an example of such a function g. If not, explain why no such function g exists.

45. $g_x(2, -1) = 3$

46. $g_y(1, 2) = -3$

47. g is a linear function

48. g is not a linear function

49. The equation $PV = nRT$ describes the relationship between the pressure P, volume V, and temperature T of n moles of an ideal gas; R is a number called the ideal gas constant. (This equation allows each variable to be expressed as a function of the other two.) Show that

$$\frac{\partial V}{\partial T} \cdot \frac{\partial T}{\partial P} \cdot \frac{\partial P}{\partial V} = -1.$$

50. The partial differential equation

$$\frac{\partial u}{\partial t} + \frac{\partial u}{\partial x} = ku$$

is used in population modeling. Here $u = u(x, t)$ is the number of individuals of age x at time t, and k is the mortality rate. Show that the function $u(x, t) = e^{\alpha x + \beta t}$ is a solution to this partial differential equation if the constants α and β satisfy the equation $\alpha + \beta = k$.

51. Suppose that $f(x, y) = x^4 g(y)$, $g(2) = 3$, and $g'(2) = -1$. Evaluate $f_x(-2, 2)$ and $f_y(-2, 2)$.

52. Suppose that $f(1, -1) = 1$, $f_x(1, -1) = 2$, $f_y(1, -1) = -3$, and $g(x, y) = 4f(x, y) + 5$. Evaluate $g_x(1, -1)$ and $g_y(1, -1)$.

53. Find an equation for the line tangent to the curve of intersection of the surface $z = f(x, y) = \sqrt{16 - x^2 - y^2}$ with the plane $y = 1$ at the point $(2, 1, \sqrt{11})$.

54. Find an equation for the line tangent to the curve of intersection of the surface $z = f(x, y) = \sqrt{16 - x^2 - y^2}$ with the plane $x = 2$ at the point $(2, 1, \sqrt{11})$.

55. Find a function $g(y)$ so that $f(x, y) = e^{cx} g(y)$, where c is a constant, satisfies the equation $f_x + f_y = 0$.

56. Let f be a linear function. Explain why f has either no stationary points or infinitely many stationary points.

13.3 LINEAR APPROXIMATION IN SEVERAL VARIABLES

Linear approximation in one variable

For a differentiable function $y = f(x)$ of one variable, the ordinary derivative can be interpreted in terms of linear approximation. For any point (x_0, y_0) on the f-graph, there is a certain line through this point—the tangent line—that best "fits" the graph near $x = x_0$. The derivative $f'(x_0)$ gives the slope of this tangent line. Knowing this makes it easy to find an equation for the tangent line in point-slope form:

$$y - y_0 = f'(x_0)(x - x_0).$$

We can also think of the tangent line as the graph of the linear function L defined as follows: ➡

Recall: $f(x_0) = y_0$.

$$y = L(x) = y_0 + f'(x_0)(x - x_0).$$

The function L is called the **linear approximation to** f **at** x_0. The name is appropriate for three good reasons:

(i) L is linear;   (ii) $L(x_0) = f(x_0)$;   (iii) $L'(x_0) = f'(x_0)$.

In short, L "agrees" with f at x_0 as closely as *any* linear function can—both L and f have the same value and the same (first) derivative.

Linear approximation in several variables

The idea of linear approximation is essentially the same for functions of two (or more) variables: Given a function $f(x, y)$ and a point (x_0, y_0) in the domain of f, we look for a *linear* function $L(x, y)$ that has the same value and partial derivatives as those of f at (x_0, y_0). In other words, we want a linear function L such that

(i) $L(x_0, y_0) = f(x_0, y_0)$;   (ii) $L_x(x_0, y_0) = f_x(x_0, y_0)$;   (iii) $L_y(x_0, y_0) = f_y(x_0, y_0)$.

In the one-variable case the graph of a linear approximation function is called a **tangent line**; for a function of two variables, the graph of the linear approximation function is called a **tangent plane**.

The geometric view Figure 1 shows the tangent plane to the surface $z = f(x, y) = 3 + \cos(x)\sin(2y)$ at the point $(0, 0, 3)$; part of the curved surface is also shown:

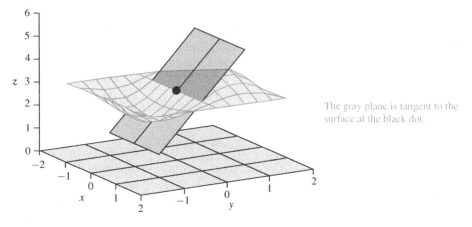

The gray plane is tangent to the surface at the black dot.

FIGURE 1
A surface and a linear approximation

Observe how closely the flat plane and the curved surface fit together near the target point $(0, 0, 3)$. Elsewhere, the surface and the tangent plane "fit" less well. In the present case, too, the tangent plane happens to *cross* the surface at the point of tangency.

The following example shows how to find a tangent plane's formula. ◂

Calculations for the plane in Figure 1 are left to the exercises.

E X A M P L E 1 Find the linear approximation function L to $f(x, y) = x^2 + y^2$ at $(x_0, y_0) = (2, 1)$. How are the graphs of f and L related? How are their contour plots related?

Solution The partial derivatives of f are $f_x(x, y) = 2x$ and $f_y(x, y) = 2y$. Thus, at our base point $(x_0, y_0) = (2, 1)$:

$$f(2, 1) = 5; \qquad f_x(2, 1) = 4; \qquad f_y(2, 1) = 2.$$

Let's find a linear function L that "matches" these values.

It turns out to be easiest to write L in the convenient form

$$L(x, y) = a(x - x_0) + b(y - y_0) + c = a(x - 2) + b(y - 1) + c$$

and then to choose appropriate values for a, b, and c. Because

$$L(x, y) = a(x - 2) + b(y - 1) + c,$$

it is easy to see ➤ that

$$L(2, 1) = c; \qquad L_x(2, 1) = a; \qquad L_y(2, 1) = b.$$

But check for yourself—especially for the derivatives.

To "match" the value and partial derivatives of f we must have $c = 5$, $a = 4$, and $b = 2$, and so

$$L(x, y) = a(x - 2) + b(y - 1) + c = 4(x - 2) + 2(y - 1) + 5.$$

Figure 2 shows both f and L:

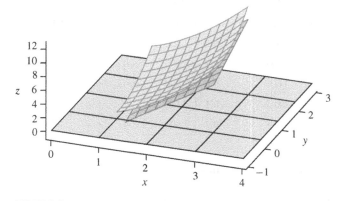

The curved surface and the plane are almost identical at this scale.

FIGURE 2
Graphs of f and L together

This picture, too, illustrates the phrases "tangent plane" and "linear approximation": The plane $z = L(x, y)$ touches the surface $z = f(x, y)$ at $(2, 1, 5)$; at this point, moreover, the flat plane "fits" the curved surface as well as possible. A closer look (Figure 3) at both functions near $(2, 1)$, this time using contour maps, shows how good the fit really is: ➤

The straight lines are contours of L.

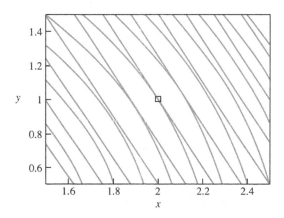

FIGURE 3
Level curves of f and L together

The two contour maps are almost identical near $(2, 1)$.

The numerical view The idea of linear approximation can be seen numerically as well as symbolically and geometrically.

EXAMPLE 2 Find the linear approximation L to $f(x, y) = 3 + \cos(x)\sin(2y)$ at $(x_0, y_0) = (0, 0)$. Compare numerical values of L and f near (x_0, y_0).

We calculated these derivatives in Example 4, page 718.

Solution Easy calculations give the needed value and partial derivatives: ←

$$f(0, 0) = 3, \qquad f_x(0, 0) = 0, \qquad \text{and} \qquad f_y(0, 0) = 2;$$

it follows that $L(x, y) = 3 + 2y$. The following tables show sample values of f and L near $(0, 0)$:

Values of $f(x, y) = 3 + \cos(x)\sin(2y)$					
y \ x	−0.2	−0.1	0.0	0.1	0.2
0.2	3.382	3.388	3.389	3.388	3.382
0.1	3.195	3.198	3.199	3.198	3.195
0.0	3.000	3.000	3.000	3.000	3.000
−0.1	2.805	2.802	2.801	2.802	2.805
−0.2	2.618	2.613	2.611	2.613	2.618

Values of $L(x, y) = 3 + 2y$					
y \ x	−0.2	−0.1	0.0	0.1	0.2
0.2	3.400	3.400	3.400	3.400	3.400
0.1	3.200	3.200	3.200	3.200	3.200
0.0	3.000	3.000	3.000	3.000	3.000
−0.1	2.800	2.800	2.800	2.800	2.800
−0.2	2.600	2.600	2.600	2.600	2.600

The numerical data show how closely L approximates f near the origin. ∎

Linear approximation: the general formula The procedure illustrated in Example 2 works the same way for any function of several variables as long as the necessary partial derivatives exist. Here are definitions for two and three variables: ←

The same idea works for any number of variables.

> **DEFINITION (Linear approximation)** Let $f(x, y)$ and $g(x, y, z)$ be functions, and suppose that all the partial derivatives mentioned below exist. The linear approximation to f at (x_0, y_0) is the function
>
> $$L(x, y) = f(x_0, y_0) + f_x(x_0, y_0)(x - x_0) + f_y(x_0, y_0)(y - y_0).$$
>
> The linear approximation to g at (x_0, y_0, z_0) is the function
>
> $$L(x, y, z) = g(x_0, y_0, z_0) + g_x(x_0, y_0, z_0)(x - x_0)$$
> $$+ g_y(x_0, y_0, z_0)(y - y_0) + g_z(x_0, y_0, z_0)(z - z_0).$$

EXAMPLE 3 Find the linear approximation to $g(x, y, z) = x + yz^2$ at $(1, 2, 3)$. Does the answer make numerical and graphical sense?

Solution Easy calculations give

$$g(1, 2, 3) = 19; \qquad g_x(1, 2, 3) = 1; \qquad g_y(1, 2, 3) = 9; \qquad g_z(1, 2, 3) = 12.$$

The linear approximation function, therefore, has the form

$$L(x, y, z) = g_x(1, 2, 3)(x - 1) + g_y(1, 2, 3)(y - 2) + g_z(1, 2, 3)(z - 3) + g(1, 2, 3)$$
$$= 1(x - 1) + 9(y - 2) + 12(z - 3) + 19.$$

To see the situation numerically, we will tabulate some values of each function:

Values of L and g near $(1, 2, 3)$						
(x, y, z)	$(1, 2, 3)$	$(1.1, 2, 3)$	$(1, 2.1, 3)$	$(1, 2, 3.1)$	$(1.1, 2.1, 3.1)$	$(3, 4, 5)$
$g(x, y, z)$	19	19.1	19.9	20.22	21.281	103
$L(x, y, z)$	19	19.1	19.9	20.2	21.2	63

The last column illustrates what happens "far" from $(1, 2, 3)$.

As the numbers illustrate, $L(x, y, z)$ and $g(x, y, z)$ are close together if, but only if, (x, y, z) is near $(1, 2, 3)$. ➡

One for each input variable and one for the output.

To plot ordinary graphs of g and L would require four dimensions. ➡ Instead we will plot, for comparison, the level surfaces $L(x, y, z) = 19$ and $g(x, y, z) = 19$, both of which pass through the base point $(1, 2, 3)$. (A **level surface** is like a level curve, i.e., a set of inputs along which a function has constant output value.) Figure 4 suggests, again, how similarly $g(x, y, z)$ and $L(x, y, z)$ behave when $(x, y, z) \approx (1, 2, 3)$.

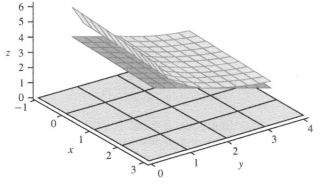

The two level surfaces resemble each other near the target point.

FIGURE 4
Level surfaces of g and L

Partial derivatives, the cross product, and the tangent plane

Here is another way to think of the tangent plane to a surface $z = f(x, y)$ at a point (x_0, y_0, z_0). Consider again the curves formed by intersecting the surface with planes either of the form $y = y_0$ or of the form $x = x_0$. Look again at the graph of $f(x, y) = 3 + \cos(x)\sin(2y)$ in Figure 2, page 718. It shows many such curves; together, these curves give the "wireframe" appearance of the surface $z = f(x, y)$.

Let's parametrize the curve of intersection between a surface $z = f(x, y)$ (with $a \leq x \leq b$ and $c \leq y \leq d$) and a plane $y = y_0$. All points on this curve are of the form $(x, y_0, f(x, y_0))$. As a vector-valued function of t, therefore, the curve can be parametrized by

$$x = t; \quad y = y_0; \quad z = f(t, y_0); \quad a \leq t \leq b.$$

At $t = x_0$, the curve has position (x_0, y_0, z_0)—that is, the curve passes through the point of interest. The velocity vector at this point is therefore

$$\mathbf{v}(x_0) = \big(1, 0, f_x(x_0, y_0) \big).$$

(The last coordinate is found by differentiating with respect to t, but t plays exactly the same the role in $f(t, y_0)$ as x does in $f(x, y_0)$.) ➡ We know that a velocity vector is tangent to its curve at the point in question. Here, the curve lies "in" the surface $z = f(x, y)$; thus, the velocity vector $\big(1, 0, f_x(x_0, y_0) \big)$ is tangent to the surface at (x_0, y_0, z_0).

The name of the variable—t, x, s, and so forth—is immaterial.

Now we replay the same game, but this time we use the curve of intersection between the surface $z = f(x, y)$ and the plane $x = x_0$. Reasoning just as before, we find another vector, $\left(0, 1, f_y(x_0, y_0)\right)$, that is also tangent to the surface at (x_0, y_0, z_0). The results are worth putting together:

FACT Let $z = f(x, y)$ define a surface, with $z_0 = f(x_0, y_0)$. Then the two vectors

$$\left(1, 0, f_x(x_0, y_0)\right) \quad \text{and} \quad \left(0, 1, f_y(x_0, y_0)\right)$$

are tangent to the surface at (x_0, y_0, z_0).

Having found *two* noncollinear vectors, each of them tangent to our surface at the "target" point, we can now write equations—in any form we like—for the tangent plane. In vector parametric form, we get

$$\mathbf{X}(s, t) = (x_0, y_0, z_0) + s\left(1, 0, f_x(x_0, y_0)\right) + t\left(0, 1, f_y(x_0, y_0)\right);$$

the scalar parametric equations are

$$x = x_0 + s; \qquad y = y_0 + t; \qquad z = z_0 + s f_x(x_0, y_0) + t f_y(x_0, y_0).$$

In either order—the answers are opposites, but both work equally well.

We can also, if we prefer, write the plane in ordinary scalar form. To find a normal vector, we can take the cross product ← of the two tangent vectors just found:

$$\mathbf{n} = \left(1, 0, f_x(x_0, y_0)\right) \times \left(0, 1, f_y(x_0, y_0)\right) = \left(-f_x(x_0, y_0), -f_y(x_0, y_0), 1\right).$$

Thus, the tangent plane has vector equation

$$\left(\mathbf{X} - (x_0, y_0, z_0)\right) \cdot \left(-f_x(x_0, y_0), -f_y(x_0, y_0), 1\right) = 0,$$

We did a little simplification.

or, in scalar form, ←

$$(x - x_0) f_x(x_0, y_0) + (y - y_0) f_y(x_0, y_0) = z - z_0.$$

See the formal definition on page 726.

The last form should look familiar; we used essentially the same formula when we defined the linear approximation $L(x, y)$. ←

EXAMPLE 4 Use the preceding Fact to describe the tangent plane to the surface $f(x, y) = 2y - x^2$ at the point $(3, 4, -1)$.

Solution It is easy to see that $f_x(3, 4) = -6$ and $f_y(3, 4) = 2$. Therefore, by Fact, the vectors $(1, 0, -6)$ and $(0, 1, 2)$ are tangent to the surface at $(3, 4, -1)$. Their cross product, which is normal to the tangent plane, is $\mathbf{n} = (6, -2, 1)$. It follows that the plane has scalar equation $6x - 2y + z = 9$. In vector parametric form, the plane is given by

$$\mathbf{X}(s, t) = (3, 4, -1) + s(1, 0, -6) + t(0, 1, 2).$$

Figure 5 gives a rough picture of the tangent plane "patch" corresponding to the parameter values $0 \le s \le 2$ and $0 \le t \le 2$:

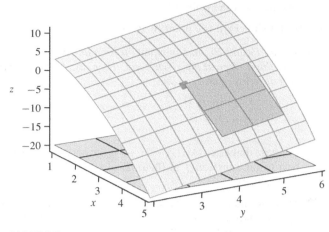

FIGURE 5
A surface and a tangent plane patch

The curved surface has equation $f(x, y) = 2y - x^2$; the plane patch is tangent at the point $(3, 4, -1)$.

BASIC EXERCISES

1. Let $f(x, y) = x^2 - 3xy + 6$.

   (a) Use the contour map of f shown in Figure 1 on page 717 to estimate the partial derivatives $f_x(1.5, 1.5)$ and $f_y(1.5, 1.5)$.

   (b) Use the formula for $f(x, y)$ to find $f_x(1.5, 1.5)$ and $f_y(1.5, 1.5)$ exactly.

   (c) Use results of part (b) to find the linear approximation $L(x, y)$ to $f(x, y)$ at $(1.5, 1.5)$.

2. Let $g(x, y) = x^2 + y^2$.

   (a) Use the contour map of g shown in Figure 4 on page 707 to estimate the partial derivatives $g_x(1, 2)$ and $g_y(1, 2)$.

   (b) Check your answers to part (a) by symbolic differentiation.

   (c) Use your answers from part (a) to find the linear approximation $L(x, y)$ to $g(x, y)$ at $(1, 2)$.

   (d) On one set of axes, plot the level curves $L(x, y) = k$ and $g(x, y) = k$ for $k = 3, 4, 5, 6, 7$. (Use the window $[0, 3] \times [0, 3]$.) What is special about the point $(1, 2)$?

3. Let $f(x, y) = xy$. This exercise explores ideas like those of Example 1.

   (a) Find the linear approximation function L to f at $(x_0, y_0) = (2, 1)$.

   (b) (Do this part by hand.) On one set of xy-axes, draw the level curves $L(x, y) = k$ for $k = 1, 2, 3, 4, 5$. On another set of axes, draw the level curves $f(x, y) = k$ for $k = 1, 2, 3, 4, 5$. (In each case, draw the curves into the square $[0, 3] \times [0, 3]$.)

   (c) How do the contour maps in part (b) reflect the fact that L is the linear approximation to f at the point $(2, 1)$? Explain briefly in words.

   (d) Use technology to plot contour maps of f and L in the window $1.8 \le x \le 2.2, 0.8 \le y \le 1.2$. (This small window is centered at $(2, 1)$.) Explain what you see.

4. Repeat Exercise 3 using the function $f(x, y) = x^2 - y^2$.

5. Let $f(x, y) = \sin x + 2y + xy$.

   (a) Find the partial derivatives $f_x(x, y)$ and $f_y(x, y)$.

   (b) Find a linear function $L(x, y) = a + bx + cy$ such that $L_x(0, 0) = f_x(0, 0)$, $L_y(0, 0) = f_y(0, 0)$, and $L(0, 0) = f(0, 0)$.

   (c) Complete the following table (report answers to four decimals).

(x, y)	(0.0)	$(0.01, 0.01)$	$(0.1, 0.1)$	$(1, 1)$
$f(x, y)$				
$L(x, y)$				

   Do the answers suggest that L approximates f closely near $(0, 0)$? Justify your answer.

   (d) Use technology to draw contour plots of both f and L on the rectangle $-1 \le x \le 1, -1 \le y \le 1$. Label several contours on each. How do the pictures reflect the fact that L approximates f closely near $(0, 0)$?

6. Let $g(x, y) = xy - 3x + 4y + 5$.

   (a) Find the partial derivatives $g_x(x, y)$ and $g_y(x, y)$.

   (b) Find a linear function $L(x, y) = a + bx + cy$ such that $L_x(0, 0) = g_x(0, 0)$, $L_y(0, 0) = g_y(0, 0)$, and $L(0, 0) = g(0, 0)$.

   (c) Complete the following table (report answers to four decimals).

(x, y)	$(-0.01, -0.01)$	$(-0.01, 0.01)$	$(0, 0)$	$(0, 01, -0.01)$	$(0.1, 0.1)$
$\|g(x, y) - L(x, y)\|$					

Do the entries suggest that L approximates g closely near $(0, 0)$? Justify your answer.

(d) Use technology to draw contour plots of both g and L on the rectangle $-1 \le x \le 1$, $-1 \le y \le 1$. Label several contours on each. How do the pictures reflect the fact that L approximates g closely near $(0, 0)$?

In Exercises 7–12, find the linear function L that linearly approximates f at the given point. (Check your answers graphically by plotting f and L near the given point.)

7. $f(x, y) = x^2 + y^2$; $(2, 1)$.

8. $f(x, y) = x^2 + y^2$; $(0, 0)$.

9. $f(x, y) = \sin x + \sin y$; $(0, 0)$.

10. $f(x, y) = \sin x \sin y$; $(0, 0)$.

11. $f(x, y) = 3x^2 - 4y^5$; $(2, 1)$

12. $f(x, y) = xy \cos(x + y)$; (π, π)

13. Let $f(x, y) = 3 + \cos(x) \sin(2y)$.

(a) Find the linear approximation function $L(x, y)$ to f at $(0, 0)$.

(b) Write a vector equation for the tangent plane at the point $(0, 0, 3)$. [HINT: First find a normal vector.]

14. Let $g(x, y) = 1 + e^x + \sin y$.

(a) Find the linear approximation function $L(x, y)$ to g at $(0, 0)$.

(b) Write a vector equation for the tangent plane at the point $(0, 0, 2)$.

In Exercises 15–18, find an equation for the plane tangent to the surface $z = f(x, y)$ at the given point, and then plot the surface and the plane together. (Finding the "right" window may take some experimenting.)

15. $z = x^2 + y^2$ at the point $(2, 1, 5)$

16. $z = x^2 - y^2$ at the point $(2, 1, 3)$

17. $z = \cos(xy)$ at the point $(0, 0, 1)$

18. $z = 1 + \sin(xy)$ at the point $(0, 0, 1)$

In Exercises 19 and 20, find an equation for a line that is perpendicular to the surface at the given point. Then, use technology to plot both the surface and the line.

19. $z = x^2 + y^2$, at $(2, 1, 5)$

20. $z = x - y^2$, at $(2, 1, 1)$

21. Suppose that $L(x, y) = 3y - 2x + 4$ is the linear approximation to the function f at $(0, 0)$. Evaluate $f(0, 0)$, $f_x(0, 0)$, and $f_y(0, 0)$.

22. Suppose that $L(x, y, z) = 2x - 3y + 4z - 5$ is the linear approximation to the function g at $(0, 0, 0)$. Evaluate $g(0, 0, 0)$, $g_x(0, 0, 0)$, $g_y(0, 0, 0)$, and $g_z(0, 0, 0)$.

23. Let $f(x, y) = 2 + \sin y$.

(a) Plot f; use the domain $-5 \le x \le 5$, $-5 \le y \le 5$. How does the shape of the graph reflect the fact that f is independent of x?

(b) Find $f_x(x, y)$ and $f_y(x, y)$. How do the answers reflect the fact that f is independent of x?

(c) Find the linear approximation function L for f at the point $(0, 0)$. How does its form reflect the fact that f is independent of x?

24. Let $g(x, y) = 3 - x^2$.

(a) Plot g; use the domain $-3 \le x \le 3$, $-3 \le y \le 3$. How does the shape of the graph reflect the fact that g is independent of y?

(b) Find $g_x(x, y)$ and $g_y(x, y)$. How do the answers reflect the fact that g is independent of y?

(c) Find the linear approximation function L for g at the point $(1, 1)$. How does its form reflect the fact that g is independent of y?

FURTHER EXERCISES

25. Are the two vectors mentioned in the Fact on page 728 ever collinear? Justify your answer.

26. Suppose that f is a function such that, $f(3, 4) = 25$, $f_x(3, 4) = 6$, $f_y(3, 4) = 8$, and $f(4, 5) = 41$.

(a) Find a linear function $L(x, y)$ that approximates f as well as possible near $(3, 4)$.

(b) Use L to estimate $f(2.9, 3.9)$, $f(3.1, 4.1)$, and $f(4, 5)$.

(c) Could f be a linear function? Justify your answer.

27. For a certain function g we know that $g(3, 4) = 5$, $g_x(3, 4) = 3/5$, $g_y(3, 4) = 4/5$, and $g(4, 5) = \sqrt{41}$.

(a) Find a linear function $L(x, y)$ that approximates g as well as possible near $(3, 4)$.

(b) Use L to estimate $g(2.9, 4.1)$ and $g(4, 5)$.

(c) Could g be a linear function? Justify your answer.

28. Suppose that $L(x, y) = 4y - 3x + 2$ is the linear approximation to the function f at $(1, 2)$. Evaluate $f(1, 2)$, $f_x(1, 2)$, and $f_y(1, 2)$.

29. Suppose that $L(x, y) = 4x + 5y - 3z - 2$ is the linear approximation to the function g at $(-1, 1, 0)$. Evaluate $g(-1, 1, 0)$, $g_x(-1, 1, 0)$, $g_y(-1, 1, 0)$, and $g_z(-1, 1, 0)$.

30. Let $f(x, y)$ be a differentiable function of two variables, let (x_0, y_0) be any point in its domain, and let $L(x, y)$ be the linear approximation to f at (x_0, y_0). Show that if f is independent of one of the variables—say, x—then so is L.

31. When two electrical resistors R_1 and R_2 are connected in parallel, the equivalent resistance R is

$$R = \left(\frac{1}{R_1} + \frac{1}{R_2} \right)^{-1}.$$

Suppose that $R_1 = 300\,\Omega$ within 6% and it is desired that $R = 75\,\Omega$ within 3%. Use a linear approximation of the function $R(R_1, R_2)$ to estimate the interval of acceptable values for R_2.

32. Find a point on the surface $x^2 + y^2 + 3z^2 = 8$ where the tangent plane is parallel to the plane $2x + y + 3z = 0$.

13.4 THE GRADIENT AND DIRECTIONAL DERIVATIVES

A function $f(x, y)$ has two partial derivatives at a point (x_0, y_0) of its domain, one for each variable. ➡ Given their common origin, it is natural to "store" both derivatives as components of a single vector

Assuming (as we do) that both partial derivatives exist

$$\nabla f(x_0, y_0) = \left(f_x(x_0, y_0), f_y(x_0, y_0) \right)$$

called the **gradient** of f at (x_0, y_0).

The preceding sections were about partial derivatives—the separate components of gradient vectors. We interpreted them in several ways: as directional rates of change, as slopes of certain curves, and as ingredients in finding linear approximation functions and tangent planes.

But what does the gradient mean as a *vector*? What do its length and direction tell us? How is the gradient related to other geometric objects we have already seen, such as level curves, tangent planes, and the surface $z = f(x, y)$? We study such questions in this section.

A vector of partial derivatives

Here is the formal definition:

> **DEFINITION (Gradient of a function at a point)** Let $f(x, y)$ be a function of two variables and (x_0, y_0) a point of its domain; assume that both partial derivatives exist at (x_0, y_0). The gradient of f at (x_0, y_0) is the plane vector
>
> $$\nabla f(x_0, y_0) = \left(f_x(x_0, y_0), f_y(x_0, y_0) \right).$$
>
> For a function $g(x, y, z)$ of three variables, the gradient is the space vector
>
> $$\nabla g(x_0, y_0, z_0) = \left(g_x(x_0, y_0, z_0), g_y(x_0, y_0, z_0), g_z(x_0, y_0, z_0) \right).$$

Notice:

- **Counting dimensions** Keeping track of dimensions can be tricky in multivariable calculus. For example, the graph of a function $f : \mathbb{R}^2 \to \mathbb{R}$ is a 2-dimensional object (a surface) hovering in three-dimensional space. To keep things in their proper spaces, it is important to remember that for any function f and any input $\mathbf{X_0}$: ➡

  We're thinking of $\mathbf{X_0}$ as a point in either $\mathbb{R}^2$ or $\mathbb{R}^3$.

  The gradient $\nabla f(\mathbf{X_0})$ is a vector with the same dimension as the domain of f.

  Indeed, gradient vectors naturally "live" in the domain of f. We will often draw them there, with the tail of $\nabla f(\mathbf{X_0})$ pinned at the point $\mathbf{X_0}$.

- **The gradient as a function** The ordinary derivative $f'(x)$ of a function of one variable is a new function of one variable. For a function $f(x, y)$, the gradient $\nabla f(x, y)$ can also be thought of as a new function—a vector-valued function—of x and y: For each domain point (x, y), the gradient recipe produces an associated vector. ⬅ Notice one new feature of the multivariable setting: Although f is scalar-valued, the gradient function ∇f is a vector-valued function sometimes known as a **vector field**.

Calculating gradients is very easy. Developing intuition for what they mean is a little more challenging; we will approach the matter through examples and pictures.

EXAMPLE 1   Let $f(x, y) = x^2 - y^2$. Calculate some gradient vectors in the vicinity of the origin. What do their lengths and directions say?

Solution   For any input (x, y), $\nabla f(x, y) = (2x, -2y)$. For instance,

$$\nabla f(0, 0) = (0, 0); \qquad \nabla f(2, 2) = (4, -4); \qquad \nabla f(-2, 4) = (-4, -8).$$

We could keep this up forever, but it is much more enlightening to *plot* the results, pinning the tail of each vector arrow at the corresponding domain point. Figure 1 is a sample; level curves of f are added for reference:

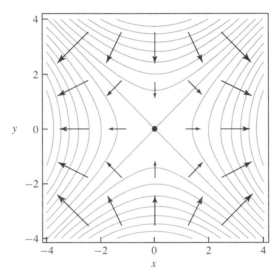

FIGURE 1
Contours and gradient vectors for $f(x, y) = x^2 - y^2$

The picture deserves a close look:

- **Just a sample**   The gradient function assigns a vector to *every* point in the domain. For obvious reasons, the picture can show only a tiny sample. But even the 25 arrows shown here give some sense of how the gradient function varies over the domain.

- **Lengths not to scale**   The gradient arrows point in the right directions, but lengths are not drawn to the same scale as shown on the axes. This is necessary to avoid having arrows overlap each other in the picture. Exactly *how* the lengths are scaled depends on the software. ⬅ In any event, arrows that should be longer *are* longer.

- **Gradient vectors point "uphill"**   Notice carefully the vectors along the coordinate axes. On the x-axis, ⬅ $f(x, y) = f(x, 0) = x^2$, and so $f(x, y)$ increases—first slowly and then faster and faster, as (x, y) moves either east or west. For similar reasons, $f(x, y)$ decreases as (x, y) moves either north or south along the y-axis.

   All of this information (and much more) appears in the picture. Along the x-axis, all gradient vectors point away from the origin and get longer as their base points get farther from the origin. On the y-axis, exactly the opposite happens. In short, gradient vectors point "uphill," and their lengths reflect how "steeply" the function increases.

- **A stationary point**   At a stationary point for f, the gradient is the zero vector; the dot at the origin represents this situation. The picture also shows the sense in which the origin is a saddle point ➤ of f: Above the origin, the surface $z = f(x, y)$ resembles a saddle with the horse heading east or west.

Not a maximum point or a minimum point

- **Gradient vectors and contour lines**   We've saved the most striking observation for last:

> *At each point (x_0, y_0) of the domain, the gradient vector is perpendicular to the level curve through (x_0, y_0).*

We will soon discuss exactly *why* this occurs, although a fully rigorous proof will have to wait a few sections. But all the pictures in this section offer strong circumstantial evidence *that* it occurs. (Intuitively, the phenomenon seems reasonable: It says, in hiking language, that a level path runs perpendicular to the steepest uphill direction.)   ■

The gradient of a linear function

A linear function, of the form $L(x, y) = ax + by + c$, has constant partial derivatives. Its gradient, therefore, is a *constant* vector: $\nabla L(x, y) = (a, b)$ for all (x, y).

This result looks especially simple and pleasing when everything is written in vector notation using the dot product. To this end, let's write $\mathbf{A} = (a, b)$ and $\mathbf{X} = (x, y)$. Then the observation above can be written as follows:

$$\text{If } L(\mathbf{X}) = \mathbf{A} \cdot \mathbf{X} + c, \qquad \text{then} \qquad \nabla L(\mathbf{X}) = \mathbf{A}.$$

The same fact—and even the same notation—holds for a linear function of three variables. Notice, too, the similarity to the analogous *one*-variable fact: $L(x) = ax + b \implies L'(x) = a$.

Plotting gradient vectors and level curves together shows what it means, geometrically, for a function to be linear. Figure 2 shows a sample of gradient vectors for the linear function $f(x, y) = 3x + 2y$:

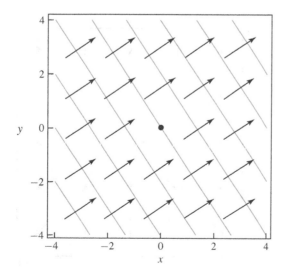

FIGURE 2

Contours and gradient vectors for $f(x, y) = 3x + 2y$

Again, the gradient vectors $(3, 2)$ look perpendicular to the level curves (lines, in this case). Indeed they are: For this function, every level line has the form $3x + 2y = k$ for some constant k. Each such line has slope $-3/2$, and so $(2, -3)$ is a tangent vector and the line is perpendicular to $(3, 2)$, as claimed.

Linear functions in three variables: gradients and level surfaces Let $L(x, y, z) = ax + by + cz + d$ be a linear function of three variables. The gradient is $\nabla L(x, y, z) = (a, b, c)$—a constant 3-vector. Let (x_0, y_0, z_0) be any point of the domain, and suppose that $L(x_0, y_0, z_0) = w_0$. Consider the level set of L passing through any point (x_0, y_0, z_0), that is, the set of points (x, y, z) such that $L(x, y, z) = ax + by + cz + d = w_0$. Since d and w_0 are constants, this set is a *plane* Π with equation $ax + by + cz = w_0 - d$. As we have seen, Π has normal vector (a, b, c). This shows that for linear functions of three variables (as for two):

The gradient vector at any point is perpendicular to the level set through that point.

Gradient vectors and linear approximations

Let $f(x, y)$ be any function and (x_0, y_0) a point of its domain. We have defined the linear approximation to f at (x_0, y_0) to be the linear function $L(x, y)$ that has the same value and the same partial derivatives as f at (x_0, y_0). That is,

$$L(x, y) = f(x_0, y_0) + f_x(x_0, y_0)(x - x_0) + f_y(x_0, y_0)(y - y_0).$$

We can write this in vector form, using the gradient vector and the notations $\mathbf{X_0} = (x_0, y_0)$ and $\mathbf{X} = (x, y)$:

$$L(x, y) = f(x_0, y_0) + \nabla f(x_0, y_0) \cdot (x - x_0, y - y_0) = f(\mathbf{X_0}) + \nabla f(\mathbf{X_0}) \cdot (\mathbf{X} - \mathbf{X_0}).$$

(For functions of three variables, the vector formula looks identical.)

As we've seen (and will return to more formally in the next section), every differentiable function of several variables—linear or not—can be closely approximated near a domain point $\mathbf{X_0}$ by its linear approximation function there. That's why the property of gradient vectors being perpendicular to level sets holds not only for linear functions but for *all* differentiable functions of several variables. Figure 3 illustrates this principle in action again—this time for the *nonlinear* function $f(x, y) = x^2 + y^2$:

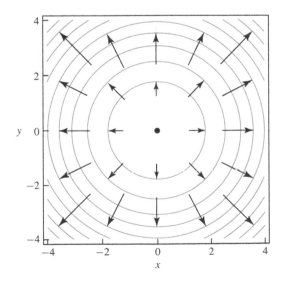

FIGURE 3
Contours and gradient vectors for $f(x, y) = x^2 + y^2$

As in earlier figures, all the gradient vectors point "uphill" and are perpendicular to the level curves.

Finding tangent planes   The perpendicularity property of gradient vectors makes it easy to find the tangent plane to a surface in $\mathbb{R}^3$.

> **EXAMPLE 2**   Find an equation for the tangent plane Π to the sphere $x^2 + y^2 + z^2 = 6$ at the point $(2, 1, 1)$. ➤
>
> *Try to picture the situation.*
>
> **Solution**   We can think of the sphere as the level surface $f(x, y, z) = 6$, where $f(x, y, z) = x^2 + y^2 + z^2$. The gradient is $\nabla f(x, y, z) = (2x, 2y, 2z)$; in particular, $\nabla f(2, 1, 1) = (4, 2, 2)$. This is a suitable normal vector for Π; an xyz-equation, therefore, is $4(x - 2) + 2(y - 1) + 2(z - 1) = 0$, or $4x + 2y + 2z = 12$.   ■

Directional derivatives

Partial derivatives tell the rates of change of a function as inputs vary in the coordinate directions. But there is nothing sacred about the coordinate directions, and it is reasonable to ask about rates of change of f as inputs vary in *any* direction. **Directional derivatives** get at just this question. The following definition works in any dimension:

> **DEFINITION (Directional derivative)**   Let f be a function, $\mathbf{X_0}$ a point of its domain, and $\mathbf{u}$ a unit vector. The derivative of f at $\mathbf{X_0}$ in the $\mathbf{u}$-direction, denoted by $D_{\mathbf{u}} f(\mathbf{X_0})$, is defined by
> $$D_{\mathbf{u}} f(\mathbf{X_0}) = \lim_{h \to 0} \frac{f(\mathbf{X_0} + h\mathbf{u}) - f(\mathbf{X_0})}{h}$$
> if the limit exists.

Notice that, with $\mathbf{u} = \mathbf{i}$, the preceding limit is the same one that defines $f_x(\mathbf{X_0})$. In fact,

$$D_{\mathbf{i}} f(\mathbf{X_0}) = f_x(\mathbf{X_0}), \qquad D_{\mathbf{j}} f(\mathbf{X_0}) = f_y(\mathbf{X_0}), \qquad \text{and} \qquad D_{\mathbf{k}} f(\mathbf{X_0}) = f_z(\mathbf{X_0}).$$

Calculating directional derivatives with the gradient   As with other derivatives, the limit definition, although crucial to *understanding* the directional derivative, is almost useless for *calculating* derivatives. Fortunately, the gradient comes to our rescue. For technical reasons, ➤ we assume that f has continuous partial derivatives at $\mathbf{X_0}$. (This is so for virtually all the functions we will encounter.)

More details follow in the next section.

> **FACT**   Let f, $\mathbf{X_0}$, and $\mathbf{u}$ be as above. Then
> $$D_{\mathbf{u}} f(\mathbf{X_0}) = \nabla f(\mathbf{X_0}) \cdot \mathbf{u}.$$

To see why the Fact is true, consider first what happens if f is a *linear* function of two variables, that is, if $f(x, y) = ax + by + c = \nabla f \cdot (x, y) + c$. In this case, ➤

Watch carefully as things drop away by subtraction.

$$\frac{f(\mathbf{X_0} + h\mathbf{u}) - f(\mathbf{X_0})}{h} = \frac{\nabla f \cdot (\mathbf{X_0} + h\mathbf{u}) - \nabla f \cdot \mathbf{X_0}}{h}$$

$$= \frac{\nabla f \cdot h\mathbf{u}}{h} = \frac{h \, \nabla f \cdot \mathbf{u}}{h}$$

$$= \nabla f \cdot \mathbf{u}.$$

This shows that the Fact holds for linear functions. Nonlinear functions are closely approximated by their linear approximations (the technical assumption stated just before the Fact guarantees this); it follows that the Fact holds for nonlinear functions as well.

Interpreting the gradient vector   Let **u** be any unit vector. A property of the dot product ← gives

Recall: For any vectors **v** *and* **w**, *we have* **v** · **w** = |**v**||**w**| cos θ.

$$D_{\mathbf{u}}f(\mathbf{X_0}) = \nabla f(\mathbf{X_0}) \cdot \mathbf{u} = |\nabla f(\mathbf{X_0})||\mathbf{u}| \cos\theta = |\nabla f(\mathbf{X_0})| \cos\theta,$$

where θ is the angle between $\nabla f(\mathbf{X_0})$ and **u**. In particular,

$$D_{\mathbf{u}}f(\mathbf{X_0}) \le |\nabla f(\mathbf{X_0})|;$$

In this case, θ = 0.

equality holds if and only if **u** points in the same direction as ∇f. ← Thus, two important properties of the gradient follow from the Fact above:

• **Steepest ascent**   The gradient ∇f points in the direction of fastest increase of f.

• **As fast as possible**   The magnitude $|\nabla f(\mathbf{X_0})|$ is the largest possible rate of change of f.

EXAMPLE 3   Find the directional derivatives of $f(x, y) = x^2 + y^2$ in various directions at $(2, 1)$. In which direction does f increase fastest? Decrease fastest?

Solution   The gradient is $\nabla f(2, 1) = (4, 2)$. In this direction, therefore, f increases at the rate of $|(4, 2)| = \sqrt{20}$ units of output per unit of input. In the opposite direction (i.e., the direction of $(-4, -2)$), $\cos\theta = -1$, and thus the directional derivative is $-\sqrt{20}$. In other directions, such as $\mathbf{u} = (1/\sqrt{2}, 1/\sqrt{2})$, the directional derivative is found from the gradient:

$$D_{\mathbf{u}}f(2, 1) = (4, 2) \cdot \left(\frac{1}{\sqrt{2}}, \frac{1}{\sqrt{2}} \right) = \frac{6}{\sqrt{2}}. \qquad \blacksquare$$

Finding functions from gradients

It's easy, as we have seen, to calculate the gradient ∇f from a given function f: We just find all the partial derivatives and collect them in a single vector. The reverse problem is something like antidifferentiation: finding a function f with prescribed partial derivatives, one for each variable. In simple cases, an answer can either be guessed and then checked for accuracy, or found through relatively easy calculations.

On the other hand, antidifferentiation problems can be difficult even in single-variable calculus, and the same is true for two or more variables. Moreover, not every pair of functions of two variables is a gradient at all. We illustrate some of the possibilities in the following example.

EXAMPLE 4   Find functions $f(x, y)$, $g(x, y)$, and $h(x, y)$, if possible, whose gradients are

$$(y, x), \qquad (y, x + 2y), \qquad \text{and} \qquad (0, x),$$

respectively. Is more than one answer possible?

Solution   With a little luck and experience we might guess $f(x, y) = xy$, and it is easy to check directly that our guess is correct.

We might also find $g(x, y)$ by guessing, but a more systematic approach is possible. Because we want $g_x(x, y) = y$, we can try to find g by antidifferentiation in x, treating y as a constant. The result is

$$g(x, y) = \int y\, dx = xy + C(y),$$

where the "constant" $C(y)$ may depend on y but not on x. For such a function g to solve our problem we need $g_y(x, y) = x + 2y$. In other words, we must have

$$g_y(x, y) = \left(xy + C(y)\right)_y = x + C'(y) = x + 2y,$$

which implies, in turn, that $C(y) = y^2$ will do. Thus, $g(x, y) = xy + y^2$ is a suitable solution.

The remaining problem, to find $h(x, y)$ with $\nabla h = (0, x)$, turns out to be impossible. The first condition, that $h_x = 0$, implies that h is independent of x and so must have the form $h(x, y) = h(y)$. The second condition, that $h_y = x$, implies that h has the form $h(x, y) = xy + C(x)$. These conditions on h are incompatible, and so no such h can exist. ∎

BASIC EXERCISES

In Exercises 1–4, draw (by hand) the gradient at each point with integer coordinates in the rectangle $[0, 2] \times [0, 2]$. Then draw the level curve that passes through each of these points.

1. $f(x, y) = (x + y)/2$

2. $f(x, y) = (x^2 - y)/2$

3. $f(x, y) = (y - x^2)/2$

4. $f(x, y) = x/2$

In Exercises 5–8, evaluate the gradient vector ∇f at the given point.

5. $f(x, y) = \sqrt{x^2 + y^2}$ at $(3, 4)$

6. $f(x, y) = \cos^2(\pi x y)$ at $(1, -1)$

7. $f(x, y, z) = \ln(xyz)$ at $(1, 2, 3)$

8. $f(x, y, z) = xe^{yz}$ at $(2, 1, 0)$

In Exercises 9–12, find the gradient at the given point. Then find the level curve through the given point and show that the gradient vector is perpendicular to the level curve at that point.
[HINT: You will need to find a tangent vector to the level curve at the given point. One way to do so is to parametrize the curve, and then use the parametrization to find a velocity vector to the curve at the given point. Another way is to use ordinary derivatives to find the slope of the curve at the given point and then to use the slope to find a suitable tangent vector. Once a tangent vector is found, check that it is perpendicular to the gradient vector.]

9. $f(x, y) = x$; $(2, 3)$

10. $f(x, y) = x^2 - y$; $(1, 1)$

11. $f(x, y) = x^2 - y$; $(2, 1)$

12. $f(x, y) = x^2 + y^2$; $(2, 1)$

In Exercises 13–18, use the method of Example 2 to find the tangent plane to the surface at the given point. Then, use technology to plot both the surface and the plane in an appropriate window.

13. $z = x^2 + y^2$ at $(2, 1, 5)$

14. $z = x^2 - 2y$ at $(3, 4, 1)$

15. $x^2 + y^2 + z^2 = 1$ at $(0, 0, 1)$

16. $x^2 + y^2 + z^2 = 1$ at $(1/\sqrt{3}, 1/\sqrt{3}, 1/\sqrt{3})$

17. $z = \cos(xy) + xe^y$ at $(1, 0, 2)$

18. $x^2 y + e^{xz} + yz = 3$ at $(0, 1, 2)$

In Exercises 19–22, find all the stationary points of f.

19. $f(x, y) = x^3 e^{-(x^2+y^2)}$

20. $f(x, y) = x^3 - 3xy + y^3$

21. $f(x, y) = x^3 + xy^2 - 12x - y^2$

22. $f(x, y) = xy(x^2 + y^2 - 1)$

In Exercises 23–26, find the derivative of f in the direction $\mathbf{v}$ at the given point $\mathbf{x}_0$.

23. $f(x, y) = \sqrt{x^2 + y^2}$; $\mathbf{v} = (\mathbf{i} + \mathbf{j})/\sqrt{2}$; $\mathbf{x}_0 = (3, 4)$

24. $f(x, y) = \sin(\pi x y)$; $\mathbf{v} = (\mathbf{i} - 2\mathbf{j})/\sqrt{5}$; $\mathbf{x}_0 = (2, 1)$

25. $f(x, y, z) = -x^2 + y^2 - z^2$; $\mathbf{v} = \mathbf{i} - \mathbf{j} + \mathbf{k}$; $\mathbf{x}_0 = (1, -1, 0)$

26. $f(x, y, z) = \ln(xy + xz + yz)$; $\mathbf{v} = \mathbf{i} + 2\mathbf{j} + 3\mathbf{k}$; $\mathbf{x}_0 = (3, 2, 1)$

In Exercises 27–30, give an example of a function whose gradient is the given function.

27. $(2x + y, x)$

28. $\left(e^{x-y}, -e^{x-y}\right)$

29. $(2x, 4y, 6z)$

30. $(1, e^z, ye^z)$

FURTHER EXERCISES

31. Let $f(x, y) = x^2 + y^2$.

  (a) Find the directional derivatives of f at $(2, 0)$ in each of the eight directions $\theta = 0$, $\theta = \pi/4$, $\theta = \pi/2, \ldots$, $\theta = 7\pi/4$. Express answers as decimal numbers.

  (b) Plot the data found in (a) as a function of θ. What general shape does the graph have?

  (c) Find a direction (is there more than one?) in which, at $(2, 0)$, f increases at the rate of 3 output units per input unit.

32. The figure below shows the gradient of a function f.

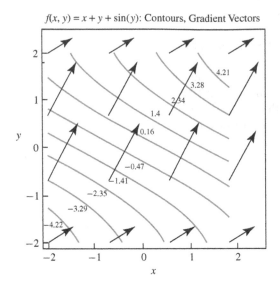

$f(x, y) = x + y + \sin(y)$: Contours, Gradient Vectors

(a) Use the diagram to estimate the partial derivatives $f_x(0, 0)$ and $f_y(0, 0)$.

(b) In fact, f is the function $f(x, y) = x + y + \sin y$. Use the formula for f to check your estimates in part (a).

33. Let $f(x, y) = ax + by + c$ be a linear function and (x_0, y_0) be a point of the domain.

(a) Write an equation for the level curve through (x_0, y_0).

(b) Show that the level curve is perpendicular to the gradient.

34. Let $f(x, y) = x^2 + y^2$ and let C be the level curve of f that passes through the point (x_0, y_0). (Figure 3 shows some contours and gradient vectors for f.)

(a) Show that the vector $(y_0, -x_0)$ is tangent to C at (x_0, y_0). [HINT: Use implicit differentiation.]

(b) Use part (a) to show that $\nabla f(x_0, y_0)$ is perpendicular to C at (x_0, y_0).

35. Let $f(x, y) = x^2 - y^2$ and let C be the level curve of f that passes through the point (x_0, y_0). (Figure 1 shows some contours and gradient vectors for f.)

(a) Show that the vector (y_0, x_0) is tangent to C at (x_0, y_0). [HINT: Use implicit differentiation.]

(b) Use part (a) to show that $\nabla f(x_0, y_0)$ is perpendicular to C at (x_0, y_0).

36. Let $g(x, y) = x^2 - 3xy + 6$ and let C be the level curve of g that passes through the point (x_0, y_0). (Figure 1 on page 717 shows some level curves of this function.)

(a) Show that if $g(x_0, y_0) \neq 6$, then (x_0, y_0) is a point on the curve described by the equation $y = (x^2 + 6 - g(x_0, y_0))/(3x)$.

(b) Suppose that $g(x_0, y_0) \neq 6$. Find a vector that is tangent to C at (x_0, y_0). [HINT: Use implicit differentiation.]

(c) Suppose that $g(x_0, y_0) = 6$ and $x_0 \neq 0$. Show that (x_0, y_0) is a point on the line $y = x/3$.

(d) Suppose that $g(x_0, y_0) = 6$ and $x_0 = 0$. Show that (x_0, y_0) is a point on the line $x = 0$.

(e) Use parts (b)–(d) to show that $\nabla g(x_0, y_0)$ is perpendicular to C at (x_0, y_0).

37. Find the minimum value of the directional derivative of $f(x, y) = 2x + \cos(xy)$ at the point $(\pi/4, 1)$. What direction corresponds to this value?

38. Find the minimum value of the directional derivative of $f(x, y) = x/y$ at the point $(4, -1)$. What direction corresponds to this value?

39. Suppose that at the point $(4, 5, 6)$ the function $f(x, y, z)$ increases most rapidly in the direction $(2, -1, -3)$ and that the rate of increase of f in this direction is 7. What is the rate of change of f at the point $(4, 5, 6)$ in the direction $(-1, 1, -2)$?

40. Suppose that at the point $(-4, 2, 4)$ the function $f(x, y, z)$ decreases most rapidly in the direction $(3, 6, 2)$ and that the rate of change of f in this direction is -5. What is the rate of change of f at the point $(-4, 2, 4)$ in the direction $(4, 0, -3)$?

41. At the point $(1, 3)$, a function $f(x, y)$ has directional derivative 2 in the direction from $(1, 3)$ to $(2, 3)$ and -2 in the direction from $(1, 3)$ to $(1, 4)$.

(a) Determine the gradient of f at $(1, 3)$.

(b) Compute the directional derivative of f in the direction from $(1, 3)$ to $(3, 6)$.

42. At the point (x_0, y_0), a function $f(x, y)$ has directional derivative 1 in the directions $\mathbf{i} + 2\mathbf{j}$ and $\mathbf{i} - 2\mathbf{j}$.

(a) Determine the gradient of f at (x_0, y_0).

(b) Compute the directional derivative of f in the direction $\mathbf{i}$.

43. Suppose that $\nabla f(x, y) = (y^2 + x, 2xy)$ for all $(x, y) \in \mathbb{R}^2$ and that $f(1, 0) = 5$. Find $f(2, 1)$.

44. Suppose that $\nabla f(x, y) = (2xy \cos(x^2 y), x^2 \cos(x^2 y))$ for all $(x, y) \in \mathbb{R}^2$ and that $f(0, 1) = 2$. Find $f(1, \pi/2)$.

45. Suppose that $f(x, y) = xy^2$. In what direction(s) from the point $(-1, 1)$ is the rate of change 2?

46. Suppose that $f(x, y) = x^2 y$. In what direction(s) from the point $(1, 2)$ is the rate of change 3?

47. Let $L(x, y)$ be the linear approximation to $f(x, y)$ at the point (x_0, y_0). Show that $D_\mathbf{u} L(x_0, y_0) = D_\mathbf{u} f(x_0, y_0)$ for every direction vector $\mathbf{u}$.

48. Show that $\nabla(fg) = f\nabla g + g\nabla f$.

49. Show that $\nabla(f/g) = (1/g^2)(g\nabla f - f\nabla g)$.

13.5 HIGHER-ORDER DERIVATIVES AND QUADRATIC APPROXIMATION

For a well-behaved function f of one variable, higher-order derivatives f'', f''',..., $f^{(17)}$,... are readily calculated. Higher derivatives have various uses and interpretations. The second derivative, f'', for instance, has an important geometric meaning: It tells how fast (and in what direction) the slope function f' varies and therefore describes the concavity of the f-graph. One use of this information is in distinguishing among various kinds of stationary points of f. Suppose, for instance, that $f'(x_0) = 0$ and that $f''(x_0) < 0$. Then the f-graph is concave down at x_0, and so f must have achieved a local *maximum* at x_0. This sort of reasoning is collected in the **second derivative test** of single-variable calculus. (We will follow this line of thought in the next section.) Another one-variable use of higher derivatives is in writing Taylor and Maclaurin polynomials to approximate a given nonpolynomial function f.

In this section and the next we will briefly review these ideas in the one-variable setting and then see how they extend to functions of several variables.

Second and higher derivatives

Functions of several variables have repeated (partial) derivatives, just like their one-variable relatives. The following calculations will not be surprising, but notice the various notations and terminology.

EXAMPLE 1   Let $f(x, y) = x^2 + xy^2$. Find all possible second partial derivatives.

Solution   The first partial derivatives are

$$f_x = \frac{\partial f}{\partial x} = 2x + y^2; \qquad f_y = \frac{\partial f}{\partial y} = 2xy.$$

Differentiating *again* gives four results:

$$f_{xx} = \frac{\partial}{\partial x}\left(\frac{\partial f}{\partial x}\right) = \frac{\partial^2 f}{\partial x^2} = 2; \qquad f_{xy} = \frac{\partial}{\partial y}\left(\frac{\partial f}{\partial x}\right) = \frac{\partial^2 f}{\partial y \partial x} = 2y;$$

and

$$f_{yx} = \frac{\partial}{\partial x}\left(\frac{\partial f}{\partial y}\right) = \frac{\partial^2 f}{\partial x \partial y} = 2y; \qquad f_{yy} = \frac{\partial}{\partial y}\left(\frac{\partial f}{\partial y}\right) = \frac{\partial^2 f}{\partial y^2} = 2x.$$

(Notice that the symbols f_{xy} and $\dfrac{\partial^2 f}{\partial y \partial x}$ mean the same thing—even though the order of symbols may seem reversed.) ∎

A matrix of second derivatives   All the second partial derivatives of a function $f(x, y)$ can be collected in a 2×2 matrix called the **second derivative** (or **Hessian**) matrix of f at (x, y):

$$f''(x, y) = \begin{pmatrix} f_{xx} & f_{xy} \\ f_{yx} & f_{yy} \end{pmatrix} = \begin{pmatrix} 2 & 2y \\ 2y & 2x \end{pmatrix}.$$

The Hessian matrix of f at $\mathbf{X_0}$ is also sometimes denoted by $H_f(\mathbf{X_0})$; it is named for the German mathematician Ludwig Otto Hesse (1811–1874). Observe some properties of the second derivative matrix:

- **Dimension**   For a function $f(x_1, x_2, \dots, x_n)$ of n variables, the Hessian is an $n \times n$ matrix. The jth entry in the ith row is $f_{x_i x_j}$, the result of differentiating first with

Do you agree? Check details.

respect to x_i and then x_j. If, say, $f : \mathbb{R}^3 \to \mathbb{R}$ is defined by $f(x, y, z) = xz^2 + yz$, then the Hessian of f is the 3×3 matrix ◄

$$f''(x, y, z) = \begin{pmatrix} f_{xx} & f_{xy} & f_{xz} \\ f_{yx} & f_{yy} & f_{yz} \\ f_{zx} & f_{zy} & f_{zz} \end{pmatrix} = \begin{pmatrix} 0 & 0 & 2z \\ 0 & 0 & 1 \\ 2z & 1 & 2x \end{pmatrix}.$$

- **Rows are gradients** Successive rows of the Hessian are gradients of the *first* partial derivatives of f. The middle row above, for instance, is the gradient $\nabla f_y(x, y, z)$.

Order of differentiation (usually) doesn't matter In both equations above, the second derivative matrix is symmetric with respect to the main diagonal. In other words,

$$f_{xy} = f_{yx}; \qquad f_{xz} = f_{zx}; \qquad f_{yz} = f_{zy};$$

the order of differentiation of "mixed partials" does not seem to matter (at least in the cases seen so far). It is an interesting fact that this phenomenon holds for *all* well-behaved functions of several variables. (It's a useful fact, too—it saves a lot of work in evaluating a Hessian.) We will state the result for a function $f(x, y)$ (though the same result holds for functions of any number of variables):

> **THEOREM 2 (Equality of cross partials)** Let $f(x, y)$ be a function; assume that the second partial derivatives f_{xy} and f_{yx} are defined and continuous on the domain of f. Then for all (x, y),
>
> $$f_{xy}(x, y) = f_{yx}(x, y).$$

It might be surprising that an integral is involved.

There are several ways to prove this theorem. One of the simplest involves a "double integral." ◄ Since we have not met this idea yet, we will defer the proof to Appendix K.

Taylor polynomials and quadratic approximation in one variable

Let $f(x)$ be a function of one variable. Recall from single-variable calculus the idea of **Taylor polynomials** for $f(x)$ at a point x_0. The polynomials

$$p_1(x) = f(x_0) + f'(x_0)(x - x_0);$$

$$p_2(x) = f(x_0) + f'(x_0)(x - x_0) + \frac{f''(x_0)}{2}(x - x_0)^2$$

are called, respectively, the first and second (or first-order and second-order) Taylor polynomials for f, based at x_0. We have seen p_1 recently—it is simply the linear approximation to f at x_0. In the same spirit, p_2 is the **quadratic approximation** to f at x_0. In general, the nth Taylor polynomial has the form

$$\begin{aligned} p_n(x) = f(x_0) + f'(x_0)(x - x_0) + \frac{f''(x_0)}{2}(x - x_0)^2 \\ + \frac{f'''(x_0)}{3!}(x - x_0)^3 + \cdots + \frac{f^{(n)}(x_0)}{n!}(x - x_0)^n, \end{aligned} \qquad (1)$$

where $f^{(n)}$ denotes the nth derivative and $n!$ is the factorial of n. The number $f^{(n)}(x_0)/n!$ is called the nth **Taylor coefficient**.

EXAMPLE 2 Let $f(x) = e^x$. Find several Taylor polynomials for f based at $x_0 = 0$. Do the same for $g(x) = \sin x$. How are the functions related to their Taylor polynomials?

Solution Using $f(x) = e^x$ makes the work especially simple. Because $f(x) = f'(x)$, we have $f^{(n)}(0) = 1$ for all n. Thus, for all n, f has Taylor polynomials of the form

$$p_n(x) = 1 + 1x + \frac{x^2}{2!} + \frac{x^3}{3!} + \cdots + \frac{x^n}{n!}.$$

A similar calculation shows that, for $g(x) = \sin x$, the first few Taylor polynomials based at $x_0 = 0$ have the form ➤

Convince yourself!

$$P_1(x) = x; \qquad P_3(x) = x - \frac{x^3}{6}; \qquad \text{and} \qquad P_5(x) = x - \frac{x^3}{6} + \frac{x^5}{120}.$$

Figure 1 shows the sine function and several of its Taylor polynomials:

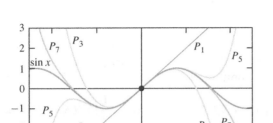

FIGURE 1
Several Taylor polynomial approximations to
$f(x) = \sin x$

The main question, of course, is how closely the Taylor polynomials approximate the sine function at and near $x_0 = 0$. ■

Matching derivatives The key property of Taylor polynomials is that they match derivatives—from order zero through order n—with their "parent" functions at the base point. ➤ If $n = 3$, for instance,

The value of a function is its "zeroth derivative."

$$p_3(x_0) = f(x_0); \qquad p_3'(x_0) = f'(x_0); \qquad p_3''(x_0) = f''(x_0); \qquad p_3'''(x_0) = f'''(x_0).$$

This is easy to check; just differentiate the right side of Equation 1. ➤ Because the function and the Taylor polynomial agree to this extent *at $x = x_0$*, they behave similarly *near $x = x_0$* as well.

Do this now; you'll see where the factorial comes from!

Taylor polynomials in several variables

The idea of Taylor polynomials makes excellent sense for functions of several variables, too. Indeed, we have already seen that, for a function $f(x, y)$, the linear approximation at $\mathbf{X_0} = (x_0, y_0)$, which we have defined by

$$L(\mathbf{X}) = f(\mathbf{X_0}) + \nabla f(\mathbf{X_0}) \cdot (\mathbf{X} - \mathbf{X_0})$$
$$= f(x_0, y_0) + f_x(x_0, y_0)(x - x_0) + f_y(x_0, y_0)(y - y_0),$$

has the same value and the same (first) partial derivatives as f at (x_0, y_0). Thus, it is reasonable to call $L(x, y)$ the first-order Taylor polynomial for f at (x_0, y_0).

The natural next step is to *second* partial derivatives: The **quadratic approximation** to f at (x_0, y_0) is the function Q defined by

$$Q(x, y) = f(x_0, y_0) + f_x(x_0, y_0)(x - x_0) + f_y(x_0, y_0)(y - y_0)$$

$$+ \frac{f_{xx}(x_0, y_0)}{2}(x - x_0)^2 + f_{xy}(x_0, y_0)(x - x_0)(y - y_0) + \frac{f_{yy}(x_0, y_0)}{2}(y - y_0)^2.$$

The formula is slightly complicated, but there are patterns to see:

- **Partial derivatives match** The definition of $Q(x, y)$ guarantees that Q and f have the same value, the same *first* partial derivatives, and the same *second* partial derivatives at (x_0, y_0). To see, for instance, that $f_{xy}(x_0, y_0) = Q_{xy}(x_0, y_0)$, we differentiate twice, once with respect to each variable. ← We get

Notice that the order doesn't matter!

$$Q_x(x, y) = f_x(x_0, y_0) + f_{xx}(x_0, y_0)(x - x_0) + f_{xy}(x_0, y_0)(y - y_0);$$

therefore $Q_{xy}(x, y) = f_{xy}(x_0, y_0)$, as claimed.

- **In vector form** The second-order coefficients in the definition of Q come from the Hessian matrix of second partial derivatives f. It isn't surprising, therefore, that the vector form of the definition of $Q(x, y)$ involves the Hessian matrix $f''(\mathbf{X_0})$:

$$Q(\mathbf{X}) = f(\mathbf{X_0}) + \nabla f(\mathbf{X_0}) \cdot (\mathbf{X} - \mathbf{X_0}) + \frac{1}{2}\left[f''(\mathbf{X_0})(\mathbf{X} - \mathbf{X_0})^t\right] \cdot (\mathbf{X} - \mathbf{X_0}).$$

(Here the notation $(\mathbf{X} - \mathbf{X_0})^t$ denotes the **transpose** of the vector $(\mathbf{X} - \mathbf{X_0})$. To "transpose" a vector means simply to write it in *column* form.)

Both dots in the preceding equation indicate ordinary dot products. There is also a *matrix* multiplication $f''(\mathbf{X_0})(\mathbf{X} - \mathbf{X_0})^t$ in the last summand; using the transpose ensures that the matrix multiplication makes good sense.

The expression is slightly complicated, to be sure. The main point, however, is the typographical similarity with the one-variable form of the second Taylor polynomial.

- **Additional variables** A similar definition holds for functions f and Q of three or more variables. We will stick mainly to two variables.

- **Appeal to higher powers** There is no need to stop with second-order Taylor polynomials. As in the one-variable case, the idea makes perfectly good sense for higher-order derivatives. The **cubic approximation** $C(x, y)$ to f at (x_0, y_0), for instance, includes all the terms in $Q(x, y)$ plus additional terms of the following forms: ←

The denominators turn out to be slightly different because some partial derivatives involve one variable and others involve two.

$$\frac{f_{xxx}(x_0, y_0)}{3!}(x - x_0)^3, \qquad \frac{f_{xxy}(x_0, y_0)}{2!}(x - x_0)^2(y - y_0),$$

$$\frac{f_{xyy}(x_0, y_0)}{2!}(x - x_0)(y - y_0)^2, \text{ etc.}$$

At some point the bookkeeping becomes excessive for humans (although *Maple*, *Mathematica*, and their relatives have no trouble); we will content ourselves with second-order approximations.

EXAMPLE 3 Find the quadratic approximation $Q(x, y)$ to $f(x, y) = xe^y$ at $(0, 0)$. How is it related to the Hessian matrix? How closely does Q appear to approximate f near $(0, 0)$?

Solution Calculating partial derivatives of f gives

$$f(0, 0) = 0; \qquad \nabla f(0, 0) = (1, 0); \qquad f''(0, 0) = \begin{pmatrix} 0 & 1 \\ 1 & 0 \end{pmatrix}.$$

Thus,

$$Q(x, y) = 0 + x + 0y + \frac{1}{2}(0x^2 + xy + yx + 0y^2) = x + xy.$$

The coefficients of the quadratic terms in Q, therefore, come directly from the Hessian matrix.

Like the linear approximation L, Q approximates f closely near $(0, 0)$. Figure 2 shows graphs of f, L, and Q for comparison. In each case, the darker surface is f:

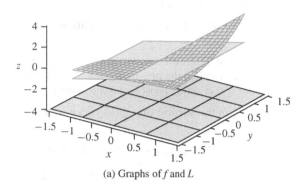

(a) Graphs of f and L

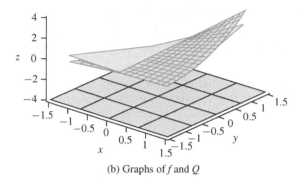

(b) Graphs of f and Q

FIGURE 2

Linear and quadratic approximations to $f(x, y) = xe^y$

A close look shows that both of the functions L and Q approximate f near the domain point $(0, 0)$. But Q, being slightly curved, does a better job of approximation than L, which is flat. This is no surprise given that Q has "extra" matching derivatives.　■

BASIC EXERCISES

In Exercises 1–4, find the second partial derivative of f with respect to each variable (e.g., f_{xx}).

1. $f(x, y) = x^3 y^4$

2. $f(t, w) = \sin(tw^2)$

3. $f(u, v) = e^{-(u^2 + v^2)}$

4. $f(r, s, t) = r^3(s + t)^2$

In Exercises 5–8, show that $f_{xy} = f_{yx}$.

5. $f(x, y) = x^4 y^3$

6. $f(x, y) = \cos(xy)$

7. $f(x, y) = e^{-(x^2 + y^2)}$

8. $f(x, y) = \arctan(y/x)$

In Exercises 9–20, find the Hessian matrix f''.

9. $f(x, y) = \sin(xy)$

10. $f(x, y) = xy$

11. $f(x, y) = \sin x + \cos(2y)$

12. $f(x, y) = x^2 + y^2$

13. $f(x, y) = x^2 - y^2$

14. $f(x, y) = x^2 y^2 + xy^3$

15. $f(x, y) = x \cos(x + y)$

16. $f(x, y) = y^2 e^{xy}$

17. $f(x, y) = Ax^2 + By^2 + Cxy + Dx + Ey + F$

18. $f(x, y) = Ax^3 + Bx^2 y + Cxy^2 + Dy^3$

19. $f(x, y, z) = Ax + By + Cz + D$

20. $f(x, y, z) = e^{(x + y^2 + z^3)}$

In Exercises 21–30, find the quadratic approximation function $Q(x, y)$ at the given point. Then, plot $|f - Q|$ near the given point to see the connection between f and Q.

21. $f(x, y) = xy$; $(0, 0)$

22. $f(x, y) = xe^y$; $(0, 2)$

23. $f(x, y) = \sin(xy)$; $(0, 0)$

24. $f(x, y) = \cos(xy)$; $(\sqrt{\pi}, \sqrt{\pi})$

25. $f(x, y) = \sin x + \cos(2y)$; $(-\pi/2, \pi/4)$

26. $f(x, y) = x^2 + y^2$; $(0, 0)$

27. $f(x, y) = x^2 - y^2$; $(0, 0)$

28. $f(x, y) = \sqrt{x^2 + y^2}$; $(3, 4)$

29. $f(x, y, z) = \sqrt{x^2 + y^2 + z^2}$; $(2, 2, 1)$

30. $f(x, y, z) = \sin(x + y + z)$; $(\pi/3, \pi/3, \pi/3)$

FURTHER EXERCISES

31. The partial differential equation $u_t = au_{xx}$, where $a > 0$ is a constant, is called the one-dimensional heat equation. Show that $u(x, t) = e^{-at} \cos x$ is a solution of this equation.

32. Show that $u(x, t) = e^{-at} \cos x + 2e^{-4at} \sin(2x)$, where $a > 0$ is a constant, is a solution of the one-dimensional heat equation $u_t = au_{xx}$.

33. The partial differential equation $u_t = a(u_{xx} + u_{yy})$, where $a > 0$ is a constant, is called the two-dimensional heat equation. Show that $u(x, y, t) = e^{-(m^2+n^2)at} \cos(mx) \sin(ny)$ is a solution of this equation.

34. Show that $u(x, y, t) = \dfrac{1}{4\pi at} e^{-(x^2+y^2)/4at}$, where $a > 0$ is a constant, is a solution of the two-dimensional heat equation $u_t = a(u_{xx} + u_{yy})$.

35. The partial differential equation $u_{xx} + u_{yy} = 0$ is called Laplace's equation. Show that $u(x, y) = \ln(\sqrt{x^2 + y^2})$ is a solution of this equation.

36. Show that $u(x, y) = \arctan(y/x)$ is a solution of Laplace's equation $u_{xx} + u_{yy} = 0$.

37. Show that $u(x, y) = e^{x^2-y^2} \sin(2xy)$ is a solution of Laplace's equation $u_{xx} + u_{yy} = 0$.

38. Show that $u(x, y) = e^{-x} \cos y$ is a solution of Laplace's equation $u_{xx} + u_{yy} = 0$.

39. The partial differential equation $u_{tt} = c^2 u_{xx}$, where $c > 0$ is a constant, is called the one-dimensional wave equation. Show that $u(x, t) = \sin(3x) \cos(3ct)$ is a solution of this equation.

40. We said in this section that $Q(x, y)$ can be written in vector form as

$$Q(\mathbf{X}) = f(\mathbf{X_0}) + \nabla f(\mathbf{X_0}) \cdot (\mathbf{X} - \mathbf{X_0})$$
$$+ \frac{1}{2} \left[f''(\mathbf{X_0})(\mathbf{X} - \mathbf{X_0})^t \right] \cdot (\mathbf{X} - \mathbf{X_0}),$$

where the dots indicate dot products and the "transpose" $(\mathbf{X} - \mathbf{X_0})^t$ in the last summand denotes a column vector. Work out the details to convince yourself of this.

41. Let $f(x, y, z)$ be a function of three variables, and let $\mathbf{X_0} = (0, 0, 0)$. Write out the formula for the quadratic approximation $Q(x, y, z)$ at $\mathbf{X_0}$. How many terms are there?

13.6 MAXIMA, MINIMA, AND QUADRATIC APPROXIMATION

Optimization—finding maximum and minimum values of a function—is as important for multivariable functions as for one-variable functions. Functions of several variables are more complicated, of course, but derivatives remain the crucial tools.

A one-variable review

If f is not differentiable, then maxima and minima may occur where $f'(x)$ fails to exist.

"Extremum" (singular) means "either maximum or minimum"; "extrema" is the plural.

We'll review the "second derivative test" in more detail later.

For a one-variable differentiable function $y = f(x)$ on an interval I, finding maximum and minimum values is relatively straightforward. Maxima and minima are found only at **stationary points**—where $f'(x) = 0$—or at the endpoints (if any) of the interval I. ◄ Usually, only a few such "candidate" points exist, and we can check directly which produces, say, the largest value of f.

A simple idea lies behind all talk of derivatives and extrema: ◄ At a local maximum or minimum point x_0, the graph of a differentiable function f must be "flat," and so $f'(x_0) = 0$. However, a stationary point x_0 might be (i) a local maximum point; (ii) a local minimum point; or (iii) neither. (The function $f(x) = x^3$ at $x = 0$ illustrates the "neither" case.)

For simple, one-variable functions, deciding which of (i)—(iii) actually holds is easy. One strategy is to check the sign of $f''(x_0)$. If, say, $f''(x_0) < 0$, then f is concave down at x_0, and f has a local maximum there. ◄ Alternatively, we might just plot f and see directly how it behaves near x_0.

Local talk   When is a maximum or minimum "local"? When is it "global"? Figure 1 shows the difference for a function defined on the domain $[-10, 10]$:

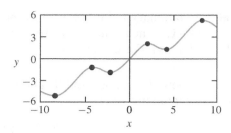

FIGURE 1

Graph of f: local vs. global extrema

All six bulleted points (three maximum points and three minimum points) correspond to *local* extrema of the function f. At each bulleted point $(x_0, f(x_0))$, $f(x_0)$ is either highest or lowest among *nearby* points $(x, f(x))$. (For present purposes, exactly *how* nearby doesn't matter.) Only the first and last points, however, are *global* extrema because they (and only they) represent the largest and smallest values of $f(x)$ among all possibilities shown in the picture. (On a larger domain interval, say $-15 \le x \le 15$, these extrema might not be global.)

Local maxima and minima are conceptually simpler than the global variety, readily recognizable on graphs, and often convenient to locate symbolically using derivatives. With all these advantages, local extrema are usually the main tools in solving optimization problems.

Functions of several variables

In optimizing functions of several variables, it is natural (just as for functions of one variable) to look especially at stationary points of the domain, where all partial derivatives are zero. Indeed, as we observed in Section 13.2, ➡ local maximum and minimum values can occur *only* at stationary points:

See page 719 for the idea of a proof.

> **FACT (Extreme points and partial derivatives)** Suppose that $f(x, y)$ has a local maximum or a local minimum at (x_0, y_0). If both partial derivatives exist, then
> $$f_x(x_0, y_0) = 0 \quad \text{and} \quad f_y(x_0, y_0) = 0.$$

But the Fact alone doesn't tell the whole story—different types of behavior are possible for a function at and near a stationary point.

Different types of stationary points Although the basic strategy for optimizing a function—find the stationary points and analyze them—is exactly the same for functions of one and of several variables, the situation is usually more complicated for functions of several variables. For one thing, finding stationary points may be harder; for another, functions of several variables can behave in more complicated ways near a stationary point. This makes multivariable optimization harder but also more interesting.

As we did for functions of one variable, we will identify three main types of stationary point for a function $f(x, y)$. (The definitions for a function $g(x, y, z)$ are almost identical.)

- **Local minimum point**   A stationary point (x_0, y_0) is a **local minimum point** for f if $f(x, y) \ge f(x_0, y_0)$ for all (x, y) near (x_0, y_0). (A little more formally: $f(x, y) \ge f(x_0, y_0)$ for all (x, y) in some rectangle surrounding (x_0, y_0).) In this case we say that f **assumes a local minimum value** at (x_0, y_0). For example, $(0, 0)$ is a local minimum point for $f(x, y) = x^2 + y^2$. ➡

Picture the graph—it's a paraboloid opening upward.

The graph is another paraboloid and opens downward.

- **Local maximum point**   A stationary point (x_0, y_0) is a **local maximum point** for f if $f(x, y) \leq f(x_0, y_0)$ for all (x, y) near (x_0, y_0). In this case we say that f **assumes a local maximum value** at (x_0, y_0). For example, $(0, 0)$ is a local maximum point for $f(x, y) = 1 - x^2 - y^2$. ◄

- **Saddle point**   A stationary point (x_0, y_0) is a **saddle point** for f if f assumes neither a local maximum nor a local minimum at (x_0, y_0).

Look in the picture for a stationary point of each type listed above.

The contour plot in Figure 2 shows that all three possibilities for stationary points can coexist in close proximity. The function in question is $f(x, y) = \cos(x)\sin(y)$: ◄

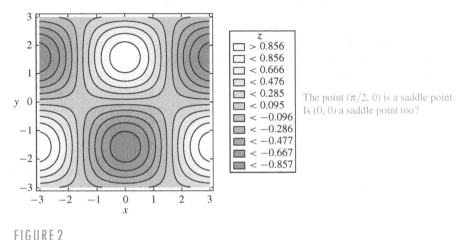

The point $(\pi/2, 0)$ is a saddle point. Is $(0, 0)$ a saddle point too?

FIGURE 2
Contour plot of $f(x, y) = \cos(x)\sin(y)$

Second derivatives and quadratic approximation

For functions of one variable, we sometimes use the second derivative to distinguish among stationary points of different types. The situation is similar—but a little more complicated—for functions of two variables. (Things become more complicated still for functions of three or more variables, but the important ideas remain the same. We will focus mainly on the two-variable case.)

Think about it "geographically." The surface $z = f(x, y)$ resembles a mountainous landscape; at a stationary point, the terrain is "level" in both the east–west and the north–south directions. As hikers know, this can happen in various ways. The point in question might be a mountain peak, the base of a bowl-shaped valley, a mountain pass between two peaks, or a point on a "terrace."

If technology is available it is sometimes possible to distinguish among different types of stationary points just by looking at a graph or a contour map. The contour plot in Figure 2, for instance, shows quite clearly the nature of the various stationary points.

But graphs are not always available, and in any case it is satisfying to relate stationary points to a function's symbolic formula. Second derivatives—in this case the Hessian matrix of second partial derivatives—turn out to be helpful, and quadratic approximation is the key tool.

Quadratic approximation in one variable   Consider first the case of a one-variable function $f(x)$. At a stationary point x_0 we have $f'(x_0)$, and so the quadratic approximation function has the form

$$Q(x) = f(x_0) + \frac{f''(x_0)}{2}(x - x_0)^2.$$

Several useful conclusions follow:

- If $f''(x_0) > 0$, then the Q-graph is a parabola opening *upward*, and so Q (and therefore f) has a local *minimum* at $x = x_0$.

- If $f''(x_0) < 0$, then the Q-graph is a parabola opening *downward*, and so Q (and therefore f) has a local *maximum* at $x = x_0$.

- If $f''(x_0) = 0$, then Q is a constant function, and so the quadratic approximation says nothing about the type of stationary point at x_0. (If the cubic term of the Taylor approximation is nonzero, then it may help determine the nature of the stationary point. But that is another story.)

The following key principle is implicit in the conclusions just listed and in most of what follows; it holds for functions of several variables, too:

> **FACT** Let f and Q be as above. If Q is not constant, then f and Q have the same type of stationary point at x_0.

This Fact means that quadratic functions are, for many other functions, simpler prototypes for checking behavior near a stationary point. It is especially important, therefore, to understand clearly how quadratic functions themselves behave near stationary points.

Stationary points and quadratic approximation

Let f be a function of two variables; to avoid technical complications we assume that all the partial derivatives in question exist and are continuous. (This ensures that, for any input point (x_0, y_0), f *has* a quadratic approximation; without that, nothing that follows makes sense.) To simplify notation and save a little space, we will use $(x_0, y_0) = (0, 0)$. ➡ Then the quadratic approximation has the form

This is only a convenience—we won't use any special properties of the origin.

$$Q(x, y) = f(0, 0) + f_x(0, 0)x + f_y(0, 0)y + \frac{f_{xx}(0, 0)}{2}x^2 + f_{xy}(0, 0)xy + \frac{f_{yy}(0, 0)}{2}y^2.$$

If $(0, 0)$ is a stationary point, then the first-order terms disappear. Here is what remains:

$$Q(x, y) = f(0, 0) + \frac{f_{xx}(0, 0)}{2}x^2 + f_{xy}(0, 0)xy + \frac{f_{yy}(0, 0)}{2}y^2.$$

Our question, then, is, How do the values of f_{xx}, f_{xy}, and f_{yy} determine the type of stationary point?

The first term above is a constant, and so what matters here are the last three terms, which have the form $Ax^2 + Bxy + Cy^2$. Let's decide how such a function behaves near $(0, 0)$; then we will return and relate the answers to second derivatives. (For later reference, $A = f_{xx}(0, 0)/2$, $B = f_{xy}(0, 0)$, and $C = f_{yy}(0, 0)/2$.)

Analyzing $Ax^2 + Bxy + Cy^2$ Let $f(x, y) = Ax^2 + Bxy + Cy^2$; then $(0, 0)$ is a stationary point of f regardless of the values of A, B, and C. To see how the *type* of stationary point depends on these values, we will study some simple but important examples. ➡ In each case, the Hessian matrix at $(0, 0)$ will be calculated for later reference.

Plot as many surfaces as possible as aids to intuition.

EXAMPLE 1 Let $f(x, y) = x^2 + y^2$. How does f behave near the stationary point $(0, 0)$? Describe the surface $z = f(x, y)$. What difference would it make if, instead, $f(x, y) = -(x^2 + y^2)$?

Solution The Hessian matrix is simple:

$$f''(0,0) = \begin{pmatrix} 2 & 0 \\ 0 & 2 \end{pmatrix}.$$

Clearly, f has a local minimum at $(0,0)$ because, for all (x,y),

$$f(x,y) = x^2 + y^2 \geq 0 = f(0,0).$$

The surface $z = x^2 + y^2$ is a **circular paraboloid** with vertex at $(0,0)$. The level curves of f are circles centered at $(0,0)$.

Trading $f(x,y) = x^2 + y^2$ for $f(x,y) = -x^2 - y^2$ turns everything upside down. The local minimum becomes a local maximum at $(0,0)$, the surface becomes a downward-opening paraboloid, and the Hessian matrix acquires negative signs. ∎

E X A M P L E 2 Let $g(x,y) = 3x^2 + 2y^2$. How does g behave near the stationary point $(0,0)$? Describe the surface $z = g(x,y)$.

Solution The difference from the preceding example is only in the positive constants 2 and 3. So again, $(0,0)$ is a local minimum point: For all (x,y),

$$g(x,y) = 3x^2 + 2y^2 \geq 0 = g(0,0).$$

This time, however, the different coefficients of x^2 and y^2 mean that the level curves (which correspond to equations $g(x,y) = 3x^2 + 2y^2 = c$) are ellipses. Figure 3 shows a contour plot of g:

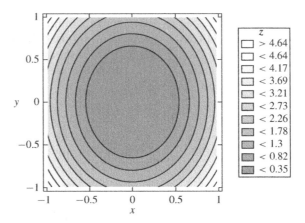

FIGURE 3

Contour plot of $g(x,y) = 3x^2 + 2y^2$

Note the factors of two.

The surface $z = g(x,y)$ is called an **elliptic paraboloid**. (See the exercises at the end of this section for more on this terminology.) The Hessian matrix is ◄

$$g''(0,0) = \begin{pmatrix} 6 & 0 \\ 0 & 4 \end{pmatrix}.$$

∎

E X A M P L E 3 Let $h(x,y) = 3x^2 - 2y^2$. How does h behave near the stationary point $(0,0)$? Describe the surface $z = h(x,y)$.

Solution Because the coefficients of x^2 and y^2 have different signs, the level curves (which correspond to equations of the form $h(x,y) = 3x^2 - 2y^2 = c$) are now hyperbolas, and the surface is called a **hyperbolic paraboloid**. Figure 4 shows a contour map:

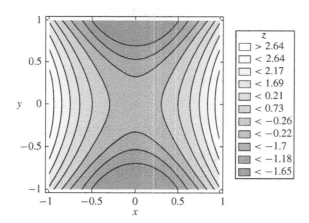

FIGURE 4
Contour plot of $h(x, y) = 3x^2 - 2y^2$

The contour map shows that, in this case, $(0, 0)$ is a **saddle point**, that is, a point that represents *both* maximum and minimum values of h, depending on direction. If we fix $x = 0$, then $h(0, y) = -2y^2$; thus, this slice of the surface is a parabola opening downward, with a *maximum* at the origin. ➡ But if we fix $y = 0$, then $h(x, 0) = 3x^2$—a parabola opening upward, with a *minimum* at the origin. The Hessian matrix is now ➡

An ant walking south to north along the surface experiences the origin as a peak. Note the minus sign.

$$h''(0, 0) = \begin{pmatrix} 6 & 0 \\ 0 & -4 \end{pmatrix}.$$

E X A M P L E 4   Let $j(x, y) = xy$. How does j behave near the stationary point $(0, 0)$? Describe the surface $z = j(x, y)$.

Solution   The function j behaves similarly to h of the preceding example. Again, level curves of j are hyperbolas, this time of the form $xy = c$, as Figure 5 reveals:

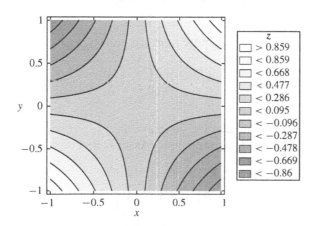

FIGURE 5
Contour plot of $j(x, y) = xy$

Slicing the surface with the plane $y = x$ gives $j(x, y) = x^2$, while if $y = -x$ then $j(x, y) = -x^2$. These "opposite" results mean that, again, $(0, 0)$ is a saddle point and that the surface is another hyperbolic paraboloid. This time the Hessian matrix is

$$j''(0, 0) = \begin{pmatrix} 0 & 1 \\ 1 & 0 \end{pmatrix}.$$

A final example illustrates an important technique we will use more generally in a moment.

EXAMPLE 5   Let $k(x, y) = x^2 + xy + y^2$. Discuss the stationary point at the origin.

Solution   Let's complete the square in x:

$$k(x, y) = x^2 + xy + y^2 = (x + y/2)^2 - y^2/4 + y^2$$

$$= (x + y/2)^2 + \frac{3}{4}y^2.$$

The last version shows that $(0, 0)$ is a minimum point since, for all (x, y),

$$k(x, y) = (x + y/2)^2 + \frac{3}{4}y^2 \geq 0 = k(0, 0).$$

The contour map (Figure 6) also shows a local minimum at the origin:

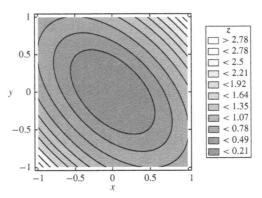

FIGURE 6
Contour plot of $k(x, y) = x^2 + xy + y^2$

This time the Hessian matrix has more nonzero entries:

$$k''(0, 0) = \begin{pmatrix} 2 & 1 \\ 1 & 2 \end{pmatrix}.$$

The general case   We will handle the general case $f(x, y) = Ax^2 + Bxy + Cy^2$ as in the preceding example, by completing the square. Assume for convenience that $A \neq 0$. ◄ Here is the symbolic calculation:

We could carry out the same calculation assuming that $B \neq 0$ or $C \neq 0$.

$$f(x, y) = Ax^2 + Bxy + Cy^2 = A\left(x^2 + \frac{B}{A}xy + \frac{C}{A}y^2\right)$$

$$= A\left(\left(x + \frac{B}{2A}y\right)^2 + \left(\frac{C}{A} - \frac{B^2}{4A^2}\right)y^2\right).$$

This shows that the type of stationary point depends on the sign of the coefficient of y^2. Notice that

$$\frac{C}{A} - \frac{B^2}{4A^2} \geq 0 \iff 4AC - B^2 \geq 0.$$

We conclude:

- If $4AC - B^2 > 0$, then f has either a local maximum or local minimum at $(0, 0)$, depending on whether $A < 0$ or $A > 0$.
- If $4AC - B^2 < 0$, then f has a saddle point at $(0, 0)$.

(In the remaining case, where $4AC - B^2 = 0$, various things can happen, and we normally look for more information.)

In terms of derivatives   If we rewrite these conclusions using partial derivatives (recalling that $2A = f_{xx}(0, 0)$, $B = f_{xy}(0, 0) = f_{yx}(0, 0)$, and $2C = f_{yy}(0, 0)$), we see that

$$4AC - B^2 = f_{xx}f_{yy} - f_{xy}^2;$$

in other words, $4AC - B^2$ is the **determinant** of the Hessian matrix

$$f''(0, 0) = \begin{pmatrix} f_{xx}(0, 0) & f_{xy}(0, 0) \\ f_{yx}(0, 0) & f_{yy}(0, 0) \end{pmatrix}.$$

Let's rewrite what we've found as a general theorem. We assume, as above, that f has continuous second partial derivatives.

THEOREM 3 (Stationary points and the Hessian matrix)   Let (x_0, y_0) be a stationary point of a function f, let $f''(x_0, y_0)$ be the Hessian matrix of f, and let

$$D = f_{xx}(x_0, y_0)f_{yy}(x_0, y_0) - \left(f_{xy}(x_0, y_0)\right)^2$$

be the determinant of $f''(x_0, y_0)$. Then

- If $D > 0$ and $f_{xx}(x_0, y_0) > 0$, then f has a local minimum at (x_0, y_0).
- If $D > 0$ and $f_{xx}(x_0, y_0) < 0$, then f has a local maximum at (x_0, y_0).
- If $D < 0$, then f has a saddle point at (x_0, y_0).
- If $D = 0$, then more information is needed.

The theorem makes many maximum–minimum calculations routine.

EXAMPLE 6   The function $f(x, y) = xy - y - 2x + 2$ has one stationary point. Find it. What type of stationary point is it?

Solution   To find the stationary point, we solve

$$\nabla f(x, y) = (y - 2, x - 1) = (0, 0);$$

clearly, this occurs (only) at the point $(1, 2)$. At this point the Hessian matrix has the form

$$H_f(1, 2) = \begin{pmatrix} 0 & 1 \\ 1 & 0 \end{pmatrix}.$$

If possible, plot the surface $z = f(x, y)$ for yourself to see the saddle.

Therefore, $D = -1$, and it follows from Theorem 3 that $(1, 2)$ is a saddle point. ■

In higher dimensions

Similar ideas can be applied to give a second derivative test, based on the Hessian matrix, for functions of three or more variables. But the derivation and statement are considerably more complicated; we omit them. In practice, moreover, it is sometimes clear from other considerations or simple experimentation how a function behaves near a stationary point. Example 7 illustrates such an ad hoc argument.

> **EXAMPLE 7** Let $f(x, y, z)$ be a function of three variables, and suppose that f has a stationary point at the origin, that is, $\nabla f(0, 0, 0) = (0, 0, 0)$. Suppose also that $f_{xx}(0, 0, 0) > 0$ and $f_{yy}(0, 0, 0) < 0$. Show that f has neither a local maximum nor a local minimum at the origin.
>
> **Solution** Consider the function $g(t) = f(t, 0, 0)$. Then $g'(0) = f_x(0, 0, 0) = 0$, and $g''(0) = f_{xx}(0, 0, 0) > 0$. Thus, $g(t)$ has a local *minimum* at $t = 0$. For similar reasons, the function $h(t) = f(0, t, 0)$ has a local *maximum* at $t - 0$. This difference means that f itself can not have either a local maximum or a minimum at $(0, 0, 0)$. ∎

Extremes on the boundary: optimization on closed regions

Recall what happens in elementary calculus for a differentiable function $f(x)$ defined on a closed interval $[a, b]$: f may assume its maximum and minimum values either at a stationary point, where $f'(x) = 0$, or at either of the endpoints $x = a$ and $x = b$.

The situation is similar for a function of two variables defined on a region, such as a rectangle or a circle, that has a definite "edge," or boundary: $f(x, y)$ may assume its maximum and minimum either at a stationary point or somewhere on the boundary of the region. We illustrate the situation, and one way to approach it, with a simple example.

> **EXAMPLE 8** Where on the rectangle $R = [-1, 1] \times [-1, 1]$ does $j(x, y) = xy$ assume its minimum and maximum values?
>
> **Solution** We saw in Example 4 that $j(x, y)$ has only one stationary point—a saddle point—in the interior of R. Therefore, the maximum and minimum values of j must occur somewhere on the boundary of R. A look at the contour plot of j (notice the symmetry in Figure 5) shows that it's enough to look along *any* boundary edge of R, such as the right edge. On this edge we have $x = 1$, and so j behaves like a function of just one variable: $j(x, y) = j(1, y) = y$. Clearly, $j(1, y) = y$ is largest at $y = 1$ and smallest at $y = -1$. We conclude, therefore, that $j(1, 1) = 1$ and $j(1, -1) = -1$ are, respectively, maximum and minimum values of j on R. ∎

BASIC EXERCISES

1. Let $j(x, y) = xy$. (Its contour map is shown in Figure 5.) To an ant walking along the surface $z = j(x, y)$ from lower left to upper right, the origin seems to be a low spot; another ant walking from upper left to lower right would experience the origin as a high spot.

   (a) An ant walks along the surface from $(0, -1)$ to $(0, 1)$. How does the ant's altitude change along the way?

   (b) Another ant walks along the surface from $(0.5, -1)$ to $(0.5, 1)$. How does the ant's altitude change along the way? Where is the ant highest? How high is the ant there?

2. See the contour map of $f(x, y) = \cos(x) \sin(y)$ in Figure 2.

   (a) The surface $z = f(x, y)$ resembles an egg carton. Where do the eggs go?

   (b) From the picture alone, estimate the coordinates of a local minimum point, a local maximum point, and a saddle point.

   (c) Use the formula $f(x, y) = \cos(x) \sin(y)$ to find (exactly) all the stationary points of f in the rectangle $R = [-3, 3] \times [-3, 3]$.

   (d) Find the maximum and minimum values of f in the rectangle $R = [-3, 3] \times [-3, 3]$.

3. Consider the function $f(x, y) = x(x-2)\sin(y)$. Here is a contour map:

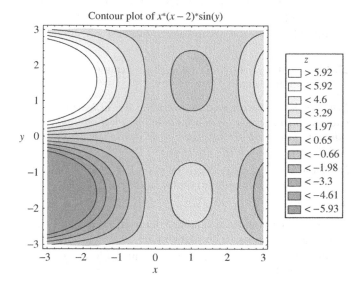

Contour plot of x*(x − 2)*sin(y)

z
> 5.92
< 5.92
< 4.6
< 3.29
< 1.97
< 0.65
< −0.66
< −1.98
< −3.3
< −4.61
< −5.93

(a) The function f has two stationary points along the line $x = 1$. Use the picture to estimate their coordinates. What type is each one?

(b) There are four stationary points inside the rectangle $R = [-3, 3] \times [-3, 3]$. Use the formula for f to find all four.

(c) The contour plot shows that f assumes its maximum and minimum values on $R = [-3, 3] \times [-3, 3]$ somewhere along the left boundary, that is, where $x = -3$. Find these maximum and minimum values. [HINT: If $x = -3$, then

$f(x, y) = f(-3, y) = 15\sin(y)$. This is a function of one variable defined for $-3 \le y \le 3$.]

4. Consider the linear function $L(x, y) = 1 + 2x + 3y$. Does L have any stationary points? If so, what type are they? If not, why not?

5. Let $f(x, y) = x^2 y + xy^2 + x + y$.
   (a) Show that $(1, -1)$ is a stationary point of the function f.
   (b) Is the point $(1, -1)$ a local maximum, a local minimum, or a saddle point of f? Justify your answer.

6. Check that Theorem 3 gives the correct information for each of the functions f, g, h, j, and k in Examples 1–5.

7. Find all stationary points of the function $f(x, y) = -x^3 + 4xy - 2y^2 + 1$.

8. Does the function $f(x, y) = 2y - \sin x$ have any stationary points? If so, find one. If not, explain why not.

In Exercises 9–16, use the formula to find all stationary points. Use Theorem 3 to classify each as a local maximum, a local minimum, or a saddle point. Then use technology (e.g., a contour or surface plot) to verify your answer.

9. $f(x, y) = -x^2 - y^2$

10. $f(x, y) = x^2 - y^2$

11. $f(x, y) = 3x^2 + 2y^2$

12. $f(x, y) = xy - y - 2x + 2$

13. $f(x, y) = x^2 + y^2 - x^2 y/2$

14. $f(x, y) = 3x^2 - 6xy + 2y^3$

15. $f(x, y) = x^3 + y^3 + 3x^2 - 3y^2 - 8$

16. $f(x, y) = x^2 - xy - y^2$

FURTHER EXERCISES

17. Consider the linear function $L(x, y) = a + bx + cy$, where a, b, and c are any constants.
   (a) The graph of L is a plane. Which planes have stationary points? For these planes, where are the stationary points?
   (b) Under what conditions on a, b, and c will L have stationary points? In this case, where are the stationary points? Reconcile your answers with those in part (a).

18. Let $f(x, y) = x^2$. The graph of f is a cylinder unrestricted in the y-direction.
   (a) Use technology to plot the surface $z = f(x, y)$. Where in the xy-plane are the stationary points? What type are they? [HINT: There's a whole line of stationary points.]
   (b) Use partial derivatives of f to find all the stationary points. Reconcile your answer with part (a).

In Exercises 19–21, give an example of a function with the specified properties. (Many examples are possible.)

[HINTS: (1) See Exercise 18 for ideas. (2) Check your answers by plotting.]

19. A function $g(x, y)$ for which every point on the x-axis is a local minimum point

20. A function $h(x, y)$ for which every point on the line $x = 1$ is a local maximum point

21. A nonconstant function $k(x, y)$ that has a local minimum at $(3, 4)$

22. Let $f(x, y) = x^2 + y^2 + kxy$.
   (a) Show that the origin is a stationary point of f regardless of the value of k.
   (b) Show that if $k < -2$ or $k > 2$, the origin is a saddle point.
   (c) Suppose that $-2 < k < 2$. Is the origin a local maximum, a local minimum, or a saddle point of f? Justify your answer.

(d) Suppose that $k = 2$. Is the origin a local maximum, a local minimum, or a saddle point of f? Justify your answer.

23. Suppose that $f(x, y)$ has a stationary point at (x_0, y_0) and that $f_{xx}(x_0, y_0)$ and $f_{yy}(x_0, y_0)$ have opposite signs. Show that (x_0, y_0) must be a saddle point.

24. Find and classify all stationary points of $f(x, y) = x^4 + y^4$. Does Theorem 3 help?

25. Let $f(x, y) = x^2 + axy + by^2$, where a and b are real constants (positive, negative, or zero).

   (a) Explain why $(0, 0)$ is a stationary point of f regardless of the values of a and b.

   (b) Under what conditions on a and b will $(0, 0)$ be a local maximum? A local minimum? A saddle point? Explain your answers carefully and give an example of each possibility.

   (c) Suppose (in this part only) that $b = a^2/4$. What sort of stationary point occurs at $(0, 0)$? Why?

26. Let $f(x, y) = x^2$; f is a cylindrical function.

   (a) Describe the shape of the f-graph near the stationary point $(0, 0)$.

(b) Show from the formula that f has a local minimum at $(0, 0)$.
   [NOTE: The local minimum is not "strict" in the sense that $f(x, y) \geq f(0, 0)$ holds, but $f(x, y) > f(0, 0)$ may not.]

(c) What does the second derivative test say about f at $(0, 0)$?

27. Show that if $f_{xx} + f_{yy} = 0$, then $f(x, y)$ cannot have a maximum or minimum.

If a continuous function of one variable has at least two local maxima, then it must also have at least one local minimum. The situation is different for functions of two variables. Exercises 28 and 29 illustrate this fact.

28. Show that $f(x, y) = (x^2 - 1)^2 + y^2$ has three stationary points—two local minima and a saddle point.

29. Let $g(x, y) = (x^2 - 1)^2 + (x^2 y - x - 1)^2$.

   (a) Show that g has stationary points at $(1, 2)$ and $(-1, 0)$ and that both are local minima.

   (b) Show that g has stationary points *only* at $(1, 2)$ and $(-1, 0)$.

13.7 THE CHAIN RULE

Note *This section assumes some basic familiarity with matrices and matrix multiplication. For a brief introduction or review, see Appendix J.*

Introduction The chain rule, the product rule, and the quotient rule are all "combinatorial" results. They tell how to find derivatives of "new" functions that are formed by combining "old" functions in various ways. (They also ensure that, under appropriate conditions, these new derivatives exist.) The new derivatives are, of course, appropriate combinations of the old derivatives and functions. ◄

But not necessarily the "obvious" combinations!

Composition: one variable vs. several

The chain rule—in any dimension—concerns derivatives of functions formed by composition. Composition is relatively simple in single-variable calculus: Composing two ordinary scalar-valued functions, say f and g, of one variable produces another scalar-valued function of one variable, $f \circ g$, defined by $f \circ g(x) = f(g(x))$. ◄ If $f(x) = \sin x$ and $g(x) = x^2$, then $f \circ g(x) = \sin(x^2)$. ◄

An equivalent notation is $(f \circ g)(x)$.

Remember: Composition isn't commutative.

Composition means exactly the same thing for functions of several variables—the output from one function is used as input to another. In multivariable calculus, however, inputs and outputs may be either scalars or vectors, and so it is important to keep careful track of each function's **domain space** and **image space**. The arrow notations in the next example can help in this regard. (The arrows point from the domain space to the image space.)

EXAMPLE 1 Consider functions $f : \mathbb{R} \to \mathbb{R}$, $g : \mathbb{R}^2 \to \mathbb{R}$, and $\mathbf{h} : \mathbb{R} \to \mathbb{R}^2$, defined by

$$f(t) = t^3; \qquad g(x, y) = x^2 + y^2; \qquad \mathbf{h}(t) = (\cos t, \sin t).$$

Note the use of boldface for the vector-valued function.

Which compositions make sense? Which do not? Why? ◄

Solution The notations $f : \mathbb{R} \to \mathbb{R}$ and $g : \mathbb{R}^2 \to \mathbb{R}$ show (even before one looks at function definitions) that $g \circ f$ can't make sense. Outputs from f are scalars; they have the wrong "type" to be inputs to g, which accepts 2-vectors. The composition $f \circ g$ *does* make sense; outputs from g are scalars, as are inputs to f. Symbolically, we have $f \circ g : \mathbb{R}^2 \to \mathbb{R} \to \mathbb{R}$ (or, simply, $f \circ g : \mathbb{R}^2 \to \mathbb{R}$) defined by

$$f \circ g(x, y) = f(x^2 + y^2) = \left(x^2 + y^2\right)^3 .$$

Similarly, $g : \mathbb{R}^2 \to \mathbb{R}$ and $\mathbf{h} : \mathbb{R} \to \mathbb{R}^2$ can be composed to give the vector-valued function $\mathbf{h} \circ g : \mathbb{R}^2 \to \mathbb{R}^2$ defined by

$$\mathbf{h} \circ g(x, y) = \mathbf{h}(x^2 + y^2) = \left(\cos(x^2 + y^2), \sin(x^2 + y^2) \right).$$

Alternatively, g and $\mathbf{h}$ can be composed in the opposite order to give an ordinary calculus-style function $g \circ \mathbf{h} : \mathbb{R} \to \mathbb{R}$ defined by

$$g \circ \mathbf{h}(t) = g(\cos t, \sin t) = \cos^2 t + \sin^2 t = 1.$$

The composite function $\mathbf{h} \circ f : \mathbb{R} \to \mathbb{R}^2$ is defined by

$$\mathbf{h} \circ f(t) = \mathbf{h}(t^3) = \left(\cos(t^3), \sin(t^3)\right);$$

it is a vector-valued function of one variable. ∎

The one-variable chain rule: multiply derivatives

Recall the one-variable chain rule:

> **FACT (Chain rule in one variable)** Let f and g be differentiable functions, with a in the domain of g. Then
> $$\left(f \circ g \right)'(a) = f'\big(g(a)\big) \cdot g'(a).$$

There are other notations for saying the same thing. ➤ If we write $y = f(u)$ and $u = g(x)$, then the chain rule can be written as

We'll see several as we go along.

$$\frac{dy}{dx}(a) = \frac{dy}{du}\big(g(a)\big) \cdot \frac{du}{dx}(a).$$

In *any* symbolic form, the key idea is this:

The derivative of the composition $f \circ g$ is a *product* of the derivatives f' and g'.

Where are the factors evaluated? What the chain rule says—multiply individual derivatives—sounds (and is) simple enough. Notice, however, that the two derivatives in the product $f'\big(g(a)\big) \cdot g'(a)$ are evaluated at different places. In the right factor, g' is evaluated at $x = a$; in the left factor, f' is evaluated at $g(a)$. ➤ A function diagram—

Not at a.

$$a \overset{g}{\longmapsto} g(a) \overset{f}{\longmapsto} f\big(g(a)\big)$$

shows why these choices make sense: The two derivatives are evaluated at the corresponding domain points.

Derivatives and linear functions: why the chain rule works Why does the chain rule work? How would one ever guess such a formula?

What follows is not (quite) a rigorous proof; the missing ingredient is a fully formal notion of approximation.

The answer is surprisingly simple—and the answer is the same for any number of variables. The key idea has two parts: ➤

Using capital letters saves other letters for later use.

- **The chain rule "works" for linear functions.** It is easy to show, right from scratch, that the chain rule holds for linear functions $f(x) = A + Bx$ and $g(x) = C + Dx$, where A, B, C, and D are all constants. ← For linear functions, the derivatives are just the coefficients of x: In this case, $f'(x) = B$ and $g'(x) = D$. The composite $f \circ g(x)$ is another linear function:

$$f \circ g(x) = f(C + Dx) = A + B(C + Dx) = A + BC + BDx.$$

Thus, $(f \circ g)'(x) = BD$, the *product* of f' and g'.

- **Differentiable functions are locally linear.** Most functions are not linear. However, every differentiable function g is *locally linear* in the sense that near any point of its domain g can be closely approximated by a linear approximation function, L_g. Similarly, f can be closely approximated near any point of *its* domain by another suitably chosen linear function, L_f. Because $f \approx L_f$ and $g \approx L_g$, it follows that $f \circ g \approx L_f \circ L_g$. In particular, $(f \circ g)'(a) = (L_f \circ L_g)'(a)$. Because the chain rule "works" for $L_f \circ L_g$, it must work for $f \circ g$, too.

Further details and some fine print appear in the following example.

EXAMPLE 2 Let g and f be differentiable functions and a a domain point of g. Use linear approximations to derive the chain rule for $(f \circ g)'(a)$.

Solution For (temporary) convenience, write $b = g(a)$. Then we have

$$a \xmapsto{g} b \xmapsto{f} f(b).$$

The linear approximation to g at $x = a$ is

$$L_g(x) = g(a) + g'(a)(x - a) = b + g'(a)(x - a);$$

notice that $L_g(a) = g(a)$ and $L_g'(a) = g'(a)$. Similarly, the linear approximation to f at $x = b$ is

$$L_f(x) = f(b) + f'(b)(x - b)$$

with $L_f(b) = f(b)$ and $L_f'(b) = f'(b)$. Composing L_g with L_f gives

$$a \xmapsto{L_g} b \xmapsto{L_f} f(b),$$

Check the second identity.

which resembles the earlier diagram. Working out the formula gives ←

$$L_f \circ L_g(x) = L_f\big(b + g'(a)(x - a)\big) = f(b) + f'(b)g'(a)(x - a).$$

Therefore, $L_f \circ L_g$ is a linear function of x with derivative $f'(b) \cdot g'(a)$. This gives the chain rule:

$$(f \circ g)'(a) = (L_f \circ L_g)'(a) = f'\big(g(a)\big) \cdot g'(a).$$

(The first equality in the line above *is* true, but a fully rigorous proof is a little beyond our scope.) ∎

The multivariable chain rule: multiply derivative matrices

The chain rule says the same thing in all dimensions: *The derivative of a composition $f \circ g$ is found by multiplying—in an appropriate sense—the derivatives of f and g.* The qualification is necessary because, for multivariable functions, derivatives are usually not single numbers but vectors or matrices. In these cases, as we will see, "multiplication" means matrix multiplication or the dot product. ← This difference aside, the multivariable chain rule is the same as in the one-variable case.

The dot product can be thought of as a special case of matrix multiplication.

Derivatives as matrices Functions of more than one variable and vector-valued functions generate a whole collection of derivatives and partial derivatives. It is convenient to store all this information in an array called the **derivative matrix** (or **Jacobian matrix**). Suppose, for instance, that $\mathbf{k}: \mathbb{R}^2 \to \mathbb{R}^2$ is defined by

$$\mathbf{k}(x, y) = \big(u(x, y), v(x, y)\big) = (x^2 + y, 2x - y^3).$$

Then the derivative matrix is

$$\mathbf{k}'(x, y) = \begin{pmatrix} \dfrac{\partial u}{\partial x} & \dfrac{\partial u}{\partial y} \\ \dfrac{\partial v}{\partial x} & \dfrac{\partial v}{\partial y} \end{pmatrix} = \begin{pmatrix} 2x & 1 \\ 2 & -3y^2 \end{pmatrix}.$$

The idea of a derivative matrix makes sense regardless of the dimensions of the domain and image spaces. ➥ We will need dimensions no higher than three, but here is the general definition: *Including dimension one*

DEFINITION (Derivative matrix) Let $\mathbf{f}: \mathbb{R}^n \to \mathbb{R}^m$ be a vector-valued function of n variables, with component functions $f_1(x_1, x_2, \ldots, x_n)$, $f_2(x_1, x_2, \ldots, x_n), \ldots, f_m(x_1, x_2, \ldots, x_n)$. Let $\mathbf{X_0}$ be a point of the domain of $\mathbf{f}$. The derivative matrix of $\mathbf{f}$ at $\mathbf{X_0}$ is the $m \times n$ matrix

$$\mathbf{f}'(\mathbf{X_0}) = \begin{pmatrix} \dfrac{\partial f_1}{\partial x_1}(\mathbf{X_0}) & \dfrac{\partial f_1}{\partial x_2}(\mathbf{X_0}) & \cdots & \dfrac{\partial f_1}{\partial x_n}(\mathbf{X_0}) \\ \dfrac{\partial f_2}{\partial x_1}(\mathbf{X_0}) & \dfrac{\partial f_2}{\partial x_2}(\mathbf{X_0}) & \cdots & \dfrac{\partial f_2}{\partial x_n}(\mathbf{X_0}) \\ \vdots & \vdots & \vdots & \vdots \\ \dfrac{\partial f_m}{\partial x_1}(\mathbf{X_0}) & \dfrac{\partial f_m}{\partial x_2}(\mathbf{X_0}) & \cdots & \dfrac{\partial f_m}{\partial x_n}(\mathbf{X_0}) \end{pmatrix}.$$

Notice that each row of the matrix $\mathbf{f}'$ is the gradient vector of a component function. Indeed, the derivative matrix is sometimes written as

$$\mathbf{f}'(\mathbf{X_0}) = \begin{pmatrix} \nabla f_1(\mathbf{X_0}) \\ \nabla f_2(\mathbf{X_0}) \\ \vdots \\ \nabla f_m(\mathbf{X_0}) \end{pmatrix}.$$

The abstract definition may look formidable, but in specific cases the calculations are easy.

EXAMPLE 3 Consider the functions ➥

$$f(t) = t^3; \qquad g(x, y) = x^2 + y^2; \qquad \mathbf{h}(t) = \big(\cos t, \sin t\big);$$

$$\mathbf{L}(x, y) = (1 + 2x + 3y, 4 + 5x + 6y).$$

Find their derivative matrices.

The first three are from Example 1.

But check them for yourself!

Solution All the derivative calculations are easy. ◂ Here are the results:

$$f'(t) = (3t^2); \qquad g'(x, y) = (2x \quad 2y);$$

$$\mathbf{h}'(t) = \begin{pmatrix} -\sin t \\ \cos t \end{pmatrix}; \qquad \mathbf{L}'(x, y) = \begin{pmatrix} 2 & 3 \\ 5 & 6 \end{pmatrix}.$$

Notice especially the size and shape of each matrix. In particular, $f'(t)$ is a 1×1 "matrix," that is, a scalar; $g'(x, y)$ is simply the gradient ∇g written as a row vector. ∎

Jacobi and the Jacobian. The Jacobian matrix and determinant are named for the German mathematician Karl Jacobi (1804–1851). Jacobi was not the first to organize derivative information in a matrix, but in his 1841 paper *De determinantibus functionalibus* Jacobi applied matrix theory to prove results about relations among sets of functions. Jacobi was a popular teacher at the University of Königsberg and is credited with having pioneered the seminar method of teaching. (For a wealth of historical information of this nature, see the History of Mathematics Archive [`http://www-groups.dcs.st-and.ac.uk/~history/`], a Web-based resource at the University of St. Andrews, in Scotland.)

Linear functions: a note on terminology Linear functions, such as $\mathbf{L}$ in the preceding example, are those in which the variables appear in powers no greater than one. The name is reasonable because graphs and level curves of linear functions are lines, planes, or other "flat" objects. In particular, we allow linear functions to have nonzero constant terms (as $\mathbf{L}$ does, above).

For the record, the word "linear" is used a little differently in linear algebra courses. There, a "linear function" or "linear transformation" is usually required to have only zero constant terms. The difference is worth noting, but it should not cause us trouble in context.

Linear functions, matrices, and derivatives. There is a close connection between linear functions (in our sense of the word) and matrices. For instance,

$$\mathbf{L}(x, y) = (1 + 2x + 3y, \, 4 + 5x + 6y)$$

Convince yourself.

says exactly the same thing as the matrix equation ◂

$$\mathbf{L}(x, y) = \begin{pmatrix} 1 \\ 4 \end{pmatrix} + \begin{pmatrix} 2 & 3 \\ 5 & 6 \end{pmatrix} \cdot \begin{pmatrix} x \\ y \end{pmatrix},$$

where the dot stands for matrix multiplication.

More generally, *every* linear function $\mathbf{L}$ has the form

$$\mathbf{L}(\mathbf{X}) = \mathbf{C} + M \cdot \mathbf{X},$$

In one variable, a linear function has the form $L(x) = c + mx$.

where M is a matrix (called the **coefficient matrix**), $\mathbf{X}$ is a vector input, $\mathbf{C}$ is a constant vector, and the dot represents matrix multiplication. ◂ Writing linear functions in matrix form has two main advantages for us:

- **Composition and matrix multiplication** Consider two linear functions given by matrix equations

$$\mathbf{L_1}(\mathbf{X}) = \mathbf{C_1} + M_1 \cdot \mathbf{X} \qquad \text{and} \qquad \mathbf{L_2}(\mathbf{X}) = \mathbf{C_2} + M_2 \cdot \mathbf{X}.$$

Then the composition $\mathbf{L_1} \circ \mathbf{L_2}$ has matrix form

$$\mathbf{L_1} \circ \mathbf{L_2}(\mathbf{X}) = \mathbf{L_1}(\mathbf{C_2} + M_2 \cdot \mathbf{X})$$
$$= \mathbf{C_1} + M_1 \cdot (\mathbf{C_2} + M_2 \cdot \mathbf{X})$$
$$= \mathbf{C_1} + M_1 \mathbf{C_2} + (M_1 \cdot M_2) \cdot \mathbf{X}.$$

(The last equation follows from algebraic properties of matrix multiplication.) The first two summands above are constant vectors, and so the composition has the general form $\mathbf{L_1} \circ \mathbf{L_2}(\mathbf{X}) = (M_1 \cdot M_2) \cdot \mathbf{X} + \mathbf{C}$. In particular:

The coefficient matrix of $\mathbf{L_1} \circ \mathbf{L_2}$ is the matrix product $M_1 \cdot M_2$.

- **Derivatives of linear functions**   The preceding example ⟿ illustrates a simple but important property of linear functions and their matrices:

See the derivative of $\mathbf{L}$.

> **FACT**   Let $\mathbf{L}$ be the linear function
> $$\mathbf{L}(\mathbf{X}) = \mathbf{C_0} + M \cdot \mathbf{X}.$$
> Then the derivative matrix is $\mathbf{L}' = M$.

This fact, combined with the italicized sentence above, tells how to find the derivative matrix of a composition of linear functions:

The derivative matrix of the composition $\mathbf{L_1} \circ \mathbf{L_2}$ is the matrix product $M_1 \cdot M_2$.

Linear approximations to functions   For a real-valued function f the linear approximation to f at $\mathbf{X_0}$ is the function

$$L(\mathbf{X}) = f(\mathbf{X_0}) + \nabla f(\mathbf{X_0}) \cdot (\mathbf{X} - \mathbf{X_0});$$

the dot represents the dot product. If $\mathbf{f} = (f_1, f_2)$ (or (f_1, f_2, f_3)) is a vector-valued function, then the linear approximation to $\mathbf{f}$ at $\mathbf{X_0}$ is the matrix analogue of the preceding equation:

$$\mathbf{L}(\mathbf{X}) = \mathbf{f}(\mathbf{X_0}) + \mathbf{f}'(\mathbf{X_0}) \cdot (\mathbf{X} - \mathbf{X_0}),$$

where $\mathbf{f}'(\mathbf{X_0})$ is now the derivative matrix ⟿ and the dot represents matrix multiplication. Notice that $\mathbf{L}$ is (like $\mathbf{f}$) a vector-valued function, such that

The rows are gradients.

$$\mathbf{L}(\mathbf{X_0}) = \mathbf{f}(\mathbf{X_0}) \qquad \text{and} \qquad \mathbf{L}'(\mathbf{X_0}) = \mathbf{f}'(\mathbf{X_0}).$$

A generic example   We have now assembled all the necessary ingredients to state the multivariable chain rule carefully. But first we present a "generic" example ⟿ (one that uses function names, not specific formulas) that suggests *why* it is that matrices and matrix multiplication prove to be natural bookkeeping devices for storing and organizing all the necessary information. ⟿ Linear approximation is the key idea.

The authors thank Reg Laursen for suggesting this calculation.

There's much information, and so organized storage is important!

EXAMPLE 4   Let $z = f(x, y)$, where $x = g(s, t)$ and $y = h(s, t)$. (To avoid distractions, we'll suppose that the functions f, g, and h are all "well behaved" in the sense that all the necessary derivatives exist.) We can think of z as a composite function of s and t:

$$z = f(x, y) = f\big(g(s, t), h(s, t)\big).$$

Use linear approximation to estimate the partial derivatives $\partial z / \partial s$ and $\partial z / \partial t$ at the point (s_0, t_0). How is matrix multiplication involved?

Solution   First, some handy notation. We will write

$$x_0 = g(s_0, t_0), \qquad y_0 = h(s_0, t_0), \qquad \text{and} \qquad z_0 = f(x_0, y_0).$$

When convenient, we will also use the vector forms

$$\mathbf{X_0} = (x_0, y_0), \qquad \mathbf{S_0} = (s_0, t_0), \qquad \text{and} \qquad \mathbf{G} = (g, h).$$

(In the last case, $\mathbf{G}$ is the function $\mathbf{G} : \mathbb{R}^2 \to \mathbb{R}^2$; the coordinate functions g and h are scalar-valued functions of two variables.)

Consider the following linear approximation formulas for z, x, and y:

$$x \approx x_0 + g_s(s_0, t_0)\,(s - s_0) + g_t(s_0, t_0)\,(t - t_0);$$
$$y \approx y_0 + h_s(s_0, t_0)\,(s - s_0) + h_t(s_0, t_0)\,(t - t_0);$$
$$z \approx z_0 + f_x(x_0, y_0)\,(x - x_0) + f_y(x_0, y_0)\,(y - y_0).$$

We use some vector notation here to reduce clutter.

The first two approximations give ◄

$$x - x_0 \approx g_s(\mathbf{S_0})(s - s_0) + g_t(\mathbf{S_0})(t - t_0) \quad \text{and} \quad y - y_0 \approx h_s(\mathbf{S_0})(s - s_0) + h_t(\mathbf{S_0})(t - t_0).$$

Substituting these into our approximation for z gives

$$z \approx z_0 + f_x(\mathbf{X_0})\big[g_s(\mathbf{S_0})(s - s_0) + g_t(\mathbf{S_0})(t - t_0)\big] + f_y(\mathbf{X_0})\big[h_s(\mathbf{S_0})(s - s_0) + h_t(\mathbf{S_0})(t - t_0)\big].$$

Rearranging terms slightly gives

$$z \approx z_0 + \big[f_x(\mathbf{X_0})g_s(\mathbf{S_0}) + f_y(\mathbf{X_0})h_s(\mathbf{S_0})\big](s - s_0) + \big[f_x(\mathbf{X_0})g_t(\mathbf{S_0}) + f_y(\mathbf{X_0})h_t(\mathbf{S_0})\big](t - t_0).$$

The line above carries the key insight: we have found a linear approximation to z in terms of the *new* variables s and t. This suggests that the coefficients of $(s - s_0)$ and $(t - t_0)$ must be the sought-after partial derivatives of z with respect to s and t:

$$\frac{\partial z}{\partial s}(\mathbf{S_0}) = f_x(\mathbf{X_0})g_s(\mathbf{S_0}) + f_y(\mathbf{X_0})h_s(\mathbf{S_0});$$
$$\frac{\partial z}{\partial t}(\mathbf{S_0}) = f_x(\mathbf{X_0})g_t(\mathbf{S_0}) + f_y(\mathbf{X_0})h_t(\mathbf{S_0}).$$

We omitted input variable names for compactness.

Rewriting these facts in matrix form ◄ gives

$$\begin{bmatrix} z_s & z_t \end{bmatrix} = \begin{bmatrix} f_x & f_y \end{bmatrix} \cdot \begin{bmatrix} g_s & g_t \\ h_s & h_t \end{bmatrix}.$$

In derivative matrix notation, it is equivalent to write simply

$$\big(f(\mathbf{G})\big)' = f' \cdot \mathbf{G}',$$

where all symbols represent matrices and the operation on the right is matrix multiplication. We have arrived.

The multivariable chain rule We can now state the chain rule in its general form.

THEOREM 4 (The multivariable chain rule) Let $\mathbf{f}$ and $\mathbf{g}$ be differentiable functions with $\mathbf{X_0}$ in the domain of $\mathbf{g}$. Then we have the matrix equation

$$(\mathbf{f} \circ \mathbf{g})'(\mathbf{X_0}) = \mathbf{f}'\big(\mathbf{g}(\mathbf{X_0})\big) \cdot \mathbf{g}'(\mathbf{X_0});$$

the dot represents matrix multiplication.

A proof sketch The idea behind a proof of Theorem 4 is much the same as for the one-variable chain rule. We approximate $\mathbf{f}$ and $\mathbf{g}$ with appropriate linear functions $\mathbf{L_f}$ and $\mathbf{L_g}$ for which the theorem "clearly" holds. ◄ Then we conclude that the theorem holds in general.

That's the conclusion drawn just before the theorem.

For $\mathbf{g}$ we have the following linear approximation at $\mathbf{X_0}$:

$$\mathbf{L_g}(\mathbf{X}) = \mathbf{g}(\mathbf{X_0}) + \mathbf{g}'(\mathbf{X_0}) \cdot (\mathbf{X} - \mathbf{X_0}).$$

Similarly, for **f** we have the following linear approximation at $\mathbf{g}(\mathbf{X}_0)$:

$$\mathbf{L_f}(\mathbf{X}) = \mathbf{f}\big(\mathbf{g}(\mathbf{X}_0)\big) + \mathbf{f}'\big(\mathbf{g}(\mathbf{X}_0)\big) \cdot (\mathbf{X} - \mathbf{X}_0).$$

The nature of the approximations $\mathbf{f} \approx \mathbf{L_f}$ and $\mathbf{g} \approx \mathbf{L_g}$ implies that $\mathbf{f} \circ \mathbf{g} \approx \mathbf{L_f} \circ \mathbf{L_g}$ and also that $(\mathbf{f} \circ \mathbf{g})'(\mathbf{X}_0) = (\mathbf{L_f} \circ \mathbf{L_g})'(\mathbf{X}_0)$. Finally, the last derivative, as we showed above, is the product of the corresponding derivative matrices. Thus,

$$(\mathbf{f} \circ \mathbf{g})'(\mathbf{X}_0) = (\mathbf{L_f} \circ \mathbf{L_g})'(\mathbf{X}_0) = \mathbf{f}'\big(\mathbf{g}(\mathbf{X}_0)\big) \cdot \mathbf{g}'(\mathbf{X}_0). \qquad \blacksquare$$

EXAMPLE 5   Consider the functions

$$\mathbf{f}(u, v) = (uv, u - v) \qquad \text{and} \qquad \mathbf{g}(x, y) = (x + y, x^2 + y^2).$$

What is $(\mathbf{f} \circ \mathbf{g})'(x, y)$? What is $(\mathbf{f} \circ \mathbf{g})'(3, 4)$?

Solution   The derivative matrices are

$$\mathbf{f}'(u, v) = \begin{pmatrix} v & u \\ 1 & -1 \end{pmatrix} \qquad \text{and} \qquad \mathbf{g}'(x, y) = \begin{pmatrix} 1 & 1 \\ 2x & 2y \end{pmatrix}.$$

Now the chain rule says:

$$(\mathbf{f} \circ \mathbf{g})'(x, y) = \begin{pmatrix} v & u \\ 1 & -1 \end{pmatrix} \cdot \begin{pmatrix} 1 & 1 \\ 2x & 2y \end{pmatrix} = \begin{pmatrix} v + 2ux & v + 2uy \\ 1 - 2x & 1 - 2y \end{pmatrix}.$$

Substituting $u = x + y$ and $v = x^2 + y^2$ gives

$$(\mathbf{f} \circ \mathbf{g})'(x, y) = \begin{pmatrix} (x^2 + y^2) + 2(x + y)x & (x^2 + y^2) + 2(x + y)y \\ 1 - 2x & 1 - 2y \end{pmatrix}.$$

To find $(\mathbf{f} \circ \mathbf{g})'(3, 4)$, we could substitute in the preceding matrix. Alternatively, we can observe that $g(3, 4) = (7, 25)$, and so

$$\mathbf{f}'(7, 25) = \begin{pmatrix} 25 & 7 \\ 1 & -1 \end{pmatrix} \qquad \text{and} \qquad \mathbf{g}'(3, 4) = \begin{pmatrix} 1 & 1 \\ 6 & 8 \end{pmatrix}.$$

Applying the chain rule again gives the sought-for result:

$$(\mathbf{f} \circ \mathbf{g})'(3, 4) = \begin{pmatrix} 25 & 7 \\ 1 & -1 \end{pmatrix} \cdot \begin{pmatrix} 1 & 1 \\ 6 & 8 \end{pmatrix} = \begin{pmatrix} 67 & 81 \\ -5 & -7 \end{pmatrix}. \qquad \blacksquare$$

EXAMPLE 6   Composite expressions sometimes appear without explicit function names. For example, suppose that u is a function of x and y, while x and y are functions of s and t. Find the partial derivatives $\partial u / \partial s$ and $\partial u / \partial t$.

Solution   The chain rule still holds, but it looks a little different. First, write

$$u = u(x, y) \qquad \text{and} \qquad \mathbf{X}(s, t) = \big(x(s, t), y(s, t)\big).$$

Then,

$$u'(x, y) = \begin{pmatrix} \dfrac{\partial u}{\partial x} & \dfrac{\partial u}{\partial y} \end{pmatrix} \qquad \text{and} \qquad \mathbf{X}'(s, t) = \begin{pmatrix} \dfrac{\partial x}{\partial s} & \dfrac{\partial x}{\partial t} \\[2mm] \dfrac{\partial y}{\partial s} & \dfrac{\partial y}{\partial t} \end{pmatrix}.$$

The chain rule says to find the matrix product

$$u'(s, t) = \left(\frac{\partial u}{\partial x} \quad \frac{\partial u}{\partial y} \right) \cdot \begin{pmatrix} \dfrac{\partial x}{\partial s} & \dfrac{\partial x}{\partial t} \\ \dfrac{\partial y}{\partial s} & \dfrac{\partial y}{\partial t} \end{pmatrix} = \left(\frac{\partial u}{\partial s} \quad \frac{\partial u}{\partial t} \right),$$

where

$$\frac{\partial u}{\partial s} = \frac{\partial u}{\partial x} \frac{\partial x}{\partial s} + \frac{\partial u}{\partial y} \frac{\partial y}{\partial s}; \qquad \frac{\partial u}{\partial t} = \frac{\partial u}{\partial x} \frac{\partial x}{\partial t} + \frac{\partial u}{\partial y} \frac{\partial y}{\partial t}.$$

(Notice the "symbolic cancellation" of ∂x and ∂y in the preceding equations.) ∎

Remembering the chain rule: a graphical aid   Remembering what the multivariate chain rule says can be challenging. Interpreting the chain rule in the notation of matrices, as we have done here, is one helpful organizing device. Figure 1 offers a graphic representation that may sometimes be helpful:

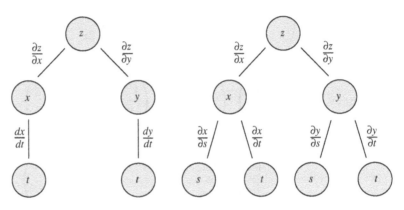

FIGURE 1
Memory aids for the chain rule

The diagram represents two possible chain-rule-related settings:

- **On the left**   The left-hand diagram depicts a function $z(x, y)$, which accepts two input variables x and y. Each of x and y is in turn a function of t. The result is a composite function $z\big(x(t), y(t)\big)$ of only one variable, t. The derivative is the sum of two products, one for each "path" that ends at t:

$$\frac{dz}{dt} = \frac{\partial z}{\partial x} \cdot \frac{dx}{dt} + \frac{\partial z}{\partial y} \cdot \frac{dy}{dt}.$$

- **On the right**   The right-hand picture shows another function $z(x, y)$, but here each of x and y accepts *two* inputs, s and t. The result is a composite function $z\big(x(s, t), y(s, t)\big)$ of the two variables s and t. The two partial derivatives z_s and z_t are, again, sums of two products, one for each "path" that ends at either s or t:

$$\frac{\partial z}{\partial s} = \frac{\partial z}{\partial x} \cdot \frac{\partial x}{\partial s} + \frac{\partial z}{\partial y} \cdot \frac{\partial y}{\partial s}; \qquad \frac{\partial z}{\partial t} = \frac{\partial z}{\partial x} \cdot \frac{\partial x}{\partial t} + \frac{\partial z}{\partial y} \cdot \frac{\partial y}{\partial t}.$$

A similar diagram can be drawn to represent derivatives of any composite function.

BASIC EXERCISES

In Exercises 1–4, write the derivative matrix for the function at the given point.

1. $\mathbf{f}(x, y) = (x + 2y + 3, 4x + 5y + 6)$; $\mathbf{X_0} = (0, 0)$

2. $\mathbf{g}(x, y, z) = (y + z, x + z, x + y)$; $\mathbf{X_0} = (1, 2, 3)$

3. $\mathbf{h}(t) = (\cos t, \sin t, t)$; $t_0 = 0$

4. $\mathbf{k}(s, t) = (1, 2, 3) + s(4, 5, 6) + t(7, 8, 9)$; $(s_0, t_0) = (1, 1)$

Let $\mathbf{f}$, $\mathbf{g}$, $\mathbf{h}$, *and* $\mathbf{k}$ *be as in Exercises 1–4. In Exercises 5–10, decide whether the given composition makes sense. If it does, find a formula for the composite function and use the chain rule to find the derivative matrix at the given point.*

5. $\mathbf{k} \circ \mathbf{f}$; $(x_0, y_0) = (0, 0)$

6. $\mathbf{f} \circ \mathbf{g}$; $(x_0, y_0, z_0) = (1, 2, 3)$

7. $\mathbf{g} \circ \mathbf{k}$; $(s_0, t_0) = (1, 1)$

8. $\mathbf{f} \circ \mathbf{k}$; $(s_0, t_0) = (1, 1)$

9. $\mathbf{g} \circ \mathbf{h}$; $t_0 = 0$

10. $\mathbf{k} \circ \mathbf{h}$; $t_0 = 0$

11. Let $f(t) = t^2 - 9t + 20$, $g(x, y) = x^2 + y^2$, and $\mathbf{X_0} = (2, 1)$.
    (a) Find L_g, the linear approximation to g at $\mathbf{X_0}$.
    (b) Find L_f, the linear approximation to f at $g(\mathbf{X_0})$.
    (c) Find formulas for $f \circ g$ and $L_f \circ L_g$.
    (d) Show that $(f \circ g)'(\mathbf{X_0}) = (L_f \circ L_g)'(\mathbf{X_0})$.
    (e) Plot $(f \circ g)$ and $(L_f \circ L_g)$ in the vicinity of $\mathbf{X} = \mathbf{X_0}$. Do the graphs "look right"? Justify your answer.

12. Repeat Exercise 11 with $f(t) = e^t$, $g(x, y) = x + y \cos(\pi xy)$, and $\mathbf{X_0} = (1, 2)$.

In Exercises 13–16, let f, g, $\mathbf{h}$, and $\mathbf{L}$ be the functions $f(t) = t^3$, $g(x, y) = x^2 + y^2$, $\mathbf{h}(t) = (\cos t, \sin t)$, and $\mathbf{L}(x, y) = (1 + 2x + 3y, 4 + 5x + 6y)$. (Their derivative matrices are calculated in Example 3.)

13. Use the chain rule to find the derivative matrix $(f \circ g)'$ (x_0, y_0).

14. Use the chain rule to find the derivative matrix $(\mathbf{h} \circ g)'$ (x_0, y_0).

15. Use the chain rule to find the derivative matrix $(g \circ \mathbf{L})'$ (x_0, y_0).
    [HINT: To avoid clashing variable names, first rewrite $g(u, v) = u^2 + v^2$.]

16. (a) Use the chain rule to find the derivative matrix $(g \circ \mathbf{h})'(t)$.
    (b) Find a formula for $g \circ \mathbf{h}(t)$ in terms of t. Differentiate (without the chain rule!) to find $(g \circ \mathbf{h})'(t)$. Compare your answer to part (a).

17. Let $f(t) = t^3$ and $\mathbf{h}(t) = (\cos t, \sin t, t)$.
    (a) Use the chain rule to find the derivative matrix $(\mathbf{h} \circ f)'(t)$.
    (b) Find a formula for $(\mathbf{h} \circ f)(t)$. Differentiate this formula to find $(\mathbf{h} \circ f)'(t)$. Compare your answer to part (a).

18. Let $g(t) = \ln(1 + t^2)$ and $\mathbf{h}(t) = (t, t^2, t^3)$.
    (a) Use the chain rule to find the derivative matrix $(\mathbf{h} \circ g)'(t)$.
    (b) Find a formula for $(\mathbf{h} \circ g)(t)$. Differentiate this formula to find $(\mathbf{h} \circ g)'(t)$. Compare your answer to part (a).

19. Let $\mathbf{f}(u, v) = (uv, u + v)$, $\mathbf{g}(x, y) = (x^2 - y^2, x^2 + y^2)$, and $\mathbf{h}(s, t) = (\mathbf{f} \circ \mathbf{g})(s, t)$. Use the chain rule to find $\mathbf{h}'(2, 1)$.

20. Let $\mathbf{f}(x, y) = (x^2 y, x^3 + y^2, y - x)$. Evaluate $\mathbf{f}'(2, -3)$.

21. Suppose that u is a function of x, y, and z and that x, y, and z are functions of r and s. Find the partial derivatives $\partial u / \partial r$ and $\partial u / \partial s$.

22. Suppose that u is a function of x and y and that x and y are functions of r, s, and t. Find the partial derivatives $\partial u / \partial r$, $\partial u / \partial s$, and $\partial u / \partial t$.

FURTHER EXERCISES

23. Give an example of a function $\mathbf{k}$ such that $\mathbf{k}'$ is a
    (a) 1×2 matrix.
    (b) 2×1 matrix.

24. Give an example of a function $\mathbf{k}$ such that $\mathbf{k}'$ is a
    (a) 3×2 matrix.
    (b) 2×3 matrix.

25. The voltage (V), current (I), and resistance (R) in an electrical circuit are related by Ohm's law: $I = V/R = I(V, R)$. Suppose that the voltage produced by an aging battery decreases at a rate of 0.1 volts/hour and that the resistance in the circuit increases at the rate of 2 Ω/hour because of heating. At what rate is the current through the circuit changing when $R = 500$ Ω and $V = 11$ volts?

26. Suppose that $f : \mathbb{R}^3 \to \mathbb{R}$ is a differentiable function, and let

$u(x, y, z) = f(x - y, y - z, z - x)$. Show that
$$\frac{\partial u}{\partial x} + \frac{\partial u}{\partial y} + \frac{\partial u}{\partial z} = 0.$$

27. The partial differential equation $u_{tt} = c^2 u_{xx}$, where $c > 0$ is a constant, is called the one-dimensional wave equation. Show that $u(x, t) = \sin(x + ct)$ is a solution of this equation.

28. Suppose that f are g are twice-differentiable functions of a single variable and $c > 0$ is a constant. Show that $u(x, t) = f(x - ct) + g(x + ct)$ is a solution of the one-dimensional wave equation $u_{tt} = c^2 u_{xx}$.

In Exercises 29–32, the position of a particle at time t is given. Find the rate of change of the distance of the particle from the origin at time $t = 1$.

29. $\mathbf{r}(t) = 3 \cos t \, \mathbf{i} + 3 \sin t \, \mathbf{j}$

30. $\mathbf{r}(t) = t\,\mathbf{i} + (1+t^2)\,\mathbf{j}$

31. $\mathbf{r}(t) = 2\cos t\,\mathbf{i} + 3\sin t\,\mathbf{j} + 4t\,\mathbf{k}$

32. $\mathbf{r}(t) = t\,\mathbf{i} + t^2\,\mathbf{j} + t^3\,\mathbf{k}$

33. Let $z = f(x, y)$ be a function of the Cartesian coordinates x and y. Show that if the variable substitutions $x = r\cos\theta$ and $y = r\sin\theta$ are used to express f in polar coordinates, then

(a) $\dfrac{\partial z}{\partial r} = f_x \cos\theta + f_y \sin\theta$

(b) $\dfrac{1}{r}\dfrac{\partial z}{\partial \theta} = -f_x \sin\theta + f_y \cos\theta$

34. Let $z = f(x, y)$ be a function of the Cartesian coordinates x and y. Show that if the variable substitutions $x = r\cos\theta$ and $y = r\sin\theta$ are used to express f in polar coordinates, then

$$\left(\frac{\partial f}{\partial x}\right)^2 + \left(\frac{\partial f}{\partial y}\right)^2 = \left(\frac{\partial z}{\partial r}\right)^2 + \frac{1}{r^2}\left(\frac{\partial z}{\partial \theta}\right)^2.$$

35. Let $z = f(x, y)$ be a function of the Cartesian coordinates x and y. Show that if the variable substitutions $x = r\cos\theta$ and

$y = r\sin\theta$ are used to express f in polar coordinates, then

$$\frac{\partial^2 f}{\partial x^2} + \frac{\partial^2 f}{\partial y^2} = \frac{\partial^2 z}{\partial r^2} + \frac{1}{r}\frac{\partial z}{\partial r} + \frac{1}{r^2}\frac{\partial^2 z}{\partial \theta^2}.$$

36. Let $z = f(x, y)$ be a function of the Cartesian coordinates x and y. Show that if the variable substitutions $x = e^t \cos\theta$ and $y = e^t \sin\theta$ are used to express z as a function of t and θ, then

$$\frac{\partial^2 z}{\partial t^2} + \frac{\partial^2 z}{\partial \theta^2} = (x^2 + y^2)\left(\frac{\partial^2 f}{\partial x^2} + \frac{\partial^2 f}{\partial y^2}\right).$$

37. Suppose that the temperature at the point (x, y) is maintained at $T(x, y) = x^2 + 4y^2$ °C. What are the maximum and minimum temperatures on the circle $x(t) = \cos t$, $y(t) = \sin t$, $0 \le t \le 2\pi$?

38. Suppose that the temperature at the point (x, y) is maintained at $T(x, y) = x^2 - xy + y^2$ °C. What are the maximum and minimum temperatures on the ellipse $x(t) = 2\cos t$, $y(t) = 3\sin t$, $0 \le t \le 2\pi$?

39. Suppose that $\nabla f(x, y, z) = (y, x, 2)$ and that $g(u, v) = f(u - v, u + 2v, u^2 - 2u)$. Find all stationary points of g.

13.8 LOCAL LINEARITY: SOME THEORY OF THE DERIVATIVE

Our discussion of multivariable derivatives has been, so far, mainly "practical." We have seen how to calculate partial derivatives and gradients, what they mean geometrically and numerically, and how to use them for such purposes as finding linear and quadratic approximations to more complicated functions, locating and classifying maxima and minima, calculating derivatives in various directions, and so on.

This section is a brief excursion into the *theory* of multivariate functions and their derivatives. We will consider somewhat more carefully than before what it means for a function to "be differentiable" in the multivariable sense or to "have a linear approximation"—phrases we've used freely but informally up to now. We won't dot every *i* or cross every *t*, but we will point out some important theoretical questions (and a pitfall or two) and suggest approaches to handling them.

For simplicity, only functions of two variables are discussed here. But this is only a convenience—the same questions arise (and answers apply) in any dimension.

Linear approximation and differentiable functions

Let $f(x, y)$ be a function and (x_0, y_0) a point of its domain. We have defined the **linear approximation** to f at (x_0, y_0) to be the linear function

$$L(x, y) = f(x_0, y_0) + f_x(x_0, y_0)(x - x_0) + f_y(x_0, y_0)(y - y_0)$$
$$= f(\mathbf{X_0}) + \nabla f(\mathbf{X_0}) \cdot (\mathbf{X} - \mathbf{X_0}).$$

(The third expression is the vector form of the second. Note that $\mathbf{X} = (x, y)$ and $\mathbf{X_0} = (x_0, y_0)$.)

Thus, to write L down, all we need from f is the value $f(x_0, y_0)$ and both partial derivatives $f_x(x_0, y_0)$ and $f_y(x_0, y_0)$. If these numbers exist, then, thanks to its definition, L has the same value and the same partial derivatives as f at (x_0, y_0). For these reasons it is natural to expect—and usually true in practice, as we've seen in examples—that $L(x, y)$ approximates $f(x, y)$ closely, not only *at* (x_0, y_0) but nearby as well.

A graphical example Things can go wrong …

EXAMPLE 1 We saw informally in Example 4, page 711, that the function

$$f(x, y) = \begin{cases} \dfrac{xy}{x^2 + y^2} & \text{if } (x, y) \neq (0, 0); \\ 0 & \text{if } (x, y) = (0, 0) \end{cases}$$

behaves badly near $(x, y) = (0, 0)$. Figure 1 suggests something of the problem near the origin.

To describe the situation more formally, find the linear approximation L to f at $(0, 0)$. Does L approximate f closely near $(0, 0)$?

Solution In a word, no—the function f has no close linear approximation near $(x, y) = (0, 0)$.

To see why, notice first that $f(x, y) = 0$ if either $x = 0$ or $y = 0$. ➔ This property implies that both $f_x(0, 0) = 0$ and $f_y(0, 0) = 0$, and so the linear approximation function at $(0, 0)$ is simply the zero function:

The formula for f guarantees this.

$$L(x, y) = f(0, 0) + f_x(0, 0)(x - 0) + f_y(0, 0)(y - 0) = 0.$$

So far so good, but here is the problem: $f(x, y)$ is *not* near zero for all inputs (x, y) near the origin. This property can be seen in Figure 1, and the formula gives further evidence.

For instance, if (x, y) lies on the line $y = x$, and $(x, y) \neq (0, 0)$, then

$$f(x, y) = \frac{xy}{x^2 + y^2} = \frac{x^2}{x^2 + x^2} = \frac{1}{2}.$$

If (x, y) is on the line $y = -x$, then

$$f(x, y) = \frac{xy}{x^2 + y^2} = \frac{-x^2}{x^2 + x^2} = -\frac{1}{2}.$$

In short, $f(x, y)$ behaves strangely near the origin, assuming constant but different values on different lines through the origin. Figure 1 illustrates this curious phenomenon: it shows the graphs of f and the linear approximation function L and how poorly the latter approximates the former.

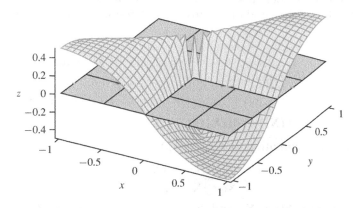

FIGURE 1
The graph of f and the plane $z = 0$

The conclusion is graphically clear: f is *not* closely approximated by L near $(x, y) = (0, 0)$ and is therefore not differentiable there. ∎

Defining differentiability precisely The preceding example illustrates the fact that, for functions of several variables, the existence of partial derivatives at (x_0, y_0) does *not* guarantee that f is continuous—let alone differentiable—at (x_0, y_0). That is why the definition we'll give below, which involves a limit, requires something more than mere existence of partial derivatives.

Recall that for a function $f(x)$ of one variable, the derivative $f'(x_0)$ is usually defined as the limit

$$\lim_{x \to x_0} \frac{f(x) - f(x_0)}{x - x_0} = f'(x_0),$$

if the limit exists. For a function $f(x, y)$ of two variables, which accepts vectors as inputs, we might first guess at something like this limit,

$$\lim_{\mathbf{X} \to \mathbf{X_0}} \frac{f(\mathbf{X}) - f(\mathbf{X_0})}{\mathbf{X} - \mathbf{X_0}},$$

obtained by replacing the old scalar inputs x and x_0 with vectors $\mathbf{X}$ and $\mathbf{X_0}$. Alas, this strategy fails; we have no way of dividing a scalar (the numerator inside the limit) by a vector (the denominator inside the limit).

We can skirt this problem in quite a clever way. Notice first that for a function of one variable, the difference quotient definition of derivative is equivalent to another statement about limits:

$$\lim_{x \to x_0} \frac{f(x) - f(x_0)}{x - x_0} = f'(x_0) \iff \lim_{x \to x_0} \frac{f(x) - f(x_0) - f'(x_0)(x - x_0)}{x - x_0} = 0.$$

The last condition, in turn, is equivalent to requiring that, for some number $f'(x_0)$,

$$\lim_{x \to x_0} \frac{f(x) - \left[f(x_0) + f'(x_0)(x - x_0) \right]}{|x - x_0|} = 0.$$

(The absolute value in the denominator is harmless here, but it will be essential in the multivariable definition to come.)

Notice the numerator in the preceding limit: It is the difference $f(x) - L(x)$, where $L(x)$ is a *linear* function. That the limit above is zero means that, as x tends to x_0, the difference $f(x) - L(x)$ tends to zero faster than the denominator $|x - x_0|$—in other words, the linear approximation $L(x)$ "fits" $f(x)$ even better than x "fits" x_0.

This last condition is key, and the multivariate definition makes it formal:

DEFINITION (The total derivative) Let $f(x, y)$ be a function and $\mathbf{X_0} = (x_0, y_0)$ a point of its domain. Let

$$L(x, y) = f(x_0, y_0) + f_x(x_0, y_0)(x - x_0) + f_y(x_0, y_0)(y - y_0)$$

$$= f(\mathbf{X_0}) + \nabla f(\mathbf{X_0}) \cdot (\mathbf{X} - \mathbf{X_0})$$

be the linear approximation to f at (x_0, y_0). If

$$\lim_{\mathbf{X} \to \mathbf{X_0}} \frac{f(\mathbf{X}) - L(\mathbf{X})}{|\mathbf{X} - \mathbf{X_0}|} = 0,$$

then f is **differentiable** at $\mathbf{X_0}$, and the vector $\nabla f(\mathbf{X_0})$ is the **total derivative** of f at $\mathbf{X_0}$.

Observe:

- **Local linearity** Here, as in the one-variable situation, the limit in the definition guarantees that the linear approximation $L(\mathbf{X})$ "fits" $f(\mathbf{X})$ even more closely near $\mathbf{X}_0$ than $\mathbf{X}$ approximates $\mathbf{X}_0$. This condition is described by the phrase: f is **locally linear** at $\mathbf{X}_0$.

- **Limit subtleties** For the limit condition in the definition to hold, the quantity inside the limit must tend to zero *regardless* of the direction through which $\mathbf{X}$ tends to $\mathbf{X}_0$. This is what's wrong with the function in Example 1: The quantity inside the limit tends to different "destinations" along different lines through the origin.

  A fully rigorous treatment of the theory of limits in several variables would have to take careful account of the infinitely many directions along which one vector quantity can tend to another. As in the one-variable case, a rigorous definition of limits involves a little Greek: Quantities δ and ϵ are defined to measure small changes in inputs and outputs and carefully balanced against each other. (Further discussion of the ϵ-δ approach appears in Appendix K.)

Continuity in several variables

Like differentiability, continuity of a multivariate function is defined through a (multivariate) limit. The definition closely resembles its one-variable counterpart:

> **DEFINITION (Continuity)** Let $f(x, y)$ be a function and $\mathbf{X}_0 = (x_0, y_0)$ a point of its domain. If
> $$\lim_{\mathbf{X} \to \mathbf{X}_0} f(\mathbf{X}) = f(\mathbf{X}_0),$$
> then f is **continuous** at $\mathbf{X}_0$.

Observe:

- **A strong limit again** Here, as in the preceding definition, the limit involved must hold in a strong sense: $f(x, y)$ must tend to $f(x_0, y_0)$ regardless of the direction through which (x, y) tends to (x_0, y_0). Although surprises *can* occur, they seldom do for ordinary calculus functions except at "obviously suspicious" points, such as where denominators vanish.

- **Differentiable functions must be continuous** As in the one-variable case, there is hope for a function f to be differentiable at a point (x_0, y_0) only when f is continuous at (x_0, y_0). This is because, without continuity, the more complicated limit in the definition of differentiability cannot exist.

EXAMPLE 2 Discuss continuity and differentiability at $(0, 0)$ of the function $f(x, y)$ in Example 1, page 765.

Solution The function is discontinuous at $(0, 0)$ because the limit $\lim\limits_{(x,y) \to (0,0)} f(x, y)$ does not exist. To see why, let (x, y) tend to $(0, 0)$ along the x-axis. Then we have $(x, y) = (x, 0)$, with $x \to 0$. But $f(x, 0) = 0$ for all x, and so $f(x, y) \to 0$ as (x, y) tends to zero in this direction.

For comparison, now let (x, y) tend to $(0, 0)$ along the line $y = x$. Then we have $(x, y) = (x, x)$ with $x \to 0$. But now $f(x, x) = 1/2$ for all x, and so $f(x, y) \to 1/2$ as (x, y) tends to zero in this direction. Thus, we have two incompatible limits at $(0, 0)$, and so the limit in question doesn't exist. We conclude that f is discontinuous and therefore also not differentiable at $(0, 0)$. ∎

Which functions have total derivatives?

Fortunately, functions like the one in Examples 1 and 2, which have partial derivatives but no total derivative, appear only rarely (and then, mainly as counterexamples) in courses like this one. The following handy Fact (which we state without proof) guarantees this for most of the functions one is likely to meet in calculus courses:

> **FACT (A condition for local linearity)**　If the partial derivatives f_x and f_y are continuous at (x_0, y_0), then f is differentiable at (x_0, y_0), and has total derivative $\nabla f(x_0, y_0)$.

This simple and easy-to-check criterion guarantees that many functions are indeed differentiable and that the gradient is in fact the total derivative.

EXAMPLE 3　Let $f(x, y)$ be the function f defined in Example 1, page 765. What does the Fact say about differentiability?

Solution　We have already seen that f is not differentiable at $(x, y) = (0, 0)$. For other values of (x, y), however, we have

$$f_x(x, y) = \frac{y^3 - x^2 y}{(x^2 + y^2)^2}; \quad f_y(x, y) = \frac{x^3 - xy^2}{(x^2 + y^2)^2}.$$

Since both partial derivatives are continuous *except* at $(x, y) = (0, 0)$ (where the denominators vanish), the Fact guarantees that f is differentiable everywhere other than at the origin.　∎

BASIC EXERCISES

1. Let f be the "pathological" function defined in Example 1.

   (a) Explain why $f(x, y)$ is constant along every line $y = mx$ through the origin (except at the origin). What is the value of f along the line $y = mx$?

   (b) Draw (by hand) the contour lines $f(x, y) = \pm 1/2$, $f(x, y) = \pm 2/5$, and $f(x, y) = \pm 3/10$. Compare them with what a computer shows. Why do you think the computer has trouble?

   (c) Let $\mathbf{u} = (1/\sqrt{2}, 1/\sqrt{2})$. Does the directional derivative $D_{\mathbf{u}} f(0, 0)$ exist? Why or why not?

   (d) In what directions $\mathbf{u}$, if any, does the directional derivative $D_{\mathbf{u}} f(0, 0)$ exist?

   (e) Plot the surface $z = \dfrac{xy}{x^2 + y^2}$ near the origin. Does the surface's shape seem right? Why do you think a computer has trouble plotting this surface?

2. Let

   $$g(x, y) = \begin{cases} \dfrac{x^2}{x^2 + y^2} & \text{if } (x, y) \neq (0, 0); \\ 0 & \text{if } (x, y) = (0, 0); \end{cases}$$

   (a) Does one or both of the partial derivatives $g_x(0, 0)$ and $g_y(0, 0)$ exist? Why or why not?

   (b) Explain why $g(x, y)$ is constant along every line $y = mx$ through the origin (except at the origin). What is the value of g along the line $y = mx$?

   (c) Draw (by hand) the contour lines $g(x, y) = 1/2$, $g(x, y) = 1/5$, and $g(x, y) = 1/10$. Compare them with what a computer shows. Why do you think the computer has trouble?

   (d) Let $\mathbf{u} = (1/\sqrt{2}, 1/\sqrt{2})$. Does the directional derivative $D_{\mathbf{u}} g(0, 0)$ exist? Why or why not?

   (e) In what directions $\mathbf{u}$, if any, does the directional derivative $D_{\mathbf{u}} g(0, 0)$ exist?

   (f) Plot the surface $z = \dfrac{x^2}{x^2 + y^2}$ near the origin. Does the surface's shape seem right? Why do you think the computer has trouble drawing this surface?

In Exercises 3–6, find the linear approximation L to f at the specified point (x_0, y_0). Then plot the quantity

$$g(x, y) = \frac{f(x, y) - L(x, y)}{\sqrt{(x - x_0)^2 + (y - y_0)^2}}$$

near this point. (This is the quantity in the limit definition of the total derivative. For a function that is differentiable at the specified point, $g(x, y)$ should tend to zero near that point.)

3. $f(x, y) = \sin(x + y)$; $(x_0, y_0) = (0, 0)$

4. $f(x, y) = \sin(xy)$; $(x_0, y_0) = (0, 0)$

5. $f(x, y) = x^2 + y$; $(x_0, y_0) = (1, 2)$

6. $f(x, y) = e^{-(x^2 + y^2)}$; $(x_0, y_0) = (1, -1)$

7. Let f be the function defined in Example 1. Use the Fact on page 768 to show that f is differentiable at $(1, 2)$ and then find its total derivative there.

8. Let g be the function defined in Exercise 2. Use the Fact on page 768 to show that g is differentiable at $(1, -1)$ and then find its total derivative there.

9. Let $h(x, y) = \sqrt{x^2 + y^2}$. Is h differentiable at the origin? Justify your answer.

10. Let $f(x, y) = \begin{cases} \dfrac{x^2 + y^2}{xy^2 - 2} & \text{if } xy^2 \neq 2 \\ 0 & \text{if } xy^2 = 2. \end{cases}$

Is f differentiable at $(2, 1)$? Justify your answer.

11. Let $L(x, y)$ be a linear function. Using the Fact on page 768, explain why L is differentiable at every point (x_0, y_0) in $\mathbb{R}^2$.

FURTHER EXERCISES

12. Let $f(x, y) = |y| \cos x$. This exercise explores the fact that the partial derivatives of a function may or may not exist at a given point.

 (a) Plot $z = f(x, y)$ over the rectangle $[-5, 5] \times [-5, 5]$. The graph suggests that there may be trouble with partial derivatives where $y = 0$, that is, along the x-axis. How does the graph suggest this? Which partial derivative (f_x or f_y) seems to be in trouble?

 (b) Use the definition to show that $f_y(0, 0)$ does not exist. In other words, explain why the limit

 $$\lim_{h \to 0} \frac{f(0, h) - f(0, 0)}{h}$$

 does not exist.

 (c) Show that $f_x(0, 0)$ *does* exist; find its value. How does the result appear on the graph?

 (d) Use the limit definition to show that $f_y(\pi/2, 0)$ *does* exist; find its value.

 (e) How does the graph reflect the result of part (d)?

 (f) Find a function $g(x, y)$ for which $g_y(0, 0)$ exists but $g_x(0, 0)$ does not. Use technology to plot its graph.

13. Let $f(x, y) = |x|y$.

 (a) By experimenting with graphs (use technology!), try to guess where $f_x(x, y)$ and $f_y(x, y)$ do exist and where they do not. (No proofs are needed.)

 (b) Use the limit definition to find $f_x(0, 0)$.

 (c) Using the limit definition, explain why $f_x(0, 1)$ does not exist.

14. Let $h(x, y) = \begin{cases} \dfrac{x^2 y + xy^2}{x^2 + y^2} & \text{if } (x, y) \neq (0, 0) \\ 0 & \text{if } (x, y) = (0, 0). \end{cases}$

 (a) Show that h is continuous at $(0, 0)$. [HINT: Use polar coordinates.]

 (b) Is h differentiable at $(0, 0)$? Justify your answer.

15. Suppose that $f_x(x, y) = xy^2$ and $f_y(x, y) = x^2 y$. Does f have a total derivative at $(0, 0)$? If so, find it. If not, explain how you know this.

16. Let f be the function defined in Example 1.

 (a) By converting to polar coordinates, show that

 $$f(r, \theta) = \begin{cases} \frac{1}{2} \sin(2\theta) & \text{if } (r, \theta) \neq (0, 0) \\ 0 & \text{if } (r, \theta) = (0, 0). \end{cases}$$

 (b) Use the formula in part (a) to show that f is not continuous at the origin.

17. Let g be the function defined in Exercise 2. Use polar coordinates to show that g is not continuous at the origin.

In this chapter we extended the general notion of derivatives, which are familiar for functions of one variable, to the multivariate setting. Some basic ideas from one-variable calculus carry over—but usually with new twists and deeper meanings. Multivariable derivatives can be interpreted as rates of change, for instance, but in slightly subtler ways than for functions of one variable.

Functions of several variables   The basic object of one-variable calculus is a function $y = f(x)$; its simplest multivariable analogue is a real-valued function $z = f(x, y)$ of two input variables. We began by studying such functions and their graphs, emphasizing similarities to, and differences from, the one-variable setting. Graphs of such functions are usually **surfaces** in $\mathbb{R}^3$ and therefore pose special problems for visualization. Alternative representations, including numerical tables and **contour maps**, may convey more clearly than ordinary graphs how such functions behave—and misbehave.

Partial derivatives   Partial derivatives are the simplest multivariable analogues of the ordinary derivative of single-variable calculus. For a function $f(x, y)$, for instance, the **partial derivative with respect to** x, denoted by $f_x(x, y)$, is found by differentiating $f(x, y)$ with respect to the variable x, treating y as a constant. Partial derivatives (like other derivatives) give information about **rates** and about **slopes**. For a given input (x_0, y_0), the partial derivative $f_x(x_0, y_0)$ describes the rate of increase of f with respect to x when $y = y_0$ is held fixed. Alternatively, $f_x(x_0, y_0)$ describes the "steepness in the x-direction" of the graph of $z = f(x, y)$ at the point (x_0, y_0, z_0). If $f(x, y)$ has a **local minimum** or a **local maximum** at a point (x_0, y_0), then *both* partial derivatives must be zero: $f_x(x_0, y_0) = f_y(x_0, y_0) = 0$.

Linear approximation in several variables   A one-variable differentiable function $f(x)$ is closely approximated near a point x_0 by a **linear** function $L(x) = ax + b$. The coefficient of x is the derivative $f'(x_0)$, and the graph of L is the **tangent line** at $x = x_0$. In the same spirit, a function $f(x, y)$ with well-behaved partial derivatives has a two-variable **linear approximation** function, of the form $L(x, y) = ax + by + c$. Here the coefficients a and b are partial derivatives ($a = f_x(x_0, y_0)$ and $b = f_y(x_0, y_0)$), and the graph of L is the **tangent plane** at the point (x_0, y_0).

The gradient and directional derivatives   The **gradient** of a function f at an input point (x_0, y_0), denoted $\nabla f(x_0, y_0)$, is the vector $(f_x(x_0, y_0), f_y(x_0, y_0))$ of partial derivatives, one for each variable. Although easily calculated, the gradient conveys a lot of information. For example, the vector $\nabla f(x_0, y_0)$ always points "uphill"—that is, in the direction of fastest increase of the function f at the point (x_0, y_0). For a given unit vector $\mathbf{u}$, the **directional derivative** $D_{\mathbf{u}} f(x_0, y_0)$ measures the rate at which $f(x, y)$ increases as (x, y) moves in the $\mathbf{u}$-direction. Although defined as a limit, the directional derivative is easily calculated using the gradient and the dot product: At any point (x_0, y_0), we have $D_{\mathbf{u}} f(x_0, y_0) = \nabla f(x_0, y_0) \cdot \mathbf{u}$, and so $D_{\mathbf{u}} f(x_0, y_0)$ is the **scalar component** of $\nabla f(x_0, y_0)$ in the direction of $\mathbf{u}$.

Higher-order derivatives and quadratic approximation   Functions of several variables permit *repeated* partial derivatives. The notation $f_{xy}(x, y)$, for instance, indicates the result of differentiating first with respect to x and then with respect to y. A striking property of such derivatives is that, for well-behaved functions, the **order of differentiation** does not matter: $f_{xy}(x, y) = f_{yx}(x, y)$. Using higher-order derivatives, we can extend the idea of linear approximation to the quadratic level. For a given function $f(x, y)$, the **quadratic approximation** function $Q(x, y)$ at (x_0, y_0) is chosen so that the Q and f have identical values, partial derivatives, and second-order derivatives at (x_0, y_0). More generally, we can find **Taylor polynomials** in several variables that approximate a given function to any desired degree in much the same way as in the one-variable case.

Maxima and minima   As in the one-variable case, optimizing functions of several variables is done using derivatives: For a differentiable function, a local maximum or minimum can occur *only* at a **stationary point**, where all of the partial derivatives vanish simultaneously. Even in one variable a stationary point may correspond to a maximum, a minimum, or neither, and "classifying" stationary points among these possibilities is an important problem. For functions of several variables, the situation is similar—but more complicated because of the extra "room" in higher dimensions. We analyzed the situation carefully for functions of two variables. For such functions, a stationary point may be a **maximum point**, a **minimum point**, or a **saddle point**. To choose among these options we looked carefully at second derivatives and at the quadratic approximation discussed earlier.

The chain rule The chain rule—regardless of dimension—concerns derivatives of new functions formed by composing old functions. The multivariable setting allows various types of compositions and therefore leads to slightly different versions of the chain rule. All versions are linked, however, by the common idea of **linear approximation**: If any two functions f and g have linear approximations L_f and L_g respectively, then $f \circ g$ has linear approximation $L_f \circ L_g$.

Local linearity: a look at theory Most of the functions normally met in a multivariable calculus course are differentiable in the sense that they permit suitable linear approximations. In this brief section we studied more carefully what such approximation means—and what can go wrong in its absence.

REVIEW EXERCISES

1. Carefully, but briefly, describe the idea of a level curve of a function $f(x, y)$.

2. Briefly but carefully say what it means for (x_0, y_0) to be a stationary point for $f(x, y)$.

3. (a) What does it mean for (x_0, y_0) to be a saddle point of $f(x, y)$?
   (b) Give an example of a function $f(x, y)$ that has a saddle point at $(0, 0)$.
   (c) Give an example of a function $g(x, y)$ that has a saddle point at $(2, 1)$.
   (d) What does the Hessian matrix $f''(x_0, y_0)$ say about saddle points?

4. Describe the direction in which the directional derivative of a function is a maximum.

5. Describe the direction in which the directional derivative of a function is a minimum.

6. Describe the direction in which the directional derivative of a function is zero.

7. Let $f(x, y) = x^2 + 2xy - 6x - 2y + 8$.
   (a) f has one stationary point (x_0, y_0). Find it.
   (b) Classify the stationary point you found in part (a) (i.e., tell what type it is and why).
   (c) Write the linear and quadratic approximation functions $L(x, y)$ and $Q(x, y)$ to f at the stationary point (x_0, y_0) you found in part (a).

8. Let $f(x, y) = 3x^2 - 12x + y^2 - 6y + 25$.
   (a) f has one stationary point (x_0, y_0). Find it.

(b) Classify the stationary point you found in part (a) (i.e., tell what type it is and why).
   (c) Write the linear and quadratic approximation functions $L(x, y)$ and $Q(x, y)$ to f at the stationary point (x_0, y_0) you found in part (a).

9. Let $f(x, y) = xy - y - 2x + 2$.
   (a) f has one stationary point (x_0, y_0). Find it.
   (b) Classify the stationary point you found in part (a) (i.e., tell what type it is and why).
   (c) Write the linear and quadratic approximation functions $L(x, y)$ and $Q(x, y)$ to f at the stationary point (x_0, y_0) you found in part (a).

10. Let $f(x, y) = xy - y + 2x - 2$.
    (a) f has one stationary point (x_0, y_0). Find it.
    (b) Classify the stationary point you found in part (a) (i.e., tell what type it is and why).
    (c) Write the linear and quadratic approximation functions $L(x, y)$ and $Q(x, y)$ to f at the stationary point (x_0, y_0) you found in part (a).

11. Let $f(x, y) = x^2 + 2y^2$, and let $\mathbf{u} = (1/\sqrt{2}, 1/\sqrt{2})$. Find $D_{\mathbf{u}}(2, 3)$, the directional derivative of f at $(2, 3)$ in the direction of $\mathbf{u}$.

12. Let $f(x, y) = xy^2$, and let $\mathbf{u} = (-1/\sqrt{2}, 1/\sqrt{2})$. Find $D_{\mathbf{u}} f(2, 3)$, the directional derivative of f at $(2, 3)$ in the direction of $\mathbf{u}$.

13. Let $f(x, y) = x + y^2$, and let $\mathbf{u} = (-1/\sqrt{2}, 1/\sqrt{2})$. Find $D_{\mathbf{u}} f(3, 2)$, the directional derivative of f at $(3, 2)$ in the direction of $\mathbf{u}$.

14. Find the linear approximation $L(x, y)$ to the function $f(x, y) = x^2 + y^2$ at $(3, 4)$.

15. Give an example of a function $f(x, y)$ that has a local maximum at the point $(0, 0)$. Explain briefly how you know that there is a local maximum there.

16. Find the quadratic approximation $Q(x, y)$ at $(0, 0)$ to the function $f(x, y) = \cos x + xy + y + 1$.

17. Find the quadratic approximation $Q(x, y)$ at $(0, 0)$ to the function $f(x, y) = e^x + xy + y^3 + 1$.

18. The function $f(x, y) = x^2 + xy - 3y$ has a stationary point at $(3, -6)$.
    (a) What type of stationary point is $(3, -6)$?
    (b) Does f have any other stationary points? Justify your answer.

19. Let $f(x, y) = x^2 + y$ and $\mathbf{u} = (1/\sqrt{2}, -1/\sqrt{2})$.
    (a) Find $\nabla f(2, 3)$.
    (b) Find $D_{\mathbf{u}} f(2, 3)$.

20. Find the quadratic approximation $Q(x, y)$ at $(0, 0)$ to the function $f(x, y) = 1 + 2x + \sin(xy)$.

In Exercises 21–27, draw the level "curves" $f(x, y) = z = k$ for $k = -1, 0, 1, 2, 3$. (In some—but not all—cases, the level sets are curves.)

21. $f(x, y) = x^2 + y^2$

22. $f(x, y) = x^2 + y^2 + 5$

23. $f(x, y) = \sqrt{x^2 + y^2}$

24. $f(x, y) = x^2$

25. $f(x, y) = \sqrt{x}$

26. $f(x, y) = x + y$

27. $f(x, y) = (x + y)^2$

In Exercises 28–32, write a formula for the functions defined in words, state the function's natural domain, and then draw the level sets $z = k$ for $k = -1, 0, 1, 2, 3$.

28. $f(x, y)$ is the distance from (x, y) to the point $(1, 2)$.

29. $f(x, y)$ is the slope of the line joining (x, y) to $(0, 0)$.

30. $f(x, y)$ is the distance from (x, y) to the line $x = 1$.

31. $f(x, y)$ is the distance from (x, y) to the circle $x^2 + y^2 = 1$.

32. $f(x, y)$ is the distance from (x, y) to the line $x + y = 1$.

33. The curvature for a plane curve $\mathbf{r}(t) = (x(t), y(t))$ is

$$\kappa = \frac{|x'y'' - y'x''|}{\left((x')^2 + (y')^2\right)^{3/2}}$$

    (a) Write a simpler formula for curvature for the case where the curve is of the form $y = f(x)$. [HINT: Let $\mathbf{r}(t) = (t, f(t))$.]

    (b) Compute the curvature at the vertex of the parabola $y = x^2$.

    (c) Show that the curvature anywhere on the circle $x^2 + y^2 = r^2$ is $1/r$. [HINT: Parametrize the circle.]

34. Compute f_x and f_y for $f(x, y) = x \sin(xy)$.

35. Let $g(x, y) = x^2 + 2y^2$. Draw the level curves $g(x, y) = 1$, $g(x, y) = 2$, and $g(x, y) = 4$.

36. Suppose $f(x, y)$ is a differentiable function, and $x = \sin t$, $y = t^2$. What does the chain rule say about df/dt?

37. Let $f(x, y) = x^2 + xy$.
    (a) Find the directional derivative of f in the direction of the gradient vector $\nabla f(1, 2)$.
    (b) Write an equation for the line tangent to the level curve through $(1, 2)$.
    (c) Write an equation for the plane tangent to the surface $z = x^2 + xy$ at the point $(1, 2, 3)$.

38. Let $f(x, y) = x^2 + y$.
    (a) Draw level curves $f(x, y) = c$ for $c = -1, 0, 1, 2, 3$.
    (b) Write an equation for the line tangent to the level curve $f(x, y) = 2$ at the point $(1, 1)$.
    (c) In which direction $\mathbf{u}$ does $f(x, y)$ increase fastest at $(1, 1)$? For this direction $\mathbf{u}$, find the directional derivative $f_{\mathbf{u}}'(1, 1)$.
    (d) Write an equation for the plane tangent to the surface $z = f(x, y)$ at $(1, 1, 2)$.

39. Let $f(x, y) = y \sin(x + y)$.
    (a) Compute ∇f.
    (b) Find the directional derivative of f at $(0, \pi)$ in the direction of the vector $(3\mathbf{i} + 4\mathbf{j})/5$.

40. Let $f(x, y) = x^2 + y$, $x(s, t) = s + t$, and $y(s, t) = st$. Use the chain rule to find $\partial f / \partial t$.

41. Is there a function $f(x, y)$ for which $\nabla f = (\cos x + \cos(x + y), \cos(x + y))$? If so, find the most general such function. If not, explain why not.

42. Let $f(x, y) = yx^2 - y$. Find all stationary points of f and determine whether each corresponds to a local maximum, a local minimum, or a saddle point.

43. Let $f(x, y) = \cos(\pi x + 2y)$. Observe that $f(3, 0) = -1$.
    (a) Use a linear approximation to estimate $f(2.8, 0.1)$.
    (b) Use a quadratic approximation to estimate $f(2.8, 0.1)$.

In Exercises 44–46, $P = (1, 2, 3)$, $Q = (4, 6, 8)$, and $f(x, y) = x^2 - y + 4$.

44. The graph of f is a surface in xyz-space. Does P lie on this graph? Justify your answer.

45. On one set of axes, draw and label the level curves of f that correspond to $z = -4$, $z = 0$, and $z = 4$.

46. Find the linear approximation function $L(x, y)$ to f at the point $(2, 1)$.

47. Here are some values of a linear function $g(x, y)$:

y \ x	−1	0	1
1	−2	3	8
0	1	6	11
−1	4	9	14

Find $g_x(0, 0)$ and $g_y(1, 1)$.

In Exercises 48–51, $f(x, y) = x^2 + y - 2$, $P = (2, 1, 3)$, and $Q = (4, 5, 6)$.

48. The graph of $z = f(x, y)$ is a surface in xyz-space. Does P lie on the graph? Does Q? Justify your answers.

49. On one set of axes, draw and label the level curves of f that correspond to $z = -4$, $z = 0$, and $z = 4$.

50. **(a)** Find $f_x(2, 1)$.
 (b) Find $f_y(2, 1)$.
 (c) Find the linear approximation function $L(x, y)$ to f at the point $(2, 1)$.

51. Does f have any local maximum or minimum points? If so, find one. If not, indicate why not.

52. Let $z = f(x, y)$ and let (x_0, y_0) be a point. State the definition (involving a limit) of $\dfrac{\partial f}{\partial x}(x_0, y_0)$.

53. Let $z = f(x, y)$, let (x_0, y_0) be a point in the xy-plane, and let $\mathbf{u}$ be a unit vector. Suppose that $\dfrac{\partial f}{\partial \mathbf{u}}(x_0, y_0) = -3$. What does this tell us about rates of change of f?

In Exercises 54–58, $f(x, y) = x^2 + xy$.

54. Find $\nabla f(2, 1)$.

55. Find an equation for the plane tangent to the surface $z = f(x, y)$ at the point $(2, 1, 6)$.

56. Find the directional derivative $D_{\mathbf{u}} f(2, 1)$, where $\mathbf{u} = (1/\sqrt{2}, -1/\sqrt{2})$.

57. **(a)** For what unit direction vector $\mathbf{u}$ is $D_{\mathbf{u}} f(2, 1)$ largest?
 (b) What is the largest possible value of $D_{\mathbf{u}} f(2, 1)$? Justify your answer.

58. Suppose that $x(t) = t^2$ and $y(t) = e^t$. Find $\dfrac{df}{dt}$. (Write the answer as a function of t alone.)

59. Let $f(x, y) = x^2 + y$ and $g(t) = \sin t$. Use the chain rule to find $(g \circ f)'(2, 1)$.

60. Consider two functions $\mathbf{f} : \mathbb{R}^2 \to \mathbb{R}^2$ and $g : \mathbb{R}^2 \to \mathbb{R}$. Let $\mathbf{X}_0 = (x_0, y_0)$ be a point in $\mathbb{R}^2$. Write the chain rule formula for the derivative $(g \circ \mathbf{f})'(\mathbf{X}_0)$. (Indicate clearly which symbols on both sides of the equation represent matrices and matrix multiplication and state the dimensions of each matrix.)

61. Suppose that (x_0, y_0) is a stationary point of $f(x, y)$, and let D be the determinant of the Hessian matrix $H_f(x_0, y_0)$. What does D tell us about the nature of the stationary point?

62. Let $f(x, y) = \sin x + \cos y$. Find the linear approximation function $L(x, y)$ to $f(x, y)$ at the point $(0, 0)$.

63. Find and classify all the stationary points of the function $f(x, y) = x^3 - 12x + y^2 - 2y$.

64. Let $f(x, y) = 2y - \sin x$. Describe and plot the level curve that passes through the point $(\pi/2, 2)$.

65. Does the function $f(x, y) = 2y - \sin x$ have any stationary points? If so, find one. If not, say briefly why not.

66. Let $f(x, y) = 3e^x + 2 \sin y$. Find the linear approximation function $L(x, y)$ to $f(x, y)$ at the point $(0, 0)$.

67. Find and classify all the stationary points of the function $f(x, y) = x^3 - 3x + y^2 - y$.

68. Let $f(x, y) = xy^2$, and let $\mathbf{u} = (-1, 0)$. Find $D_{\mathbf{u}} f(2, 3)$, the directional derivative of f at $(2, 3)$ in the direction of $\mathbf{u}$.

69. Let $f(x, y) = \cos x + xy + y + 1$.
 (a) Find the linear approximation $L(x, y)$ to $f(x, y)$ at $(0, 0)$.
 (b) Find the quadratic approximation $Q(x, y)$ to $f(x, y)$ at $(0, 0)$.

70. Consider two functions $\mathbf{f} : \mathbb{R}^3 \to \mathbb{R}^2$ and $g : \mathbb{R}^2 \to \mathbb{R}$. Let $\mathbf{X}_0 = (x_0, y_0, z_0)$ be a point in $\mathbb{R}^3$. Write the chain rule formula for the derivative $(g \circ \mathbf{f})'(\mathbf{X}_0)$. (Indicate clearly which symbols on both sides of the equation represent matrices and matrix multiplication and state the dimensions of each matrix.)

71. Let $f(x, y) = 3x^2 - 6xy + 2y^3$. Find and classify all the stationary points of f.

72. Let $f(x, y) = \sin(x) + \cos(2y)$.
 (a) Find the directional derivative of f at $(1, 2)$ in the direction of the vector $(3, 4)$.
 (b) Find $p_2(x, y)$, the second-order Taylor approximation to $f(x, y)$ at $(0, 0)$.
 (c) Find a stationary point of f. (Any one is OK.) Use the Hessian to decide what type it is.

73. Let S be the surface $z = f(x, y) = x^2 + y^2$.
 (a) If (a, b) is any point in the xy-plane, then $(a, b, a^2 + b^2)$ is on S. Write an equation for the plane tangent to S at the point $(a, b, a^2 + b^2)$. (The answer will involve a and b, of course.)
 (b) Where does the plane you found in the previous part intersect each of the three axes? Justify your answers.

74. Let $f(x, y) = x^2 + y^2$. Show that for any (a, b) the level curve of f through (a, b) is perpendicular to the gradient vector $\nabla f(a, b)$.

75. Let $f(x, y) = x - 2y^2$ and $P = (1, 2)$.
 (a) Find $\nabla f(1, 2)$.
 (b) Find an equation of the level curve of f that passes through the point P.

(c) Show that the $\nabla f(1, 2)$ is perpendicular to the curve in part (b) at P.

76. Let $f(x, y) = x - 2y^2$ and $\mathbf{X}_0 = (1, 2)$.
  (a) Find the directional derivatives $D_{\mathbf{u}} f(\mathbf{X}_0)$ for $\mathbf{u} = \mathbf{i}$, for $\mathbf{u} = \mathbf{j}$, and $\mathbf{u} = (3\mathbf{i} + 4\mathbf{j})/5$.
  (b) Find a direction vector $\mathbf{u}$ for which $D_{\mathbf{u}} f(\mathbf{X}_0) = 0$.

77. Find and classify all the stationary points of $f(x, y) = x^3 + 2y^3 + 6x^2 - 9y^2$. [HINT: There are four stationary points in all.]

78. Let $\mathbf{f}(u, v) = (uv^2, u - v))$, $\mathbf{g}(x, y) = (x^2 - y, x^2 + y)$, and $\mathbf{h}(x, y) = (\mathbf{f} \circ \mathbf{g})(x, y)$. Use the chain rule to find $\mathbf{h}'(2, 1)$.

79. Let $f(x, y) = e^{xy} + y^3$.
  (a) Find $\nabla f(x, y)$.
  (b) Find a scalar equation for the plane tangent to the surface $z = f(x, y)$ at the point $(0, 1, 2)$.

80. Suppose that the level curves of the function $h(x, y)$ are described by equations of the form $y = x^2 + k$, where k is a real number. Furthermore, suppose that $h(2, 1) = 3$ and $h(1, 2) = 7$.
  (a) Show that $h(1, -2) = 3$.
  (b) Evaluate $h(0, -3)$ and $h(0, 1)$.

INTEGRALS

14.1 MULTIPLE INTEGRALS AND APPROXIMATING SUMS

The two most important ideas of calculus—of one *or* several variables—are the derivative and the integral. In studying multivariable calculus up to now, we have seen the derivative idea in many forms and settings, including partial derivatives, derivatives of vector-valued functions, gradients, directional derivatives, and linear and quadratic approximation.

Now we study integrals, the other main idea of our subject. Like derivatives, integrals have various meanings and interpretations in different multivariable settings. In this chapter our main focus is on **multiple integrals**, that is, integrals of real-valued functions of two or three variables, integrated over regions in two- or three-dimensional space. ➡ We will study how multiple integrals are defined, how to calculate (or approximate) their values, and (last, but certainly not least) what they tell us in various circumstances.

Multiple integrals can also be defined in dimensions higher than three.

Integrals and approximating sums

All integrals—single, double, triple, or whatever—are defined to be certain limits of **approximating sums** (also known, sometimes, as **Riemann sums**). This important idea is always studied in single-variable calculus, but it may be quickly (and, perhaps, gratefully) forgotten. Readers whose memories are vague on this score have an excellent excuse: Although integrals are *defined* as limits of approximating sums, they are usually *calculated* in an entirely different way, using antiderivatives. Here is a typical calculation: ➡

$$\int_0^1 x^2\, dx = \frac{x^3}{3}\bigg]_0^1 = \frac{1}{3}.$$

There's not a Riemann sum in sight.

This method of evaluating an integral—find an antiderivative for the integrand and plug in the endpoints—works just fine thanks to the fundamental theorem of calculus.

So why bother at all with approximating sums? Here are two good reasons:

Antiderivative trouble The antiderivative method depends on finding a convenient antiderivative of the integrand. Unfortunately, not every function, even in single-variable calculus, has an "elementary" antiderivative, that is, an antiderivative with a symbolic formula built from standard "elements." The simple-looking function $f(x) = \sin(x^2)$, for

Spend a moment looking for an antiderivative formula. Nothing works.

instance, has no elementary antiderivative. ◄ The best we can do with the integral

$$I = \int_0^1 \sin(x^2)\,dx,$$

The integral can also be estimated using infinite series. See Example 3, page 801.

therefore, is to approximate it, perhaps with an approximating sum using (say) the left rule, the midpoint rule, or the trapezoid rule. For the record, approximating I with a trapezoid-rule sum with 10 subdivisions gives $I \approx 0.311$. ◄

But not always—sometimes it's hard or impossible to find antiderivatives.

What integrals mean   The fundamental theorem often makes *calculating* integrals easy, ◄ but approximating sums may illustrate more clearly what the answers *mean*. Figure 1, for instance, illustrates the sense in which a midpoint-rule sum with four subdivisions approximates the area bounded by the curve $y = x^2$ from $x = 0$ to $x = 1$:

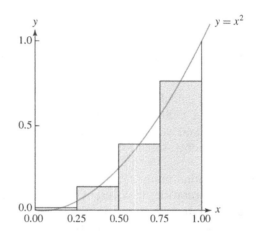

FIGURE 1

A midpoint sum estimate for
$\int_0^1 x^2\,dx : M_4 = 21/64 \approx 0.328$

The one-variable case: a review   The basic idea of an integral as a limit of approximating sums is much the same for functions of two (or more) variables as for functions of one variable. Let's review the main single-variable ideas and notations that arise on the way to defining the integral of a function f over an interval $[a, b]$. The first step is to form Riemann sums:

> **DEFINITION**   Let the interval $[a, b]$ be partitioned into n subintervals by any $n+1$ points
>
> $$a = x_0 < x_1 < x_2 < \cdots < x_{n-1} < x_n = b;$$
>
> let $\Delta x_i = x_i - x_{i-1}$ denote the width of the ith subinterval. Within each subinterval $[x_{i-1}, x_i]$, choose any point c_i. The sum
>
> $$\sum_{i=1}^n f(c_i)\Delta x_i = f(c_1)\Delta x_1 + f(c_2)\Delta x_2 + \cdots + f(c_n)\Delta x_n$$
>
> is called a **Riemann sum with n subdivisions** for f on $[a, b]$.

Left-rule, right-rule, and midpoint-rule approximating sums all fit this definition. Each of these sums is built from a **regular partition** of $[a, b]$ (i.e., one with subintervals of equal length) and some consistent scheme for choosing the **sampling points** c_i. (For left, right, and midpoint sums we choose each c_i as the left endpoint, the right endpoint, or the midpoint of the ith subinterval, respectively.)

Graphical and numerical intuition suggests that, as the number n of subdivisions tends to infinity, all of these approximating sums should converge to some fixed number. Geometrically speaking, moreover, this number measures the signed area ➔ bounded by the f-graph from $x = a$ to $x = b$.

The formal definition of the integral makes these ideas precise:

"Signed" means that areas under the x-axis are considered to be negative.

> **DEFINITION** Let the function f be defined on the interval $[a, b]$. The **integral of f over** $[a, b]$, denoted $\int_a^b f(x)\,dx$, is the number to which all Riemann sums S_n tend as n tends to infinity and as the widths of all subdivisions tend to zero. In symbols:
>
> $$\int_a^b f(x)\,dx = \lim_{n\to\infty} S_n = \lim_{n\to\infty} \sum_{i=1}^n f(c_i)\,\Delta x_i,$$
>
> if the limit exists.

Honesty dictates a brief admission: The limit in the definition, taken at face value, is a slippery customer. Understanding every ramification of permitting arbitrary partitions and sampling points, for example, can be tricky. Fortunately, these issues need not trouble us for the well-behaved functions (e.g., continuous functions) we typically meet in single-variable and multivariable calculus. For such functions, almost any respectable sort of approximating sum does what we expect—it approaches the true value of the integral as n tends to infinity.

Sums and integrals in two variables Most of the differences between single-variable integrals and multivariable integrals are technical rather than theoretical. Indeed, the definitions of

$$\iint_R f(x, y)\,dA \qquad \text{and} \qquad \int_a^b f(x)\,dx$$

are almost identical. Here, f is a function of two variables, and R is a region—in the simplest case, a rectangle—in the xy-plane. On the other hand, the mechanics of evaluating these two types of integrals by antidifferentiation are quite different. ➔

First, let's list the ingredients that go into defining the **double integral** $\iint_R f(x, y)\,dA$. Look, in each case, for points of similarity and difference with the one-variable situation.

We'll get to that in the next section.

- **R, the region of integration** In one variable the region of integration is always an interval $[a, b]$ in the domain of f; this is implicit in the notation $\int_a^b f(x)\,dx$. In two variables, by contrast, the region of integration, denoted by R, may be almost *any* two-dimensional subset of the plane. In the simplest case R is a rectangle $[a, b] \times [c, d]$.

- **Why the dA?** The symbol "dA" in the double integral resembles the "dx" that appears in single integrals. The "A" reminds us of area, and "dA" suggests a small increment of area.

- **Partitions** In one variable we partition an interval $[a, b]$ by cutting it (perhaps unevenly) into smaller intervals, with endpoints $a = x_0 < x_1 < x_2 < \cdots < x_n = b$. The "size" of the ith subinterval is simply its length, Δx_i.

In two variables we do something similar: We chop the plane region R into m smaller regions $R_1, R_2, R_3, \ldots, R_m$, perhaps of different sizes and shapes. The "size" of a subregion R_i is now its *area* denoted by ΔA_i.

In practice—whatever the number of variables—it is usually convenient to choose the partition in some consistent way. In one variable, a **regular partition** (one with equal-length subintervals) is simplest. An analogous procedure works in two variables if R is a rectangle $[a, b] \times [c, d]$: Cut R by an n-by-n grid in each direction, producing n^2 rectangular subregions in all.

- **Approximating sums**   In one variable, an approximating sum has the form

$$f(c_1)\Delta x_1 + f(c_2)\Delta x_2 + \cdots + f(c_n)\Delta x_n = \sum_{i=1}^{n} f(c_i)\Delta x_i,$$

where each c_i is a sample point chosen from the ith subinterval.

A two-variable approximating sum is similar. In each subregion R_i we choose a sampling point $P_i = (x_i, y_i)$ and then form the approximating sum

$$S_m = f(P_1)\Delta A_1 + f(P_2)\Delta A_2 + \cdots + f(P_m)\Delta A_m = \sum_{i=1}^{m} f(P_i)\Delta A_i,$$

where ΔA_i is the area of R_i.

EXAMPLE 1   Consider the function $f(x, y) = x + y$ on the rectangle $R = [0, 4] \times [0, 4]$. Calculate an approximating sum S_{16} for the integral $\iint_R f(x, y)\, dA$; use $4^2 = 16$ square subregions, each with area $\Delta A_i = 1$. In each subregion, sample the integrand f at the corner closest to the origin.

Solution   Figure 2 illustrates the idea. The base of each square "block" is one of the 16 subregions of R; the height of each block is determined at the corner nearest the origin.

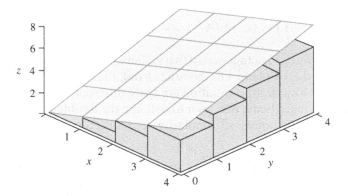

FIGURE 2
The surface $z = x + y$ and an approximating sum

Figure 2 shows that the approximating sum (the total volume of all 16 blocks) approximates the volume of the solid bounded on top by the surface $z = f(x, y)$ and whose base is the rectangle $R = [0, 4] \times [0, 4]$. This volume is the value of the integral $\iint_R (x + y)\, dA$.

The following table shows values $f(x, y)$ at the 16 sampling points:

y \ x	0	1	2	3
3	3	4	5	6
2	2	3	4	5
1	1	2	3	4
0	0	1	2	3

Since each subregion has area 1, the approximating sum is simply the sum of all 16 table entries, or 48. In symbols:

$$S_{16} = 48 \approx \iint_R (x+y)\, dA.$$

Figure 2 also shows that S_{16} *underestimates* the true volume—as would any approximating sum S_m for this integral with sampling points at the corners closest to the origin. Still, we expect these approximating sums S_m to converge to the "true" volume as m increases to infinity. A table of values supports this guess:

m	2^2	4^2	8^2	16^2	24^2	32^2	100^2
S_m	32.00	48.00	56.00	60.00	61.33	62.00	63.36

The values of S_m seem to be creeping up toward a number around 64. Figure 3 illustrates another approximating sum for the same integral.

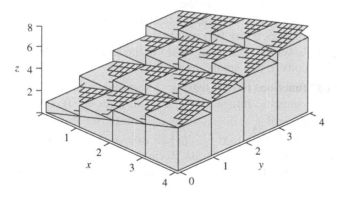

FIGURE 3

The surface $z = x + y$ and a midpoint Riemann approximation

Here sampling points are chosen at the *midpoints* of respective subrectangles, and this time the approximating sum has value 64. In fact, 64 turns to be the exact value of the integral, and a close look suggests that overestimates and underestimates in the approximating sum cancel each other out. (We will calculate the same integral in the next section, using other methods.) ∎

EXAMPLE 2 Consider again the double integral $I = \iint_R f(x, y)\, dA$, where $f(x, y) = x + y$ and $R = [0, 4] \times [0, 4]$. Calculate—this time by hand—an approximating sum S_4 with four equal subdivisions (two in each direction). In each square subregion, evaluate f at the corner *farthest* from the origin. Does the sum underestimate or overestimate the integral?

Solution   All four subregions are squares with edge length 2 and therefore area 4. The sampling points are $P_1 = (2, 2)$, $P_2 = (4, 2)$, $P_3 = (2, 4)$, and $P_4 = (4, 4)$, and the desired approximating sum is

$$S_4 = \sum_{i=1}^{4} f(P_i) \Delta A_i = 4 \cdot 4 + 6 \cdot 4 + 6 \cdot 4 + 8 \cdot 4 = 96.$$

This sum *overestimates* I because, in each subrectangle, the integrand function is sampled at the point where it takes its largest possible value.

The integral as a limit

We defined the one-variable integral $\int_a^b f(x)\,dx$ as the limit of approximating sums S_n, as n tends to infinity and the widths of all subdivisions tend to zero. The double integral $\iint_R f(x, y)\,dA$ is defined in a similar way.

Before giving a detailed definition of the integral we mention two technical subtleties.

- **Diameter**   A useful definition requires that the *size* of subregions of R should go to zero as the *number* of subregions tends to infinity. If the subregions are all, say, squares, then there is no problem—as the number of squares increases the size of each square automatically decreases. Trouble could arise, however, if the subdivisions are extremely "oblong." This would happen if, say, we subdivided the rectangle $[0, 1] \times [0, 1]$ only in the x-direction, leaving the y-interval untouched. One technical "fix" that avoids such possible trouble is to define the **diameter** of a subregion to be the largest possible distance between any two points in the subregion. Requiring the diameters of *all* subregions to tend to zero—as we do in the following definition—automatically forces the subregions to be "small" in all directions.

- **"Well-behaved" functions have integrals**   Even in the one-variable setting, not every function f *has* a sensible integral over a given interval. If, say, f is unbounded on $[a, b]$, then the integral $\int_a^b f(x)\,dx$ does not exist. But if f is continuous on $[a, b]$, then the theory guarantees that $\int_a^b f(x)\,dx$ does exist, and so we can "do calculus" with such functions and their integrals. Because the familiar elementary functions are continuous (and even differentiable) wherever they are defined, existence of integrals is seldom a live question in single-variable calculus, and we make little fuss over it.

  The situation is similar in the multivariable setting. If a function f of two or more variables is *continuous* on a domain R, then f is also *integrable* on such a domain, and (multivariate) calculus can proceed. What this bland statement slightly conceals is that the concept of continuity is somewhat subtler for functions of several variables than for functions of one variable. Nevertheless, nearly every function seen in a multivariable calculus course is continuous except perhaps at "obviously" suspicious points of its domain such as where a denominator vanishes or a symbolic formula makes no sense. We will consider almost exclusively functions that are "well behaved" in the sense that they and (if needed) their derivatives are continuous on the domains in question.

See the preceding paragraph for more on "good behavior."

The formal definition   The following definition is adequate for the well-behaved functions $f(x, y)$ studied in this book:

> **DEFINITION** Let the function $f(x, y)$ be defined on the region R, and let S_m be an approximating sum with m subdivisions, as described above. Suppose that S_m tends to a number I as m tends to infinity and the diameters of all subdivisions tend to zero. Then I is the **double integral of f over R**, and we write
>
> $$I = \iint_R f(x, y)\, dA = \lim_{m \to \infty} \sum_{i=1}^{m} f(P_i) \Delta A_i.$$

As in the one-variable setting, the limit definition—although crucial to understanding integrals and often useful for approximating them—almost never lends itself to calculating integrals exactly. For this purpose, fortunately, there are methods based on antidifferentiation. ➡

We'll see some in the next section.

A special case: integrating constant functions   Constant functions are easy to integrate in one variable. The same is true in several variables, thanks to the definition just given. Thus, let R be a region in the plane, let $f(x, y) = k$ be any constant function, and consider the integral $I = \iint_R f(x, y)\, dA$. Since $f(x, y) = k$ for all (x, y), *every* approximating sum for I has the same value:

$$S_m = \sum_{i=1}^{m} f(P_i) \Delta A_i = \sum_{i=1}^{m} k \, \Delta A_i = k \sum_{i=1}^{m} \Delta A_i = k \times \text{area of } R.$$

As the limit of approximating sums, the integral itself must have the same value. The result is worth remembering:

$$\text{If } f(x, y) = k, \text{ then } \iint_R f(x, y)\, dA = k \times \text{area of } R.$$

EXAMPLE 3   Evaluate $\displaystyle\iint_R 3\, dA$, where R is the region inside the unit circle.

Solution   By the preceding observation,

$$\iint_R 3\, dA = 3 \times \text{area of } R = 3\pi.$$

∎

Triple sums and triple integrals

The idea of integral can be extended to three (and even higher) dimensions. In this section we will consider only the simplest case: the **triple integral** of a function $g(x, y, z)$ over a **rectangular parallelepiped** $R = [a, b] \times [c, d] \times [e, f]$. ➡ Such an integral is denoted by

To put it more humbly, a "brick."

$$\iiint_R g(x, y, z)\, dV;$$

note that dV, which suggests *volume*, replaces the dA (for *area*) that appears in double integrals.

Triple integrals, like single and double integrals, are defined formally as limits of approximating sums. An approximating sum in *three* dimensions is formed by subdividing a rectangular solid region R into m smaller rectangular solid subregions R_i with *volume*

ΔV_i, choosing a sampling point P_i in each subregion, and then evaluating the sum

$$\sum_{i=1}^{m} f(P_i)\Delta V_i.$$

See the remarks about diameter at the beginning of this section.

The triple integral is formally defined as the limit of such sums as the number of subregions tends to infinity and the diameter ⬅ of all subregions tends to zero.

> **EXAMPLE 4**   Consider the triple integral $I = \iiint_R g(x, y, z)\, dV$, where $g(x, y, z) = x + y + z$ and $R = [0, 2] \times [0, 2] \times [0, 2]$. Calculate an approximating sum S_8 with eight equal subdivisions (two in each direction). In each subregion, evaluate g at the corner nearest to the origin.
>
> **Solution**   All eight subregions are cubes with edge length 1 and, therefore, volume 1. The sampling points are
>
> $$P_1 = (0, 0, 0), \quad P_2 = (1, 0, 0), \quad P_3 = (0, 1, 0), \quad P_4 = (1, 1, 0),$$
> $$P_5 = (0, 0, 1), \quad P_6 = (1, 0, 1), \quad P_7 = (0, 1, 1), \quad P_8 = (1, 1, 1),$$
>
> and the approximating sum is
>
> $$S_8 = \sum_{i=1}^{8} g(P_i)\Delta V_i = 0 + 1 + 1 + 2 + 1 + 2 + 2 + 3 = 12.$$ ∎

Properties of multivariate integrals

Many properties of double and triple integrals are similar to those of single-variable integrals—and they hold for similar reasons. Regardless of dimension, integrals are limits of approximating sums, and so integrals "inherit" basic properties of sums. ⬅

Rigorous proofs of these properties appeal to properties of sums.

Theorem 1 collects several basic properties of double integrals (but they hold for triple integrals, too).

THEOREM 1 (New integrals from old)   Let $f(x, y)$ and $g(x, y)$ be continuous functions on a domain R in the xy-plane; let k denote a constant. Then

(i) **(Sum rule)** $\iint_R (f(x, y) \pm g(x, y))\, dA = \iint_R f(x, y)\, dA \pm \iint_R g(x, y)\, dA.$

(ii) **(Constant multiple rule)** $\iint_R kf(x, y)\, dA = k \iint_R f(x, y)\, dA.$

(iii) **(Smaller integrand, smaller integral)**   If $f(x, y) \le g(x, y)$ for all x, y in R, then $\iint_R f(x, y)\, dA \le \iint_R g(x, y)\, dA.$

(iv) **(Splitting the domain)**   If $R = R_1 \cup R_2$, where the domains R_1 and R_2 overlap only at their edges, then $\iint_R f(x, y)\, dA = \iint_{R_1} f(x, y)\, dA + \iint_{R_2} f(x, y)\, dA.$

Theorem 1 is simple but surprisingly useful. We exploit it twice in the following example.

EXAMPLE 5 Consider the integrals $I_1 = \iint_R x\, dA$, $I_2 = \iint_R (2+3x)\, dA$, and

$I_3 = \iint_R \left(2 + \sin(x^2 + y^2)\right) dA$, where R is the rectangle $[0,2] \times [0,3]$ in the xy-plane.
Assuming that $I_1 = 6$, ➤ find I_2. Find a reasonable estimate for I_3.

*We'll explain why $I_1 = 6$
shortly; note that R has area 6.*

Solution Theorem 1 says that

$$I_2 = \iint_R (2+3x)\, dA$$
$$= \iint_R 2\, dA + \iint_R 3x\, dA = \iint_R 2\, dA + 3\iint_R x\, dA \quad \text{(using Theorem 1)}$$
$$= 2 \times \text{area of } R + 3 \times 6 = 30. \quad \text{(the first integrand is constant)}$$

The integrand in I_3 is too complicated to handle exactly, but we can bound the
integrand above and below: ➤

The sine never exceeds 1.

$$1 \le 2 + \sin(x^2 + y^2) \le 3.$$

Now Theorem 1(iii) gives

$$\iint_R 1\, dA \le \iint_R \left(2 + \sin(x^2+y^2)\right) dA \le \iint_R 3\, dA.$$

The left- and right-hand integrands are constants, and so we have $6 \le I_3 \le 18$. ∎

Interpreting multiple integrals

Integrals can be interpreted geometrically, physically, or in other ways. Following are
several possibilities; notice, especially, that the appropriate units for an integral depend on
the interpretation at hand.

Double integrals and volume The simplest geometric interpretation of a single-variable
integral $\int_a^b f(x)\, dx$ is in terms of *area*: If $f(x) \ge 0$, then $\int_a^b f(x)\, dx$ measures the area of
the region having vertical sides and bounded above by the curve $y = f(x)$ and below by
the interval $[a,b]$ in the x-axis. In a similar vein, if $f(x,y) \ge 0$ for (x,y) in R, then the
double integral $\iint_R f(x,y)\, dA$ measures the *volume* of the three-dimensional solid with
sides perpendicular to the xy-plane and bounded above by the surface $z = f(x,y)$ and
below by the region R in the xy-plane. If the units of distance in x and y are centimeters,
then units of volume are cubic centimeters.

Double integrals and area: a special case If R is a region in the xy-plane and g happens
to be the constant function $g(x,y) = 1$, then (as the preceding paragraph says) the integral
$\iint_R 1\, dA$ represents the volume of the solid S bounded below by R and having *constant*
height 1. Recall, however, that the volume of any such "cylindrical" solid S is the *area* of
the base times the height. Therefore, in this very special case, the volume of S and the area
of R happen to have the same numerical value. That is,

$$\iint_R 1\, dA = \text{area of } R$$

for any plane region R. If the units of distance in x and y are centimeters, then units of
area are square centimeters.

Using known volumes and areas The geometric interpretations just given for double
integrals are often used to calculate unknown areas and volumes. Sometimes the reasoning
goes the other way: "known" volumes or areas are used to evaluate integrals.

EXAMPLE 6 Find $\iint_R 3\, dA$, where R is the region in the xy-plane between the line $y = x$ and the curve $y = x^2$.

Solution Combining the constant multiple rule and the area interpretation gives

$$\iint_R 3\, dA = 3 \iint_R 1\, dA = 3\, \text{area of } R.$$

Sketch R to convince yourself.

The area of R is readily found by *single-variable* methods: ◄

$$\text{area of } R = \int_0^1 \left(x - x^2 \right) dx = \frac{1}{6},$$

from which we conclude that the $\iint_R 3\, dA = 1/2$. ∎

We used this integral in Example 5.

EXAMPLE 7 Evaluate $\iint_R x\, dA$, where R is the rectangle $[0, 2] \times [0, 3]$, by interpreting it geometrically. ◄.

Solution The integral is the volume of a "block" of the general form shown in Figure 4:

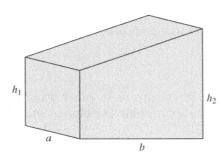

The base is an $a \times b$ rectangle; the height tapers linearly from h_1 to h_2.

FIGURE 4
A simple block

Two such blocks could form a brick with dimensions $a, b,$ and $h_1 + h_2$.

A close look shows that the block's volume is $a \times b \times (h_1 + h_2)/2$—the base area times the "average height." ◄

For the integral in question the "block" has dimensions $a = 3, b = 2, h_1 = 0$, and $h_2 = 2$, and so the volume—and hence the integral—has value 6. ∎

Triple integrals and volume There is no "room" in three-dimensional space even to *plot* a function $w = g(x, y, z)$—four variables would be needed. For this reason, interpreting triple integrals geometrically is usually difficult or impossible.

There is one important exception to this rule: For reasons similar to those for double integrals and area, integrating the *constant* function $g(x, y, z) = 1$ over a solid region R in xyz-space measures the *volume* of the region R. In symbols:

$$\iiint_R 1\, dV = \text{volume of } R.$$

If the units of distance in x, y, and z are centimeters, then units of volume are cubic centimeters.

Density and mass Both double and triple integrals can often be interpreted physically, in the language of density and mass. This view has the special advantage of making sense for both double and triple integrals.

For a *double* integral $\iint_R f(x, y) \, dA$, one thinks of the plane region R as a flat plate with variable **density**; at any point (x, y), the value $f(x, y)$ gives the density, which is measured in appropriate units (such as grams per square centimeter). From this viewpoint, the double integral $\iint_R f(x, y) \, dA$ measures the total mass (in grams) of the plate R.

For a *triple* integral $\iiint_R g(x, y, z) \, dV$, one imagines a solid region R with variable density—at any point (x, y, z), the function value $g(x, y, z)$ is the solid's density, which is measured in appropriate units (e.g., grams per cubic centimeter). From this viewpoint, the triple integral $\iiint_R g(x, y, z) \, dV$ is the total mass (in grams) of the solid R.

Average value of a function   Let $f(x, y)$ be a function defined on a plane region R, with total area $A(R)$. To form an approximating sum S_n for $I = \iint_R f(x, y) \, dA$, we chop R into n small subregions R_i, each with area ΔA_i. Then we calculate the sum

$$S_n = f(P_1)\Delta A_1 + f(P_2)\Delta A_2 + \cdots f(P_n)\Delta A_n,$$

where the P_i are any convenient sampling points inside R_i. Thus, S_n can be thought of as a **weighted sum** of n output values of f, with the inputs P_i scattered around R and each output value "weighted" by the size of the subregion it represents. Therefore, the quotient $S_n/A(R)$ can reasonably be thought of as an approximate **average value** of f over R. Taking this idea to the limit, and recalling that the area of R can also be calculated as an integral, we obtain the definition. ➡

An analogous definition holds for the average value of a one-variable function.

> **DEFINITION**   Let $f(x, y)$ be defined on a region R, with area $A(R)$. The **average value** of f over R is the ratio
>
> $$\frac{\iint_R f(x, y) \, dA}{A(R)},$$
>
> or, equivalently,
>
> $$\frac{\iint_R f(x, y) \, dA}{\iint_R 1 \, dA}.$$

The same reasoning applies to functions of three variables. If $g(x, y, z)$ is defined on a 3-D region R, with volume $V(R)$, then the average value of g over R is given by

$$\frac{\iiint_R g(x, y, z) \, dV}{V(R)} = \frac{\iiint_R g(x, y, z) \, dV}{\iiint_R 1 \, dV}.$$

EXAMPLE 8   Discuss average values in relation to the integrals of Example 1 and Example 4.

Solution   Example 1 concerns the function $f(x, y) = x + y$ over the square region $R = [0, 4] \times [0, 4]$, so $A(R) = 16$. We calculated several approximating sums for $I = \iint_R f(x, y) \, dA$, with values ranging from 32 to 96. These approximations produce, in turn, estimates for the average value of f over R ranging from $32/16 = 2$ to $96/16 = 6$. Because the exact value of I is 64, the "correct" average value of f over R is 4.

Example 4 concerns the function $g(x, y, z) = x + y + z$ and the region $R = [0, 2] \times [0, 2] \times [0, 2]$, which has volume 8. The (crude) approximating sum $S_8 = 12$ leads, therefore, to the estimate $12/8 = 1.5$ for the average value. In fact, the exact value of $I = \iiint_R g \, dA$ can be shown to be 24; ➡ this implies a true average value of $24/8 = 3$ for g over R. ∎

We show this in the next section.

General advice

Multivariable integrals can be interpreted in many ways, depending on the situation, but is there any consistent theme?

One useful strategy for interpreting the information a given integral conveys is to think first about approximating sums. If approximating sums for an integral estimate a quantity, such as the mass of a solid object, then the integral measures the same quantity. We illustrate this line of reasoning in the next example.

EXAMPLE 9   The audience at a public event (a night at the opera, say, or a rock concert) may either be evenly spread through the available space, or it may be tightly packed near the stage and less crowded farther away.

Of course, people are counted in whole numbers, not fractions, and so any effort to model the situation with calculus can only be approximate. Despite this, calculus methods can and do produce useful estimates.

Suppose that at a certain event the function $f(x, y)$ describes the "crowd density," in people per square meter, at the point x meters east and y meters north of some point on the stage, and let $R = [0, 50] \times [-15, 15]$. What information does the integral $I = \iint f(x, y)\, dA$ convey?

Solution   An approximating sum for the integral I has the form

$$f(P_1)\, \Delta A_1 + f(P_2)\, \Delta A_2 + \cdots + f(P_n)\, \Delta A_n.$$

Each summand is the product of a density (in people per unit area) and an area, and so each summand approximates the number of people in a small area ΔA_i. Therefore, the approximating sum estimates the total number of people in a rectangular region extending 50 meters east, 15 meters north, and 15 meters south of the stage. The integral, in turn, can also be interpreted as gauging the total number of people in the region.   ◾

A word of caution   With so many possible interpretations for integrals, a final warning may be in order:

The integral is defined as a number, *say 5—not as a geometric or physical quantity.*

Although it is often useful and enlightening to *interpret* an integral in the language of area, volume, mass, or some other quantity, these are *only* interpretations, not intrinsic properties of an integral. In particular, there is nothing illogical about interpreting an integral $\iint_R f(x, y)\, dA = 5$ today as the volume of a certain solid, tomorrow as the mass of a plane region with variable density, the day after that in terms of average value, and next week as a crowd estimate.

BASIC EXERCISES

1. Let $I = \iint_R \sin(x)\sin(y)\, dA$, where $R = [0, 1] \times [0, 1]$.

    (a) Compute the double midpoint sum $S_{n^2} \approx I$ with $n = 3$ (i.e., 9 subdivisions in all).

    (b) Compute the double midpoint sum $S_{n^2} \approx I$ with $n = 10$ (i.e., 100 subdivisions in all) for I.

2. Let $I = \iiint_R xyz\, dV$, where $R = [0, 4] \times [0, 4] \times [0, 4]$.

    (a) Compute the triple midpoint sum $S_{n^3} \approx I$ with $n = 2$ (i.e., 8 subdivisions in all) for I.

    (b) Compute the triple midpoint sum $S_{n^3} \approx I$ with $n = 4$ (i.e., $4^3 = 64$ subdivisions in all).

3. Let $I = \iint_R f(x, y)\, dA$, where $R = [0, 4] \times [0, 4]$.

    (a) Use the contour map of f shown below to evaluate a double midpoint sum $S_{n^2} \approx I$ with $n = 4$ (i.e., 16 subdivisions in all).

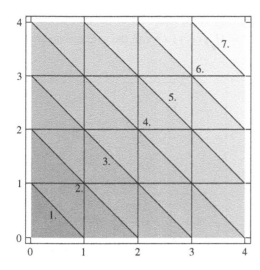

(b) Use the fact that $f(x, y) = x + y$ to check your answer to part (a).

(c) The double midpoint sum for I computed in part (a) is, in fact, the exact value of the integral I. How does the symmetry of the contour map show this?

4. Let $I = \iint_R f(x, y)\, dA$, where $R = [0, 4] \times [0, 4]$.

(a) Use the contour map of f shown below to evaluate a double midpoint sum $S_{n^2} \approx I$, with $n = 4$ (i.e., 16 subdivisions in all).

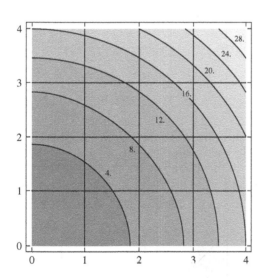

(b) Use the fact that $f(x, y) = x^2 + y^2$ to check your answer to part (a).

(c) Does your answer to part (a) underestimate the true value of I? Justify your answer.

Exercises 5–7 are about the situation described in Example 9.

5. Suppose the event is a free concert, the hall is full, the area in question is completely filled with seats, and each seat in the hall occupies a rectangle 1 meter by 0.8 meters. Find a

formula for $f(x, y)$ and calculate $I = \iint_R f(x, y)\, dA$. What does the answer mean in appropriate units?

6. Suppose the event is a rock concert in an open (seatless) area and that the crowd density decreases linearly from 3 people per square meter at the "front" of the hall (where $x = 0$) to 1 person per square meter at the "back" of the hall (where $x = 50$). Find a formula for $f(x, y)$; then estimate $I = \iint_R f(x, y)\, dA$ using a midpoint sum with 25 subdivisions (5 in each direction). What does the answer mean in appropriate units?

7. Suppose the event is a sold-out opera and that the hall has two classes of seats: high-priced seats in the "left half" of the hall (where $x \leq 25$) and low-priced seats in the "right" half of the hall (where $x \geq 25$). For their extra money, the higher-paying customers get larger seats, occupying 1.2 square meters; cheaper seats occupy only 0.8 square meters. Express the total number of people in the hall as the sum of two integrals over different rectangular domains. What is the total number of people?

8. (a) Let $g(x, y)$ be the depth (in inches) of snow at the point x miles east and y miles north of downtown Frostbite Falls, Minnesota. Let $R = [-50, 50] \times [-50, 50]$ and suppose that $\iint_R g(x, y)\, dA = 60{,}000$. What does the integral measure? What are the units?

(b) Let $g(x, y)$ and R be as in the preceding part. Evaluate $\dfrac{\iint_R g(x, y)\, dA}{\iint_R 1\, dA}$. What does the answer mean about snow?

9. Suppose that the function $f : \mathbb{R}^2 \to \mathbb{R}$ has the values shown in the following table.

y \ x	1	2	3	4
4	9	10	11	12
3	7	8	9	10
2	5	6	7	8
1	3	4	5	6

(a) Use a midpoint approximating sum with four subdivisions (two in each direction) to estimate $\iint_R f(x, y)\, dA$, where $R = [1, 5] \times [1, 5]$.

(b) Use a midpoint approximating sum with four subdivisions (two in each direction) to estimate $\iint_R f(x, y)\, dA$, where $R = [0, 4] \times [0, 4]$.

(c) The sum of all 16 table entries is 120. Use this fact to calculate an approximating sum with 16 subdivisions for the integral $\iint_R f(x, y)\, dA$, where $R = [0, 4] \times [0, 4]$. At which point in each subrectangle is f evaluated?

(d) The sum of all 16 table entries is 120. Use this fact to calculate a 16-term midpoint approximating sum for an appropriate integral $\iint_R f(x,y)\,dA$. (State the domain R and the value of the approximation.)

10. Suppose that the function $f : \mathbb{R}^2 \to \mathbb{R}$ has the values shown in the following table.

y \ x	1.5	2.0	2.5	3.0
3.0	4	6	9	6
2.5	6	9	7	5
2.0	4	8	6	4
1.5	3	5	5	7

(a) Use a midpoint approximating sum with four subdivisions (two in each direction) to estimate $\iint_R f(x,y)\,dA$, where $R = [1,3] \times [1,3]$.

(b) The sum of all 16 table entries is 94. Use this fact to calculate an approximating sum with 16 subdivisions for the integral $\iint_R f(x,y)\,dA$, where $R = [1,3] \times [1,3]$. At which point in each subrectangle is f evaluated?

(c) Use a midpoint approximating sum to estimate $\iint_R f(x,y)\,dA$, where $R = [2,3] \times [1,3]$. (Use as many table entries as possible.)

(d) The sum of all 16 table entries is 94. Use this fact to calculate a 16-term midpoint approximating sum for an appropriate integral $\iint_R f(x,y)\,dA$. (State the domain R and the value of the approximation.)

11. The figure below shows values of a function $f(x,y)$ at the midpoint of each of nine subdivisions of the rectangle $R = [0,3] \times [0,3]$. Use the data to estimate the integral $I = \iint_R f(x,y)\,dA$ and the average value of f on R.

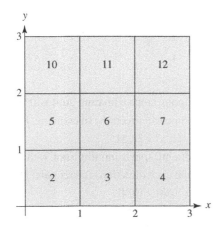

12. The figure below shows values of a function $g(x,y)$ at the midpoints of each of several (unequal) subdivisions of the rectangle $R = [0,3] \times [0,3]$. Use the data to estimate the integral $I = \iint_R g(x,y)\,dA$ and the average value of g on R.

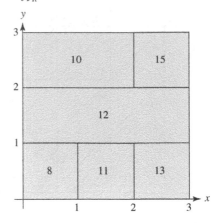

13. The figure below shows values of a function $h(x,y)$ at the midpoints of each of several (unequal) subdivisions of the rectangle $R = [0,3] \times [0,3]$. Use the data to estimate the integral $I = \iint_R h(x,y)\,dA$ and the average value of h on R.

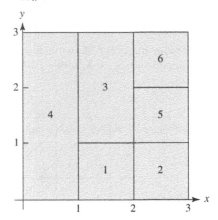

14. The figure below shows values of a function $k(x,y)$ at points inside several (nonrectangular) subdivisions of the rectangle $R = [0,3] \times [0,3]$. Use the data to estimate the integral $I = \iint_R k(x,y)\,dA$ and the average value of k on R.

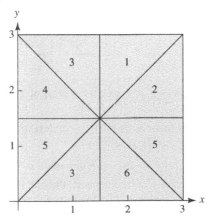

In Exercises 15–18, use geometry to evaluate $\iint_R 1\, dA$, where R is the region of the xy-plane enclosed by

15. the rectangle with vertices at $(-1, -2)$, $(-1, 1)$, $(3, 1)$, and $(3, -2)$.

16. the triangle with vertices at $(-2, 0)$, $(4, 0)$, and $(1, 3)$.

17. the polygon with vertices at $(1, -1)$, $(3, 1)$, $(5, -1)$, and $(3, -3)$.

18. the circle $x^2 + y^2 = 1$.

In Exercises 19–22, use geometry to evaluate $\iiint_R 1\, dV$, where

19. $R = [-2, 1] \times [3, 6] \times [-4, 5]$.

20. R is the region inside the cylinder $x^2 + y^2 = 4$ between $z = -3$ and $z = 5$.

21. R is the region inside the sphere $x^2 + y^2 + z^2 = 4$.

22. R is the region where $0 \le \sqrt{x^2 + y^2} \le z \le 4$.

23. Let R denote the region of the xy-plane enclosed by the circle $x^2 + y^2 = 1$. Use geometry to explain why $\iint_R \sqrt{1 - x^2 - y^2}\, dA = 2\pi/3$. [HINT: $x^2 + y^2 + z^2 = 1$ is the equation of a sphere in xyz-space.]

24. Let R denote the region of the xy-plane enclosed by the circle $x^2 + y^2 = 9$. Use geometry to evaluate $\iint_R \left(3 - \sqrt{x^2 + y^2}\right) dA$.

FURTHER EXERCISES

25. Let $I = \iint_R f(x, y)\, dA$, where $f(x, y) = x^2 + y$ and $R = [0, 1] \times [0, 2]$.

   (a) Estimate I by calculating a Riemann sum using four subrectangles and the value of f at the lower left corner of each subrectangle.

   (b) How does the estimate in part (a) compare with the exact value of I? Justify your answer.

26. Let $I = \iint_R f(x, y)\, dA$, where $f(x, y) = x + y^2$ and $R = [0, 3] \times [0, 6]$.

   (a) Estimate I by calculating a Riemann sum using nine subrectangles and the value of f at the upper right corner of each subrectangle.

   (b) How does the estimate in part (a) compare with the exact value of I? Justify your answer.

27. Let $I = \iint_R (x^2 + y)\, dA$, where $R = [0, 1] \times [0, 2]$.

   (a) Explain why $I \ge 0$.

   (b) Explain why $I \le 6$.

   (c) Estimate I by calculating a double midpoint sum with four subdivisions (two in each direction).

28. Explain why $4 \le \iint_R e^{x^2 + y^2}\, dA \le 30$, where $R = [-1, 1] \times [-1, 1]$.

29. Let $R = [1, 2] \times [3, 4]$. Explain why $\iint_R (x^2 + y^2)\, dA \le \iint_R (x^3 + y^3)\, dA$.

30. Let $R = [0, 1] \times [0, 1]$. Is it true that $\iint_R (x^2 + y^2)\, dA \le \iint_R (x^3 + y^3)\, dA$? Justify your answer.

31. Let $I = \iint_R xy^2\, dA$. Give an example of a rectangular region R for which $I < 0$.

32. Let $I = \iint_R f(x, y)\, dA$, where $R = [-2, 2] \times [-2, 2]$. Give an example of a nonconstant function f for which $I < 0$.

In Exercises 33–36, explain why the value of the double integral is zero. [HINT: Exploit the symmetry of the region of integration.]

33. $\iint_R (2x - y)\, dA$, where R is the disk of radius 3 centered at the origin.

34. $\iint_R x^3 \cos y\, dA$, where $R = [-1, 1] \times [-1, 1]$.

35. $\iint_R \sin(xy)\, dA$, where R is the triangle with vertices at $(-1, 0)$, $(1, 0)$, and $(0, 1)$.

36. $\iint_R e^{-x} y\, dA$, where R is the triangle with vertices at $(0, 2)$, $(0, -2)$, and $(2, 0)$.

37. Evaluate $\iint_R (2 + (y - x)^3)\, dA$, where R is the triangle with vertices at $(0, 1)$, $(1, 0)$, and $(1, 1)$.

38. Evaluate $\iint_R (1 + \sin(x + y))\, dA$, where R is the region with vertices at $(0, 1)$, $(1, 0)$, $(0, -1)$, and $(-1, 0)$.

In Exercises 39–44, $R_1 = [0, 4] \times [0, 4]$, $R_2 = [1, 3] \times [2, 3]$, and R_3 is the region inside R_1 and outside R_2.

39. Evaluate $\iint_{R_1} 1\, dA$.

40. Evaluate $\iint_{R_3} 1\, dA$.

41. Evaluate $\iint_{R_1} (5 - x)\, dA$.

42. Evaluate $\iint_{R_2} (1 + 2y)\, dA$.

43. Suppose that $\iint_{R_3} f(x, y)\, dA = 42$ and $\iint_{R_2} f(x, y)\, dA = 27$. Evaluate $\iint_{R_1} f(x, y)\, dA$.

44. Suppose that $\iint_{R_3} f(x, y)\, dA = 42$ and $\iint_{R_1} f(x, y)\, dA = 27$. Evaluate $\iint_{R_2} f(x, y)\, dA$.

In Exercises 45–48, R_1 is the disk of radius 5 with center at the origin, R_2 is the polygon with vertices at $(-1, 3)$, $(-3, 1)$, $(-1, -1)$, and $(1, 1)$, and R_3 is the region inside R_1 and outside R_2.

45. Suppose that the average value of the function f over the region R_1 is 13. Evaluate $\iint_{R_1} f(x, y)\, dA$.

46. Suppose that the average value of the function g over the region R_2 is 13. Evaluate $\iint_{R_2} g(x, y) \, dA$.

47. Suppose that the average value of the function h over the region R_3 is 13. Evaluate $\iint_{R_3} h(x, y) \, dA$.

48. Suppose that the average value of the function f over the region R_1 is 13 and that the average value of f over the region R_2 is 7. Find the average value of f over the region R_3.

49. Let R_1 denote the region $[0, 5] \times [-4, 4]$, R_2 the region $[0, 5] \times [0, 4]$, and R_3 the region $[-5, 0] \times [-4, 0]$. Suppose that $\iint_{R_2} f(x, y) \, dA = 10$, that $\iint_{R_3} f(x, y) \, dA = 24$, and that $f(x, y) = -f(-x, y)$. Evaluate $\iint_{R_1} f(x, y) \, dA$.

50. Explain why $\left| \iint_R f(x, y) \, dA \right| \le \iint_R |f(x, y)| \, dA$.

14.2 CALCULATING INTEGRALS BY ITERATION

The previous section was about *defining* multivariable integrals as limits of approximating sums. This section is about *calculating* multivariable integrals, using antidifferentiation. Approximating sums are conceptually simple and (with technology) easy to calculate. But approximating sums are only approximate; to evaluate integrals exactly, we would like an appropriate version of the single-variable antiderivative method ← suitably modified for multivariable use.

It works thanks to the fundamental theorem of calculus.

Iteration: how it works

In math-speak, "iterate" means "repeat."

The key idea is to integrate a multivariable function *one variable at a time*, treating other variables as constants. The process is called **iterated integration**. ← To start, here is an example to show *how* it works; we will see *why* it works in a moment.

Let's put it to rest at last.

Attaching variable names to the limits of integration is optional, but doing so can help remind us which variable is involved.

E X A M P L E 1   Let $f(x, y) = x + y$ and $R = [0, 4] \times [0, 4]$. Find $\iint_R f(x, y) \, dA$ by iterated integration. (We discussed this integral—repeatedly—in the preceding section.) ←

Solution   We integrate first in x, treating y as a constant. ← Because both x and y range from 0 to 4, these numbers appear repeatedly as limits of integration:

$$\iint_R f(x, y) \, dA = \int_{y=0}^{y=4} \left(\int_{x=0}^{x=4} (x + y) \, dx \right) dy$$

$$= \int_{y=0}^{y=4} \left(\frac{x^2}{2} + xy \right]_{x=0}^{x=4} \right) dy \qquad \text{(antidifferentiating in } x\text{)}$$

$$= \int_{y=0}^{y=4} (8 + 4y) \, dy \qquad \text{(plugging in endpoints)}$$

$$= 8y + 2y^2 \Big]_{y=0}^{y=4} = 64.$$

Observe some features of the calculation:

- **Work from inside out**   Iterated integrals are calculated from the inside out. In the preceding calculation, the "inner" integral—done first—involves x; the "outer" integral involves y. Observe also the nested parentheses in the *Maple* command below.

- **A familiar answer**   The final answer, 64, is familiar—we estimated it repeatedly in the preceding section.
- **The answer as volume**   The answer means that the solid having straight vertical walls and bounded below by $R = [0, 4] \times [0, 4]$ and above by the plane $z = x + y$ has a volume of 64 cubic units. Figure 1 shows the solid:

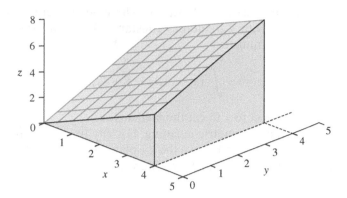

FIGURE 1

A solid bounded above by $z = x + y$

- **Checking answers with technology**   Is the answer 64 geometrically reasonable? Is it symbolically correct? Technology (such as *Maple* and *Mathematica*) can help with both questions. Figure 1 shows that the answer 64 is reasonable. As a further check, here is *Maple*'s version of the symbolic calculation above:

```
> int( int( x+y, x=0..4), y=0..4 );
                    64
```

- **Either order works**   There is nothing sacred about integrating first in x and then in y. For well-behaved functions we can integrate in either order—and both orders give the same answer. (We will return to this question below.)   ∎

Nested integral notation   As the preceding calculation suggests, double and triple integrals are usually calculated by integrating in one variable at a time. For this reason, such integrals are often written in "nested" form with the limits of integration indicating the domain of integration. If, say, the domain R is a rectangle $[a, b] \times [c, d]$, we will often write

$$\int_a^b \int_c^d f(x, y)\, dy\, dx \quad \text{or} \quad \int_c^d \int_a^b f(x, y)\, dx\, dy, \quad \text{not} \quad \iint_R f(x, y)\, dA.$$

Notice especially the difference between the first two forms. Working "from inside out" means that in the first form we integrate first in y and then in x; in the second form we reverse the order of integration. ➡

We'll see soon that both orders give the same result!

For a *triple* integral $\iiint_R g(x, y, z)\, dV$ over a parallelepiped $R = [a, b] \times [c, d] \times [e, f]$ in xyz-space we might write, say,

$$\int_a^b \int_c^d \int_e^f g(x, y, z)\, dz\, dy\, dx \qquad \text{or} \qquad \int_c^d \int_e^f \int_a^b g(x, y, z)\, dx\, dz\, dy,$$

using whichever order of integration is more convenient.

Cross-sectional area: an intermediate function The inner integral in Example 1 was calculated with respect to x with y treated as constant. The result was an intermediate function, g, of y alone; the formula is

$$g(y) = \int_0^4 (x + y)\, dx = 8 + 4y.$$

We integrated $g(y)$ with respect to y to get the final answer.

The function g has a nice geometric meaning: For any fixed y_0 in $[0, 4]$, $g(y_0)$ is the area under the curve $z = f(x, y_0)$ and above the xy-plane. In other words, $g(y)$ is the *area* of the cross section of the solid obtained by slicing with the plane $y = y_0$. Figure 2 shows the picture for $y_0 = 2$:

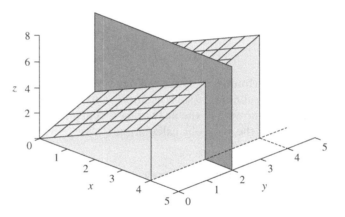

FIGURE 2

A cross section with the plane $y = 2$

Since $g(y) = 8 + 4y$, we get $g(2) = 16$; this is the area of the part of the plane inside the solid. As y runs from $y = 0$ to $y = 4$, $g(y)$ measures the area "swept out" by planes parallel to the one shown.

This area increases with y, as Figure 2 shows.

In fact, the notion of a cross-sectional area function is not a new idea. We took the same approach in single-variable calculus when calculating volumes of simple solids by ordinary integration. Back then we put it like this:

FACT Suppose that a solid lies with its base on the xy-plane, between the vertical planes $x = a$ and $x = b$. For all x in $[a, b]$, let $A(x)$ denote the area of the cross section at x, perpendicular to the x-axis. If $A(x)$ is a continuous function, then

$$\text{volume} = \int_a^b A(x)\, dx.$$

Why the method works

Why does the iteration method—integrate in one variable and then the other—work? The preceding Fact offers some geometric insight—at least for integrals that can be thought of as measuring volumes.

But not every integral can or should be thought of in this way. A better reason why iteration works—a reason that makes sense in *any* dimension—is based on approximating sums. We will describe the idea in two dimensions, but everything transfers readily to three (and higher) dimensions. The main idea is that an approximating sum for a double integral can be grouped either along "rows" or along "columns." We explain first with an example and then give a more general argument.

EXAMPLE 2   Let $f(x, y) = 2 - x^2 y$, $R = [0, 1] \times [0, 1]$ and $I = \iint_R f(x, y)\, dA$. Find the exact value of I by iterated integration. Then estimate I using a midpoint-rule approximating sum S_{100} with 100 equal subdivisions (10 in each direction).

Solution   Figure 3 shows the solid whose volume is given by I:

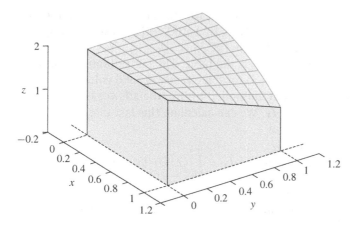

FIGURE 3
The solid under $z = 2 - x^2 y$ over $[0, 1] \times [0, 1]$

It is easy to calculate I *exactly*, by iteration. This time we integrate first in y: ➤   *Check each easy step.*

$$I = \int_{x=0}^{y=1} \left(\int_{y=0}^{y=1} (2 - x^2 y)\, dy \right) dx \qquad \text{(the inner integral involves } y\text{)}$$

$$= \int_{x=0}^{x=1} \left(2y - \frac{x^2 y^2}{2} \Big]_{y=0}^{y=1} \right) dx \qquad \text{(by the fundamental theorem)}$$

$$= \int_0^1 \left(2 - \frac{x^2}{2} \right) dx = \frac{11}{6}.$$

Finding the sum S_{100} takes a little more work. First we tabulate values of f at the midpoints of all 100 subrectangles of $[0, 1] \times [0, 1]$. For later use, column sums are at the bottom: ➤

Table entries and column sums are rounded for brevity.

Sample values of $f(x, y) = 2 - x^2 y$										
$\frac{x}{y}$	**0.05**	**0.15**	**0.25**	**0.35**	**0.45**	**0.55**	**0.65**	**0.75**	**0.85**	**0.95**
0.95	2.00	1.98	1.94	1.88	1.81	1.71	1.60	1.47	1.31	1.14
0.85	2.00	1.98	1.95	1.90	1.83	1.74	1.64	1.52	1.39	1.23
0.75	2.00	1.98	1.95	1.91	1.85	1.77	1.68	1.58	1.46	1.32
0.65	2.00	1.99	1.96	1.92	1.87	1.80	1.73	1.63	1.53	1.41
0.55	2.00	1.99	1.97	1.93	1.89	1.83	1.77	1.69	1.60	1.50
0.45	2.00	1.99	1.97	1.94	1.91	1.86	1.81	1.75	1.68	1.59
0.35	2.00	1.99	1.98	1.96	1.93	1.89	1.85	1.80	1.75	1.68
0.25	2.00	1.99	1.98	1.97	1.95	1.92	1.89	1.86	1.82	1.77
0.15	2.00	2.00	1.99	1.98	1.97	1.95	1.94	1.92	1.89	1.87
0.05	2.00	2.00	2.00	1.99	1.99	1.98	1.98	1.97	1.96	1.95
sum	20.0	19.9	19.7	19.3	19.0	18.4	17.9	17.2	16.3	15.4

Here, therefore, the approximating sum is the average of all 100 table entries.

Because each subrectangle has area 1/100, the sum S_{100} can be found by adding all 100 function values and multiplying by 1/100. ◀ The result turns out to be 1.83—not far from the exact answer, $11/6 \approx 1.8333$.

Now consider any *column* in the table. The last column, for instance, contains numbers of the form $f(0.95, y)$ for ten equally spaced values of y, and $\Delta y = 0.1$. Therefore, the column sum multiplied by 0.1 (the answer is 1.54) is a Riemann sum for the integral $\int_0^1 f(0.95, y)\,dy$. We can calculate this last integral directly using the symbolic rule for f. We get

$$\int_0^1 f(0.95, y)\,dy = \int_0^1 (2 - 0.95^2 y)\,dy = 1.54875,$$

which is not far from the Riemann sum. Tabulating results for the other columns shows the same pattern. The Riemann sum associated to each column closely approximates the corresponding y-integral: ◀

Integrals are rounded to two decimals.

Column sums and y-integrals										
x	**0.05**	**0.15**	**0.25**	**0.35**	**0.45**	**0.55**	**0.65**	**0.75**	**0.85**	**0.95**
column sum	20.0	19.9	19.7	19.3	19.0	18.4	17.9	17.2	16.3	15.4
$\int_0^1 f(x, y)\,dy$	2.00	1.99	1.97	1.94	1.90	1.85	1.79	1.72	1.64	1.55

The calculations are tedious, but the moral is clear: Whether adding up approximating sums or calculating integrals, we can work one variable at a time. ■

A general argument Let's summarize the ideas of Example 2 to convince ourselves that the integral $I = \iint_R f(x, y)\,dA$ can reasonably be calculated by iteration. To do so, suppose we are given a rectangle $R = [a, b] \times [c, d]$ in the xy-plane, and a function f defined on R.

To begin we write an approximating sum for I. First we subdivide both $[a, b]$ and $[c, d]$ into n equal subintervals with respective lengths $\Delta x = (b - a)/n$ and $\Delta y = (d - c)/n$. This produces a grid of n^2 subrectangles R_{ij}, where $1 \leq i, j \leq n$; each rectangle has area $\Delta x \, \Delta y$. Let (x_i, y_j) be the midpoint of the (i, j)th rectangle; these midpoints will serve as sampling points for an approximating sum S_{n^2} for I. Figure 4 shows the idea:

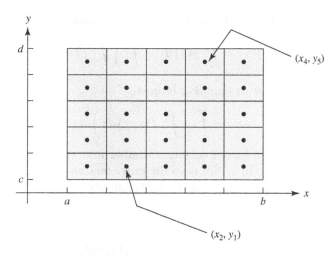

Sampling points for an approximating sum

With all ingredients in place we can write out the approximating sum: →

There are n^2 summands in all.

$$S_{n^2} = \sum_{i,j=1}^{n} f(x_i, y_j)\, \Delta x\, \Delta y \qquad \text{(both } i \text{ and } j \text{ run from 1 to } n)$$

$$= f(x_1, y_1)\, \Delta x\, \Delta y + \cdots + f(x_n, y_n)\, \Delta x\, \Delta y.$$

We can group and add the summands in any order or pattern we like. Here is one convenient pattern: →

It's OK to factor Δy out of each row because Δy is common to all summands.

$$S_{n^2} = (\ f(x_1, y_1)\Delta x + f(x_2, y_1)\Delta x + \cdots + f(x_n, y_1)\Delta x\ )\ \Delta y$$
$$+(\ f(x_1, y_2)\Delta x + f(x_2, y_2)\Delta x + \cdots + f(x_n, y_2)\Delta x\ )\ \Delta y$$
$$+\cdots$$
$$+(\ f(x_1, y_j)\Delta x + f(x_2, y_j)\Delta x + \cdots + f(x_n, y_j)\Delta x\ )\ \Delta y$$
$$+\cdots$$
$$+(\ f(x_1, y_n)\Delta x + f(x_2, y_n)\Delta x + \cdots + f(x_n, y_n)\Delta x\ )\ \Delta y.$$

Here is the first of two key points:

> *The sum inside parentheses on each line above is a Riemann sum with n subdivisions for a* single-variable *integral in x.*

Specifically, the sum on the first line is a Riemann sum for the integral $\int_a^b f(x, y_1)\, dx$; the sum on the second line approximates $\int_a^b f(x, y_2)\, dx$, and so on. Now, if n is large, then all of these sums are close to the integrals they approximate. Thus, for large n we have

$$S_{n^2} \approx \left(\int_a^b f(x, y_1)\, dx\right) \Delta y + \left(\int_a^b f(x, y_2)\, dx\right) \Delta y + \cdots + \left(\int_a^b f(x, y_n)\, dx\right) \Delta y.$$

Here is the second key observation:

> *The sum on the right above is a Riemann sum with n subdivisions for the integral $\int_c^d g(y)\, dy$, where $g(y) = \int_a^b f(x, y)\, dx$ and $g(y)$ is sampled at the n points y_1, y_2, $\ldots, y_n$.*

This should sound reasonable, and it is indeed true for continuous integrands. But a rigorous proof depends on technical properties of the integrand function.

For large n, this sum, too, is near the integral it approximates. ← In other words,

$$S_{n^2} \approx \sum_{j=1}^{n} g(y)\,\Delta y \approx \int_c^d g(y)\,dy = \int_c^d \left(\int_a^b f(x, y)\,dx \right) dy.$$

The desired conclusion is now in sight: For large n,

$$I \approx S_{n^2} \approx \int_c^d \left(\int_a^b f(x, y)\,dx \right) dy.$$

We conclude that we can, indeed, evaluate a double integral I over a rectangle R by integrating first in x and then in y. Nor does the *order* matter—we could just as well have reversed the roles of x and y throughout the argument above.

Iterated integrals in three variables

Iteration works for a triple integral defined on a rectangular parallelepiped in xyz-space in essentially the same way as just seen in two dimensions.

Check each step.

EXAMPLE 3   Let $f(x, y, z) = x + y + z$, $R = [0, 2] \times [0, 2] \times [0, 2]$, and $I = \iiint_R f(x, y, z)\,dV$. Calculate I exactly, by iteration. (Compare Example 4, page 782, where we calculated a crude approximating sum for I.) What does the answer mean?

Solution   We integrate in each of the three variables in turn: ←

$$\iiint_R f(x, y, z)\,dV = \int_0^2 \left(\int_0^2 \left(\int_0^2 (x + y + z)\,dx \right) dy \right) dz \quad \text{(nesting parentheses)}$$

$$= \int_0^2 \left(\int_0^2 \left(\frac{x^2}{2} + xy + xz \right]_0^2 \right) dy \right) dz \quad \text{(antidifferentiate in } x\text{)}$$

$$= \int_0^2 \left(\int_0^2 (2 + 2y + 2z)\,dy \right) dz \quad \text{(plug in endpoints)}$$

$$= \int_0^2 \left(2y + y^2 + 2yz \right]_0^2 \right) dz \quad \text{(antidifferentiate in } y\text{)}$$

$$= \int_0^2 (8 + 4z)\,dz \quad \text{(an ordinary integral, in } z\text{)}$$

$$= 24.$$

Here the average value happens to be $f(1, 1, 1)$, the value of f at the "center" of the cube.

What the answer means depends on our point of view. If we think of the integrand $f(x, y, z)$ as the density (in, say, grams per cubic centimeter) of the solid R at the point (x, y, z), then the integral tells the *mass* of the solid (in grams). If, instead, we think in terms of average value over a cube with volume 8, then the answer means that $f(x, y, z)$ has average value $24/8 = 3$ on the cube. ←

BASIC EXERCISES

In Exercises 1–20, use iterated integration to evaluate the integral.

1. $\iint_R \sin(x)\sin(y)\,dA$; $R = [0, \pi/2] \times [0, \pi/2]$.

2. $\iint_R \sin(x + y)\,dA$; $R = [-\pi/2, \pi/2] \times [0, \pi]$.

3. $\iint_R (x^2 + y^2)\,dA$; $R = [0, 4] \times [0, 4]$.

4. $\iint_R (3x^2 y + 4xy^2)\,dA$; $R = [-1, 1] \times [-2, 3]$.

5. $\iint_R \sqrt{x + y + 1}\,dA$; $R = [0, 1] \times [0, 1]$.

6. $\iint_R \cos(x + y)\,dA$; $R = [0, \pi/2] \times [0, \pi/2]$.

7. $\iint_R \dfrac{1}{x+y}\, dA;\ R = [1, 2] \times [2, 3].$

8. $\iint_R \dfrac{x}{1+y}\, dA;\ R = [0, 1] \times [0, 1].$

9. $\iint_R e^{x+2y}\, dA;\ R = [1, 2] \times [2, 3].$

10. $\iint_R xe^{xy}\, dA;\ R = [0, 1] \times [0, 1].$

11. $\iint_R (1 + 2x + 3y)^4\, dA;\ R = [-1, 1] \times [0, 1].$

12. $\iint_R \dfrac{1}{(1+x+y)^2}\, dA;\ R = [0, 1] \times [0, 2].$

13. $\iint_R y\sin(xy)\, dA;\ R = [0, 1] \times [0, \pi].$

14. $\iint_R \dfrac{1}{xy}\, dA;\ R = [1, 10] \times [10, 100].$

15. $\iiint_R x\, dV;\ V = [0, 1] \times [0, 2] \times [0, 3].$

16. $\iiint_R y\, dV;\ V = [0, 1] \times [0, 2] \times [0, 3].$

17. $\iiint_R xyz\, dV;\ R = [0, 3] \times [0, 2] \times [0, 1].$

18. $\iiint_R (xy + yz)\, dV;\ R = [1, 2] \times [2, 3] \times [3, 4].$

19. $\iiint_R (x^2 y + yz)\, dV;\ R = [0, 1] \times [2, 4] \times [1, 3].$

20. $\iiint_R xye^{y+z}\, dV;\ R = [0, 2] \times [0, 1] \times [0, 3].$

FURTHER EXERCISES

In Exercises 21–26, $R = [a, b] \times [c, d]$, and $f(x)$ and $g(y)$ are functions such that $\displaystyle\int_a^b f(x)\, dx = 29$ and $\displaystyle\int_c^d g(y)\, dy = 37$. Use this information to evaluate each integral.

21. $\iint_R f(x)g(y)\, dA$

22. $\iint_R (f(x) + g(y))\, dA$

23. $\iint_R (yf(x) + xg(y))\, dA$

24. $\iint_R (x + g(y))\, dA$

25. $\iiint_T (f(x) - g(y))\, dV$, where $T = [a, b] \times [c, d] \times [-1, 2].$

26. $\iiint_T (1 + f(x)g(y))\, dV$, where $T = [a, b] \times [c, d] \times [-3, 2].$

27. Give a geometric interpretation of $\displaystyle\int_1^4 f(x, y)\, dx.$

28. Suppose that a and b are real numbers such that $a < b$. Give a geometric interpretation of $\displaystyle\int_a^b f(x, y)\, dy.$

29. Let $R = [a, b] \times [c, d]$ and suppose that $f(x, y) = g(x)h(y)$. Use the Riemann sum definition of the integral to explain why

$$\iint_R f(x, y)\, dA = \left(\int_a^b g(x)\, dx\right) \cdot \left(\int_c^d (y)\, dy\right).$$

14.3 INTEGRALS OVER NONRECTANGULAR REGIONS

We have seen in preceding sections how to define and calculate integrals of the form $I = \iint_R f(x, y)\, dA$, where $f(x, y)$ is a well-behaved function of two variables and $R = [a, b] \times [c, d]$ is a rectangle. In this case we can calculate I by integrating first in one variable and then in the other—either order of integration will do: ➥

$$I = \iint_R f(x, y)\, dA = \int_a^b \left(\int_c^d f(x, y)\, dy\right) dx = \int_c^d \left(\int_a^b f(x, y)\, dx\right) dy.$$

We'll usually omit parentheses in iterated integrals; note that they are evaluated "from inside out."

(Similar formulas hold for triple integrals $\iiint_R f(x, y, z)\, dV$, where R is a three-dimensional, rectangular solid domain.)

Integrals over rectangular domains are relatively easy to evaluate. In practice, however, we often want to integrate functions over nonrectangular plane domains bounded by curves rather than by straight lines.

As we will see, many integrals over nonrectangular domains can be calculated by iteration—first with respect to one variable and then with respect to the other—much as for integrals over rectangular domains. But some extra care and "setup" is needed.

Why nonrectangular domains matter: Examples

Integrals over nonrectangular domains arise naturally and often. In Section 14.1, for example, we described several applications and interpretations of multiple integrals: Volume, area, mass, average value, and even crowd sizes can all be calculated or estimated using integrals over appropriate function domains. In practice, of course, the domains in question need not be rectangular. ← We illustrate uses of integrals over nonrectangular domains with two types of examples.

The real value of calculus lies in its ability to handle a wide variety of functions and domains.

EXAMPLE 1 Imagine the region shown in Figure 1 as a thin, flat plate cut from cardboard or some other stiff material of constant thickness. The top edges are straight lines, and the bottom curve is the parabola $y = x^2$. Where is the region's **center of mass**—the point at which the cutout would balance on the point of a pin? ←

If the lamina were hung from a thread fastened at the center of mass, the lamina would hang horizontally.

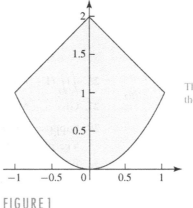

The lower edge is a parabola; the upper edges are the lines $y = 2 + x$ and $y = 2 - x$.

FIGURE 1
A simple plane lamina

Solution It is clear from the region's left–right symmetry that the center of mass lies somewhere along the y-axis. But at what y-value? At the "midpoint," $y = 1$? Or is the value somewhere below $y = 1$ because more of the region lies below than above $y = 1$? We will answer these questions later in this section, using integrals. ←

But guess an answer now.

Center of mass: the simplest case A thin plate like the one in Figure 1 is called a **plane lamina** in mathematical jargon; the adjective "plane" applies because, for present purposes, the figure is essentially two-dimensional. ← If the plate is cut from a homogeneous material, such as cardboard or plastic of constant thickness and density, then the center of mass depends only on the geometry of the lamina. In this simplest case, the **center of mass** of a region R in the xy-plane is defined as the point $(\bar{x}, \bar{y})$, where

In effect, we ignore the (small) thickness dimension.

$$\bar{x} = \frac{\iint_R x\,dA}{\iint_R 1\,dA}; \qquad \bar{y} = \frac{\iint_R y\,dA}{\iint_R 1\,dA}. \tag{1}$$

(Note that $\bar{x}$ and $\bar{y}$ are simply the *average values* of x and y over the region R.) These equations determine the "balance point" described in Example 1. ← We will use these equations to find the desired balance point exactly—once we know how to calculate integrals over nonrectangular regions.

The reasons for this physical property are physical, not mathematical.

We can avoid integration, admittedly, in a few very special, highly symmetric cases. For a postcard or a compact disk, for instance, it seems obvious (and it is true) that the center of mass is also the geometric center. But finding centers of mass in most other cases requires integration—usually over nonrectangular domains.

Center of mass: more general cases In this section we consider only the simple plane laminas described just above. We remark here, however, that an important variant of the same problem concerns plane laminas whose density varies from place to place in the region R. This occurs if, for instance, the lamina has variable thickness or is made, say, partly from gold and partly from lead. This view is also useful in modeling other situations such as the (uneven) distribution of population across a country. In this case the center of mass is defined much as in Equations 1:

$$\bar{x} = \frac{\iint_R x\,\rho(x,y)\,dA}{\iint_R \rho(x,y)\,dA}; \qquad \bar{y} = \frac{\iint_R y\,\rho(x,y)\,dA}{\iint_R \rho(x,y)\,dA}, \tag{2}$$

where $\rho(x,y)$ is the density, in appropriate units, at the point (x,y). ➧ (The Greek letter ρ is pronounced "row.") If $\rho(x,y)$ is not too complicated, then the integrals in Equations 2 are not much harder to calculate than those in Equations 1.

A typical unit of density is grams per square centimeter

Iterated integration over nonrectangular domains

In the real-world setting of Example 5, we had no explicit formula either for the rainfall function or for the curve that defines the island's boundary. In more "mathematical" settings, like that of Example 1, the ingredients of an integral $\iint_R f(x,y)\,dA$ are indeed given symbolically—both the integrand f and the domain R are described by formulas or equations in x and y. Formulas for functions are old, familiar friends; the live question is to describe the domain R symbolically—and to use that description to calculate integrals exactly.

What types of domains? Domains in the xy-plane may have extremely complicated shapes, as a look at any world atlas will attest. ➧ For many purposes, however, it is enough to consider nonrectangular domains of two main types:

Imagine an integral over, say, the fjord-fringed nation of Norway.

- bounded above and below by curves and on the left and right by straight lines;
- bounded on the left and right by curves and on the top and bottom by straight lines.

Figure 2 shows examples of both kinds.

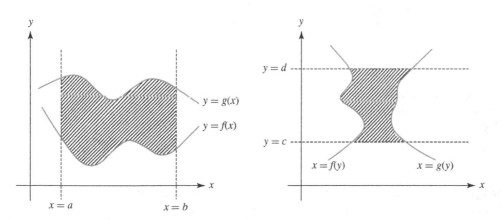

FIGURE 2
Two basic regions for integration

Iterated integrals over basic regions Integrals over regions like those just described can be calculated by iteration in much the same way as for integrals over rectangles. We illustrate with examples.

EXAMPLE 2 Find $I = \iint_R (2-x)\,dA$, where R is the triangular plane region between the curves $y = 0$ and $y = x$, and $x = 1$.

Solution The integral gives the volume of the solid in Figure 3(a); Figure 3(b) shows the domain of integration:

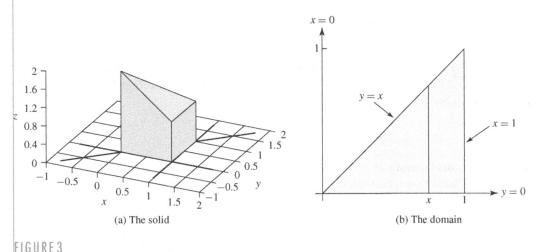

(a) The solid

(b) The domain

FIGURE 3
Integration over a nonrectangular domain

The domain R is basic in the sense just described: It is bounded by vertical lines $x = 0$ and $x = 1$ on the left and right and by the curves $y = 0$ and $y = x$ on the bottom and top. As with integrals over rectangles, we can integrate this one by iteration. The *outer* integral in x runs from $x = 0$ to $x = 1$. The upper and lower limits of the *inner* integral, however, depend on x: For a given x, the variable y runs from from $y = 0$ to $y = x$. Observe how the limits of integration in the following calculation describe R:

$$I = \iint_R (2-x)\,dA = \int_{x=0}^{x=1} \int_{y=0}^{y=x} (2-x)\,dy\,dx$$

$$= \int_{x=0}^{x=1} \left(2y - xy\right]_{y=0}^{y=x}\right) dx = \int_{x=0}^{x=1} \left(2x - x^2\right) dx = \frac{2}{3}.$$

Notice the main new feature of this calculation: The inner integral has *variable* limits of integration, signifying that the domain of integration has varying vertical "heights" that depend on x.

Changing the order of integration We said in Section 14.2 that, in calculating integrals over rectangles, we could integrate either first in x or first in y. The same property holds for integrals over nonrectangular domains—although, as determined by the shape of the domain, one order of integration may be easier or more convenient than the other. For regions bounded by a few lines or familiar curves, either order of integration is often equally feasible. We illustrate the process by redoing Example 2.

EXAMPLE 3 Calculate—again—the integral $I = \iint_R (2 - x)\, dA$ of Example 2, but now integrate first in x and then in y.

Solution This time we think of R as bounded below and above by the horizontal lines $y = 0$ and $y = 1$. For given y between 0 and 1, the region R begins at the left boundary line $x = y$ and ends at the right boundary line $x = 1$. These formulas lead to the following integral:

$$I = \iint_R (2 - x)\, dA = \int_{y=0}^{y=1} \int_{x=y}^{x=1} (2 - x)\, dx\, dy$$

$$= \int_{y=0}^{y=1} \left(2x - \frac{x^2}{2} \right]_{x=y}^{x=1} \right) dy$$

$$= \int_{y=0}^{y=1} \left(\frac{3}{2} - 2y + \frac{y^2}{2} \right) dy = \frac{2}{3},$$

which is the answer we found above in Example 2. ∎

Not always so simple Matters are not always as simple as Example 3 might suggest. As determined by the domain of integration, the integrand, or both, one order of integration may permit much easier calculations than another. ➡

The exercises give some examples.

Three-dimensional domains of integration can also be more challenging—both to visualize and to describe symbolically. We will return to three-dimensional integrals and develop some additional techniques for handling them, later in this chapter.

General formulas Example 2 illustrates how to integrate by iteration over *any* region R of either type illustrated in Figure 2. If $h(x, y)$ is any function defined on such a region, then the integral $\iint_R h(x, y)\, dA$ can be written in one of the following forms:

$$\int_{x=a}^{x=b} \int_{y=f(x)}^{y=g(x)} h(x, y)\, dy\, dx; \qquad \int_{y=c}^{y=d} \int_{x=f(y)}^{x=g(y)} h(x, y)\, dx\, dy.$$

The challenging part of evaluating such integrals is often in the setup, not the calculation itself. ➡

If technology is available, the calculation may be completely trivial.

Beyond the basics: more general regions Not every domain of interest has either of the basic shapes in Figure 2. The plane lamina in Figure 1, for example, has one curve on the bottom but *two* straight lines (with equations $y = 2 + x$ and $y = 2 - x$) on top. This small difficulty is easy to get around. The left and right halves of the lamina (we will call them R_1 and R_2, respectively) are basic in the sense at hand; we just integrate over each half separately and add the results.

EXAMPLE 4 Use the tools just developed to find the center of mass of the lamina in Example 1.

Solution Note that R_1 and R_2 correspond, respectively, to the x-intervals $[-1, 0]$ and

[0, 1]. Hence, the needed integrals are set up as follows:

$$\iint_R y\, dA = \iint_{R_1} y\, dA + \iint_{R_2} y\, dA$$

$$= \int_{x=-1}^{x=0} \int_{y=x^2}^{y=2+x} y\, dy\, dx + \int_{x=0}^{x=1} \int_{y=x^2}^{y=2-x} y\, dy\, dx;$$

$$\iint_R 1\, dA = \int_{x=-1}^{x=0} \int_{y=x^2}^{y=2+x} 1\, dy\, dx + \int_{x=0}^{x=1} \int_{y=x^2}^{y=2-x} 1\, dy\, dx.$$

The calculations themselves are now routine exercises. The results are

$$\iint_{R_1} y\, dA = \iint_{R_2} y\, dA = \frac{16}{15}; \qquad \iint_{R_2} 1\, dA = \iint_{R_2} 1\, dA = \frac{7}{6}.$$

For other integrands, the left- and right-hand integrals might be different.

(The left- and right-hand integrals are equal in this case because of the symmetry of the region and of the integrand.) ← These results and Equations 1 now give

$$\bar{y} = \frac{\iint_R y\, dA}{\iint_R 1\, dA} = \frac{32/15}{14/6} = \frac{32}{35}.$$

An even easier calculation (or physical intuition) shows that $\bar{x} = 0$. Thus, the center of mass is at $(0, 32/35) \approx (0, 0.91)$, which is a little *below* the half-height $y = 1$. ■

Estimating nonrectangular integrals

Explicit formulas are often unavailable for integrals over nonrectangular plane regions, especially in real-world applications. Such cases call for estimates—both of the integrand function and of areas in the domain of integration.

EXAMPLE 5 A research meteorologist measures rainfall depths on a small island using rain gauges placed at the centers of 1-km squares. Figure 4 shows rainfall readings, in centimeters, following a brief thunderstorm. (The black dot represents a lighthouse just offshore.)

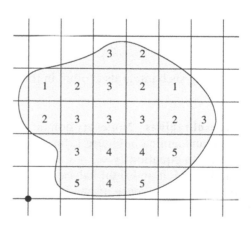

FIGURE 4
Rainfall over an island

Under these rustic conditions we can only estimate.

What total volume of water did the storm drop on the island? What was the average rainfall depth on the island? ←

Solution The answers can be understood using integrals over an island-shaped region in the plane. If $r(x, y)$ represents the rainfall depth at the point x km east and y km north of the black-dot lighthouse, then the total rainfall over the island is given, in appropriate units of volume, by the integral

$$\text{total volume} = \iint_R r(x, y)\, dA,$$

where R, the region of integration, is the island itself.

Without explicit formulas, *exact* calculation is impossible. But the picture suggests a natural estimation strategy: Let the gridlines determine a partition of R, and let the rainfall numbers represent samples of the function r within each subdivision. These data determine an approximating sum for the integral in question:

$$r(P_1)\Delta A_1 + r(P_2)\Delta A_2 + \cdots + r(P_n)\Delta A_n = \sum_{i=1}^{n} r(P_i)\Delta A_i,$$

where P_i is the center point, ΔA_i is the area of the ith subdivision, and n is the total number of subdivisions. In this case most of the areas ΔA_i are 1 km^2. To handle the other subdivisions, we simply estimate the area, the rainfall depth, or both. In the southwest corner, for instance, we might estimate the rain depth as 4 cm and the shaded area as 0.1 km^2. Thus, calculating the approximating sum (starting at lower left and working across rows) gives something like the following: ➡

The total value depends on our particular estimates.

$$4 \cdot 0.1 + 5 \cdot 0.7 + 4 \cdot 0.8 + 5 \cdot 0.6 + \cdots + 3 \cdot 0.7 + 2 \cdot 0.6 + 2 \cdot 0.2 = 56.8.$$

Thus, our numerical estimate is $\iint_R r(x, y)\, dA \approx 56.8$. Because the units of area are square kilometers and units of depth are centimeters, the (rather peculiar) units of our integral estimate are cm-km^2, and a little calculator work shows that our total rainfall estimate reduces to $56.8 \times 10^4 = 568{,}000$ cubic meters.

The average rainfall depth over the island is simply the average value of the rainfall function $r(x, y)$ over the region R. As discussed in Section 14.1, average value is also calculated as an integral by

$$\text{average value} = \frac{\iint_R f(x, y)\, dA}{A(R)},$$

where $A(R)$ is the total area of the region of integration. Here, an eyeball estimate suggests $A(R) \approx 20$, and so

$$\text{average rainfall} \approx \frac{56.8}{20} = 2.84\,\text{cm},$$

which looks consistent with the general size of numbers in Figure 4. ■

BASIC EXERCISES

1. Show by integration that the center of mass of the lamina in Example 1 lies on the y-axis (i.e., $\bar{x} = 0$).

2. Find the center of mass of the right half R_2 of the lamina in Example 1.

3. Find the center of mass of the lower part (below $y = 1$) of the lamina in Example 1.

4. Show by integration that the center of mass of a plane rectangle $[a, b] \times [c, d]$ is the center point $\left(\dfrac{a+b}{2}, \dfrac{c+d}{2}\right)$.

5. Let $f(x, y) = x + y$ and let R be the triangle with vertices $(0, 0), (6, 0),$ and $(6, 6)$. The following figure shows two ways of cutting this triangle into smaller subdivisions. All of the black dots have integer coordinates.

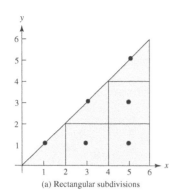

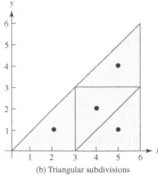

(a) Rectangular subdivisions   (b) Triangular subdivisions

(a) Use the indicated subdivisions in part (a) of the figure to calculate an approximating sum for $\iint_R f(x, y)\, dA$; evaluate the integrand at the black dot in each subdivision.

(b) Use the indicated subdivisions in part (b) of the figure to calculate an approximating sum for $\iint_R f(x, y)\, dA$; evaluate the integrand at the black dot in each subdivision.

(c) Calculate the integral $\iint_R f(x, y)\, dA$ exactly as an iterated integral.

6. Repeat Exercise 5 with $f(x, y) = y^2$.

7. In the situation of Example 5, estimate the total volume of rain that fell on the eastern half of the island (to the right of the middle vertical grid line).

8. In the situation of Example 5, estimate the average depth of rainfall on the eastern half of the island (to the right of the middle vertical grid line).

In Exercises 9–12, calculate the nonrectangular integral using an iterated integral in which the inner integral is with respect to y and the outer integral is with respect to x.

9. $\iint_R xy\, dA$; R is the region bounded by the curves $y = x^2$ and $y = 1$.

10. $\iint_R (x + y)\, dA$; R is the region bounded by the curves $y = x$ and $y = x^2$.

11. $\iint_R x\, dA$; R is the region bounded by the curves $y = x^2$ and $y = \sqrt{x}$.

12. $\iint_R 1\, dA$; R is the region bounded by the curves $y = \ln x$, $y = 0$, and $x = e$.

13–16. Evaluate the double integrals in Exercises 9–12 using iterated integrals in which the inner integral is with respect to x and the outer integral is with respect to y.

17. Let $f(x, y) = x$, and let R be the plane region bounded by the curves $y = e^x$, $y = 0$, $x = 0$, and $x = 1$.
   (a) Calculate $I = \iint_R f(x, y)\, dA$ by integrating first in y and then in x.
   (b) Calculate $I = \iint_R f(x, y)\, dA$ by integrating first in x and then in y. [HINT: First split the region R into two simpler pieces; each simpler piece should be bounded on the left by one curve and on the right by another.]

18. Let $f(x, y) = xy$, and let R be the plane region bounded by the curves $y^2 = 2x + 6$ and $y = x - 1$.
   (a) Calculate $I = \iint_R f(x, y)\, dA$ by integrating first in x and then in y.
   (b) Calculate $I = \iint_R f(x, y)\, dA$ by integrating first in y and then in x.

19. Find the volume of the region under the surface $z = x^2 + y^2$ and above the region in the xy-plane bounded by the curves $y = x^2$ and $x = y^2$.

20. Find the volume of the region under the surface $z = xe^y$ and above the triangle with vertices at $(0, 0)$, $(2, 2)$, and $(4, 0)$.

FURTHER EXERCISES

21. Let $y = f(x)$ be a function, with $f(x) \geq 0$ if $a \leq x \leq b$; let R be the plane region bounded by the curves $y = f(x)$, $y = 0$, $x = a$, and $x = b$.
   (a) What does single-variable calculus say about the area of R?
   (b) We have claimed that the double integral $I = \iint_R 1\, dA$ gives the area of R. Use an iterated integral to reconcile this formula with the one in part (a).

22. Let $x = g(y)$ be a function with $g(y) \geq 0$ if $c \leq y \leq d$; let R be the plane region bounded by the curves $x = g(y)$, $x = 0$, $y = c$, and $y = d$.
   (a) What does single-variable calculus say about the area of R?
   (b) We have claimed that the double integral $I = \iint_R 1\, dA$ gives the area of R. Use an iterated integral to reconcile this formula with the one in part (a).

23. Let R be the region under the graph of $y = 2 + \cos x$ between $x = 1$ and $x = 3$.
   (a) Write a single-variable integral for the area of R.
   (b) Write an iterated integral for the area of R.

24. Let R be the region between the graphs of $y = 1 + e^x$ and $y = \cos x$ between $x = -3$ and $x = 2$.
   (a) Write a single-variable integral for the area of R.
   (b) Write an iterated integral for the area of R.

25. Find the center of mass of the triangle with vertices at $(-1, 1)$, $(0, 2)$, and $(1, 1)$.

26. Consider the triangle with vertices at $(0, 1)$, $(0, -1)$, and (a, b), where (a, b) is any point in the xy-plane with $a > 0$. (Every triangle is similar to a triangle of this form.)

(a) Show by calculating an appropriate integral that the triangle has center of mass at $(a/3, b/3)$. [HINT: Write equations for the "top" and "bottom" edges of the triangle; use these lines in an iterated integral.]

(b) The **centroid** of a triangle with vertices (a_1, b_1), (a_2, b_2), and (a_3, b_3) is defined to be the point
$$\frac{(a_1, b_1) + (a_2, b_2) + (a_3, b_3)}{3} = \left(\frac{a_1 + a_2 + a_3}{3}, \frac{b_1 + b_2 + b_3}{3}\right).$$
(This is the "average" of the three vertices.) Explain why the center of mass of the triangle in part (a) is also the centroid.

27. Let D be the "half-disk" region bounded below by the x-axis and above by the unit circle $x^2 + y^2 = 1$. Find the center of mass $(\bar{x}, \bar{y})$ of D.

28. Find the center of mass of the region R_k bounded above by the line $y = 1$ and below by the curve $y = x^k$, where k is any even positive integer. (Requiring k to be even ensures that the region makes good sense.) The symmetry of R_k implies that the center of mass is at some point $(0, \bar{y})$ on the y-axis.

(a) Find $\bar{y}$ in terms of k.

(b) What happens as $k \to \infty$? Could the result have been predicted geometrically?

In Exercises 29–34, sketch the region of integration and write the iterated integral with the order of integration reversed.

29. $\displaystyle\int_0^1 \int_{x^2}^1 f(x, y) \, dy \, dx$.

30. $\displaystyle\int_0^2 \int_1^{e^x} f(x, y) \, dy \, dx$.

31. $\displaystyle\int_0^1 \int_0^{\sqrt{1-y^2}} f(x, y) \, dx \, dy$.

32. $\displaystyle\int_0^{\ln 2} \int_{e^y}^2 f(x, y) \, dx \, dy$.

33. $\displaystyle\int_0^{\pi/2} \int_{2x/\pi}^{\sin x} f(x, y) \, dy \, dx$.

34. $\displaystyle\int_0^1 \int_{x^2}^{x^{1/3}} f(x, y) \, dy \, dx$.

The iterated integrals in Exercises 35–42 are difficult to evaluate in the order given. Write each iterated integral with the order of integration reversed and then evaluate the integral.

35. $\displaystyle\int_0^1 \int_{2x}^2 x\sqrt{1 + y^3} \, dy \, dx$.

36. $\displaystyle\int_0^1 \int_{x/2}^{1/2} e^{y^2} \, dy \, dx$.

37. $\displaystyle\int_0^9 \int_{\sqrt{y}}^3 \sin(x^3) \, dx \, dy$.

38. $\displaystyle\int_0^1 \int_{e^x}^e \frac{1}{\ln y} \, dy \, dx$.

39. $\displaystyle\int_0^1 \int_{\sqrt{x}}^1 \sqrt{1 + y^3} \, dy \, dx$.

40. $\displaystyle\int_0^2 \int_{x^2}^4 x \cos(y^2) \, dy \, dx$.

41. $\displaystyle\int_0^1 \int_y^{\sqrt{y}} \frac{\sin x}{x} \, dx \, dy$.

42. $\displaystyle\int_0^1 \int_0^1 \sin(e^x) \, dx \, dy + \int_1^e \int_{\ln y}^1 \sin(e^x) \, dx \, dy$.

14.4 DOUBLE INTEGRALS IN POLAR COORDINATES

Good and bad integrals

What makes a double integral $I = \iint_R f(x, y) \, dA$ hard to calculate? Both f and R can play a role: If either one is complicated or messy to describe, then I may be correspondingly ugly. Example 1 illustrates both "good" and "bad" integrals.

EXAMPLE 1 Discuss
$$I_1 = \iint_{R_1} x^2 \, dA \qquad \text{and} \qquad I_2 = \iint_{R_2} \sqrt{x^2 + y^2} \, dA,$$

where R_1 is the rectangle $[0, 1] \times [0, 2\pi]$ and R_2 is the region inside the unit circle $x^2 + y^2 = 1$. ➡

Sketch R_1 and R_2 for yourself.

Solution The first integral is easily calculated:
$$I_1 = \int_{x=0}^{x=1} \left(\int_{y=0}^{y=2\pi} x^2 \, dy \right) dx = \int_0^1 \left(x^2 y \right]_0^{2\pi} \right) dx = \int_0^1 2\pi x^2 \, dx = \frac{2\pi}{3}.$$

The ingredients of I_2 are more complicated to describe. ➡ The circular region R_2 can be thought of as bounded by the curves $y = \sqrt{1 - x^2}$ and $y = -\sqrt{1 - x^2}$ on the top and bottom and by the lines $x = -1$ and $x = 1$ on the left and right. Now we can write I_2 in

Polar coordinates will simplify the work.

iterated form:

$$I_2 = \int_{x=-1}^{x=1} \left(\int_{y=-\sqrt{1-x^2}}^{y=\sqrt{1-x^2}} \sqrt{x^2 + y^2} \, dy \right) dx.$$

Even Maple has trouble with it.

The integral looks—and is—complicated. ◄ Just to get started on the inner integral, we would need an antiderivative formula like

$$\int \sqrt{y^2 + p^2} \, dy = \frac{1}{2} \left(y\sqrt{y^2 + p^2} + p^2 \ln \left| y + \sqrt{y^2 + p^2} \right| \right),$$

which might be found in an integral table. Faced with this prospect, we retreat. But only temporarily—we shall soon return. ■

What went wrong, and what to do   The integral I_2 in Example 1 led to an unpleasant calculation in x and y for two reasons:

No thanks to the square root.

(i) the integrand, $\sqrt{x^2 + y^2}$, has a complicated antiderivative in x or y; ◄

No thanks, again, to square roots.

(ii) the domain of integration, although geometrically simple, is messy to describe in rectangular coordinates. ◄

In polar coordinates, on the other hand, both the integrand and the domain have simple, uncluttered formulas. The integrand is

$$f(x, y) = \sqrt{x^2 + y^2} = r.$$

The domain of integration is, in the polar sense, much like a rectangle: It is defined by the inequalities

$$0 \le r \le 1 \quad \text{and} \quad 0 \le \theta \le 2\pi.$$

In this case, apparently, both the integrand f and the domain R are better described in polar than in rectangular coordinates. It seems reasonable, therefore, that the double integral I_2 should be calculated using polar rather than rectangular coordinates.

That hunch is correct. We will see in this section how to calculate double integrals in polar form, using r and θ as opposed to Cartesian (i.e., rectangular) form, which uses x and y. Integrals such as I_2, in which the integrand, the domain of integration, or both are simplest in polar form, are natural candidates for polar treatment.

Polar "rectangles"   A rectangle in Cartesian coordinates is defined by two inequalities of the form

$$a \le x \le b; \quad c \le y \le d;$$

each of the coordinates x and y ranges through an interval. A **polar rectangle** is defined by two similar inequalities:

$$a \le r \le b; \quad \alpha \le \theta \le \beta;$$

again, each of the coordinates r and θ ranges through an interval. Figure 1 shows "generic" pictures of both types of rectangle:

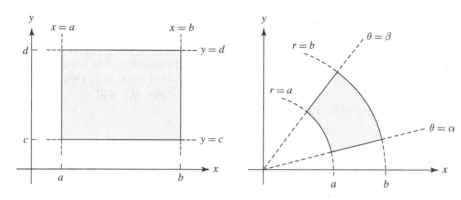

Cartesian and polar rectangles

For polar integrals, as for rectangular ones, rectangles (in the appropriate sense) are the simplest regions over which to integrate.

Polar integration — how it works

A double integral $I = \iint_R f(x, y)\, dA$ in rectangular coordinates, where $R = [a, b] \times [c, d]$ is an ordinary rectangle, is written in iterated form as follows: ➡

This time we integrate first in y.

$$\iint_R f(x, y)\, dA = \int_a^b \int_c^d f(x, y)\, dy\, dx.$$

(Here and below we omit some parentheses—the inner integral is always done first.)

Now consider a double integral $I = \iint_R g(r, \theta)\, dA$ in polar coordinates, where R is the polar rectangle defined by inequalities

$$a \le r \le b \quad \text{and} \quad \alpha \le \theta \le \beta,$$

and $g(r, \theta)$ is a function defined on R. The following Fact gives the appropriate integral formula:

FACT (Double integrals in polar coordinates)   Let the integrand g and the region R be as above. Then

$$\iint_R g(r, \theta)\, dA = \int_{\theta=\alpha}^{\theta=\beta} \int_{r=a}^{r=b} g(r, \theta)\, r\, dr\, d\theta.$$

The formula requires some comment:

Trading x and y for r and θ   Any function $f(x, y)$ can be "traded" for an equivalent function $g(r, \theta)$ by using the relations

$$x = r\cos\theta \qquad \text{and} \qquad y = r\sin\theta.$$

The same method works for *equations* in x and y. The equation $x = y$, for example, says in polar coordinates that $r\cos\theta = r\sin\theta$, or, equivalently, that $\tan\theta = 1$. This polar equation describes the same line as the original Cartesian equation. We will use these principles below when evaluating polar integrals.

We won't go into great depth, but we'll give some informal justification soon.

A useful mnemonic   Compare the formulas given in this section for integrating in rectangular and polar coordinates. An important difference between the two has to do with the "dA" expression. The full mathematical story is much deeper, ← but as a first quick aid to memory the following formulas are handy:

$$dA = dx\,dy; \qquad \text{(for cartesian coordinates)}$$
$$dA = r\,dr\,d\theta. \qquad \text{(for polar coordinates)}$$

That extra factor of r ...   Why is that mysterious "extra" r needed in the polar formula $dA = r\,dr\,d\theta$? Why not simply $dA = dr\,d\theta$?

We owe the reader an explanation—and we will honor that debt soon, when we discuss *why* the formula works. First, however, let's see *how* it works.

EXAMPLE 2   Let R_2 be the region inside the unit circle. Use polar coordinates to calculate the troublesome integral $I_2 = \iint_{R_2} \sqrt{x^2 + y^2}\,dA$ from Example 1.

Solution   First we write all our data in polar form. For the integrand, we have $f(x, y) = \sqrt{x^2 + y^2} = r = g(r, \theta)$. For the domain of integration, we translate the Cartesian equation $x^2 + y^2 = 1$ into its (simpler!) polar form, $r = 1$. The rest is easy:

$$\iint_{R_2} \sqrt{x^2 + y^2}\,dA = \int_{\theta=0}^{\theta=2\pi} \int_{r=0}^{r=1} r\,dA \qquad \text{(the domain is a "rectangle")}$$

$$= \int_{\theta=0}^{\theta=2\pi} \int_{r=0}^{r=1} r^2\,dr\,d\theta \qquad \text{(because } dA = r\,dr\,d\theta)$$

$$= \int_{\theta=0}^{\theta=2\pi} \left.\frac{r^3}{3}\right]_0^1 d\theta \qquad \text{(integrating first in } r)$$

$$= \int_{\theta=0}^{\theta=2\pi} \frac{1}{3}\,d\theta = \frac{2\pi}{3}. \qquad \text{(integrating in } \theta)$$

Notice, especially, the similarity to the integral I_1 of Example 1—I_1 and I_2 turned out to have the same value. This is no accident, of course. After rewriting in polar coordinates, I_2 turned out to be the same integral in r and θ as I_1 is in x and y.   ∎

Polar integrals—what they mean

Polar integrals have exactly the same interpretations as any other double integrals. ➔ As determined by the situation and on our point of view, an integral might represent a volume, the area of a plane region, the mass of a thin plate, or many other things.

We discussed several possible interpretations of integrals in Section 14.1.

For example, both the integrals I_1 and I_2 of Example 1, page 805, can be interpreted as volumes of solids. For I_2, the solid lies above the unit disk and below the surface $z = \sqrt{x^2 + y^2}$; see Figure 2:

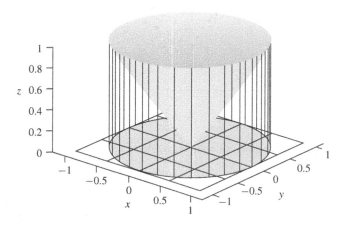

FIGURE 2
A polar solid

As we just calculated, this figure has volume $2\pi/3 \approx 2.094$ cubic units. ➔

Does this number seem reasonable from the picture?

Polar integration—why it works

Why does the polar integration formula work? Where, especially, does the r in $dA = r\,dr\,d\theta$ come from?

All properties of integrals—whether in Cartesian, polar, or any other form—stem ultimately from properties of the approximating sums that are used to define integrals. For any function f defined on a region R, we have

$$\iint_R f\,dA = \lim_{m \to \infty} \sum_{i=1}^{m} f(P_i)\Delta A_i,$$

where ΔA_i is the area of the ith subregion of R, and P_i is a sampling point chosen inside this subregion.

If $R = [a, b] \times [c, d]$ is a Cartesian rectangle, then it is natural to subdivide R into smaller rectangles, each with sides Δx and Δy. Any such rectangle has area $\Delta A_i = \Delta x\,\Delta y$. In the limit that defines the integral, therefore, $dA = dx\,dy$.

If R is a *polar* rectangle, the picture is a little different. In this case, a "polar grid" is the natural way to subdivide R. Figure 3 shows the idea; each bulleted point represents a sampling point "centered" in its respective subdivision.

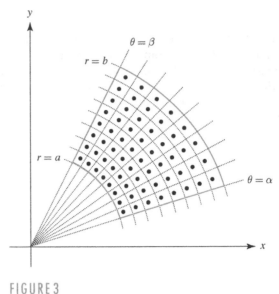

FIGURE 3

A polar grid on a polar rectangle

Notice:

Similar subregions All subregions correspond to the same $\Delta\theta$ (the angle between any two adjacent dotted radial lines) and the same Δr (the radial distance from one arc to the next).

But not identical Similar as they are, the subregions shown are not identical. Here is the key point:

> *In a polar grid, the subregions have different sizes. The area depends not only on $\Delta\theta$ and Δr but also on r; larger values of r produce larger subregions.*

This fact explains the difference between polar and Cartesian integrals—and hints at why that extra r is needed.

The area of one subregion Each subregion described above is a small polar rectangle, with polar dimensions $\Delta\theta$ and Δr, inner radius r, and outer radius $r + \Delta r$. An important property of such a polar rectangle is that

$$\text{area} = \Delta A_i = \frac{r + (r + \Delta r)}{2} \, \Delta r \, \Delta \theta.$$

(We leave verification of this straightforward fact to the exercises.) The first factor on the right is crucial—it represents the *average* radius of the given subregion, that is, the r-coordinate of the ith bulleted midpoint (r_i, θ_i) shown above. Therefore,

$$\wedge A_i = r_i \, \Delta r \, \Delta \theta.$$

This equation is what we have been waiting for; it shows that, for a polar rectangular region, a midpoint-rule approximating sum has the form

$$\sum_{i=1}^{m} f(r_i, \theta_i) \, \Delta A_i = \sum_{i=1}^{m} f(r_i, \theta_i) \, r_i \, \Delta r \, \Delta \theta.$$

The integral itself, therefore, has the limiting form $\int_R f(r, \theta) r \, dr \, d\theta$.

Polar integrals over nonrectangular regions

Polar integrals, like Cartesian integrals, can be taken over nonrectangular regions. The method is similar, too.

> **FACT** Let R be the region bounded by the radial lines $\theta = \alpha$ and $\theta = \beta$, by an "inner" curve $r = r_1(\theta)$, and by an "outer" curve $r = r_2(\theta)$. ("Inner" and "outer" are understood relative to the origin.) Let $g(r, \theta)$ be a function defined on R. Then
>
> $$\iint_R g \, dA = \int_{\theta=\alpha}^{\theta=\beta} \int_{r=r_1(\theta)}^{r=r_2(\theta)} g(r, \theta) r \, dr \, d\theta.$$

Figure 4 shows two examples, one "generic" and one specific:

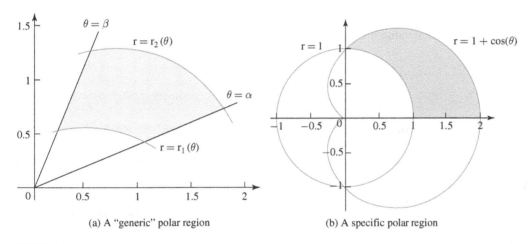

(a) A "generic" polar region   (b) A specific polar region

FIGURE 4
Two nonrectangular polar regions

EXAMPLE 3 Find $\displaystyle\iint_R \frac{1}{\sqrt{x^2 + y^2}} \, dA$, where R is the shaded region in Figure 4(b). Then find the area of the shaded region.

Solution Note that R is bounded outside by the cardioid $r = 1 + \cos\theta$ and inside by the circle $r = 1$, and that it lies between the positive coordinate axes, which have polar forms $\theta = 0$ and $\theta = \pi/2$. Also

$$f(x, y) = \frac{1}{\sqrt{x^2 + y^2}} = \frac{1}{r} = g(r, \theta),$$

and so the preceding Fact gives

$$\iint_R \frac{1}{x^2 + y^2} \, dA = \int_0^{\pi/2} \int_{r=1}^{r=1+\cos\theta} \frac{1}{r} r \, dr \, d\theta = \int_0^{\pi/2} \int_{r=1}^{r=1+\cos\theta} dr \, d\theta$$

$$= \int_0^{\pi/2} \cos\theta \, d\theta = 1. \qquad \text{(cancel in } r \text{, then integrate twice)}$$

To find the area of R, we apply the familiar formula

$$\text{area of } R = \iint_R 1 \, dA,$$

but we calculate the integral in polar form:

$$\iint_R 1 \, dA = \int_{\theta=0}^{\theta=\pi/2} \int_{r=1}^{r=1+\cos\theta} 1\, r \, dr \, d\theta$$

$$= \int_{\theta=0}^{\theta=\pi/2} \frac{r^2}{2}\bigg]_1^{1+\cos\theta} d\theta$$

$$= \frac{1}{2} \int_{\theta=0}^{\theta=\pi/2} \left((1+\cos\theta)^2 - 1 \right) d\theta.$$

But it's not really difficult.

The last integral takes a little effort by hand, ⟵ but *Maple* has no trouble:

```
> 1/2 * int( (1+cos(t))^2-1, t=0 .. Pi/2 );
                    1 + Pi/8
```

which has decimal value around 1.4.

BASIC EXERCISES

1. Let R be the polar rectangle defined by $a \le r \le b$ and $\alpha \le \theta \le \beta$.

    (a) Show that the area of R is $\dfrac{a+b}{2}(b-a)(\beta-\alpha)$.

    (b) Use part (a) to show that a polar rectangle with dimensions Δr and $\Delta\theta$ and inner radius r has area $\dfrac{r+r+\Delta r}{2}\,\Delta r\, \Delta\theta$.

2. Let $f(x, y) = y$, let R be the upper half of the region inside the unit circle $x^2 + y^2 = 1$, and let $I = \iint_R f \, dA$.

    (a) Calculate I as an iterated integral in rectangular coordinates with the inner integral in y.

    (b) Calculate I as an iterated integral in rectangular coordinates with the inner integral in x.

    (c) Calculate I as an iterated integral in polar coordinates.

3. Use formulas for the volumes of cones and cylinders to explain why

$$I_2 = \iint_{R_2} \sqrt{x^2 + y^2}\, dA = \frac{2\pi}{3},$$

where R_2 is the region inside the unit circle $r = 1$. (See Figure 2.)

4. Let I_1 be the integral defined in Example 1.

    (a) Sketch the solid whose volume is given by I_1.

    (b) Evaluate I_1 using an iterated integral where the inner integral is with respect to x rather than y.

In Exercises 5–8, use a polar double integral to find the area of the region R. (Draw each region first.)

5. R is the region inside the cardioid $r = 1 + \sin\theta$.

6. R is the region inside the curve $r = 2 - \cos(2\theta)$.

7. R is the region bounded by $y = x$, $y = 0$, and $x = 1$. [HINT: First write the boundary equations in polar form.]

8. R is the region bounded by the circle of radius $1/2$ centered at $(0, 1/2)$. [HINT: First write a Cartesian equation for the circle, then change it to polar form.]

In Exercises 9–14, write the given integral as an iterated integral in polar coordinates.

9. $\displaystyle\int_0^1 \int_0^{\sqrt{1-y^2}} (x^2 + y^2)\, dx \, dy$.

10. $\displaystyle\int_{-a}^0 \int_{-\sqrt{a^2-x^2}}^{\sqrt{a^2-x^2}} xy \, dy \, dx$.

11. $\displaystyle\int_0^1 \int_{x^2}^x xy \, dy \, dx$.

12. $\displaystyle\int_0^1 \int_{y^2}^{\sqrt{y}} (x + y)\, dx \, dy$.

13. $\displaystyle\int_{-1}^1 \int_{1-\sqrt{1-y^2}}^{1+\sqrt{1-y^2}} (x^2 + y^2)\, dx \, dy$.

14. $\displaystyle\int_0^2 \int_{-\sqrt{2x-x^2}}^{\sqrt{2x-x^2}} (x + y)\, dy \, dx$. [HINT: $2x - x^2 = 1 - (x-1)^2$.]

In Exercises 15–20, evaluate the given integral.

15. $\displaystyle\int_0^1 \int_0^{\sqrt{1-x^2}} e^{x^2+y^2} \, dy \, dx$.

16. $\displaystyle\int_{-1}^1 \int_0^{\sqrt{1-y^2}} \sqrt{x^2 + y^2}\, dx \, dy$.

17. $\int_0^1 \int_0^{\sqrt{1-x^2}} \frac{1}{\sqrt{1+x^2+y^2}}\, dy\, dx.$

18. $\int_0^2 \int_0^{\sqrt{4-y^2}} e^{x^2+y^2}\, dx\, dy.$

19. $\int_0^1 \int_0^{\sqrt{x}} \frac{1}{\sqrt{x^2+y^2}}\, dy\, dx.$

20. $\int_0^1 \int_x^1 \frac{1}{(1+x^2+y^2)^{3/2}}\, dy\, dx.$

FURTHER EXERCISES

In Exercises 21–30, evaluate the double integral using polar coordinates.

21. $\iint_R \frac{1}{\sqrt{x^2+y^2}}\, dA$, where R is the region inside the cardioid $r = 1 + \sin\theta$ and above the x-axis.

22. $\iint_R \frac{x^2}{x^2+y^2}\, dA$, where R is the region between the circles $x^2 + y^2 = 1$ and $x^2 + y^2 = 4$.

23. $\iint_R (x^2+y^2)\, dA$, where R is the region in the first quadrant where $x^2 + y^2 \le 2y$.

24. $\iint_R \frac{1}{\sqrt{x^2+y^2}}\, dA$, where R is the quarter of the unit disk in the first quadrant.

25. $\iint_R \frac{1}{x^2+y^2}\, dA$, where R is the region between the circles $x^2 + y^2 = 4$ and $x^2 + y^2 = 9$.

26. $\iint_R x\, dA$, where R is the region inside the circle $r = 2$ and outside the cardioid $r = 1 + \cos\theta$.

27. $\iint_R \sqrt{x^2+y^2}\, dA$, where R is the region enclosed by the four loops of the rose $r = \cos(2\theta)$.

28. $\iint_R y\, dA$, where R is the region in the first quadrant bounded by the curves $x^2 + y^2 = 1$ and $x^2 + y^2 = 4$.

29. $\iint_R (x^2+y^2)\, dA$, where R is the region defined by $0 \le r \le \theta \le \pi$.

30. $\iint_R \sqrt{x^2+y^2}\, dA$, where R is the triangle with vertices at $(0,0)$, $(3,0)$, and $(3,3)$.

31. Find the area of the region inside the cardioid $r = 1 + \cos\theta$ and above the x-axis.

32. Find the area of the region $0 \le r \le \theta \le \pi$.

33. Find the area of the region inside the curve $r = 1 + \cos(2\theta)$ and outside the circle $r = 1$.

34. Find the area of the region inside the cardioid $r = 1 + \cos\theta$ and outside the cardioid $r = 1 - \cos\theta$.

35. Find the area of the region inside the circle $r = 2\cos\theta$ and outside the circle $r = 1$.

36. Find the volume of the solid under the surface $z = 1 - x^2 - y^2$ and above the xy-plane. [HINT: First decide where the surface hits the xy-plane.]

37. Find the volume of the conical solid under the surface $z = 1 - \sqrt{x^2+y^2}$ and above the xy-plane. [HINT: First decide where the surface hits the xy-plane.]

38. Find the volume V of a sphere with radius R using a double integral expressed in polar coordinates.

39. Find the volume of the solid bounded above by the surface $z = \sqrt{x^2+y^2}$ and below by the circle $0 \le (x-1)^2 + y^2 \le 1$.

40. Find the volume of the solid bounded by the paraboloids $z = x^2 + y^2$ and $z = 2 - (x^2 + y^2)$.

41. Find the volume of the solid that is inside the paraboloid $z = 4 - x^2 - y^2$, above the xy-plane, and outside the cylinder $x^2 + y^2 = 1$.

42. Find the volume of the solid that is inside the paraboloid $z = 4 - x^2 - y^2$, above the xy-plane, and outside the cylinder $x^2 + 2y^2 = 2y$.

14.5 TRIPLE INTEGRALS

Triple integrals are defined in the same way as single and double integrals. ➡ For a function $f(x, y, z)$ defined on a solid region S in xyz-space, the integral is defined to be a limit of approximating sums:

We discussed triple integrals briefly in Section 14.1.

$$\iiint_S f\, dV = \lim_{m\to\infty} \sum_{i=1}^m f(P_i)\Delta V_i.$$

The sum on the right is formed by subdividing S into m small subregions $S_1, S_2, \ldots, S_m$ with the ith subregion having volume ΔV_i. ➡ A sampling point P_i is chosen within each subregion S_i, and the approximating sum is calculated. Notice, in particular, that in the

The S_i may have different volumes.

sum the contribution of each value $f(P_i)$ is "weighted" by the volume of its respective subdivision—larger subdivisions receive larger weights.

Dividing each edge into n pieces produces n^3 subdivisions.

Pros and cons of the definition   Approximating sums are not hard to calculate (perhaps with help from computing)—*if* the number of subdivisions is modest. But that's a big "if" for functions of three variables: Subdividing a three-dimensional cube rapidly generates thousands of subregions. ← Approximating sums, moreover, do no more than their name implies: They approximate integrals rather than evaluate them exactly.

It is therefore useful—for both practical and theoretical reasons—to have exact, antiderivative-based methods. In this section we survey several useful techniques for evaluating triple integrals by antidifferentiation. In the following section we introduce two new coordinate systems for $\mathbb{R}^3$ much as we did with polar coordinates in the plane. Choosing the "right" coordinate system for a given integral can simplify the calculation considerably.

Triple integrals, plain and fancy

Ordinary rectangles are nicest for double integrals.

Triple integrals are easiest to calculate if the region of integration S is a rectangular solid, or **parallelepiped**, of the form $S = [a, b] \times [c, d] \times [e, f]$. ← ("Brick" is a less pretentious synonym.) In this case we can integrate f by iteration, one variable at a time—without any fuss over variable limits of integration. See Example 4, page 782 for a straightforward example.

Integrals are harder to evaluate by iteration if the domain of integration is not rectangular. Very irregular domains may be beyond our powers, but many useful domains in $\mathbb{R}^3$ fit a standard pattern that we *can* readily handle. We will call such domains **basic**. ←

"Basic" is useful shorthand, not formal jargon.

Basic domains in two variables   We have already identified some basic domains in the plane:

- a region bounded *above* by a curve $y = f_2(x)$, *below* by a curve $y = f_1(x)$, and on the left and right by vertical lines $x = a$ and $x = b$;
- a region bounded on the *right* by a curve $x = f_2(y)$, on the *left* by a curve $x = f_1(y)$, and on the bottom and top by horizontal lines $y = c$ and $y = d$;
- a polar region bounded on the *outside* by a curve $r = f_2(\theta)$ and on the *inside* by a curve $r = f_1(\theta)$ and lying between the radial lines $\theta = \alpha$ and $\theta = \beta$.

For any domain R of one of these types we can readily integrate a function $f(x, y)$ or $g(r, \theta)$ by iteration using one of the following forms:

$$\int_{x=a}^{x=b} \int_{y=f_1(x)}^{y=f_2(x)} f(x, y) \, dy \, dx; \tag{1}$$

$$\int_{y=c}^{y=d} \int_{x=f_1(y)}^{x=f_2(y)} f(x, y) \, dx \, dy; \tag{2}$$

$$\int_{\theta=\alpha}^{\theta=\beta} \int_{r=f_1(\theta)}^{x=f_2(\theta)} g(r, \theta) \, r \, dr \, d\theta. \tag{3}$$

Basic domains in three variables   Some domains in three variables are almost as convenient for integration as ordinary rectangular solids. For simplicity, we will concentrate in this section on one basic type:

the solid S bounded above by a surface $z = f_2(x, y)$, below by a surface $z = f_1(x, y)$ and lying directly above (or below) a region R in the xy-plane.

(In the next section we will introduce new coordinate systems that will help us handle other three-dimensional domains.) Figure 1 shows a generic picture of such a domain.

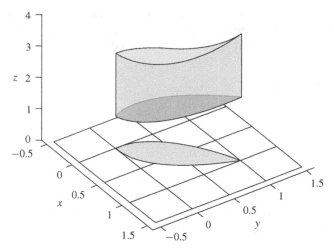

The solid domain has top and bottom surfaces and lies above a "shadow" region R in the xy-plane.

FIGURE 1
A basic domain in three variables

Domains of this type deserve some comment:

- **The shadow** The plane region R referred to just above can be thought of as the "shadow" cast by S in the xy-plane if a light shines straight downward, parallel to the z-axis. ▶ The idea is similar to that in two dimensions, where the region between two curves in the xy-plane casts a "shadow" either on the x-axis or on the y-axis. (See Figure 2, page 799.)

  If S lies below the xy-plane, then the light shines upward.

- **Changing directions** The shadow region R discussed here is described as lying in the xy-plane rather than, say, the yz-plane. Is there something special about the xy-plane?

  The short answer is no: We chose the xy-plane mainly for convenience and ease of visualization. On the other hand, the choice is not really as arbitrary as it may seem; in most real problems we can simply rename the variables as needed.

The iteration recipe for a triple integral over a basic region closely resembles Formulas 1–3 above:

> **FACT** Let S be a basic region as described above with shadow R in the xy-plane, and let $f(x, y, z)$ be defined on S. Then
> $$\iiint_S f(x, y, z)\, dV = \iint_R \int_{z=f_1(x,y)}^{z=f_2(x,y)} f(x, y, z)\, dz\, dA.$$

The Fact says that, if we integrate first in z, we can trade the original *triple* integral (over S) for a *double* integral (over the "shadow" region R). This is often a good trade because we already have sharp tools for handling double integrals.

In particular, if the shadow R has one of our basic two-dimensional forms, then the inner integral, too, can be found in iterated form. If, say, R is bounded above and below by curves $y = h(x)$ and $y = g(x)$, and on the left and right by $x = a$ and $x = b$, then we get the triply nested integral

$$\iiint_S f(x, y, z)\, dV = \int_{x=a}^{x=b} \int_{y=g(x)}^{y=h(x)} \int_{z=f_1(x,y)}^{z=f_2(x,y)} f(x, y, z)\, dz\, dy\, dx,$$

which we integrate from inside out. The process is easier than the impressive symbolic formula might suggest.

E X A M P L E 1   The region S in Figure 1 is bounded above by the surface $z = 3 + x^2 y$ and below by the surface $z = 1 + 3x - 2y$; it lies above the plane region R bounded by the curves $y = x^2$ and $y = \sqrt{x}$. Express the integral $\iiint_S 2x\, dV$ in iterated form; then find its value.

S o l u t i o n   By the preceding Fact, we have

$$\iiint_S z\, dV = \iint_R \int_{z=1+3x-2y}^{z=3+x^2 y} 2x\, dz\, dA.$$

Technology relieves the tedium; we show only the result.

The inner integral is easy, but a little tedious, to find: ←

$$\iiint_S z\, dV = \iint_R \left(4x + 2x^3 y - 6x^2 + 4xy\right) dA.$$

Observe the progress already made: We reduced the original *triple* integral to a relatively straightforward *double* integral. We evaluate the remaining integral by iteration in x and y: ←

We omit the calculation.

$$\iiint_S z\, dV = \int_{x=0}^{x=1} \int_{y=x^2}^{y=\sqrt{x}} \left(4x + 2x^3 y - 6x^2 + 4xy\right) dy\, dx = \frac{83}{168}.$$

To save a little writing we could have written the original integral in triply nested form:

$$\iiint_S z\, dV = \int_{x=0}^{x=1} \int_{y=x^2}^{y=\sqrt{x}} \int_{z=1+3x-2y}^{z=3+x^2 y} 2x\, dz\, dy\, dx.$$

Simpler basic domains   Example 1 required somewhat complicated symbolics because of the shape of the solid domain S. In practice, triple integrals often involve simpler domains.

E X A M P L E 2   The region S in Figure 2 is bounded above by the surface $z = 1 + x^2$, below by the xy-plane, and lying above the plane region R bounded by the curves $y = 0$ and $y = 1 - x$. Evaluate the integral $\iiint_S 2x\, dV$.

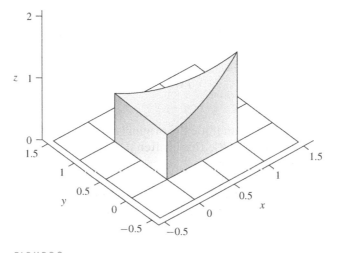

Here the "shadow" region R is also the lower surface of S.

FIGURE 2
A simpler domain in three variables

Solution The domain's simplicity is reflected in the integral's setup and calculation:

$$\iiint_S 2x \, dV = \int_{x=0}^{x=1} \int_{y=0}^{y=1-x} \int_{z=0}^{z=1+x^2} 2x \, dz \, dy \, dx$$

$$= \int_{x=0}^{x=1} \int_{y=0}^{y=1-x} \left(2x + 2x^3\right) dy \, dx,$$

$$= \int_0^1 \left(2x(1-x) + 2x^3(1-x)\right) dx = \frac{13}{30}.$$

Finding the shadow Sometimes (in three dimensions as in two) the shadow region R needs to be calculated from the given data.

EXAMPLE 3 Let S be the region bounded below by the surface $z = x^2 + y^2$ and above by the surface $z = 4$. Calculate $I_1 = \iiint_S z \, dz$ and $I_2 = \iiint_S 1 \, dz$.

Solution Figure 3 shows the domain of integration S (it lies between the two surfaces) and its shadow R below.

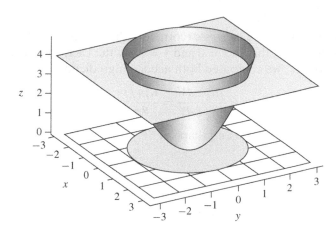

FIGURE 3
The region between $z = x^2 + y^2$ and $z = 4$

To find the shadow region R, we need to find where the surfaces intersect. We do so by setting the surface equations equal, that is, $x^2 + y^2 = 4$. Thus, the surfaces intersect above the circle $x^2 + y^2 = 4$ in the xy-plane, and the shadow region R lies inside this circle. Now we can find I_1:

$$\iiint_S z \, dV = \iint_R \int_{z=x^2+y^2}^{z=4} z \, dz \, dA = \iint_R \left(8 - \frac{(x^2+y^2)^2}{2}\right) dA.$$

The last integral naturally deserves polar treatment: ⬈

Check the result, perhaps with technology.

$$\int_0^{2\pi} \int_0^2 \left(8 - \frac{r^4}{2}\right) r \, dr \, d\theta = \frac{64\pi}{3} \approx 67.$$

A similar calculation gives I_2:

$$\iiint_S 1 \, dV = \iint_R \int_{z=x^2+y^2}^{z=4} 1 \, dz \, dA = \int_0^{2\pi} \int_0^2 \left(4 - r^2\right) r \, dr \, d\theta = 8\pi.$$

Understanding triple integrals: volume, average value, and center of mass Triple integrals, like integrals in lower dimensions, have geometric and physical meaning. We have already remarked that an integral of the form $\iiint_S 1\, dV$ gives the **volume** of S and that, for any integrand $f(x, y, z)$,

$$\frac{\iiint_S f(x, y, z)\, dV}{\text{volume of } S} = \frac{\iiint_S f(x, y, z)\, dV}{\iiint_S 1\, dV} = \textbf{average value of } f \text{ on } S.$$

We can also interpret some integrals physically, using the **center of mass**. We have already seen for double integrals that the center of mass of a *plane* region R (regarded as having constant density) has coordinates $(\bar{x}, \bar{y})$ that are simply the average values of the functions x and y over the region R. The three-dimensional situation is analogous:

> **FACT** Let S be a solid region in $\mathbb{R}^3$ with constant density. The center of mass of S is the point $(\bar{x}, \bar{y}, \bar{z})$ whose coordinates are the average values over S of the functions x, y, and z, respectively.

EXAMPLE 4 Locate the center of mass of the solid S described in Example 3 and shown in Figure 3.

Solution The symmetry of S suggests (correctly) that the center of mass lies somewhere on the z-axis; that is, $\bar{x} = 0$ and $\bar{y} = 0$. By the Fact, $\bar{z}$ is the average value of z over S. By good luck, we computed both needed ingredients in Example 3:

$$\bar{z} = \frac{\iiint_S z\, dV}{\iiint_S 1\, dV} = \frac{64\pi/3}{8\pi} = \frac{8}{3},$$

which is plausible given the size and shape of S. ∎

BASIC EXERCISES

1. Let T be the tetrahedron bounded by the planes $x = 0$, $y = 0$, $z = 0$, and $x + 2y + 3z = 6$.
 (a) Express the volume of T as a triple iterated integral.
 (b) Express the volume of T as a double iterated integral.
 (c) Find the volume of T.

2. Let R denote the tetrahedron with vertices at $(0, 0, 0)$, $(a, 0, 0)$, $(0, b, 0)$, and $(0, 0, c)$.
 (a) Express the volume of R as a triple iterated integral.
 (b) Express the volume of R as a double iterated integral.
 (c) Find the volume of R.
 (d) Evaluate $\iiint_R x\, dV$.

3. Suppose that $I = \iiint_R f(x, y, z)\, dV = \int_0^1 \int_0^x \int_0^1 f(x, y, z)\, dz\, dy\, dx$.
 (a) Sketch the region of integration R.
 (b) Let $f(x, y, z) = xy^2z$. Evaluate I.

4. Suppose that $I = \iiint_R f(x, y, z)\, dV = \int_0^1 \int_0^y \int_{x^2+y^2}^{x+y} f(x, y, z)\, dz\, dx\, dy$.
 (a) Sketch the region of integration R.
 (b) Let $f(x, y, z) = x + y$. Evaluate I.

5. Suppose that $I = \iiint_R f(x, y, z)\, dV = \int_0^1 \int_0^z \int_0^1 f(x, y, z)\, dx\, dy\, dz$.
 (a) Sketch the region of integration R.
 (b) Let $f(x, y, z) = xy$. Evaluate I.

6. Suppose that $I = \iiint_R f(x, y, z)\, dV = \int_0^1 \int_0^{\sqrt{1-z^2}} \int_0^x f(x, y, z)\, dy\, dx\, dz$.
 (a) Sketch the region of integration R.
 (b) Let $f(x, y, z) = xy$. Evaluate I.

7. Suppose that $I = \iiint_R f(x, y, z)\, dV$, where R is defined by $1 \leq z \leq 2$, $0 \leq y \leq x$, and $1 \leq x \leq 2$.
 (a) Sketch the region of integration R.
 (b) Let $f(x, y, z) = (y + z)/x$. Evaluate I.

8. Suppose that $I = \iiint_R f(x, y, z)\, dV$, where R is the region above the triangle with vertices $(0, 0, 0)$, $(1, 0, 0)$, and $(0, 2, 0)$ and below the surface $z = 3$.
 (a) Sketch the region of integration R.
 (b) Let $f(x, y, z) = \dfrac{x + y}{z + 1}$. Evaluate I.

9. Suppose that $I = \iiint_R f(x, y, z)\, dV$, where R is defined by $0 \le z \le x + 2y^2$, $x \ge 0$, $y \ge 0$, and $x + y \le 1$.

(a) Sketch the region of integration R.

(b) Let $f(x, y, z) = x$. Evaluate I.

10. Suppose that $I = \iiint_R f(x, y, z)\, dV$, where R is the region above the triangle in the xy-plane described by $0 \le x \le 2$ and $0 \le y \le 2 - x$ and below the cone $z = 3\sqrt{x^2 + y^2}$.

(a) Sketch the region of integration R.

(b) Let $f(x, y, z) = xz$. Evaluate I.

11. Suppose that $I = \iiint_R f(x, y, z)\, dV$, where R is the region between the parabolic cylinders $z = y^2$ and $z = 8 - y^2$, for $0 \le x \le 5$.

(a) Sketch the region of integration R.

(b) Let $f(x, y, z) = z$. Evaluate I.

12. Suppose that $I = \iiint_R f(x, y, z)\, dV$, where R is the region bounded by the paraboloids $z = x^2 + y^2$ and $z = 1 - (x^2 + y^2)$.

(a) Sketch the region of integration R.

(b) Let $f(x, y, z) = x + y + z$. Set up an iterated integral for I and evaluate it.

13. Suppose that $I = \iiint_R f(x, y, z)\, dV$, where R is the region bounded by the cone $z = \sqrt{x^2 + y^2}$ and the paraboloid $z = 2 - (x^2 + y^2)$.

(a) Sketch the region of integration R.

(b) Let $f(x, y, z) = x + y + z$. Evaluate I.

14. Suppose that $I = \iiint_R f(x, y, z)\, dV$, where R is the region bounded by the plane $x = 1$, the cylinder $x = e^z$, the plane $y = z$, and the cylinder $y^2 + z^2 = 8$.

(a) Sketch the region of integration R.

(b) Let $f(x, y, z) = 1/x$. Evaluate I.

15. Suppose that $I = \iiint_R f(x, y, z)\, dV$, where R is the region defined by the inequalities $z^2 \le x \le z$ and $y^2 + z^2 \le 1$.

(a) Sketch the region of integration R.

(b) Let $f(x, y, z) = x + y$. Evaluate I.

14.6 MORE TRIPLE INTEGRALS: CYLINDRICAL AND SPHERICAL COORDINATES

We have already seen for double integrals how polar coordinates can simplify integrals in which the integrand or the domain of integration, or both, have a polar structure. In this section we consider two new coordinate systems that offer similar advantages for triple integrals.

Cylindrical coordinates

As we saw earlier, polar coordinates in the plane radically simplify the formulas used to describe certain regions and functions. **Cylindrical coordinates** represent a natural extension of the same idea to $\mathbb{R}^3$. The idea is simple: Use r, θ, and z (rather than x, y, and z) to describe points in three-dimensional space. Figure 1 shows the the the idea.

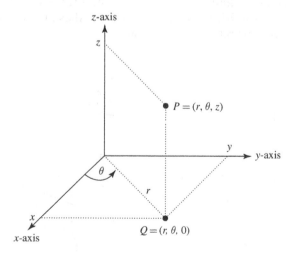

FIGURE 1
Cylindrical coordinates

The various coordinates are related to each other just as for polar coordinates. Look for the following relations in the picture:

$$x = r\cos\theta; \quad y = r\sin\theta; \quad \sqrt{x^2 + y^2} = r.$$

(For simplicity we will assume that $r \geq 0$ when working with cylindrical coordinates.)

Why the name? Simple shapes   Cylindrical coordinates deserve the name because they are so well suited to describing circular cylinders and related shapes. Here are some samples:

Cylinders   In cylindrical coordinates, the graph of the equation $r = a$, for any fixed $a > 0$, is an infinite circular *cylinder* of radius a, centered along the z-axis.

Horizontal planes   The graph of the equation $z = a$ is the horizontal plane at height a above (or below) the xy-plane. (Such planes have exactly the same equations in Cartesian coordinates.)

Vertical planes   The graph of the equation $\theta = a$ is a vertical plane through the origin, perpendicular to the xy-plane and makes angle a radians with the x-axis.

Cones   The graph of the equation $z = mr$ is a cone centered on the z-axis, with vertex at the origin; m measures the slope of the sides of the cone.

Integration in cylindrical coordinates

Recall the formula for double integrals in polar coordinates. For a polar rectangle R, we have seen that

$$\iint_R f(x, y)\, dA = \iint_R g(r, \theta)\, r\, dr\, d\theta.$$

For short, $dA = r\, dr\, d\theta$; dA is called the **area element in polar coordinates**.

The corresponding formula for cylindrical coordinates is similar. If S is a region in cylindrical form,

$$\iiint_S f(x, y, z)\, dV = \iiint_S g(r, \theta, z)\, r\, dr\, d\theta\, dz.$$

For short, $dV = r\, dr\, d\theta\, dz$ is the **volume element in cylindrical coordinates**.

The form of the volume element should not be too surprising given what we have seen for polar coordinates. For cylindrical coordinates, the idea is to subdivide the region of integration into small "cylindrical rectangles" and use these to construct approximating sums. Figure 2 shows *one* cylindrical subdivision:

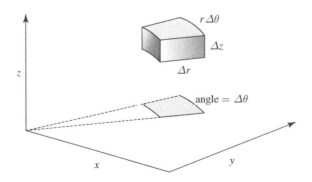

FIGURE 2
A small cylindrical "rectangle"

We saw in the previous section that one such rectangle has base area approximately $r \, \Delta r \, \Delta \theta$. Since the thickness is Δz, it follows that

$$\text{volume of one box} \approx r \, \Delta r \, \Delta \theta \, \Delta z.$$

Thus, any approximating sum in cylindrical coordinates has the form

$$\sum_{i=1}^{m} g(r_i, \theta_i, z_i) \, \Delta V_i \approx \sum_{i=1}^{m} g(r_i, \theta_i, z_i) r_i \, \Delta r \, \Delta \theta \, \Delta z.$$

In the limit, therefore,

$$\iiint_S f(x, y, z) \, dV = \iiint_S g(r, \theta, z) \, r \, dr \, d\theta \, dz.$$

Cylindrical coordinates are the natural choice when the region of integration or the integrand (or both) have simpler formulas or descriptions in terms of these coordinates.

EXAMPLE 1 Derive the classical formula for the volume of a cone with height h and radius a.

Solution The desired cone C has cylindrical equation $z = rh/a$. ➡ The part we want lies below $z = h$ and above $z = rh/a$. Let's calculate the volume in cylindrical coordinates:

Convince yourself of this; note that $z = h$ when $r = a$.

$$\text{volume} = \iiint_C 1 \, dV = \int_0^{2\pi} \int_{r=0}^{r=a} \int_{z=rh/a}^{z=h} r \, dz \, dr \, d\theta.$$

It is a routine matter to calculate the iterated integral on the right; the answer is $\pi a^2 h / 3$. ∎

Spherical coordinates

Spherical coordinates offer yet another way of describing three-dimensional space. Spherical coordinates are denoted $\rho, \theta,$ and ϕ. The first coordinate measures three-dimensional distance from the origin, and the other two coordinates are angular (θ is the same angle as in the polar and cylindrical case). The general picture is shown in Figure 3:

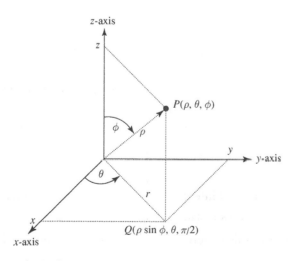

FIGURE 3

Spherical coordinates

A close and careful look at the picture reveals various relations among spherical and other

Do you see why they hold?

coordinates. These are especially important: ←

$$z = \rho \cos\phi; \quad r = \rho \sin\phi.$$

Combining these relations with those already known from polar coordinates gives

$$x = r\cos\theta = \rho\sin\phi\cos\theta; \quad y = r\sin\theta = \rho\sin\phi\sin\theta; \quad z = \rho\cos\phi.$$

These relations let us convert any function $f(x, y, z)$ in Cartesian coordinates into a new function $g(\rho, \theta, \phi)$ in spherical coordinates.

Integration in spherical coordinates   Triple integrals in spherical coordinates have their own special form. If S is a spherical region and $f(x, y, z)$ is defined on S, then

$$\iiint_S f(x, y, z)\, dV = \iiint_S g(\rho, \theta, \phi)\, \rho^2 \, \sin\phi \, d\rho \, d\theta \, d\phi.$$

In short, $dV = \rho^2 \sin\phi \, d\rho \, d\theta \, d\phi$ is the **volume element in spherical coordinates**.

Explaining why the volume element has precisely this form is similar to the argument given above for cylindrical coordinates. We omit some details, but Figure 4 contains all the necessary ingredients. In particular, the lengths of three edges of the "box" are marked; their product gives the volume element.

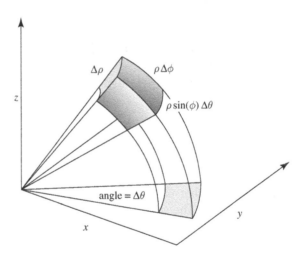

FIGURE 4
A small spherical "rectangle"

Figure 4 gives (after some very careful examination) a good estimate to the volume of one small spherical rectangle:

$$\text{volume of one box} \approx \rho\,\Delta\phi\,\rho\,\sin\phi\,\Delta\theta\,\Delta\rho = \rho^2\,\sin\phi\,\Delta\rho\,\Delta\theta\,\Delta\phi.$$

The integral formula follows as a result.

> **E X A M P L E 2**   Use a spherical integral to verify the classical formula $V = \dfrac{4}{3}\pi a^3$ for the volume of a solid sphere of radius a.
>
> **Solution**   Let S be the solid sphere; it is defined by three inequalities in spherical coordinates:
>
> $$0 \le \phi \le \pi; \quad 0 \le \theta \le 2\pi; \quad 0 \le \rho \le a.$$

Thus, the sphere is, in essence, a "spherical rectangle," and the integral calculation is straightforward (though a little complicated):

$$\text{volume} = \iiint_S 1 \, dV$$

$$= \int_{\phi=0}^{\phi=\pi} \int_{\theta=0}^{\theta=2\pi} \int_{\rho=0}^{\rho=a} \rho^2 \sin\phi \, d\rho \, d\theta \, d\phi \qquad \text{(note the order of nesting)}$$

$$= \int_{\phi=0}^{\phi=\pi} \sin\phi \, d\phi \cdot \int_{\theta=0}^{\theta=2\pi} d\theta \cdot \int_{\rho=0}^{\rho=a} \rho^2 \, d\rho \qquad \text{(separating the variables)}$$

$$= 2 \cdot 2\pi \cdot \frac{a^3}{3} = \frac{4}{3}\pi a^3,$$

as desired. ∎

BASIC EXERCISES

In Exercises 1–6, sketch the surface described in cylindrical coordinates.

1. $z = -r$.

2. $z = 1 - r$.

3. $\theta = 1$.

4. $r = \theta$, for $0 \le \theta < 2\pi$.

5. $z = r^2$.

6. $z = 1 - r^2$.

In Exercises 7–10, write a formula in cylindrical coordinates for the given surface.

7. The surface with Cartesian equation $x = 1$.

8. A cylinder of radius 3, centered on the z-axis.

9. A sphere of radius 3, centered at the origin.

10. The surface with Cartesian equation $x^2 - x + y^2 - y = 0$. Describe the surface in words.

In Exercises 11–16, describe in words the surface associated with the equation given in spherical coordinates and then give a Cartesian equation for the surface.

11. $\rho = a$, for any $a > 0$.

12. $\rho \sin\phi = 3$.

13. $\phi = \pi/4$.

14. $\rho \sin\phi \sin\theta = 4$.

15. $\rho = 3 \sec\phi$.

16. $\theta = \pi/4$.

17. (a) Express the point $(x, y, z) = (2, 3, 1)$ in cylindrical coordinates.

   (b) Express the point $(x, y, z) = (2, 3, 1)$ in spherical coordinates.

18. (a) Express the point $(r, \theta, z) = (2, 3, 1)$ in Cartesian coordinates.

   (b) Express the point $(\rho, \theta, \phi) = (2, 3, 1)$ in Cartesian coordinates.

19. Verify the calculation in Example 1.

20. Verify the calculation in Example 2.

In Exercises 21–26, use cylindrical coordinates to find the volume of the region R.

21. R is a cylinder of radius a and height h.

22. R is the region bounded above by the plane $z = x + y$, below by the xy-plane, and on the sides by the cylinder $r = 1$.

23. R is the region in the first octant bounded by the cylinder $x^2 + y^2 = 1$ and the plane $z = x$.

24. R is the region bounded below by the surface $z = x^2 + y^2$ and above by the plane $z = x + y$.

25. R is the region under the cone $z = r$, above the xy-plane, and inside the cylinder $r = 2$.

26. R is a sphere of radius a centered at the origin.

In Exercises 27–32, use cylindrical coordinates to evaluate the given integral.

27. $\iiint_C z \, dV$, where C is the cylinder of radius a and height h.

28. $\iiint_S x \, dV$, where S is the region bounded above by the plane $z = x + y$, below by the xy-plane, and on the sides by the cylinder $r = 1$.

29. $\iiint_S y \, dV$, where S is the region in the first octant bounded by the cylinder $x^2 + y^2 = 1$ and the plane $z = x$.

30. $\iiint_R z \, dV$, where R is the region bounded below by the surface $z = x^2 + y^2$ and above by the plane $z = x + y$.

31. $\iiint_S z \, dV$, where S is the region under the cone $z = r$, above the xy-plane, and inside the cylinder $r = 2$.

32. $\iiint_S x \, dV$, where S is a sphere of radius a centered at the origin.

33. Give a geometric description of the solid whose volume is given by the triple integral $\int_0^{2\pi} \int_1^2 \int_0^5 r \, dz \, dr \, d\theta$.

34. The volume of a solid is $\int_0^2 \int_0^{\sqrt{2x-x^2}} \int_{-\sqrt{4-x^2-y^2}}^{\sqrt{4-x^2-y^2}} dz \, dy \, dx$.

   (a) Describe the solid by giving equations in Cartesian coordinates for the surfaces that form its boundary.

   (b) Rewrite the integral in cylindrical coordinates.

35. Use spherical coordinates to find the volume of the region above the cone $z = \sqrt{x^2 + y^2}$ and below the sphere $x^2 + y^2 + z^2 = 1$.

36. Rewrite the integral $\int_0^{2\pi} \int_0^1 \int_0^{\sqrt{4-r^2}} r^2 \, dz \, dr \, d\theta$ as an iterated integral in

   (a) rectangular coordinates;

   (b) spherical coordinates.

37. A hemispherical bowl of radius 5 cm is filled with water to within 3 cm of the top. Set up an integral for the volume of water in the bowl.

38. A hemisphere of radius 6 inches is truncated to make a bowl of volume 99π cubic inches. Find the height of the bowl.

39. Let R be the "ice cream cone" region enclosed by a sphere of radius 2 centered at the origin and the cone $z = \sqrt{x^2/3 + y^2/3}$. Write the triple integral $\iiint_R y \, dV$ as an iterated integral in

   (a) rectangular coordinates

   (b) cylindrical coordinates

   (c) spherical coordinates

14.7 MULTIPLE INTEGRALS OVERVIEWED; CHANGE OF VARIABLES

In this chapter so far we have encountered several approaches to multiple integrals. In this final section, we first compare these ideas of the integral by reviewing some of the approaches and methods we've seen. Then we link some apparently different approaches to the integral by presenting a general change-of-variables formula that permits us to transform one integral to another.

Integrals reviewed

Different types of integrals In elementary calculus there is only one main type of integral. For a function $f : \mathbb{R} \to \mathbb{R}$ and an interval $[a, b]$, the familiar integral $I = \int_a^b f(x) \, dx$ is a certain real number, defined as a limit of approximating sums. Although I has various possible geometric, physical, and numerical interpretations (as area under the f-graph, as displacement if f represents velocity, as a weighted average of f-values, and so on), they are nothing more than different ways of looking at the same mathematical object.

In multivariable calculus the possibilities for integrals are much broader. Integrands may be functions of one or more variables, and they may be either scalar- or vector-valued functions. The sets over which functions are integrated may also be of various types, including ordinary intervals, one-dimensional curves in $\mathbb{R}^2$ or $\mathbb{R}^3$, planar regions in $\mathbb{R}^2$, solid regions in $\mathbb{R}^3$, and two-dimensional surfaces in $\mathbb{R}^3$. We have seen several such possibilities, with symbolic forms such as these:

$$\int_a^b \mathbf{v}(t) \, dt; \quad \iint_R f(x, y) \, dA; \quad \iint_R g(r, \theta) r \, dr \, d\theta; \quad \iiint_S h(r, \theta, z) r \, dr \, d\theta \, dz.$$

In Chapter 16 we will see still more integrals, including **line integrals** and **surface**

integrals, in which the domains of integration are, respectively, either curves or surfaces in $\mathbb{R}^2$ or $\mathbb{R}^3$.

Derivatives, integrals, and fundamental theorems A first course in calculus usually culminates with the **fundamental theorem of calculus** (FTC), which relates derivatives and integrals. In its simplest ➤ form, the fundamental theorem says that, under appropriate conditions,

But most important.

$$\int_a^b f'(x)\,dx = f(b) - f(a).$$

The theorem says that, in a certain sense, single-variable integration "undoes" single-variable differentiation, and vice versa.

As one might guess, the ideas behind the fundamental theorem can be extended and reinterpreted to fit the multivariable setting. For instance, a version of the **fundamental theorem for line integrals** says that

$$\int_\gamma \nabla f \cdot d\mathbf{X} = f(\mathbf{b}) - f(\mathbf{a}), \tag{1}$$

where $f(x, y)$ is a function of two variables, and γ is a curve joining $\mathbf{a}$ to $\mathbf{b}$ in $\mathbb{R}^2$. Much remains to be explained, of course, including the precise meaning of the line integral on the left. (We discuss line integrals and their properties in Chapter 16.) But the general similarity to the ordinary fundamental theorem should be apparent. In each case, some sort of integral is applied to some sort of derivative of f, ➤ and the answer involves f itself.

The gradient ∇f is a type of derivative of f.

Integrating functions of one variable Our first goal is to review the various types of integrals encountered so far. The simplest involve only one input variable. Single-variable integrals such as

$$\int_0^1 x^2\,dx = \left.\frac{x^3}{3}\right]_0^1 = \frac{1}{3}$$

are the stock in trade of elementary calculus; the integrand is a scalar-valued function of one variable and the domain of integration is an interval. Integrals of this basic type do arise in multivariable calculus, but often as byproducts to some higher-dimensional problem. Arclength calculations are one good example.

E X A M P L E 1   The position function

$$\mathbf{r}(t) = (\cos t, \sin t, t)$$

describes a helix (spiral) in $\mathbb{R}^3$. Find the arclength of one "turn"; think of t as time.

S o l u t i o n   Position is a *vector*-valued function of time t, but the speed at time t, given by

$$|\mathbf{r}'(t)| = |(-\sin t, \cos t, 1)| = \sqrt{\sin^2 t + \cos^2 t + 1} = \sqrt{2},$$

is *scalar*-valued. (In this case, the speed function happens to be constant.) One turn of the helix takes 2π units of time, and so the arclength is given by the ordinary integral

$$\int_0^{2\pi} |\mathbf{r}'(t)|\,dt = \int_0^{2\pi} \sqrt{2}\,dt = 2\sqrt{2}\pi \approx 8.89$$

units of distance. ▣

We have also seen how to integrate vector-valued functions of one variable over an interval in the domain. An important use of such integrals is in modeling phenomena of motion.

But check our work.

EXAMPLE 2   A particle moves in the xy-plane. At time t, its velocity vector is

$$\mathbf{v}(t) = (5, -32t).$$

Find $\int_0^6 \mathbf{v}(t)\,dt$. What does the answer mean about the particle's motion?

Solution   The calculation is routine; ← we integrate in each coordinate separately:

$$\int_0^6 \mathbf{v}(t)\,dt = \left(\int_0^6 5\,dt, \int_0^6 -32t\,dt \right) = (30, -576).$$

More interesting than the numerical answer is what it tells: the particle's *displacement* over the time interval $0 \le t \le 6$. If $\mathbf{p}(t)$ is the particle's position at time t, then $\mathbf{p}'(t) = \mathbf{v}(t)$, and

$$\int_0^6 \mathbf{v}(t)\,dt = \mathbf{p}(6) - \mathbf{p}(0).$$

Thus, if $\mathbf{v}(t)$ represents velocity in units of feet per second, the calculation means that, over the 6-second interval, the object moves 30 feet to the right and 576 feet downward.   ■

Multiple integrals   We have also studied **multiple integrals** in which the integrand is a scalar-valued function of two or three variables and the domain of integration is a subset of $\mathbb{R}^2$ or $\mathbb{R}^3$. ← In the simplest case, the integrand is a function of two variables, and the domain of integration is a rectangle. In this happy event, the integral can be calculated by "iteration," that is, by integrating in one variable and then the other.

The name "multiple integral" is standard but a little vague; it doesn't say whether inputs, outputs, or both are "multiple."

EXAMPLE 3   Let $f(x, y) = 2 - x^2 y$, $R = [0, 1] \times [0, 1]$, and $I = \iint_R f(x, y)\,dA$. Calculate I as an iterated integral. What does the answer mean?

Solution   We will integrate first in x and then in y:

$$I = \int_0^1 \left(\int_0^1 (2 - x^2 y)\,dx \right) dy = \int_0^1 \left(2x - \frac{x^3 y}{3} \Big]_0^1 \right) dy$$

$$= \int_0^1 \left(2 - \frac{y}{3} \right) dy = \frac{11}{6}.$$

And gives the same answer!

The other order of integration is possible, too. ← That calculation appears, along with a useful picture, in Example 2, page 793.

Geometrically, the integral gives the volume of the solid bounded below by the domain of integration, R, and above by the surface $z = f(x, y)$. Alternatively, the integral can be thought of as the *average value* of f over the rectangle $[0, 1] \times [0, 1]$, which has area 1. (The answer 11/6, a little less than 2, accords well with the picture and with the table of representative values of f.)   ■

Techniques like those just shown can be used to integrate a function $f(x, y, z)$ over a rectangular solid region in $\mathbb{R}^3$.

Integrals over nonrectangles; polar coordinates   If a domain of integration is not rectangular, then various strategies are used as determined by the situation. One possibility is to use *variable* limits of integration on the inner integral. Another possibility arises when the domain of integration is more conveniently represented in polar coordinates r and θ. Things work best when the integrand also lends itself to polar form.

EXAMPLE 4 Let R be the region inside the unit circle, and let $f(x, y) = x^2 + y^2$. Use polar coordinates to calculate $I = \iint_R f(x, y)\, dA$. What does the answer mean geometrically?

Solution First we write everything in polar form, using the following change-of-variable equations:

$$x = r\cos\theta; \quad y = r\sin\theta; \quad x^2 + y^2 = r^2; \quad dA = r\, dr\, d\theta.$$

(The last identity is called the **area element** in polar coordinates.) For the integrand, we have $f(x, y) = x^2 + y^2 = r^2$. The domain of integration is a "polar rectangle" described by the inequalities

$$0 \le r \le 1; \quad 0 \le \theta \le 2\pi.$$

Thus, rewritten in polar coordinates, the integral I becomes

$$I = \int_0^{2\pi} \int_0^1 r^2 r\, dr\, d\theta = \int_0^{2\pi} \int_0^1 r^3\, dr\, d\theta.$$

Integrating in r and then θ gives $I = \pi/2 \approx 1.57$. Geometrically, the integral measures the volume of the region under the paraboloid $z = x^2 + y^2$ and above the unit disk $x^2 + y^2 \le 1$. ∎

Cylindrical and spherical coordinates Similar change-of-variable strategies work for triple integrals when the domain of integration, the integrand, or both are best described in cylindrical coordinates (r, θ, z) or spherical coordinates (ρ, θ, ϕ). In each case, a **volume element** formula applies. The volume elements are, respectively,

$$dV = r\, dr\, d\theta\, dz \quad \text{for cylindrical coordinates,}$$

and

$$dV = \rho^2 \sin\phi\, d\rho\, d\theta\, d\phi \quad \text{for spherical coordinates.}$$

A typical calculation in spherical coordinates follows.

EXAMPLE 5 Calculate $I = \iiint_P \dfrac{1}{x^2 + y^2 + z^2}\, dV$, where P is the part of the unit ball $x^2 + y^2 + z^2$ that lies in the first octant.

Solution The first octant in $\mathbb{R}^3$ is the set of points whose spherical coordinates satisfy $0 \le \theta \le \pi/2$ (this ensures that $x \ge 0$ and $y \ge 0$) and $0 \le \phi \le \pi/2$ (this ensures that $z \ge 0$). Points that are also inside the unit ball satisfy $0 \le \rho \le 1$. Thus, P is—from the spherical coordinate point of view—a rectangular solid. The integrand is also best converted to spherical coordinates:

$$\frac{1}{x^2 + y^2 + z^2} = \frac{1}{\rho^2}.$$

Now everything is ready:

$$\begin{aligned}
I &= \iiint_P \frac{1}{x^2 + y^2 + z^2}\, dV \\
&= \int_0^{\pi/2} \int_0^{\pi/2} \int_0^1 \frac{1}{\rho^2} \rho^2 \sin\phi\, d\rho\, d\theta\, d\phi \qquad \text{(substituting as above)} \\
&= \int_0^{\pi/2} \sin\phi\, d\phi \cdot \int_0^{\pi/2} d\theta \cdot \int_0^1 d\rho = \frac{\pi}{2}.
\end{aligned}$$

∎

The general change-of-variables formula

The most mysterious ingredients in all of the foregoing examples, in which we changed from Cartesian to other coordinates, are the area and volume elements in polar, cylindrical, and spherical coordinates:

$$dA = r\,dr\,d\theta, \qquad dV = r\,dr\,d\theta\,dz, \qquad \text{and} \qquad dV = \rho^2 \sin\phi\,d\rho\,d\theta\,d\phi.$$

In fact, all three of these formulas are instances of a more general principle known as the change-of-variables formula. It describes how integrals change when one set of coordinate variables (such as x and y) are exchanged for another (such as r and θ).

Changing variables in one dimension The change-of-variables formula is best understood as the multivariable version of integration by substitution in elementary calculus. An example will help fix the ideas and notation. Notice that, although our use of the symbols u and x may seem "reversed" compared to what is usual in elementary calculus, the reversal is only in nomenclature; the ideas below are exactly the ones familiar from elementary calculus.

> **EXAMPLE 6** Use substitution to calculate $\int_0^{\pi/2} 3\sin^2 u\,\cos u\,du$.
>
> **Solution** The substitutions $x = \sin u$ and $dx = \cos u\,du$ seem natural. Indeed, they work just fine:
>
> $$\int_{u=0}^{u=\pi/2} 3\sin^2 u\,\cos u\,du = \int_{x=0}^{x=1} 3x^2\,dx = 1.$$
>
> Notice the change in limits of integration: Since $x = \sin u$, the u-interval $I_u = [0, \pi/2]$ corresponds to the x-interval $I_x = [0, 1]$.
>   The first equality is the main point. If we write $f(x) = 3x^2$ and $x = x(u) = \sin u$, then the first equality has the form
>
> $$\int_{I_u} f\big(x(u)\big)\frac{dx}{du}\,du = \int_{I_x} f(x)\,dx.$$
>
> We will see analogues of this equation in several variables.   ■

A magnification factor Notice especially the "extra" factor dx/du on the left in the preceding equation. In fact, the whole substitution method can be summarized succinctly as follows: ←

$$\frac{dx}{du}\,du = dx.$$

Perhaps even too succinctly—it packs a lot into a small space.

The main points of this equation are that, because of the relationship between u and x, a tiny change du in u produces a corresponding change dx in x, and that du and dx are related as the equation indicates. In effect, dx/du plays the role of a "magnification factor," either enlarging or diminishing the effect of small changes in u.

Changing several variables The change-of-variables recipe for multiple integrals uses similar ingredients. For double integrals, the usual setup is to start with an integral $\iint_{R_{xy}} f(x, y)\,dA$, where f is a function and R_{xy} is a region in the xy-plane. In addition, we are given (or somehow find) two functions relating x and y (the "old" coordinates) to u and v (the "new" coordinates):

$$x = x(u, v) \qquad \text{and} \qquad y = y(u, v).$$

We assume that these functions "map" the uv-plane (or part of it) into the xy-plane in such a way that some region R_{uv} in the uv-plane corresponds to R_{xy}, the original domain of integration in the xy-plane.

One reason to carry out such a change of variables is to simplify the *domain* of integration: Ideally, the uv-region R_{uv} will be simpler or more familiar than the xy-region R_{xy}. Another possibility is that the change of variables simplifies the *integrand* function. Sometimes both types of simplification occur—as they did in Example 5. (For one-variable integrals, by contrast, domains are usually just finite intervals, and so only the integrand normally benefits from simplification.)

So far, the one- and two-variable change-of-variable setups are similar. An important difference, however, is that, instead of the ordinary derivative dx/du, we now use the **Jacobian determinant**, that is, the determinant of the derivative matrix of the mapping $(u, v) \mapsto (x, y)$:

$$\frac{\partial(x, y)}{\partial(u, v)} = \det \begin{pmatrix} x_u & x_v \\ y_u & y_v \end{pmatrix}.$$

(The peculiar-looking "fraction" on the left is just a mnemonic shorthand.) It turns out that the determinant above plays the same role for the mapping $(u, v) \mapsto (x, y)$ as does the derivative dx/du in the one-variable setting: For any input (u, v), the absolute value $|\partial(x, y)/\partial(u, v)|$ is the factor by which the mapping *magnifies areas* near (u, v). In particular, the mapping transforms a tiny rectangle at (u, v), with area $\Delta u \, \Delta v$, to another tiny near-rectangle at (x, y) with area approximately

$$\Delta x \, \Delta y \approx \left| \frac{\partial(x, y)}{\partial(u, v)} \right| \Delta u \, \Delta v.$$

We can now state the general theorem. Some technical hypotheses are added, but the final equation is the main point:

> **THEOREM 2 (Change of variables in double integrals)**   Let coordinates (x, y) and (u, v) be related as above. We assume that all derivatives in question exist and are continuous, that the mapping $(u, v) \mapsto (x, y)$ is one-to-one, and that the regions R_{uv} and R_{xy} correspond to each other under the mapping. Then
>
> $$\iint_{R_{xy}} f(x, y) \, dA_{xy} = \iint_{R_{uv}} f\big(x(u, v), y(u, v)\big) \left| \frac{\partial(x, y)}{\partial(u, v)} \right| dA_{uv}$$
>
> $$= \iint_{R_{uv}} f\big(x(u, v), y(u, v)\big) \left| \frac{\partial(x, y)}{\partial(u, v)} \right| du \, dv.$$

EXAMPLE 7   What does the theorem say about polar coordinates in double integrals?

Solution   Using r and θ in place of u and v gives $x = r \cos\theta$ and $y = r \sin\theta$. Thus, the Jacobian matrix is   *The names aren't sacred.*

$$\begin{pmatrix} x_r & x_\theta \\ y_r & y_\theta \end{pmatrix} = \begin{pmatrix} \cos\theta & -r \sin\theta \\ \sin\theta & r \cos\theta \end{pmatrix};$$

it has determinant r. Thus the theorem says that

$$\iint_{R_{xy}} f(x, y) \, dA = \iint_{R_{r\theta}} f\big(x(r, \theta), y(r, \theta)\big) r \, dr \, d\theta;$$

this is the familiar area element for polar coordinates. ∎

The dirty details Fully carrying out the change-of-variable mechanics can be messy symbolically, but it is not too bad in the case of a linear change of variables. The following example illustrates the process—with some details left to the reader.

EXAMPLE 8 Find $\iint_{R_{xy}} x\, dA$, where R is the shaded region shown in Figure 1(b).

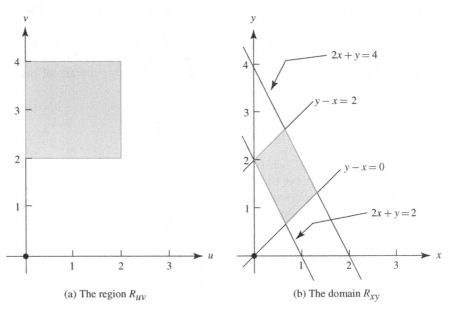

(a) The region R_{uv} (b) The domain R_{xy}

FIGURE 1
Corresponding regions R_{uv} and R_{xy}

Solution   As Figure 1(b) shows, the region R_{xy} is defined by the inequalities

$$0 \le y - x \le 2 \qquad \text{and} \qquad 2 \le 2x + y \le 4.$$

These inequalities suggest that we try the change of variables

$$u = y - x \qquad \text{and} \qquad v = 2x + y;$$

then the xy-domain R_{xy} corresponds to the *rectangular uv*-domain R_{uv} shown in Figure 1(a), for which the corresponding inequalities are simply

$$0 \le u \le 2 \qquad \text{and} \qquad 2 \le v \le 4.$$

So far so good—but to use Theorem 2 we need to write x and y in terms of u and v. Here this is easy to do, by simultaneously solving the following equations: ◄

Check the details for yourself.

$$u = y - x \qquad \text{and} \qquad v = 2x + y \quad \Longrightarrow \quad x = \frac{v - u}{3} \qquad \text{and} \qquad y = \frac{2u + v}{3}.$$

From this we get

$$\left| \frac{\partial(x, y)}{\partial(u, v)} \right| = \begin{vmatrix} -1/3 & 1/3 \\ 2/3 & 1/3 \end{vmatrix} = \frac{1}{3},$$

and we are finally ready to apply Theorem 2:

$$\iint_{R_{xy}} x \, dA_{xy} = \iint_{R_{uv}} x(u, v) \cdot \frac{1}{3} dA_{uv}$$

$$= \int_{R_{uv}} \frac{v - u}{9} dA_{uv} \qquad \text{(since } x = (v - u)/3)$$

$$= \frac{1}{9} \int_{v=2}^{v=4} \int_{u=0}^{u=2} (v - u) \, du \, dv \qquad \text{(in iterated form)}$$

$$= \frac{8}{9}. \qquad \text{(by routine computation)}$$

■

In three variables The same formula holds for triple integrals in three variables, except that the Jacobian determinant has the form $\partial(x, y, z)/\partial(u, v, w)$. ➡ The work is potentially quite messy, but in many cases of real interest the change of coordinates is simple. This is especially true when the change of variable amounts merely to "stretching" in one or more dimensions.

The Jacobian matrix is 3×3.

EXAMPLE 9 Let a, b, and c be positive numbers. Find the volume of the ellipsoid $\frac{x^2}{a^2} + \frac{y^2}{b^2} + \frac{z^2}{c^2} \leq 1$ in $\mathbb{R}^3$.

Solution If we use the "new" coordinates

$$u = \frac{x}{a}, \quad v = \frac{y}{b}, \quad \text{and} \quad w = \frac{z}{c},$$

then the ellipsoid E in xyz-space corresponds to the unit ball B in uvw-space; it has formula $u^2 + v^2 + w^2 \leq 1$ and (known) volume $4\pi/3$. To use the change-of-variable formula, we need the Jacobian determinant of the mapping $(u, v, w) \mapsto (x, y, z)$. In this case it is easy to find. The preceding formulas give

$$x(u, v, w) = au; \quad y(u, v, w) = bv; \quad z(u, v, w) = cw.$$

Thus, the Jacobian matrix is diagonal, with constant determinant ➡

Check the easy calculation.

$$\frac{\partial(x, y, z)}{\partial(u, v, w)} = abc.$$

The change-of-variable formula now finishes our problem:

$$\text{volume of E} = \iiint_E 1 \, dV_{xyz} = \iiint_B 1 \left| \frac{\partial(x, y, z)}{\partial(u, v, w)} \right| dV_{uvw}$$

$$= abc \iiint_R 1 \, dV_{uvw} = \frac{4\pi}{3} abc.$$

Notice, finally, that the result has a very pleasing geometric meaning. At every point of uvw-space, the mapping $(u, v, w) \mapsto (x, y, z) = (au, bv, cw)$ "magnifies" distances by factors of a, b, and c, respectively, in the three coordinate directions; the corresponding effect on volumes is to magnify by the factor abc. ■

EXAMPLE 10 This example explores the fact that the Jacobian determinant of a mapping $(u, v) \mapsto (x, y)$ describes an area magnification factor. In each case below, let R_{uv} be the unit square $0 \leq u \leq 1; 0 \leq v \leq 1$; it has area 1. Let R_{xy} be the image of R_{uv} in the xy-plane under the mapping $(u, v) \mapsto (x, y)$. In each case following: (i) describe R_{xy};

(ii) find the area of R_{xy}; (iii) find the Jacobian determinant of the given mapping. (In each case, R_{xy} is a parallelogram, and so it is enough to find its corners.)

(a) $x = 2u,\ y = v$;   (b) $x = 1 + 2u,\ y = 1 + 3v$;   (c) $x = 1 + 3u + v,\ y = 1 + u + 2v$.

Solution   Figure 2 shows the images R_{xy} of the unit square R_{uv} for each of the three mappings:

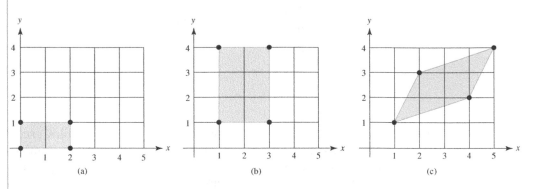

FIGURE 2
Images of the unit square for three mappings

In case (a) the image region R_{xy} is the rectangle $[0, 2] \times [0, 1]$, and so the magnification factor is 2. The derivative matrix is $\begin{bmatrix} 2 & 0 \\ 0 & 1 \end{bmatrix}$; its Jacobian determinant is 2.

In case (b) the image region R_{xy} is the rectangle $[1, 3] \times [1, 4]$ with area 6; this is also the magnification factor. The derivative matrix is $\begin{bmatrix} 2 & 0 \\ 0 & 3 \end{bmatrix}$; its Jacobian determinant is 6.

In case (c) the image region R_{xy} is the (nonrectangular) parallelogram shown. A close look at the grid shows that the shaded area, which is also the magnification factor, is 5. Here the derivative matrix is $\begin{bmatrix} 3 & 1 \\ 1 & 2 \end{bmatrix}$; its Jacobian determinant is 5. ∎

The idea of the proof   Approximating sums are the idea behind the proof of the change-of-variables theorem. Recall that any double integral $\iint_{R_{uv}} g(u, v)\, dA$ is defined as a limit of approximating sums of the form

$$\lim_{n \to \infty} \sum_{i=1}^{n} g(u_i, v_i)\, A(R_i),$$

where $A(R_i)$ is the area of a small uv-rectangle containing (u_i, v_i).

In the case at hand, suppose that

$$g(u, v) = f\big(x(u, v), y(u, v)\big) \left| \frac{\partial(x, y)}{\partial(u, v)} \right|.$$

Then the approximating sum above becomes

$$\lim_{n \to \infty} \sum_{i=1}^{n} f\big(x(u_i, v_i), y(u_i, v_i)\big) \left| \frac{\partial(x, y)}{\partial(u, v)}(u_i, v_i) \right| A(R_i).$$

By the earlier remark about area magnification, this sum is approximately equal to

$$\lim_{n \to \infty} \sum_{i=1}^{n} f(x_i, y_i)\, A(R'_i),$$

where $A(R'_i)$ is the area of a small xy-rectangle R'_i containing (x_i, y_i). But this last sum approximates the integral $\iint_{R_{xy}} f(x, y)\, dA$. We have arrived at last.

1. In Example 1 we calculated the length of one turn of a certain spiral in $\mathbb{R}^3$. The same curve can be parametrized by

$$\mathbf{r}(t) = \left(\cos(t^2), \sin(t^2), t^2 \right); \quad 0 \le t \le \sqrt{2\pi}.$$

Recalculate the arclength using this parametrization.

2. Let $\mathbf{r}(t) = (t + \sin t, 1 + \cos t)$.

   (a) Plot $\mathbf{r}(t)$.

   (b) Find (exactly, by antidifferentiation) the length of one arch of the curve. [HINT: $2 + 2\cos t = 4\cos^2(t/2)$. To simplify the computation, find the length from $t = 0$ to $t = \pi$ and double the value.]

3. Repeat Exercise 2 with $\mathbf{r}(t) = (t/2 + \sin t, 1 + \cos t)$. (This curve represents the path traced by a point on a wheel that slips as it rolls.) In this case you will need to estimate the final integral; use the midpoint rule with ten subdivisions.

4. Consider the situation described in Example 2. Assume that the units of time and distance are seconds and feet, respectively.

   (a) Find the distance traveled by the particle over the 6-second interval.

   (b) Find the vector-valued acceleration function $\mathbf{a}(t)$. What is the appropriate unit of measurement?

   (c) Calculate $\int_0^6 \mathbf{a}(t)\,dt$. What does the answer mean about the particle's motion?

5. Consider the integral $I = \iint_R x^2 y\,dA$, where R is the region bounded by the curves $y = x^2$ and $y = 1$.

   (a) Calculate I by integrating first in y and then in x.

   (b) Calculate I by integrating first in x and then in y.

6. Let $f(x, y) = y$, and let R be the plane region bounded by the curves $y = \ln x$, $y = 0$, and $x = e$.

   (a) Calculate $I = \iint_R f(x, y)\,dA$ by integrating first in y, then in x.

   (b) Calculate $I = \iint_R f(x, y)\,dA$ by integrating first in x, then in y.

7. Consider the integral I of Example 4. We said there that the integral gives the volume of the solid region (call it S) under the paraboloid $z = x^2 + y^2$ and above the unit disk $x^2 + y^2 \le 1$.

   (a) The solid region S described above is a solid of revolution. Explain why — that is, tell which plane region must be revolved around which axis to produce S.

   (b) Express the volume of S as an integral in one variable.

8. Consider the ellipse E in the xy-plane defined by $\dfrac{x^2}{a^2} + \dfrac{y^2}{b^2} \le 1$, where a and b are positive numbers.

   (a) Set up an elementary calculus-style integral for the area of E.

   (b) Use the idea of Example 9 and an appropriate change of variable to find the area of E.

   (c) How does the area calculated in part (b) behave as $a \to b$?

9. (For students with determinant experience) As in Example 7, find an appropriate determinant to explain the volume element formula $dV = r\,dr\,d\theta\,dz$ for cylindrical coordinates.

10. (For students with determinant experience) As in Example 7, find an appropriate determinant to explain the volume element formula $dV = \rho^2 \sin\phi\,d\rho\,d\theta\,d\phi$ for spherical coordinates.

Exercises 11–14 explore the fact that the Jacobian determinant of a mapping $(u, v) \mapsto (x, y)$ describes an area magnification factor. (See also Example 10.) In each exercise, let R_{uv} be the unit square $0 \le u \le 1$, $0 \le v \le 1$; it has area 1. Let R_{xy} be the image of R_{uv} in the xy-plane under the mapping $(u, v) \mapsto (x, y)$. In each part following, (i) sketch R_{xy}, (ii) find the area of R_{xy}, and (iii) find the Jacobian determinant of the given mapping. [NOTE: In each case, R_{xy} is a parallelogram, and so it is enough to find its corners.]

11. $x = u$, $y = 2v$.

12. $x = 2 + 3u$, $y = 4 + 5v$.

13. $x = u + v$, $y = u - v$.

14. $x = 1 + 2u + 3v$, $y = 4 + 5u + 6v$. [HINT: Use the cross product to find the area of the parallelogram.]

15. Let $I = \iint_{R_{xy}} xy\,dA$, where R_{xy} is the region bounded by $y = 0$, $y = 1$, $y = x$, and $y = x - 2$.

    (a) Evaluate I in xy-coordinates.

    (b) Let $u = x - y$ and $v = y$. Draw the region R_{xy} and the corresponding region R_{uv}.

    (c) Use the change of variables in part (b) to evaluate I.

16. Let $I = \iint_{R_{xy}} (x + y)\,dA$, where R_{xy} is the region bounded by $3x - 2y = 4$, $3x - 2y = -2$, $x + y = -1$, and $x + y = 2$.

    (a) Evaluate I in xy-coordinates.

    (b) Let $u = 3x - 2y$ and $v = x + y$. Draw the region R_{xy} and the corresponding region R_{uv}.

    (c) Use the change of variables in part (b) to evaluate I.

17. Let $I = \iint_{R_{xy}} e^x\,dA$, where R_{xy} is the region bounded by $y = x$, $y = 2x$, $x + y = 4$, and $x + y = 0$.

    (a) Evaluate I in xy-coordinates.

    (b) Let $u = x + y$ and $v = x - y$. Draw the region R_{xy} and the corresponding region R_{uv}.

    (c) Use the change of variables in part (b) to evaluate I.

18. Let $I = \iint_{R_{xy}} \dfrac{(x + y)^2}{x - y}\,dA$, where R_{xy} is the region bounded by $x - y = 1$, $x - y = 5$, $x = 0$, and $y = 0$.

    (a) Evaluate I in xy-coordinates.

(b) Let $u = x + y$ and $v = x - y$. Draw the region R_{xy} and the corresponding region R_{uv}.

(c) Use the change of variables in part (b) to evaluate I.

19. Let D be the region in xyz-space defined by the inequalities $1 \le x \le 2, 0 \le xy \le 2$, and $0 \le z \le 1$. Write the triple integral $\iiint_D (x^2 y + 3xyz) \, dV$ as an iterated integral in the coordinates $u = x$, $v = xy$, and $w = 3z$.

20. Let $g(x, y) = (u, v) = (-x + y, x + 2y)$, let R be a region in the xy-plane, and let T be the image of R under the mapping g. Furthermore, suppose that $f(x, y) > 0$ for all $(x, y) \in \mathbb{R}^2$. Is the statement

$$\iint_R f(x, y) \, dA < \iint_T f(x(u, v), y(u, v)) \, dA$$

true? Justify your answer.

SUMMARY

In Chapter 13 we extended the familiar idea of derivative to the multivariate setting. In this chapter we did the same thing for integrals. Although integrals are defined in much the same way as in elementary calculus, the multivariate setting requires new interpretations and permits new applications.

Multiple integrals and approximating sums   The integral of any function over any domain, in any number of variables, is a certain limit of **approximating sums**. For a function $f(x, y)$ of two variables, such a sum has the form

$$S_m = f(P_1) \Delta A_1 + f(P_2) \Delta A_2 + \cdots + f(P_m) \Delta A_m = \sum_{i=1}^{m} f(P_i) \Delta A_i,$$

where the P_i are points chosen in a region of integration (let's call it R) and ΔA_i is the area of a **subregion** R_i formed by subdividing R into m smaller pieces. Under appropriate assumptions on f and the subregions R_i, these sums approach the **double integral** $\iint_R f(x, y) \, dA$ as m tends to infinity. We began by illustrating and investigating such sums and their connections with integrals. Then we explained how the properties and interpretations of integrals themselves are inherited from similar properties of approximating sums.

Calculating integrals by iteration   Although *defined* as limits of sums, integrals are most easily *evaluated* using antiderivatives. Doing so for functions of several variables involves **iterated integration**: evaluating an integral by repeated antidifferentiation, one variable at a time. The process is especially simple when we integrate a well-behaved function over a **rectangular domain** $R = [a, b] \times [c, d]$. In this case, we can write

$$\iint_R f(x, y) \, dA = \int_a^b \int_c^d f(x, y) \, dy \, dx = \int_c^d \int_a^b f(x, y) \, dx \, dy.$$

(The last equality says that we can integrate the two variables in either order.) As a rough shorthand, $dA = dx \, dy = dy \, dx$.

Integrals over nonrectangular regions   Integrals must often be calculated over nonrectangular domains, such as plane regions bounded by curves, circles, or oblique lines. Such integrals, like those over rectangular domain, can be calculated by iteration, but special care must be taken for the geometry of the domain in question. We showed how to set up and calculate such integrals—sometimes with help from technology.

Double integrals in polar coordinates   As determined by the domain R and the function f, an integral $\iint_R f(x, y) \, dA$ may be easier to calculate in polar than in rectangular coordinates. This occurs, for instance, when R or f, or both, are simply described in polar coordinates. Changing from rectangular to polar coordinates transforms an integral $\iint_R f(x, y)\, dA$ to the new form $\iint_R g(r, \theta)\, dA$, or, in iterated form, $\iint_R g(r, \theta)\, r\, dr\, d\theta$. As a rough shorthand, we have $dA = r\, dr\, d\theta$, an equation explained and justified in Section 14.4.

Triple integrals   A function $g(x, y, z)$ of three variables can be integrated over a solid region S in three-dimensional space. We defined and illustrated the **triple integral** $\iiint_S g(x, y, z)\, dV$ and explored some of its properties. If the region S is "rectangular," of the form $S = [a, b] \times [c, d] \times [e, f]$, then the integral can be evaluated, as in the two-dimensional case, by iteration. In shorthand form, $dV = dx\, dy\, dz$.

Cylindrical and spherical coordinates   Just as polar coordinates simplify certain double integrals, the **cylindrical** and **spherical** coordinate systems simplify certain triple integrals. Cylindrical coordinates are usually denoted r, θ, and z, with r and θ related to x and y as in the polar system. Spherical coordinates are denoted ρ, θ, and ϕ, with θ as before, but ρ and ϕ are defined in reference to a sphere centered at the origin. We described properties and advantages of these special coordinate systems and derived the useful shorthand formulas

$$dV = r\, dr\, d\theta\, dz \quad \text{and} \quad dV = \rho^2 \sin\phi\, d\rho\, d\theta\, d\phi,$$

the first for cylindrical and the second for spherical coordinates.

Change of variables   Double and triple integrals come in many forms. Linking all of our integral formulas—regardless of the choice of coordinate system—is the general **change-of-variable** formula, Theorem 2. This theorem, a sophisticated analogue of the u-substitution method of elementary calculus, tells how to transform one integral into another by trading one set of variables for another.

REVIEW EXERCISES

1. Testing new seed, a farmer divides a 9-acre plot as shown in the figure below. The numbers indicate one year's crop yield in bushels of corn per acre for each field subdivision. The smallest squares have area 1 acre. (This implies, strangely enough, that each unit in the x- or y-direction represents about 209 feet.) Use the data to find the total yield of corn on the 9-acre plot and the average yield per acre. Interpret the results in terms of functions and integrals.

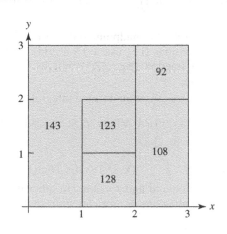

2. Find $\iint_D (x^2 + y^2 - 6xy)\, dA$, where D is the region between the curves $x^2 + y^2 = 1$ and $x^2 + y^2 = 4$.

3. Find the volume of the region between the paraboloids $z = 3 - 2x^2 - 2y^2$ and $z = x^2 + y^2$.

4. Find $\iiint_E (x + y + z)\, dV$, where E is the solid tetrahedron bounded by the coordinate planes and the plane $x + 2y + 3z = 6$.

5. Let a, b, and c be positive constants. Find the volume of the tetrahedron bounded by the three coordinate planes and the plane $ax + by + cz = 1$. [HINT: The problem can be done either as a double or as a triple integral.]

6. Use cylindrical coordinates to evaluate the integral $\iiint_C (x^2 + y^2)\, dV$, where C is the cylinder of radius 1 and height 1 with base on the xy-plane.

7. Use cylindrical coordinates to evaluate the integral $\iiint_C (x^2 + y^2)\, dV$, where C is the region under the cone $z = r$ above the xy-plane and inside the cylinder $r = 1$.

8. Let C be the solid region inside the cylinder $r = 1/2$ above the xy-plane and below the unit sphere $x^2 + y^2 + z^2 = 1$. (The

region looks like a cylinder with a rounded top.) Calculate $\iiint_C 2z \, dV$.

9. Evaluate $\iint_R \dfrac{x+y}{x^2+y^2} \, dA$, where R is the cardioid with polar equation $r = 1 + \cos\theta$.

10. Evaluate $\iiint_R z \, dV$, where R is the region above the cone $z = r$ and below the plane $z = 3$.

11. Evaluate $\iiint_R z^2 \, dV$, where R is the first octant part of the region inside the unit sphere $x^2 + y^2 + z^2 = 1$.

12. Set up an integral in polar coordinates that gives the area inside the cardioid $r = 1 + \cos\theta$.

13. Calculate $\iint_R y \, dA$, where R is the top half of the region inside the unit circle.

14. Use cylindrical coordinates to find $\iiint_R z \, dV$, where R is the region inside the cylinder $r = 1$ above the xy-plane and below the plane $z = 1$.

15. Calculate $\iint_R x \, dA$, where R is the right half of the region inside the unit circle $x^2 + y^2 = 1$.

16. Use cylindrical coordinates to find $\iiint_R z^2 \, dV$, where R is the solid region inside the cylinder $r = 1$ above the xy-plane and below the plane $z = 1$.

17. Set up an integral in spherical coordinates that gives the volume of the region (i) inside the sphere $x^2 + y^2 + z^2 = 1$ and (ii) in the first octant (where x, y, and z are all nonnegative).

18. Calculate $\iint_R (x^2 + y^2) \, dA$, where R is the upper half of the region inside the unit circle $x^2 + y^2 = 1$.

19. Use cylindrical coordinates to calculate $\iiint_R (x^2 + y^2 + 3z^2) \, dV$, where R is the solid region inside the cylinder $r = 1$ above the xy-plane and below the plane $z = 1$.

20. Calculate the integral $I = \iint_R (x + y + 1) \, dA$, where R is the rectangle $[0, 1] \times [0, 1]$. This integral gives the volume of a certain solid region; describe this solid briefly in words.

21. Let R_{xy} be a region in the xy-plane. Let R_{uv} be the region in the uv-plane that corresponds to R_{xy} when the change of variable $x = x(u, v)$, $y = y(u, v)$ is made. State the change of variable formula (i.e., the formula that relates an integral in the variables x and y to an integral in the variables u and v).

22. Write an integral for the volume of the solid bounded by the planes $x = 0$, $y = 0$, $z = 0$, $x + y + z = 1$.

23. Evaluate $\int_0^1 \int_0^{2y} \int_0^x dz \, dx \, dy$.

24. Set up an interated integral for the volume bounded above by the paraboloid $z = 4 - x^2 - y^2$ and below by the parabolic cylinder $z = 2 + y^2$.

25. Set up an integral in cylindrical coordinates for the "ice cream cone" bounded below by the surfaces $z^2 = 3(x^2 + y^2)$ and $z = 0$ and above by the sphere $x^2 + y^2 + z^2 = 1$.

26. Set up an integral in spherical coordinates for the volume in the previous exercise.

27. Find the volume of the region bounded above by the cylinder $z = 4 - y^2$ and below by the elliptic paraboloid $z = 2x^2 + y^2$.

28. Evaluate $\int_0^3 \int_{x=\sqrt{y}}^{x=2} xy \, dx \, dy$.

29. Find the mass of the region bounded below by the paraboloid $z = x^2 + y^2$ and above by the plane $z = 1$ if the density at each point is given by $f(x, y, z) = z$.

30. Let $I = \iint_R x \, dA$, where $R = [0, 2] \times [0, 2]$. Calculate a midpoint-rule approximating sum S_4 for I with 4 subdivisions (i.e., use a 2×2 grid).

31. Let $f(x, y) = \sin x + \cos y$.

(a) Evaluate $\iint_R f(x, y) \, dA$, where R is the square $[0, \pi/2] \times [0, \pi/2]$.

(b) The integral in the previous part gives the volume of a certain solid region. Describe this region carefully in words (or draw and label a picture).

32. Let R be the region described by the inequalities $0 \le r \le 1$ and $\pi/4 \le \theta \le \pi/2$, and let $I = \iint_R x \, dA$.

(a) Write I as an iterated integral in xy-coordinates.

(b) Write I as an iterated integral in polar coordinates.

(c) Evaluate I.

33. Calculate a midpoint approximating sum with four subdivisions (2 in each direction) for the double integral $\iint_R (x^2 + y^2) \, dA$, where R is the square $[0, 1] \times [0, 1]$.

34. Calculate a midpoint approximating sum with four subdivisions (two in each direction) for the integral $\iint_R f(x, y) \, dA$, where R is the square $[0, 2] \times [0, 2]$ and $f(x, y) = y$.

35. Let I be the integral $\iint_R (3x^2 + 2y) \, dA$, where $R = [0, 3] \times [0, 3]$.

(a) Compute a double midpoint sum with with a total of nine subdivisions (i.e., use a 3×3 grid). State the answer as a fraction (e.g., 22/3, not in rounded form (e.g., 7.667).

(b) Evaluate I exactly as an iterated integral.

36. Evaluate $\iint_R f(x, y) \, dA$, where $f(x, y) = \sin(x) + \cos(2y)$ and $R = [0, 1] \times [0, 1]$.

37. Calculate $I = \int_{-1}^1 \int_{x=y^2}^{x=y+2} (x + y) \, dx \, dy$. Then reverse the order of integration and find I again. (In other words, find I again by calculating one or more integrals that end in $dy \, dx$

rather than $dx\,dy$.) Draw pictures as necessary to illustrate your reasoning.

38. Find the volume of the solid under the surface $z = \sqrt{x^2 + y^2}$ and above the circle $x^2 + y^2 = 2y$.

39. Find the volume of the solid that lies inside the sphere $x^2 + y^2 + z^2 = 25$ and outside the cylinder $x^2 + y^2 = 9$.

40. Find the volume of the solid that is inside the sphere $x^2 + y^2 + z^2 = 4$ and outside the cylinder $x^2 + y^2 = 2y$.

41. Find the volume of the solid bounded by the upper surface of the sphere $x^2 + y^2 + z^2 = 4$ and the paraboloid $z = 4 - x^2 - y^2$.

42. Find the volume of the solid bounded by the lower surface of the sphere $x^2 + y^2 + z^2 = 4$ and the paraboloid $z = 4 - x^2 - y^2$.

43. Find the area of the region inside the cardioid $r = 1 + \sin\theta$ and outside the circle $r = 1$.

44. Find the volume of the solid that lies below the paraboloid $z = x^2 + y^2$ and above the disk $x^2 + y^2 \le 9$.

45. Find the volume of the solid that lies under the plane $6x + 4y + z = 12$ and above the circular region defined by $x^2 + y^2 = y$.

46. Find the volume of the solid bounded by the cone $z^2 = x^2 + y^2$ and the cylinder $x^2 + y^2 = x$.

47. Evaluate $\displaystyle\int_0^1 \int_{z^2}^{z} \int_0^{\ln z} y e^x \, dx \, dy \, dz$.

48. Evaluate $\displaystyle\iiint_R f(x, y, z) \, dV$, where R is the portion of the sphere $x^2 + y^2 + z^2 = 1$ that lies in the first octant (i.e., $x \ge 0$, $y \ge 0$).

Mass and Center of Mass

Density and mass   It is often useful to think of a quantity, such as mass, as "distributed" within a region in the xy-plane or in xyz-space. In the plane, for instance, we saw in Section 14.3 how to use integrals to find the center of mass of a region R considered as a thin plate (**plane lamina** is the technical term) with constant density. We met the three-dimensional analogue of the same problem in Section 14.5, where we integrated to find centers of mass of some simple solid regions also regarded as having constant density.

A cubic centimeter of silver weighs around 10 g; the same volume of gold weighs almost 20 g.

Density measures the mass per unit area or unit volume of a material. The numerical value of density depends on the units of measurement—grams per cubic centimeter, say, or kilograms per cubic meter. But, no matter the units, gold is about twice as "dense" as silver. ◄

If a plane lamina or a solid has *constant* density, then its mass and center of mass depend only on the object's shape and size. In useful applications, however, the density of an object *varies* from point to point. This occurs if, for example, an object is made partly of silver and partly of gold. In such cases we can describe density with a nonnegative function $\rho(x, y)$ (or $\rho(x, y, z)$), which takes different values for different inputs (x, y) or (x, y, z).

We found a similar one-dimensional formula in earlier work.

For an object of constant density, the mass is simply the product density × volume. The formula for the mass of an object with *variable* density is therefore not unexpected: ◄

FACT (Measuring mass)   Suppose that mass is distributed over a plane region R or a solid S, with variable density ρ. Then the total mass of R or S is given (in appropriate units) by

$$\iint_R \rho(x, y)\, dA \quad \text{or} \quad \iiint_S \rho(x, y, z)\, dV.$$

EXAMPLE 1   One copy of the "brick" $[0, 1] \times [0, 2] \times [0, 3]$ in $\mathbb{R}^3$ has constant density 10 grams per unit of volume. A second copy of the brick has variable density $\rho(x, y, z) = 10x$ grams per unit of volume. Find the mass of each brick.

Solution   The first brick has volume 6 and density 10, and so its mass is 60 g. The second brick's mass is found by integration:

$$\text{mass} = \int_0^1 \int_0^2 \int_0^3 10x\, dz\, dy\, dx = 30;$$

thus, the second brick has half the mass of the first.   ∎

PROBLEM 1   A third and fourth brick shaped like those in Example 1, have density functions $\rho(x, y, z) = 10y$ and $\rho(x, y, z) = 10z$, respectively. Find the mass of each.

Centers of mass and weighted averages   The center of mass of a plane or solid region can be thought of as the region's "balance point": If we lift the object by a thread attached at the center of mass, the object should hang motionless, without twisting or turning.

In the case of constant density we have seen that the coordinates $(\bar{x}, \bar{y})$ or $(\bar{x}, \bar{y}, \bar{z})$ of the center of mass are simply the **average values** of the coordinate functions x, y, and (if necessary) z over the region in question. For a plane region R, for example, we have

$$\bar{x} = \frac{\iint_R x \, dA}{\iint_R 1 \, dA}; \qquad \bar{y} = \frac{\iint_R y \, dA}{\iint_R 1 \, dA}.$$

The integral formulas in the case of nonconstant density are similar, but they involve **weighted average values** of the coordinate functions. Naturally enough, the density function ρ provides the weighting:

FACT (Finding the center of mass) Suppose that mass is distributed over a plane region R or a solid region S with variable density ρ. The centers of mass of R and S, written $(\bar{x}, \bar{y})$ and $(\bar{x}, \bar{y}, \bar{z})$, respectively, have coordinates given by weighted average values of the coordinate functions. For instance,

$$\bar{x} = \frac{\iint_R x \, \rho(x, y) \, dA}{\iint_R 1 \, \rho(x, y) \, dA} \qquad \text{or} \qquad \bar{x} = \frac{\iiint_S x \, \rho(x, y, z) \, dV}{\iiint_X 1 \, \rho(x, y, z) \, dV};$$

similar formulas hold for $\bar{y}$ and $\bar{z}$.

Observe:

- The denominator in each equation is the total mass of the object.
- If the density ρ happens to be a constant, then ρ cancels from both numerator and denominator. In this case $\bar{x}$, $\bar{y}$, and $\bar{z}$ are ordinary average values of x, y, and z.

EXAMPLE 2 Find the centers of mass of the two bricks in Example 1.

Solution For each brick we have

$$\bar{x} = \frac{\int_0^1 \int_0^2 \int_0^3 x \, \rho(x, y, z) \, dz \, dy \, dx}{\text{total mass}}.$$

(We found the denominators in Example 1.) For the first brick, we get

$$\bar{x} = \frac{\int_0^1 \int_0^2 \int_0^3 x \cdot 10 \, dz \, dy \, dx}{60} = \frac{1}{2},$$

and similar easy calculations give $\bar{y} = 1$, $\bar{z} = 3/2$. Thus, the first brick's center of mass is simply its geometric center. ⟶

For the second brick, we have

This should not be surprising.

$$\bar{x} = \frac{\int_0^1 \int_0^2 \int_0^3 x \cdot 10x \, dz \, dy \, dx}{30} = \frac{2}{3};$$

similar calculations give $\bar{y} = 1$, $\bar{z} = 3/2$. We see, in particular, that the brick's variable density "moves" the center of mass away from the geometric center in the direction of greater density. ∎

PROBLEM 2 Find the centers of mass of the third and fourth bricks described earlier with density functions $\rho(x, y, z) = 10y$ and $\rho(x, y, z) = 10z$, respectively.

PROBLEM 3 A fifth brick, shaped like those in Example 1, has variable density $\rho(x, y, z) = 10(x + y)$ grams per unit of volume. Find its mass and center of mass.

PROBLEM 4   A sixth brick, shaped like those in Example 1, has variable density $\rho(x, y, z) = 10(y-1)^2$ grams per unit of volume. Find its mass and center of mass. Why is the center of mass the ordinary center in this case?

PROBLEM 5   Let T_1 be the triangle bounded by the lines $y = x$, $y = -x$, and $y = 1$; assume T_1 has constant density. Find the center of mass of T_1. (NOTE: It is easily shown that $\bar{x} = 0$, so just find $\bar{y}$.)

PROBLEM 6   Let T_2 be the triangle bounded by the lines $y = x$, $y = -x$, and $y = 1$; assume T_2 has density function $\rho(x, y) = y$. Find the center of mass of T_2. (NOTE: It is easily shown that $\bar{x} = 0$, so just find $\bar{y}$.)

PROBLEM 7   Let C_1 be the solid region above the cone $z = r$ and below the plane $z = 1$ (we are using cylindrical coordinates). Assume C_1 has constant density. Find the center of mass of C_1. (NOTE: It is easily shown that $\bar{x} = 0$ and $\bar{y} = 0$, so just find $\bar{z}$.)

PROBLEM 8   Let C_2 be the solid region above the cone $z = r$ and below the plane $z = 1$; suppose C_2 has nonconstant density function $\rho(x, y, z) = z$. Find the center of mass of C_2. (NOTE: Here $\bar{x} = 0$ and $\bar{y} = 0$, so just find $\bar{z}$.)

OTHER TOPICS

15.1 LINEAR, CIRCULAR, AND COMBINED MOTION

This section builds on ideas developed in Section 12.6.

In Section 12.6 we studied how the calculus of vectors and vector-valued functions can be used to model phenomena of motion. Here we continue and extend that study, with emphasis on some especially important specific types of motion.

Motions that arise in practical applications can often be thought of as combinations of simpler, more basic motions. A point on a moving bicycle wheel, for instance, follows a path that is determined by two simpler types of motion: the rotation of the wheel and the linear motion of the bicycle itself. (We'll model this situation in detail later in this section.) To model such motions, therefore, it makes sense first to study some simpler, elementary motions and then see how to combine them—using vector algebra—to form more complex, "compound" motions.

Modeling compound motions

A two-part industrial robot arm might appear as in Figure 1:

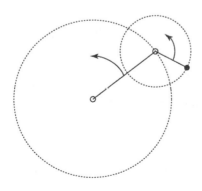

FIGURE 1
The arm of a simple robot

The long arm is pinned at the center of the large circle, the short section is pinned to the end of the long arm, and each arm rotates separately (driven by a motor, perhaps) about its fixed end. What path does the black dot (which might represent a tool at the end of the compound arm) follow as the two arms rotate? What are the velocity and acceleration of the black dot at any given time? (Controlling acceleration is practically important because acceleration is proportional to force; excessive acceleration could damage the machinery.)

At first blush these questions seem quite difficult. The answers depend, after all, on several variables: the lengths of the two arms and the speed and direction of each rotation. We will see, however, that the calculus of vector-valued functions makes modeling such motions quite straightforward. Our strategy will be to break the compound arm's motion into two simpler motions, analyze each separately, and then, quite literally, reassemble the parts.

Figure 2 shows two different paths the black dot might follow as determined by the data. Axes cross at the fixed point of the large arm. We will explain later how the paths are drawn mathematically.

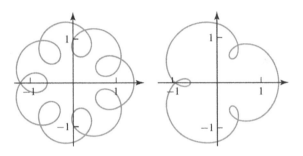

FIGURE 2
Two robot paths

Modeling linear motion Motion in a straight line, with constant speed, is called **uniform linear motion**. This type of motion is easily modeled by using linear position functions. As we saw in Chapter 12, the linear position function

$$\mathbf{p}(t) = (x_0, y_0) + t(a, b)$$

describes a particle that (i) starts at position (x_0, y_0) at time $t = 0$, (ii) moves in the direction of the velocity vector (a, b), and (iii) has constant speed $\sqrt{a^2 + b^2}$. Choosing the constants judiciously lets us model any instance of uniform linear motion.

EXAMPLE 1 Suppose that a particle moves linearly at constant speed from $(1, 2)$ at $t = 0$ to $(5, 6)$ at $t = 3$. Find the position, velocity, and acceleration functions. What is the particle's speed?

Solution The vector $\mathbf{v} = (4, 4) = (5, 6) - (1, 2)$ points in the right direction; it joins the point $(1, 2)$ to the point $(5, 6)$. To arrange that the trip take 3 seconds, we use velocity vector $(4, 4)/3$. The desired position function, therefore, is

$$\mathbf{p}(t) = (1, 2) + t\left(\frac{4}{3}, \frac{4}{3}\right);$$

the velocity and acceleration are

$$\mathbf{v}(t) = \left(\frac{4}{3}, \frac{4}{3}\right) \qquad \text{and} \qquad \mathbf{a}(t) = (0, 0).$$

The speed is $\sqrt{4^2 + 4^2}/3 = \sqrt{32}/3 \approx 1.89$ units per second. ▪

Modeling circular motion   Constant-speed motion along a circular path is called **uniform circular motion**. The following special cases are most useful:

FACT   For any center point (x_0, y_0) and any radius $R > 0$, the position function

$$\mathbf{p}(t) = (x_0, y_0) + (R\cos t, R\sin t)$$

models *counterclockwise* uniform circular motion about (x_0, y_0) with radius R, at constant speed R. The position function

$$\mathbf{p}(t) = (x_0, y_0) + (R\cos t, -R\sin t)$$

models *clockwise* uniform circular motion at constant speed R.

All parts of the Fact are easy to verify; see the exercises. Using the Fact—and a few handy tricks—we can model all sorts of uniform circular motion. We illustrate several by example.

EXAMPLE 2   **(Different speeds)** A particle has position function

$$\mathbf{p}(t) = (x_0, y_0) + \big(R\cos(at), R\sin(at)\big),$$

where $a > 0$ is a positive constant. What difference does the a make?

Solution   The a makes no difference to the particle's path—the particle still travels at constant speed along the circle of radius R about (x_0, y_0). Velocity and acceleration are affected, however. Differentiation gives

$$\mathbf{v}(t) = a\big(-R\sin(at), R\cos(at)\big),$$

and it follows that $|\mathbf{v}(t)| = aR$. Similarly,

$$\mathbf{a}(t) = \mathbf{v}'(t) = a^2\big(-R\cos(at), -R\sin(at)\big).$$

Thus, the effect of a is to multiply the speed by a and the magnitude of acceleration by a^2.   ■

EXAMPLE 3   **(Changing speed)** A particle moves counterclockwise with constant speed s around the circle of radius R, centered at (x_0, y_0). Find the position function.

Solution   The Fact says that the position function

$$\mathbf{p}(t) = (x_0, y_0) + (R\cos t, R\sin t)$$

has constant speed R. Example 2 shows that, to *change* the speed from R to s, we multiply t by s/R to obtain the new position function

$$\mathbf{p}(t) = (x_0, y_0) + \big(R\cos(ts/R), R\sin(ts/R)\big).$$

Now it is easy to see that $\mathbf{p}(t)$ has speed s, as desired.   ■

EXAMPLE 4   A particle starts at $(2, 3)$ at $t = 0$. It moves at constant speed 1 around the circle centered at $(2, 0)$ until it reaches the point $(-1, 0)$. ➝ Find the particle's position function.

Draw your own picture.

Solution   The circle has radius 3, and so $\mathbf{p}(t) = (2, 0) + (3\cos t, 3\sin t)$ describes a particle starting from the east pole $(5, 0)$ at $t = 0$. To start at angle $\pi/2$, we could use the position function

$$\mathbf{p}(t) = (2, 0) + \big(3\cos(t + \pi/2), 3\sin(t + \pi/2)\big).$$

But this position function has speed 3. To arrange for speed 1 we can divide t by 3 to get

$$\mathbf{p}(t) = (2, 0) + \big(3\cos(t/3 + \pi/2), 3\sin(t/3 + \pi/2)\big).$$

The particle arrives at $(-1, 0)$ when $t = 3\pi/2$. ◾

Circular motion, acceleration, and centripetal force   Suppose that a particle moves with uniform speed s about a circle of radius R centered at the origin. Then, by the work in preceding Examples we have

$$\mathbf{p}(t) = \big(R\cos(st/R), R\sin(st/R)\big).$$

Differentiation (twice, with respect to t) produces the acceleration function

$$\mathbf{a}(t) = \frac{s^2}{R^2}\big(-R\cos(st/R), -R\sin(st/R)\big) = -\frac{s^2}{R}\big(\cos(st/R), \sin(st/R)\big).$$

This harmless-looking calculation has an important physical meaning:

> *For uniform circular motion with speed R, the acceleration vector always points toward the center of the circle and has magnitude s^2/R.*

It may seem surprising that constant-speed circular motion generates *any* acceleration. But Newton's second law of motion asserts that acceleration is proportional to force— and we've all felt the force (called **centripetal force**) generated when we swing an object around on a string or when we drive around a tight curve in a car. The italicized remark says, moreover, that this force is *inversely* proportional to the radius and *directly* proportional to the square of the speed. The automotive lesson is clear: Take tight corners slowly.

Combining motions, adding functions

Many interesting motions, such as that of the robot in Figure 1, turn out to be simple combinations of uniform linear and circular motion. It is a pleasant physical fact that these combinations of basic motions can be modeled simply by *adding* (in the vector sense) the vector-valued functions that represent the simpler motions.

EXAMPLE 5   Consider again the jointed-arm robot in Figure 1. Assume that (i) the long arm has length 1; (ii) the short arm has length 0.4; (iii) the long arm rotates uniformly counterclockwise, making 1 rotation in 2π seconds; and (iv) the short arm rotates uniformly counterclockwise, making 8 rotations in 2π seconds. Find the position function for the black dot.

Solution   The end of the long arm undergoes uniform circular motion, with radius 1, center $(0, 0)$, and unit speed, and so it has position function

$$\mathbf{p}_{\text{long}}(t) = (\cos t, \sin t).$$

The short arm, if it were pinned at the origin, would have position function

$$\mathbf{p}_{\text{short}}(t) = \big(0.4\cos(8t), 0.4\sin(8t)\big).$$

The robot combines the two motions by moving the center of the short arm's rotation to the end of the long arm. This amounts simply to *adding* the position functions as vectors, that is,

$$\mathbf{p}(t) = \mathbf{p}_{\text{short}}(t) + \mathbf{p}_{\text{long}}(t) = \left(\cos(t) + 0.4\cos(8t), \sin(t) + 0.4\sin(8t)\right).$$

Plotting this function in parametric form for $0 \leq t \leq 2\pi$ produced Figure 2(a). ▪

Cycloids: combining circular and linear motions Combining uniform linear and circular motions produces another class of curves.

EXAMPLE 6 Combine the uniform linear motion

$$\mathbf{p}_{\text{linear}}(t) = (t, 1)$$

and the uniform (clockwise) circular motion

$$\mathbf{p}_{\text{circular}}(t) = (\cos t, -\sin t).$$

What does the combination $\mathbf{p}(t) = (t + \cos t, 1 - \sin t)$ represent?

Solution Figure 3 shows the motion for $0 \leq t \leq 4\pi$:

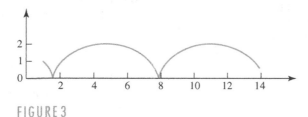

FIGURE 3
A cycloid: Combined linear and circular motion

The curve is called a **cycloid**; it represents the path traced by a point on the rim of a rolling wheel. We investigate cycloids and their relatives further in the exercises. (See also the Interlude at the end of this chapter on cycloids and cycloid-like curves.) ▪

BASIC EXERCISES

1. In each part below, find a position function that models the given uniform linear motion. Then use technology to plot the particle's path as a parametric curve.
   (a) The particle starts at $(0, 0)$ at $t = 0$ and travels to $(1, 2)$ at constant speed 1.
   (b) The particle starts at $(1, 2)$ at $t = 0$ and travels to $(0, 0)$ at constant speed 1.
   (c) The particle starts at $(1, 2)$ at $t = 0$ and travels to $(5, 6)$ at constant speed 100.
   (d) The particle is at $(1, 2)$ at time $t = 1$ and at $(5, 6)$ at time $t = 10$.

2. This exercise confirms the Fact on page 843. Let $\mathbf{p}(t) = (x_0, y_0) + (R\cos t, R\sin t)$.
   (a) Show that $|\mathbf{p}(t) - (x_0, y_0)| = R$ for all t.
   (b) Show that $\mathbf{p}$ has constant speed R.
   (c) Find $\mathbf{v}(0)$. How does the answer reflect the counterclockwise direction?

3. In this exercise let $\mathbf{p}(t) = (1, 2) + (3\cos t, -3\sin t)$ describe a particle's position for $-\pi/2 \leq t \leq \pi/2$.
   (a) Describe the particle's path geometrically. In which direction does it travel?
   (b) Show that $\mathbf{p}$ has constant speed 3.
   (c) Find $\mathbf{v}(0)$. What does the answer say about the direction of travel?

4. In each part below, give a position function that models the uniform circular motion described. Use technology to plot the position curve
   (a) with center $(0, 0)$, radius 3, speed 1, and counterclockwise direction; one full rotation starting from the east pole,
   (b) with center $(0, 0)$, radius 3, speed 2, and clockwise direction, one half rotation, starting from the east pole,
   (c) with center $(1, 2)$, passing through $(4, 5)$, speed 1, counterclockwise direction, one full rotation starting from $(4, 5)$.

5. Redo Example 5, but assume that the small arm rotates counterclockwise four times (rather than eight times) as fast as the larger arm. Find a position function; plot it for $0 \le t \le 2\pi$.

6. Redo Example 5, but assume that the small arm rotates clockwise (rather than counterclockwise) four times as fast as the larger arm. Find the position function; plot it for $0 \le t \le 2\pi$. Try to explain the different appearance from the previous problem.

7. Redo Example 5, but this time assume that the large arm rotates counterclockwise four times as fast as the smaller arm. Find a position function; plot it for $0 \le t \le 2\pi$.

8. Consider a robot like that of Example 5, but assume (i) the long arm has length 2; (ii) the short arm has length 1; (iii) the long arm rotates uniformly counterclockwise, making 1 rotation in 2π seconds; and (iv) the short arm rotates uniformly counterclockwise, making 4 rotations in 2π seconds.

   (a) Find the position function for the black dot.

   (b) Find the velocity function for the black dot.

   (c) Find the speed function for the black dot. Use it to find or estimate the length of one "cycle." (NOTE: You will probably have to use a numerical method to estimate the arclength integral.)

   (d) Find the acceleration function for the black dot.

9. This exercise is about Example 6.

   (a) Find the velocity function $\mathbf{v}(t)$. Use it to find points at which the velocity is $(0, 0)$. How do these points appear on the graph?

   (b) Find the speed function $s(t)$. Use it to find (exactly!) the length of one arch of the curve.

   (c) Show that the acceleration has constant magnitude.

10. (a) Suppose that a bicycle with tires of radius a moves at a constant speed v_0 along a straight road. Assume that the road is the x-axis and that the bicycle moves to the right. Explain why the point on the tire at the origin at time $t = 0$ is at the point $\mathbf{p}(t) = \left(v_0 t - a \sin(v_0 t/a), a - a \cos(v_0 t/a)\right)$ at time $t > 0$. [HINT: Think of the motion of the point on the tire as a combination of sliding and spinning.]

   (b) Suppose that a bicycle with tires of radius a moves at a constant speed v_0 along a straight road. At time $t = 0$ an insect starts crawling from the center of the wheel along a spoke at constant linear speed s_0. Find an expression for the location of the insect at time t. (Assume that the road is the x-axis and that the bicycle moves to the right.)

   (c) Suppose that a bicycle with tires of radius a moves at a constant speed v_0 up a hill. Assume that the road is the line $y = mx$ and that the bicycle moves to the right. Find an expression for the location at time $t > 0$ of the point on the tire that is at the origin at time $t = 0$.

11. Let $\mathbf{p}(t)$ and $\mathbf{q}(t)$ represent uniform linear motions, at constant speed 1, from $(1, 2)$ to $(2, 3)$ and from $(5, 6)$ to $(6, 7)$, respectively.

   (a) Write position formulas in vector form for $\mathbf{p}(t)$ and $\mathbf{q}(t)$. (In each case, let $t = 0$ correspond to the starting point.)

   (b) Explain why the position function $\mathbf{p}(t) + \mathbf{q}(t)$ describes a uniform linear motion. What are the initial and final points? What is the speed?

   (c) Write a position function $r(t)$ for uniform linear motion at constant unit speed from $(6, 8)$ to $(10, 12)$. (Let $t = 0$ correspond to the starting point.)

12. Let $\mathbf{p}(t)$ and $\mathbf{q}(t)$ represent uniform linear motions, at constant speed 1, from (x_1, y_1) to (x_2, y_2) and from (x_3, y_3) to (x_4, y_4), respectively. (Assume that $(x_1, y_1) \ne (x_2, y_2)$ and $(x_3, y_3) \ne (x_4, y_4)$.)

   (a) Write position formulas in vector form for $\mathbf{p}(t)$ and $\mathbf{q}(t)$. (In each case, let $t = 0$ correspond to the starting point.)

   (b) Explain why the position function $\mathbf{p}(t) + \mathbf{q}(t)$ describes a uniform linear motion. What are the initial and final points? What is the speed?

   (c) Write a position function $r(t)$ for uniform linear motion at constant unit speed from $(x_1 + x_3, y_1 + y_3)$ to $(x_2 + x_4, y_2 + y_4)$. (Let $t = 0$ correspond to the starting point.)

13. Consider a robot like that discussed in this section but with three linked arms of lengths 4, 2, and 1 unit, respectively. Suppose that the long arm rotates once in 2π seconds, the medium arm rotates four times in 2π seconds, and the short arm rotates eight times in 2π seconds.

   (a) As in Example 5, write separate position functions $\mathbf{p}_{\text{long}}$, $\mathbf{p}_{\text{medium}}$, and $\mathbf{p}_{\text{short}}$ for the three arms, imagining that all are pinned at the origin.

   (b) Use technology to plot the combined position function $\mathbf{p} = \mathbf{p}_{\text{long}} + \mathbf{p}_{\text{medium}} + \mathbf{p}_{\text{short}}$. (This function represents the position of the end of the small arm if the big arm is pinned at the origin.)

   (c) Calculate the velocity and acceleration vectors for $\mathbf{p}(t)$ at $t = 0$. Can the answers be "seen" in the curve?

   (d) Calculate the velocity and acceleration vectors for $\mathbf{p}(t)$ at $t = \pi/4$.

14. Let $\mathbf{p}(t) = (\cos t, \sin t)$ and let $\mathbf{q}(t) = \left(a \cos(bt), a \sin(bt)\right)$, where a and b are positive constants. Note that both $\mathbf{p}$ and $\mathbf{q}$ represent uniform curricular motion about the origin.

   (a) Find the (constant) speed of each of $\mathbf{p}$ and $\mathbf{q}$.

   (b) Suppose that $b = 1$. Show that $\mathbf{p} + \mathbf{q}$ represents uniform circular motion. Find the center, the radius, and the speed.

   (c) Suppose that $b \ne 1$. Explain why $\mathbf{p} + \mathbf{q}$ does not represent uniform circular motion. [HINT: Show that the $|\mathbf{p} + \mathbf{q}|$ is not constant.]

15. Show that $\mathbf{p}(t)$ represents uniform linear motion if and only if $\mathbf{p}'(t) = \mathbf{v}(t) = (a, b)$, where a and b are constants. What happens in the special case that $(a, b) = (0, 0)$?

16. Show that $\mathbf{p}(t)$ represents uniform linear motion if and only if $\mathbf{p}''(t) = \mathbf{a}(t) = (0, 0)$.

Cycloids and Epicycloids

This Interlude builds on and extends ideas developed in Sections 12.6 and 15.1.

About cycloids A cycloid is the path traced by a point on the rim of a wheel (we will assume it has radius 1) as it rolls, without slipping, along a line (we'll take it to be the *y*-axis). A cycloid can be modeled as the vector sum of a uniform *linear* motion (of the center of the wheel) and a uniform *circular* motion (of the point on the rim around the center of the wheel). It follows that the cycloid has a vector-valued position function of the general form

$$\mathbf{p}(t) = \text{linear} + \text{circular} = (at, 1) + (\cos(bt), -\sin(bt)),$$

where *a* and *b* are constants.

PROBLEM 1 Use technology to plot some curves of the form above, for several values of *a* and *b*. How do the values of *a* and *b* affect the shape of the curve? How are the values of *a* and *b* related to the physical situation?

PROBLEM 2 Suppose that the wheel rolls to the *right*, without slipping, at a constant rate of 1 unit per second. What is the formula for $\mathbf{p}(t)$? In particular, why does $-\sin$ appear rather than sin?

PROBLEM 3 What situation does the formula $\mathbf{p}(t) = (t, 1) + (\cos(2t), -\sin(2t))$ represent? (HINT: The answer involves slipping; 2 could be called the **slipping coefficient**.) Draw the curve. Where on the curve, if anywhere, does the point on the rim of the wheel move to the *left*?

PROBLEM 4 Explain why the wheel does not slip if $a > 0$ and $\mathbf{p}(t) = (at, 1) + (\cos(at), -\sin(at))$.

About epicycloids An **epicycloid** (sometimes called a **roulette**) is the curve traced out by a point *P* on the rim of a wheel as the wheel rolls (with or without slipping) around the outside of another wheel, say of radius *R*. We assume for simplicity that the outer wheel has radius 1. Figure 1(a) shows the apparatus; Figure 1(b) shows one possible curve (together with the inside wheel).

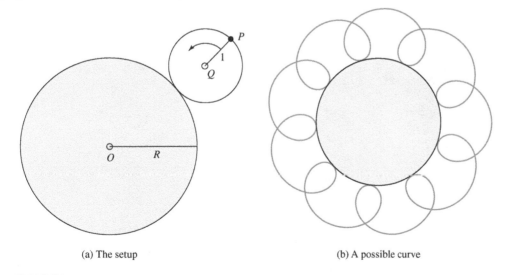

(a) The setup   (b) A possible curve

FIGURE 1
Creating an epicycloid

A little thought reveals that, in this situation, the motion of P is the vector sum of two uniform circular motions: (i) the motion of Q (the center of the small wheel) around 0 (the center of the large wheel) and (ii) the motion of P around Q. Thus, an equation of the following form describes the motion:

$$\mathbf{p}(t) = \text{motion (i)} + \text{motion (ii)} = (R+1)\big(\cos(at), \sin(at)\big) + \big(\cos(bt), \sin(bt)\big). \qquad (1)$$

PROBLEM 5 Explain why the motion formula Equation 1 "works." Why does $R+1$ appear in it?

PROBLEM 6 Suppose that $a = 1$, $R = 3$, and $b = 2$. Find all the cusps—that is, the places where the point P has zero velocity. How do these points appear on the epicycloid?

PROBLEM 7 Give a formula for, and draw with technology, an epicycloid with five cusps.

PROBLEM 8 Suppose that $a = 1$. Under what conditions on the values of R and b in Equation 1 will the wheel roll without slipping? (HINT: The answer is $b = R - 1$. Try to explain why.)

15.2 NEW CURVES FROM OLD

This section builds on ideas developed in Sections 12.6 and 12.7.

In this section we use vector-based tools—projections, length, the dot product, perpendicularity, and so on—to construct and manipulate plane curves more efficiently than we could do otherwise. Our main goal will be to building interesting and attractive new curves out of old ones. Some of these "new" curves are, in fact, quite ancient. The **conchoid of Nicomedes**, for instance, is named for a Greek mathematician of the third century B.C. who used it in studying such classical problems as trisecting angles and duplicating a cube.

The basic object: a curve in vector form   In this section our basic object of study will be a parametrized plane curve C. We will describe such a curve either in vector form

$$\mathbf{r}(t) = \big(x(t), y(t)\big); \qquad a \le t \le b,$$

or in parametric form

$$x = x(t); \qquad y = y(t); \qquad a \le t \le b.$$

The differences between these forms are mainly cosmetic, and we will toggle freely back and forth between them. When using the vector version, we will think of $\mathbf{r}(t)$ as the position vector with tail at the origin and head at $\big(x(t), y(t)\big)$. As t varies, the arrowhead "sweeps out" the curve C, as shown in Figure 1.

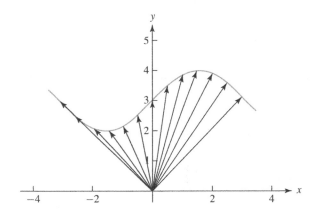

FIGURE 1
Position vectors tracing out a curve

The notation $\mathbf{r}(t)$ suggests "radius"—as it should, because all the position vectors emanate "radially" from the origin.

We studied curves in vector form in Section 12.5. In this section we combine the dot product (studied in Section 12.7) with the vector idea to produce new curves from old.

Rotations

The dot product and its connection to angles can be used to rotate curves in the plane.

Rotating vectors It is often useful, given a vector $\mathbf{v} = (a, b)$, to find the new vector $\mathbf{w}$ formed by rotating $\mathbf{v}$ counterclockwise through an angle α. The following Fact tells exactly how to do this.

FACT (How to spin a vector) Let $\mathbf{v} = (a, b)$ be any nonzero vector, and let α be an angle, in radians. The vector

$$\mathbf{w} = (a \cos\alpha - b \sin\alpha, \, a \sin\alpha + b \cos\alpha)$$

is the result of rotating $\mathbf{v}$ counterclockwise through α radians.

How was the formula for $\mathbf{w}$ found? We will leave that good question to the exercises, but it is easy to check that $\mathbf{w}$ has the claimed property. To see this, let θ be the angle between $\mathbf{v}$ and $\mathbf{w}$. The Fact asserts that $\theta = \alpha$, and the dot products is the key to seeing why. Basic calculations give

$$\mathbf{v} \cdot \mathbf{w} = (a^2 + b^2)\cos\alpha; \qquad \mathbf{v} \cdot \mathbf{v} = a^2 + b^2; \qquad \mathbf{w} \cdot \mathbf{w} = a^2 + b^2.$$

Combining these results with the identity $\mathbf{v} \cdot \mathbf{w} = |\mathbf{v}||\mathbf{w}| \cos\theta$ gives

$$(a^2 + b^2)\cos\alpha = \mathbf{v} \cdot \mathbf{w} = |\mathbf{v}||\mathbf{w}|\cos\theta = \sqrt{a^2 + b^2}\sqrt{a^2 + b^2}\cos\theta.$$

Thus, $\cos\alpha = \cos\theta$, as claimed.

Rotating curves The preceding Fact tells how to rotate any given vector through any given angle. Rotating $\mathbf{v} = (1, 2)$ through $\pi/3$ radians, for example, gives the vector

$$\mathbf{w} = \big(1\cos(\pi/3) - 2\sin(\pi/3), \, 1\sin(\pi/3) + 2\cos(\pi/3)\big)$$

$$= \left(\frac{1}{2} - \sqrt{3}, \, \frac{\sqrt{3}}{2} + 1\right) \approx (-1.23, 1.86).$$

Applying the Fact to the vector formula for a *curve* gives a more interesting result.

EXAMPLE 1 Consider the curve C given in vector form by

$$\mathbf{r}(t) = (t, 3 + \sin t); \qquad -5 \le t \le 5.$$

Using $\alpha = \pi/3$, apply the rotation formula to the position vector $\mathbf{r}(t)$ to form a new curve $\mathbf{r}_\alpha(t)$. What happens?

Solution For any angle α the the rotation formula gives

$$\mathbf{r}_\alpha(t) = \big(t\cos\alpha - (3 + \sin t)\sin\alpha, \, t\sin\alpha + (3 + \sin t)\cos\alpha\big).$$

With $\alpha = \pi/3$ this formula becomes (after simplification)

$$\mathbf{r}_\alpha(t) = \frac{1}{2}\left(t - (3 + \sin t)\sqrt{3},\ t\sqrt{3} + 3 + \sin t\right)$$

Both curves $\mathbf{r}$ and $\mathbf{r}_\alpha$ appear in Figure 2 (along with four additional rotations of the original curve, added for aesthetic appeal):

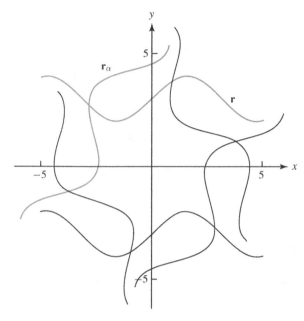

FIGURE 2
Rotating a curve

The curve $\mathbf{r}_\alpha$ results from rotating the curve $\mathbf{r}$ through $\pi/3$ radians, with center at the origin.

The picture shows what we expect: rotating every position vector on $\mathbf{r}$ through $\pi/3$ has the same effect on the curve itself.

New curves from old: conchoids

There are many ways to construct new curves from old; as just seen, rotation is one possibility. Another possibility, also based on the dot product, concerns **conchoids**. Such curves are constructed as follows:

> We start with a given curve C, a point P_0, and a fixed positive number k. For each point P on C, another point Q is determined by moving k units of distance "outward" along the line joining P_0 and P. The set of all such points Q is a **conchoid**.

Figure 3(a) illustrates the general idea; it shows several corresponding pairs P and Q. As P moves along the original curve C, the corresponding points Q sweep out the conchoid. Figure 3(b) shows the conchoid curve constructed from a circle centered on the y-axis. (The circled points correspond to each other.)

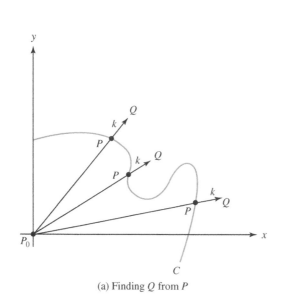

(a) Finding Q from P

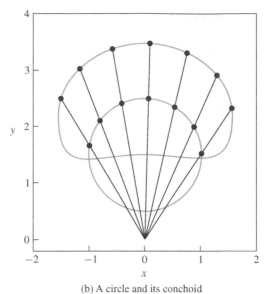

(b) A circle and its conchoid

FIGURE 3
Constructing a conchoid

A formula for the conchoid Let's find a vector description of the conchoid constructed from a given curve C expressed in vector form as $\mathbf{p}(t) = (x(t), y(t))$, for $a \le t \le b$. For simplicity, we will assume that $P_0 = (0, 0)$ and $k = 1$. Now, as Figure 3 shows, the conchoid curve is constructed simply by *extending* each vector $\mathbf{p}(t)$ by one unit of length—that is, by adding to $\mathbf{p}(t)$ the *unit* vector $\mathbf{p}(t)/|\mathbf{p}(t)|$. The result is the desired vector equation for the conchoid:

$$\mathbf{q}(t) = \mathbf{p}(t) + \frac{\mathbf{p}(t)}{|\mathbf{p}(t)|} = \mathbf{p}(t)\left(1 + \frac{1}{|\mathbf{p}(t)|}\right). \tag{1}$$

If, for instance, $\mathbf{p}(t)$ describes a circle centered at the origin with radius R, then $|\mathbf{p}(t)| = R$ is constant; thus, the right-hand factor in Equation 1 is the constant $1 + 1/R$, and the resulting conchoid is simply circle of radius $R + 1$. If $\mathbf{p}(t)$ is a straight line, the resulting curve is a **conchoid of Nicomedes**.

BASIC EXERCISES

1. This exercise is about the Fact on page 850. Use the same notations as given there for $\mathbf{v}$ and $\mathbf{w}$.
   (a) Show that $\mathbf{v} \cdot \mathbf{w} = (a^2 + b^2)\cos\alpha$.
   (b) Show that $|\mathbf{w}| = |\mathbf{v}|$.
   (c) Let $\mathbf{u} = (a\cos\alpha + b\sin\alpha, -a\sin\alpha + b\cos\alpha)$. Show that $|\mathbf{u}| = |\mathbf{v}|$ and that the angle between $\mathbf{u}$ and $\mathbf{v}$ is α.
   (d) How are $\mathbf{u}$ and $\mathbf{w}$ related to each other and to $\mathbf{v}$?

2. Use rotation and other such tricks to find vector formulas for each of the following types of curves.
   (a) The parabola $y = x^2$, but rotated $\pi/4$ radians counterclockwise.
   (b) A sine curve, but rotated so as to wiggle from northwest to southeast.

   (c) The cardioid $r = 1 + \cos\theta$, but rotated $45°$ clockwise. [HINT: First write the cardioid in parametric form.]

3. Find the distance from the point P to the line ℓ in each case following
   (a) from $P(2, 3)$ to the line ℓ through $(0, 0)$ with direction vector $\mathbf{i}$,
   (b) from $P(2, 3)$ to the line ℓ through $(1, 1)$ with direction vector $(-1, 1)$,
   (c) from $P(2, 3)$ to the line ℓ through $(1, 1)$ and $(0, 2)$,
   (d) from the origin to the line with Cartesian equation $y = mx + b$; assume that $b \ne 0$. (Could you do this without vectors?).

4. Consider the curve C given by

$$\mathbf{r}(t) = (3\cos t, \sin t); \quad 0 \le t \le 2\pi.$$

   (a) Give an equation in x and y for C.

   (b) Find and plot the new curve C_1 obtained by rotating C $\pi/4$ radians about the origin.

   (c) Find and plot the new curve C_2 obtained by rotating C $\pi/2$ radians about the origin. Give an equation in x and y for C_2.

5. Use the general conchoid formula given in this section to find (that is, give a vector equation for) and plot each conchoid below. In all parts, let $P_0 = (0, 0)$ and $k = 1$.

   (a) The conchoid based on $\mathbf{p}(t) = (2\sin t, 2\cos t)$, $0 \le t \le 2\pi$.

   (b) The conchoid based on the straight line $x = 1$. [HINT: First write a vector equation for this line.]

   (c) The conchoid based on the spiral $x = t\cos t$, $y = t\sin t$, $0 \le t \le 4\pi$.

6. Let $\mathbf{p}(t)$ and $\mathbf{q}(t)$ be position functions for two smooth curves; suppose that the parameter interval for both curves is $0 \le t \le 1$ and that both curves begin and end at the same point—that is, $\mathbf{p}(0) = \mathbf{q}(0)$ and $\mathbf{p}(1) = \mathbf{q}(1)$.

One way to form a new "hybrid" curve from $\mathbf{p}$ and $\mathbf{q}$ is with the formula $\mathbf{r}(t) = (1 - t^2)\mathbf{p}(t) + t^2\mathbf{q}(t)$, $0 \le t \le 1$.

   (a) Find $\mathbf{r}(t)$ when $\mathbf{p}(t) = (t, t)$ and $\mathbf{q}(t) = (t, t^2)$. Plot all three curves $\mathbf{p}$, $\mathbf{q}$, and $\mathbf{r}$ for $0 \le t \le 1$. (Do this part by hand.)

   (b) Find $\mathbf{r}(t)$ when $\mathbf{p}(t) = (t, t^2)$ and $\mathbf{q}(t) = (t, t^3)$. Plot all three curves $\mathbf{p}$, $\mathbf{q}$, and $\mathbf{r}$ for $0 \le t \le 1$. (Do this part by hand.)

   (c) Find $\mathbf{r}'(t)$ in terms of $\mathbf{p}(t)$, $\mathbf{q}(t)$, $\mathbf{p}'(t)$, and $\mathbf{q}'(t)$.

   (d) Show that $\mathbf{r}(0) = \mathbf{p}(0)$ and $\mathbf{r}(1) = \mathbf{q}(1)$.

   (e) Show that $\mathbf{r}'(0) = \mathbf{p}'(0)$ and $\mathbf{r}'(1) = \mathbf{p}'(1)$.

7. Let $\mathbf{p}$, $\mathbf{q}$, and $\mathbf{r}$ be defined as in the previous exercise. (In particular, the parameter interval is always $0 \le t \le 1$.)

   (a) Let $\mathbf{p}$ and $\mathbf{q}$ represent the line segment from $(0, 0)$ to $(2, 4)$ and the parabolic arc $y = x^2$ between the same two points. Find formulas for $\mathbf{p}$, $\mathbf{q}$, and $\mathbf{r}$; plot all three curves. (Be sure to use the parameter interval $0 \le t \le 1$.)

   (b) Let $\mathbf{p}$ represent the counter clockwise circular arc from $(1, 0)$ to $(-1, 0)$ on the unit circle; let $\mathbf{q}$ represent the straight line from $(1, 0)$ to $(-1, 0)$. Find formulas for $\mathbf{p}$, $\mathbf{q}$, and $\mathbf{r}$; plot all three curves.

15.3 CURVATURE

This section builds on ideas developed in Sections 12.6 and 12.7.

Anyone who has driven a winding road, even on flat terrain and at moderate speeds, knows the strain that is put on the engine, the brakes, and the passengers. The tightest curves, with the smallest "turning radius," are worst—negotiating them spills coffee, shifts cargo, and throws passengers into each others' laps.

   Curvature, which we define and calculate in this section, is a mathematical measure of how fast a plane curve "turns" near a given point. ➡ The automotive analogy shows that there are good practical reasons for measuring curvature, such as designing roads and tracks of various kinds.

The curve might be a road.

   For an arbitrary curve, such as the parabola $y = x^2$, we will see that curvature varies from point to point: It is largest at the vertex and becomes smaller far away from the vertex, where the parabola becomes more and more like a straight line. On a circle, by contrast, the curvature is the same at all points: *The smaller the radius, the larger the curvature.* On a straight line, the curvature is everywhere zero. Our first task will be to *define* curvature, using vector ingredients, in a way that captures these intuitive properties.

Defining curvature

Let a curve C be given by a position function $\mathbf{p}(t) = (x(t), y(t))$. We will define the curvature of C at the point $\mathbf{p}(t_0) = (x(t_0), y(t_0))$. First, recall that, for any t, the velocity vector $\mathbf{v}(t) = (x'(t), y'(t))$ is *tangent* to C at the point $\mathbf{p}(t)$. Our definition of curvature, therefore, will measure how fast the velocity vector turns with respect to distance traveled along the curve. For this purpose, the *length* of the velocity vector (i.e., the speed of the curve) is immaterial—we are interested only in the *direction*. With this in mind we write the velocity vector in the form

$$\mathbf{v}(t) = s(t)\mathbf{T}(t) = \sqrt{x'(t)^2 + y'(t)^2} \frac{\left(x'(t), y'(t)\right)}{\sqrt{x'(t)^2 + y'(t)^2}},$$

where $s(t)$ is the speed at time t (a scalar quantity) and $\mathbf{T}(t)$ is called the **unit tangent vector** to C at t. (The name is appropriate because $\mathbf{T}(t)$ is a unit vector and points in the direction of the curve at t.)

The curvature at $t = t_0$ tells how fast the tangent direction, given by $\mathbf{T}(t)$, varies "with respect to arclength," that is, with respect to distance traveled along the curve. Finding $|\mathbf{T}'(t_0)|$, the magnitude of the derivative $\mathbf{T}'(t_0)$, is a good first step. It tells how much the tangent direction changes per unit of *time*. To find the rate of change of the tangent direction with respect to *arclength*, we divide by the speed of the curve at $t = t_0$. ← The formal definition captures these ideas:

Does this seem reasonable? Convince yourself that it is.

> **DEFINITION**   Let the smooth curve C be defined by the position function $\mathbf{p}(t) = \left(x(t), y(t)\right)$. The **curvature** of C at the point $\mathbf{p}(t_0)$ is the quantity
>
> $$\frac{|\mathbf{T}'(t_0)|}{s(t_0)}$$
>
> wherever $s(t_0) \neq 0$.

We will see by examples that the definition makes good sense.

Curvature of a line   For a linear position function, the tangent direction $\mathbf{T}(t)$ is a *constant* vector, and so $\mathbf{T}'(t) = (0, 0)$. It follows that (as expected) a line has (constant) curvature zero.

Curvature of a circle   If C is a circle of radius R, then C has position function

$$\mathbf{p}(t) = (x_0, y_0) + R(\cos t, \sin t).$$

It follows that

$$\mathbf{v}(t) = R(-\sin t, \cos t), \qquad s(t) = R, \qquad \text{and} \qquad \mathbf{T}(t) = (-\sin t, \cos t).$$

Thus, for every t,

$$\text{curvature} = \frac{|\mathbf{T}'(t)|}{s(t)} = \frac{|(-\cos t, -\sin t)|}{R} = \frac{1}{R}.$$

The result is as expected: The curvature has the same value at every point on a circle, and the larger the radius the smaller the curvature.

Radius of curvature   If C has curvature K at a point P, then we say that C has **radius of curvature** $1/K$ at P. The preceding calculation shows that this definition makes sense for circles. We will see below what it means geometrically.

Different speeds, same curvature Curvature is an "intrinsic" property of a curve. It should not depend on such "extrinsic" factors as the speed of a particular parametrization; ➡ the denominator in the preceding definition above ensures this.

A road's "curviness" does not depend on how fast cars drive.

Suppose, for example, that the circle of radius R mentioned above is parametrized at twice the speed with the parametrization

$$\mathbf{p}(t) = (x_0, y_0) + R\big(\cos(2t), \sin(2t)\big).$$

This new and speedier parametrization gives

$$\mathbf{v}(t) = 2R\big(-\sin(2t), \cos(2t)\big); \qquad s(t) = 2R; \qquad \mathbf{T}(t) = \big(-\sin(2t), \cos(2t)\big).$$

Thus, for all t,

$$\text{curvature} = \frac{\big|\mathbf{T}'(t)\big|}{s(t)} = \frac{\big|2(-\cos(2t), -\sin(2t))\big|}{2R} = \frac{1}{R},$$

just as we found earlier.

Calculating curvature

The definition above explains what curvature *is*, but the formula is complicated. The following Fact offers a more convenient formula.

FACT Let the curve C be defined by the position function $\mathbf{p}(t) = \big(x(t), y(t)\big)$. If the speed $s(t) \neq 0$, then

$$\text{curvature} = \frac{|x'(t)y''(t) - y'(t)x''(t)|}{\big(\sqrt{x'(t)^2 + y'(t)^2}\big)^3}.$$

More succinctly: curvature $= |x'y'' - y'x''|/s^3$.

One can prove the Fact by straightforward but tedious calculation, starting from the definition. A shorter but slightly more subtle calculation is outlined in the exercises. Our main concern is with *using* the formula, and so we omit details here.

EXAMPLE 1 Discuss the curvature of the parabola $y = x^2$. Where is the curvature most and least? How does the curvature behave as x tends to ∞?

Solution We can parametrize the parabola as $\mathbf{p}(t) = (t, t^2)$. Then $x' = 1$, $x'' = 0$, $y' = 2t$, and $y'' = 2$, and so applying the formula gives

$$\text{curvature} = \frac{2}{\big(\sqrt{1 + 4t^2}\big)^3}.$$

Thus, the curvature has its largest value, 2, at the origin, where $t = 0$; the radius of curvature there is $1/2$. At the point $(1, 1)$, where $t = 1$, the curvature is $2/5\sqrt{5} \approx 0.18$;

the radius of curvature there is the reciprocal $5\sqrt{5}/2 \approx 5.59$. Figure 1 shows what these radii of curvature mean geometrically:

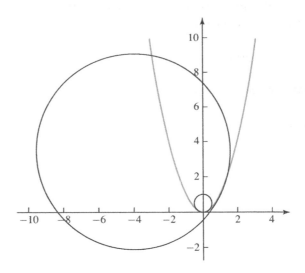

FIGURE 1
Radii of curvature at two points on a parabola

At each point P in question, a circle is drawn. The circle is (i) tangent to the curve at P and (ii) has radius equal to the radius of curvature at P. (The center of each circle is on the line perpendicular to C at P.) Such a circle is called the **osculating circle** to C at P because it "kisses" the curve at the given point. ◄ The curvature formula shows, too, that the curvature tends to zero as t tends to ∞. Equivalently, the osculating circles get larger and larger as $t \to \infty$. ▪

Mathematicians love vivid metaphors.

BASIC EXERCISES

1. Show that the line ℓ through $(1, 2)$ in the direction of $(3, 4)$ has constant curvature zero. [HINT: First write a position function for ℓ; then use the definition of curvature.]

2. Suppose that C is the graph of a function $y = f(x)$. Show that, in this case, the curvature of C at any point $(x, f(x))$ has the form
$$\text{curvature} = \frac{|f''(x)|}{\left(\sqrt{1 + f'(x)^2}\right)^3}.$$

3. Explain geometrically why the parabola $y = x^2$ should have the same curvature at the symmetric points (t, t^2) and $(-t, t^2)$. Does the formula guarantee this? How?

4. Use technology to draw your own version of Figure 1. (The point is to find suitable parametrizations for both osculating circles.)

5. Find the curvature of the curve $y = x^3$ at the point $(0, 0)$. Explain your answer geometrically.

6. Find the curvature of the curve $y = x^3$ at the point $(1, 1)$. Use technology to draw both the curve and the osculating circle there.

7. The following steps lead (through considerable symbolic calculation) to the curvature formula in the Fact on page 855. (Alas, there is no really easy way to derive the formula.)

   Let C be given by $\mathbf{p}(t) = (x(t), y(t))$. To find the curvature K at a point P, suppose that the tangent vector at P makes angle ϕ with the positive x-axis. Let $l(t)$ be the arclength along C measured from some fixed point P_0 (it doesn't really matter which point is fixed) to $\mathbf{p}(t)$. By definition, the curvature is the absolute rate of change of ϕ with respect to arclength l. That is,
$$K = \left|\frac{d\phi}{dl}\right|.$$

We will show in steps how to calculate this derivative.

(a) For all t with $dl/dt \neq 0$,

$$\frac{d\phi}{dl} = \frac{d\phi/dt}{dl/dt}.$$

Explain why.

(b) Explain why $dl/dt = s(t) = \sqrt{x'(t)^2 + y'(t)^2}$.

(c) If $x'(t) \neq 0$, then

$$\tan(\phi(t)) = \frac{y'(t)}{x'(t)}.$$

Explain why this equation holds. [HINT: Draw a typical tangent vector $(x'(t), y'(t))$.]

(d) Differentiate both sides of the preceding equation to show that

$$\sec^2(\phi(t))\frac{d\phi}{dt} = \frac{x'(t)y''(t) - y'(t)x''(t)}{x'(t)^2}.$$

(e) Use the preceding equation to show that

$$\frac{d\phi}{dt} = \frac{x'(t)y''(t) - y'(t)x''(t)}{x'(t)^2 + y'(t)^2}.$$

[HINT: Use the facts (i) $\sec^2(\phi(t)) = 1 + \tan^2(\phi(t))$ and (ii) $\tan(\phi(t)) = (y'(t)/x'(t))$.]

(f) Show that the Fact's formula is true. [HINT: Use the formulas just derived for $d\phi/dt$ and dl/dt.]

15.4 LAGRANGE MULTIPLIERS AND CONSTRAINED OPTIMIZATION

This section builds on ideas developed through Section 13.4, on gradients and directional derivatives.

Let $f(x, y)$ be a function defined on a domain in $\mathbb{R}^2$; suppose we want to find maximum and minimum values of f. We have seen in earlier sections where to look. Stationary points of f are the candidates, and the second derivative test helps us sort out the possibilities. If, say, $f(x, y) = x^2 - 4x + 2y^2$, then $\nabla f(x, y) = (2x - 4, 4y)$; thus, $(2, 0)$ is the only stationary point, and it is easy to see that f has a local minimum there.

Sometimes one is interested in finding maximum or minimum values of a function subject to some "constraint" on legal inputs. For instance, we might ask for the largest and smallest values of $f(x, y)$ when (x, y) is constrained to lie on the circle $x^2 + y^2 = 9$. Solving the gradient equation $\nabla f = (0, 0)$, as above, will do us no good—the constraint circle doesn't pass through the one stationary point $(2, 0)$. As we'll see, gradients are *still* the right tools, but they need to be used in a way appropriate to the present problem.

Naming the problem; a graphical example

Paying close attention to jargon now will save trouble later. The general problem at hand is called **constrained optimization**; the function to be maximized or minimized is called the **objective function**; and the restriction on inputs is described by a **constraint equation**. (Sometimes constraints are given by inequalities or by more than one equation.) Looking at the situation graphically will suggest the role that gradients play.

EXAMPLE 1 Optimize $f(x, y) = x^2 - 4x + 2y^2$ subject to the constraint $x^2 + y^2 = 9$.

Solution The constraint describes a circle C in the xy-plane; to "see" the objective function in the same picture, it is natural to look at level curves of f near C, as shown in Figure 1:

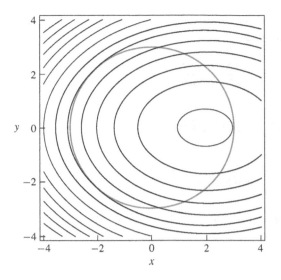

FIGURE 1
Contour plot of $z = x^2 - 4x + 2y^2$

The picture deserves a close look:

The objective function   Level curves of f are ellipses; they are all centered at $(2, 0)$, where f has a local *minimum*. Larger ellipses, therefore, represent larger values of f.

The constraint   Inputs (x, y) that satisfy the constraint are points that lie on the circle C. If we imagine an ant walking along the surface $z = f(x, y)$ above the curve C, our problem is to decide where the ant's "altitude"—described by the level curves—is largest and smallest.

Do you agree? Look especially carefully near the west pole of the circle.

A graphical answer   A close look along C suggests special attention to four points on the circle: $(-3, 0)$ and $(3, 0)$, where $y = 0$; and $(-2, \sqrt{5})$ and $(-2, -\sqrt{5})$, where $x = -2$. At these points, our wandering ant appears to experience highs and lows of altitude; between these points, the ant travels either uphill or down. ← Checking numerical values of f at these points supports these impressions:

$$f(3, 0) = -3; \qquad f(-3, 0) = 21; \qquad f(-2, \sqrt{5}) = f(-2, -\sqrt{5}) = 22.$$

We'd estimate, then, that f assumes its (constrained) maximum value, 22, at the points $(-2, \pm\sqrt{5})$ and its minimum value, -3, at $(3, 0)$.

A useful observation   At each special point P just mentioned, a crucial property holds:

The level curve of f through P is tangent to the constraint curve. ■

After the Italian-French mathematician Joseph-Louis Lagrange (1736–1813)

The exercises present a third technique.

The condition just mentioned—suitably interpreted in terms of gradients—is the main idea of this section. We will see how to use this property, called the **Lagrange condition**, ← to solve constrained optimization problems symbolically. First, however, let's acknowledge that other approaches are possible for the problem above; one follows. ←

EXAMPLE 2   Redo Example 1—this time by parametrizing the constraint curve.

Solution   The circle $x^2 + y^2 = 9$ is easily parametrized; one way is to let

$$\mathbf{X}(t) = (x(t), y(t)) = (3\cos t, 3\sin t); \quad 0 \le t \le 2\pi.$$

Constraining (x, y) to lie on the circle means, in effect, optimizing the composite function →

Check our algebra.

$$h(t) = f(\mathbf{X}(t)) = 9\cos^2 t - 12\cos t + 18\sin^2 t = 9\sin^2 t - 12\cos t + 9,$$

for t in $[0, 2\pi]$. This reduces the problem to an old familiar form; the rest is routine. Because

$$h'(t) = 18\sin t\cos t + 12\sin t = \sin t\,(18\cos t + 12),$$

we have $h'(t) = 0$ if either $\sin t = 0$ (i.e., $y = 0$) or $\cos t = -2/3$ (i.e., $x = -2$); these are the same candidate points as we found earlier.

Observe also that (according to the chain rule) the key derivative is a dot product:

$$h'(t) = \nabla f(\mathbf{X}(t)) \cdot \mathbf{X}'(t).$$

At any maximum or minimum point we must have $h'(t) = 0$; this means, geometrically, that at all such points the gradient ∇f is *perpendicular* to $\mathbf{X}'$, the tangent vector to the constraint curve. This is another way of stating the Lagrange condition mentioned earlier because, at any point $P(x, y)$, the gradient of f is perpendicular to the level curve of f through P. (We saw this important fact in Section 13.4, on gradients.) ■

Gradients and the Lagrange condition

The main point so far, illustrated by the last two examples, is as follows:

> *At a maximum or minimum point for a constrained optimization problem, the gradient of the objective function must be perpendicular to the constraint set.*

This complicated-sounding condition is easiest to understand and use with the help of level curves. The key fact, which links all the main ideas, is the connection between gradients and level sets: →

Section 13.4 has more details on this important idea.

> **FACT (Gradients and level sets)** Let $g(x, y)$ be a differentiable function, with (x_0, y_0) a point in the domain of g. Let C be the level curve of g that passes through (x_0, y_0). If $\nabla g(x_0, y_0) \neq (0, 0)$, then $\nabla g(x_0, y_0)$ is perpendicular to C at (x_0, y_0).

Proof The proof is a nice application of the chain rule. Suppose that the curve C is parametrized by a vector-valued function $\mathbf{X}(t)$ with $\mathbf{X}(t_0) = (x_0, y_0)$. Then $\mathbf{X}'(t_0)$ is a vector tangent to C at (x_0, y_0). (We're assuming the slightly more technical fact that such a parametrization must exist because of our assumption that $\nabla g(x_0, y_0) \neq (0, 0)$.) Because g is constant on the curve C, the composite function $g(\mathbf{X}(t))$ is constant in t. Therefore, by the chain rule,

$$0 = \frac{d}{dt}\big(g(\mathbf{X}(t))\big)(t_0) = \nabla g(x_0, y_0) \cdot \mathbf{X}'(t_0).$$

This means that $\nabla g(x_0, y_0)$ is perpendicular to $\mathbf{X}'(t_0)$ and therefore also to C. ■

In higher dimensions Though we'll work mainly in two variables, the preceding Fact holds just as well for functions of three (or even more) variables except that, in these cases, the level set is a *surface*, not a curve. For example, the level set $g(x, y, z) = x^2 + y^2 + z^2 = 1$ is a sphere in $\mathbb{R}^3$. The Fact says (and it's not hard to "see") that, at any point (x, y, z) on the sphere, the gradient vector $(2x, 2y, 2z)$ is perpendicular to the sphere.

The gradient of the constraint function The constraint in an optimization problem is usually described by an equation. If we write the constraint equation in the form $g(x, y) = 0$, where $g(x, y)$ is a *function*, then the constraint curve becomes the level curve $g(x, y) = 0$. The preceding Fact says, moreover, that at any point (x, y) on this level curve, the gradient vector $\nabla g(x, y)$ of the constraint function is either the zero vector or is perpendicular to the level curve.

The gradient of the objective function Let (x_0, y_0) be a point on the curve $g(x, y) = 0$, and suppose that the objective function $f(x, y)$ assumes either a local maximum or a local minimum value at (x_0, y_0) (compared with nearby points on the constraint curve). Then the following useful fact holds:

The gradient $\nabla f(x_0, y_0)$ is perpendicular to the constraint curve at (x_0, y_0).

Figure 1 illustrates this fact. At all four "candidate" points, the level curves of f are *parallel* to the constraint curve, and so the gradient ∇f must be *perpendicular* to the constraint curve. The calculation at the end of Example 2 explains why this fact holds.

Lagrange multipliers

The facts above, taken together, say something striking and useful: If (x_0, y_0) is a local maximum or local minimum point for the constrained optimization problem, then *both* vectors $\nabla f(x_0, y_0)$ and $\nabla g(x_0, y_0)$ are perpendicular to the level curve $g(x, y) = 0$. It follows, therefore, that these vectors must be parallel to each other, that is, scalar multiples. The formal result follows; we state it for two variables, but it holds in any dimension.

> THEOREM 1 (Lagrange multipliers) Let $f(x, y)$ and $g(x, y)$ be continuously differentiable functions, and consider the problem of optimizing $f(x, y)$ subject to the constraint $g(x, y) = 0$. If f assumes a constrained local maximum or local minimum value at (x_0, y_0), then for some scalar λ,
>
> $$\nabla f(x_0, y_0) = \lambda \nabla g(x_0, y_0).$$
>
> (The scalar λ is called a **Lagrange multiplier**.)

In simple cases, Theorem 1 makes quick work of finding constrained maxima and minima.

EXAMPLE 3 Optimize $f(x, y) = x + y$ subject to the constraint $x^2 + y^2 = 9$.

Solution We write $g(x, y) = x^2 + y^2 - 9$; then the constraint becomes $g(x, y) = 0$, as in the theorem. (This trick *always* works; notice that the constant 9 is "absorbed" into the function g.) Then $\nabla f(x, y) = (1, 1)$ and $\nabla g(x, y) = (2x, 2y)$.

The theorem says that a constrained maximum or minimum occurs, if at all, at a point (x, y) for which

$$\nabla f(x, y) = \lambda \nabla g(x, y)$$

holds for some scalar λ. In addition, we have the constraint equation $g(x, y) = 0$. In our setting, this means

$$(1, 1) = \lambda(2x, 2y) \quad \text{and} \quad x^2 + y^2 = 9.$$

The gradient equation counts for two!

This adds up to three equations ← in the three unknowns x, y, and λ.

There are many ways to solve these equations. One is to observe first that

$$(1, 1) = \lambda(2x, 2y) \Longrightarrow x = y;$$

plugging this into the constraint equation gives $2x^2 = 9$, or $x = \pm 3/\sqrt{2}$. (We need not bother to solve for λ; we care about x and y.) Our candidate points, therefore, are $(3/\sqrt{2}, 3/\sqrt{2})$ and $(-3/\sqrt{2}, -3/\sqrt{2})$. Checking values of f gives $f(3/\sqrt{2}, 3/\sqrt{2}) = 3\sqrt{2}$, a constrained maximum, and $f(-3/\sqrt{2}, -3/\sqrt{2}) = -3\sqrt{2}$, a constrained minimum.

Figure 2 suggests the same conclusion:

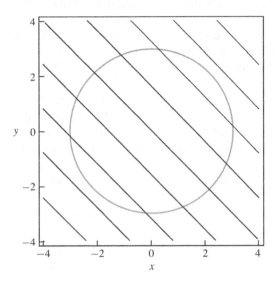

FIGURE 2
Contour plot of $z = x + y$

As in Example 1, contour lines of f are parallel to the constraint curve at the maximum and minimum points.

Caveats

Theorem 1 is often useful, but several remarks and cautions are in order. It is especially important to understand what Theorem 1 does *not* say.

Necessary but not sufficient The theorem says that the Lagrange condition $\nabla f = \lambda \nabla g$ is *necessary* for (x_0, y_0) to be a constrained optimum point—but not, in itself, sufficient. (In the same spirit, $f'(x_0) = 0$ is necessary, but not always sufficient, for x_0 to be a local maximum or minimum point.) At a point that satisfies the condition, f might assume a constrained maximum, a constrained minimum, or neither. Deciding among these possibilities can be hard; pictures may help.

There may be no solution Not every constrained optimization problem has a solution. Even in this case, however, the Lagrange condition may be of use.

EXAMPLE 4 Optimize $f(x, y) = x + y$ subject to the constraint $g(x, y) = y = 0$.

Solution It is clear that $f(x, 0) = x$ can be arbitrarily large positive or negative, and so f has neither a constrained maximum nor a constrained minimum.

Notice, however, that here $\nabla f(x, y) = (1, 1)$ and $\nabla g(x, y) = (0, 1)$. Thus, the Lagrange condition says $(1, 1) = \lambda(0, 1)$, which is clearly impossible. Therefore, by Theorem 1, no constrained maxima or minima exist. ∎

When does a solution exist? The optimization problem in the last example had no solution. The difficulty there was that the constraint set $y = 0$ is unbounded, and so $f(x, y)$ was free to blow up in magnitude.

General theory (a little beyond our scope in this book) guarantees, however, that if f and g are differentiable functions, and the constraint set $g(x, y) = 0$ is bounded, then f does indeed assume (finite) constrained maximum and minimum values. In this case, the theorem guarantees that these values must occur where the Lagrange condition is satisfied.

If the constraint set is unbounded, as in Example 3, then the objective function may or may not assume constrained maximum or minimum values—it depends on the objective function, and there is no simple rule for deciding which is the case.

The extra unknown is λ.

Solving may be difficult For a function $f(x, y)$ and a constraint $g(x, y) = 0$, the Lagrange condition and the constraint equation produce a total of three equations—not necessarily linear equations—in three unknowns. ← For functions of three variables, four unknowns are involved. Solving such systems can be very difficult, or even impossible. Fortunately, many problems of interest lead to relatively simple systems of equations. That the particular value of λ usually does not matter may also save some effort.

> **EXAMPLE 5** Optimize $f(x, y, z) = x + y + z$ subject to the constraint $g(x, y, z) = x^2 + y^2 + z^2 - 3 = 0$.
>
> **Solution** Here the constraint set—a sphere of radius $\sqrt{3}$—is finite, and so we know that constrained maximum and minimum values do exist. With these data the Lagrange condition says
>
> $$\nabla f = (1, 1, 1) = \lambda \nabla g = \lambda(2x, 2y, 2z).$$
>
> It follows (as in a previous calculation) that $x = y = z$. Plugging this into the constraint equation gives
>
> $$x^2 + y^2 + z^2 - 3 = 0 \implies 3x^2 = 3 \implies x = \pm 1.$$
>
> Thus, the candidate points are $(1, 1, 1)$ and $(-1, -1, -1)$—a constrained maximum point and a constrained minimum point, respectively. ∎

BASIC EXERCISES

1. In each part, first use the Lagrange multiplier method to find constrained maximum and minimum values, if they exist. Then redo the exercise by elementary calculus methods (i.e., use the constraint equation to rewrite the objective function as a one-variable function).

   (a) $f(x, y) = xy$, subject to $g(x, y) = x + y - 1 = 0$.

   (b) $f(x, y) = x + y$, subject to $g(x, y) = xy - 1 = 0$.

2. Use the Lagrange multiplier method to find (if they exist) constrained maximum and minimum values in each part below. If no such maxima or minima exist, explain why not. At each constrained maximum or minimum point, calculate both ∇f and ∇g.

   (a) $f(x, y) = x - y$, subject to $g(x, y) = x^2 + y^2 - 1 = 0$.

   (b) $f(x, y) = xy$, subject to $g(x, y) = x^2 + y^2 - 1 = 0$.

   (c) $f(x, y) = x^2 + y^2$, subject to $g(x, y) = x + y - 2 = 0$.

   (d) $f(x, y, z) = 2x + y + z$, subject to $g(x, y, z) = x^2 + y^2 + z^2 - 6 = 0$.

3. Here is yet another way to optimize $f(x, y) = x^2 - 4x + 2y^2$ subject to the constraint $x^2 + y^2 = 9$ (Example 1). Carry out the details as described below.

   (a) Substitute $y^2 = 9 - x^2$ into the objective function to produce a new function h of one variable, x.

   (b) On what x-interval should $h(x)$ be optimized? Why?

   (c) Use single-variable methods to maximize and minimize $h(x)$ on the appropriate x-interval. Check that your results agree with those of Example 1.

4. The following picture shows level curves of $f(x, y) = x + 2y$ and also the curve $g(x, y) = 4x^2 + 9y^2 - 36 = 0$. This exercise is about optimizing f subject to the constraint $g(x, y) = 0$.

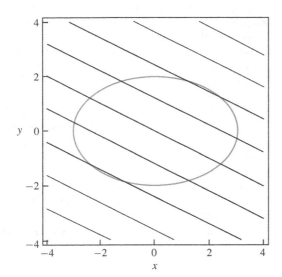

(a) Using the picture alone, estimate the points at which constrained maxima and minima occur and the values of f at these points.

(b) Use the Lagrange multiplier condition to check your work in the previous part.

5. The picture below shows several level curves of $f(x, y) = x + 2y$. This exercise is about maximizing and minimizing f subject to the constraint $g(x, y) = x^2 + y^2 - 9 = 0$.

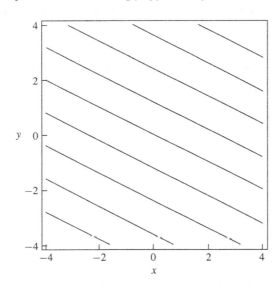

(a) Carefully draw the constraint set $g(x, y) = 0$ into the picture. Label some contour lines with their z-values.

(b) Using the picture alone, estimate the points at which constrained maxima and minima occur and the values of f at these points.

(c) Use the Lagrange multiplier condition to check your work in the previous part.

6. The picture below shows several level curves of $f(x, y) = x^2 + xy + y^2$. This exercise is about minimizing f subject to the constraint $g(x, y) = x + y - 2 = 0$. (There is no constrained maximum.)

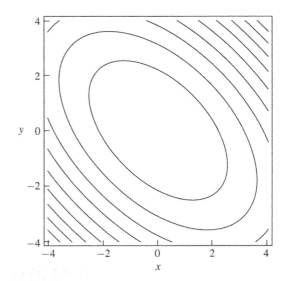

(a) Carefully draw the constraint set $g(x, y) = 0$ into the picture. Label some contour lines with their z-values.

(b) Using the picture alone, estimate the points at which the constrained minimum occurs and the values of f at this point.

(c) Use the Lagrange multiplier condition to check your work in the previous part.

7. Farmer Brown has 100 feet of fence and wants to enclose as much area as possible in a rectangular pig pen. Help Farmer Brown. (Use the Lagrange multiplier method to reinterpret and solve this tired old standard.)

8. Farmer Jones has 100 feet of fence and wants to enclose as much area as possible in (for some reason) a right-triangular pig pen. (Use Lagrange's method to assist this eccentric agriculturalist.)

9. Consider the problem of optimizing $f(x, y) = 2x + 3y$ subject to the constraint $4x + 5y = 6$. (If $g(x, y) = 4x + 5y - 6$, then the constraint is $g(x, y) = 0$.)

(a) What does the equation of Theorem 1 say in this case?

(b) Does f have a constrained maximum or minimum? Why or why not?

10. Let a and b be constants (not both zero), and consider the problem of optimizing $f(x, y) = ax + by$ subject to the constraint $g(x, y) = 4x + 5y - 6 = 0$

(a) For what values of a and b can $f(x, y)$ have a constrained maximum or minimum value? [HINT: For which a and b does the equation $\nabla f = \lambda \nabla g$ have solutions?]

(b) What does Theorem 1 say if $f(x, y) = 400x + 500y$? What are the constrained maximum and minimum values of $f(x, y)$ in this case?

11. Consider the problem of optimizing $f(x, y) = 2x + 3y$ subject to the linear constraint $ax + by = c$; here a, b, and c are constants, and $(a, b) \neq (0, 0)$.

    (a) For what values of a and b can $f(x, y)$ have a constrained maximum or minimum value? [HINT: For which a and b does the equation $\nabla f = \lambda \nabla g$ have solutions?]

    (b) What does Theorem 1 say if the constraint equation is $4x + 6y = 13$? What are the constrained maximum and minimum values of $f(x, y)$ in this case?

12. Consider the problem of optimizing $f(x, y, z) = ax + by + cz + d$ subject to the linear constraint $x + 2y + 3z = 0$; here a, b, c, and d are constants.

    (a) Does f have a constrained maximum or minimum if $f(x, y, z) = 2x + 3y + 4z + 5$? Why or why not?

    (b) Does f have a constrained maximum or minimum if $f(x, y, z) = x + 2y + 3z + 4$? Why or why not? If so, find the maximum or minimum value.

13. A solid circular cylinder of height h and radius r has volume $V(r, h) = \pi r^2 h$ and surface area $A(r, h) = 2\pi r h + 2\pi r^2$.

    (a) An old standard calculus problem is to maximize the volume $V(r, h)$ subject to the "fixed-area constraint" $A(r, h) = A_0$, where A_0 is some positive constant. Assume that this constrained maximum exists (as indeed it does). Use Lagrange multipliers to show that the constrained maximum occurs when $h = 2r$. [HINT: Write the Lagrange multiplier equation and simplify it to obtain the result.]

    (b) Another old standard calculus problem is to minimize the surface area $A(r, h)$ subject to the "fixed-volume constraint" $V(r, h) = V_0$, where V_0 is some positive constant. Assume that this constrained maximum exists (as indeed it does). Use Lagrange multipliers to show that the constrained maximum occurs when $h = 2r$. [HINT: Write the Lagrange multiplier equation and simplify it to obtain the result.]

    (c) How are the answers to the preceding parts related?

15.5 IMPROPER MULTIVARIABLE INTEGRALS

This section assumes knowledge of double integrals through Section 14.4, on polar integration.

Improper integrals in one variable Integrals of single-variable functions are called improper if either (i) the domain of integration or (ii) the integrand itself is unbounded. The following examples—presented as reminders—illustrate both possibilities and the language used to describe them.

EXAMPLE 1 Let $I_1 = \int_1^\infty \dfrac{1}{x}\,dx$ and $I_2 = \int_1^\infty \dfrac{1}{x^2}\,dx$. Does each integral converge or diverge? If either converges, find its limit.

Solution Figure 1 shows both integrands:

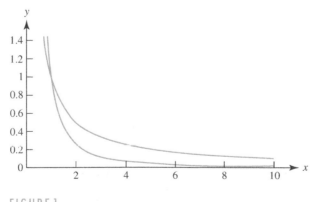

Both functions die out, but one more rapidly than the other.

FIGURE 1
Two decreasing integrands: $y = 1/x$ and $y = 1/x^2$

Both I_1 and I_2 are improper because the domain of integration is the infinite interval $[1, \infty)$. Nevertheless, one or both of the areas bounded by the integrands may be finite.

(The graphs suggest that I_2 has a better chance to converge.) To decide, we consider integrals over *finite* intervals $[1, t]$ and take limits as $t \to \infty$:

$$I_1 = \lim_{t \to \infty} \int_1^t \frac{dx}{x} = \lim_{t \to \infty} \ln x \Big]_1^t = \lim_{t \to \infty} \ln t = \infty;$$

$$I_2 = \lim_{t \to \infty} \int_1^t \frac{dx}{x^2} = \lim_{t \to \infty} \frac{-1}{x} \Big]_1^t = \lim_{t \to \infty} \left(1 - \frac{1}{t}\right) = 1.$$

We conclude that I_1 **diverges** to infinity, while I_2 **converges** to 1. ■

In the following example the integrands, rather than the domains of integration, are unbounded. ➦

Plot the integrands for yourself to suggest which has the better chance to converge.

E X A M P L E 2   Let $I_1 = \int_0^1 \frac{1}{x}\, dx$ and $I_2 = \int_0^1 \frac{1}{\sqrt{x}}\, dx$. Does each integral converge or diverge? If either converges, find its limit.

Solution   Again we consider an appropriate limit of proper integrals; this time t approaches the "problem point" 0 from above:

$$I_1 = \lim_{t \to 0^+} \int_t^1 \frac{dx}{x} = \lim_{t \to 0^+} \ln x \Big]_t^1 = \lim_{t \to 0^+} -\ln t = -\infty;$$

$$I_2 = \lim_{t \to 0^+} \int_t^1 \frac{dx}{\sqrt{x}} = \lim_{t \to 0^+} 2\sqrt{x} \Big]_t^1 = \lim_{t \to 0^+} 2 - 2\sqrt{t} = 2.$$

Thus, I_1 **diverges** to infinity and I_2 **converges** to 2. ■

Improper integrals in two variables   Two-dimensional integrals may be improper for either or both of the two reasons just illustrated, and a similar strategy applies to handling them: Express the desired improper integral as an appropriate limit of *proper* integrals calculated over smaller domains that converge in the limit to the full domain.

In the one-dimensional case these smaller domains are simply intervals. In Example 1, for instance, the full domain of integration was $[1, \infty)$, and we found limits of integrals over $[1, t]$ as $t \to \infty$. In Example 2 the full domain of integration was $[0, 1]$, and we found limits of integrals over $[t, 1]$ as $t \to 0^+$.

For two-dimensional integrals the situation may be more complicated. Domains may have complicated shapes, for instance, and so it may be difficult to find an appropriate collection of convenient smaller domains that converge in the appropriate sense to the full domain of integration. Integrands, too, may interact in complicated ways with the properties of domains of integration. We can (and will) avoid such difficulties by considering only nonnegative integrands over geometrically simple domains.

E X A M P L E 3   Discuss $I_1 = \iint_R \frac{1}{x^2 y^2}\, dA$ and $I_2 = \iint_R \frac{1}{xy}\, dA$, where R is the "infinite rectangle" $[1, \infty) \times [1, \infty)$.

Solution   The integrand is improper because the domain R is infinite. But $R = [1, \infty) \times [1, \infty)$ is in a natural sense the limit as $t \to \infty$ of domains $R_t = [1, t] \times [1, t]$, and so we consider the following limit for I_1: ➦

Check details of the calculation.

$$\lim_{t \to \infty} \iint_{R_t} \frac{dA}{x^2 y^2} = \lim_{t \to \infty} \int_{x=1}^{x=t} \int_{y=1}^{y=t} \frac{1}{x^2 y^2}\, dy\, dx = \lim_{t \to \infty} \left(1 - \frac{1}{t}\right)^2 = 1.$$

Thus, I_1 converges to the limit 1.

We approach I_2 the same way:

$$\lim_{t \to \infty} \int_1^t \int_1^t \frac{1}{xy}\, dy\, dx = \lim_{t \to \infty} (\ln t)^2 = \infty;$$

thus, I_2 diverges to ∞.

Figure 2 suggests why I_1 and I_2 behave differently:

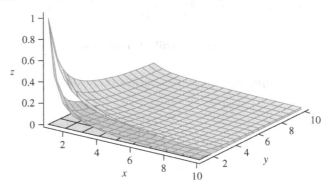

FIGURE 2

Two decreasing integrands: $z = \dfrac{1}{xy}$ and $z = \dfrac{1}{x^2 y^2}$

We see that, although both integrands approach zero as x and y tend to infinity, one of them (the integrand in I_1) dies out faster than the other thanks to the higher powers of x and y. ∎

Improper integrals, areas, and volumes   Ordinary integrals are often interpreted geometrically, as *areas* in the one-dimensional case and as *volumes* in two dimensions. The same interpretations apply to improper integrals. Each of the integrals I_1 and I_2 in Example 1, for instance, can be thought of as measuring the area under a curve, above the x-axis, and to the right of the line $x = 1$. Each of these regions is infinitely *long*, but has rapidly decreasing vertical *height*. The total *area* may therefore either remain finite, as in I_2, or diverge to infinity, as in I_1.

A similar geometric interpretation in two dimensions is as follows: If $f(x, y) \geq 0$, then an integral $\iint_R f(x, y)\, dA$ measures the *volume* of the region bounded below by the domain R and above by the graph of f. If the integral is improper, then this region is unbounded in at least one dimension, and the key question of convergence is whether the total volume is finite or infinite. For example, Figure 3 shows part of the graph of $f(x, y) = \exp(-x^2 - y^2)$:

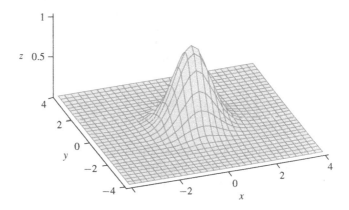

FIGURE 3

The hat-shaped graph of $f(x, y) = \exp(-x^2 - y^2)$

The figure suggests that the function dies out quickly away from its peak at the origin, and so it seems possible that the total volume under the surface and above the entire xy-plane is finite. We resolve the question in the following example.

EXAMPLE 4 Does the improper integral $I = \iint_{\mathbb{R}^2} e^{-(x^2+y^2)}\, dA$ converge or diverge? (Here $\mathbb{R}^2$ is the entire xy-plane.) If the integral converges, what is the limit?

Solution Both the symmetry in Figure 3 and the symbolic form of the integrand suggest we use polar coordinates. Thus, we consider the full domain $\mathbb{R}^2$ as the limit of finite *circular* domains R_t, where R_t is the region inside the circle of radius t about the origin. Then we have

$$I = \lim_{t \to \infty} \iint_{R_t} e^{-(x^2+y^2)}\, dA;$$

the integral converges if (but only if) a finite limit exists. The integral above is easily calculated in polar coordinates. Here R_t is the polar "rectangle" defined by $0 \le \theta \le 2\pi$ and $0 \le r \le t$, and an easy calculation gives

$$\iint_{R_t} e^{-(x^2+y^2)}\, dA = \int_0^{2\pi} \int_0^t e^{-r^2} r\, dr\, d\theta = \pi \left(1 - e^{-t^2}\right).$$

Thus, the limit as $t \to \infty$ is clearly π; this is also the total volume bounded by the xy-plane and the cymbal-shaped surface. ∎

Improper integrals with unbounded integrands As in the one-dimensional case, the *integrand* of a double integral may be unbounded, and in such cases the integral may either converge or diverge. We handle such integrals just as the reader might expect: We choose appropriate smaller domains R_t over which the integral at hand is proper, and take limits as R_t approaches R.

EXAMPLE 5 Figure 4 shows how $f(x, y) = \dfrac{1}{\sqrt{x^2+y^2}}$ behaves over the unit disk R, where $0 \le r \le 1$ and $0 \le \theta \le 2\pi$.

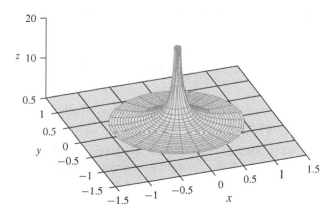

FIGURE 4
Graph of $f(x, y) = 1/\sqrt{x^2+y^2}$

The integrand spikes upward near the origin, but does it enclose a finite volume?

Solution   The question boils down to whether the improper integral $\iint_R f(x, y)\, dA$ converges or diverges. Because the impropriety occurs at the origin, it is natural to use washer-shaped smaller domains R_t with inner radius t that converges to 0. These R_t are the polar rectangles defined by $0 \leq \theta \leq 2\pi$ and $t \leq r \leq 1$. Thus,

$$\iint_R \frac{dA}{\sqrt{x^2 + y^2}} = \lim_{t \to 0^+} \iint_{R_t} \frac{dA}{\sqrt{x^2 + y^2}},$$

and thanks to the polar-friendly forms of both domain and integrand, the inner integral is easily calculated:

$$\iint_{R_t} \frac{dA}{\sqrt{x^2 + y^2}} = \int_0^{2\pi} \int_t^1 \frac{1}{\sqrt{r^2}}\, r\, dr\, d\theta = 2\pi \cdot (1 - t),$$

which clearly approaches 2π as t approaches zero.   ■

Probability calculations: using improper integrals   As in the one-dimensional case, improper *double* integrals are often applied to calculations in probability. ◄ In the two-dimensional case we imagine a continuous random variable X (such as the landing point of a dart aimed at the origin) that takes values in the xy-plane. A function $p(x, y)$ is called a **probability density function** (or **pdf**) for X if, for any region R in the plane, we have

Section 10.3 discusses probabilistic applications of one-dimensional improper integrals.

$$\iint_R p(x, y)\, dA = \text{probability that } X \text{ lies in } R.$$

Because probabilities are measured on a scale from 0 to 1 and because the value of X is always *somewhere* in the plane, a pdf must have two basic properties:

(i) $p(x, y) \geq 0$ for all (x, y);   (ii) $\iint_{\mathbb{R}^2} p(x, y)\, dA = 1.$

We illustrate uses of these ideas and properties by example.

EXAMPLE 6   A darts player aims at the origin in the xy-plane. Thanks to imperfect aim, a not-quite-steady hand, and random breezes, the dart's landing position is modeled by the pdf

$$p(x, y) = \frac{1}{2\pi} \exp\left(-\frac{x^2 + y^2}{2}\right).$$

Find the probability that a given dart lands inside the unit circle $x^2 + y^2 = 1$ (where $0 \leq r \leq 1$ in polar coordinates).

Solution   The desired probability is found by integrating the pdf $p(x, y)$ over the region R inside the unit circle. Polar coordinates clean up both R and $p(x, y)$, and the desired integral becomes

$$\iint_R p(x, y)\, dA = \frac{1}{2\pi} \int_0^{2\pi} \int_0^1 e^{-r^2/2}\, r\, dr\, d\theta = 1 - e^{-1/2} \approx 0.393.$$

(We substituted $u = -r^2/2$ in the last integral.) We see, therefore, that about 39% of darts land inside the unit circle.   ■

BASIC EXERCISES

1. In the setting of Example 6, show that $\iint_{\mathbb{R}^2} p(x, y)\, dA = 1$ (as required for a pdf).

2. The function $p(x, y)$ in Example 6 is "independent of θ" in the sense that $p(x, y)$ can be written in polar form as $e^{-r^2/2}$. What does the absence of θ mean about the dart-throwing situation?

3. In the setting of Example 6, find the probability that the dart lands inside the circle $x^2 + y^2 = 9$.

4. In the setting of Example 6, find the probability that the dart lands in the first quadrant.

5. In the setting of Example 6, find the radius a such that the dart lands inside the circle $x^2 + y^2 = a^2$ with probability $1/2$.

6. Show that the improper integral $I = \iint_R e^{-(x+y)}\, dA$ converges, where $R = [0, \infty) \times [0, \infty)$. (Let $R_t = [0, t] \times [0, t]$, integrate over R_t, and take the limit as $t \to \infty$.)

7. Show that the improper integral $I = \iint_R \frac{1}{x^2 y}\, dA$ diverges, where $R = [1, \infty) \times [1, \infty)$.

8. Show that the improper integral $I = \iint_R \frac{1}{x^2 y \sqrt{y}}\, dA$ converges, where $R = [1, \infty) \times [1, \infty)$.

9. Show that the improper integral $I = \iint_{\mathbb{R}^2} \frac{1}{(1+x^2)(1+y^2)}\, dA$ converges, where $\mathbb{R}^2$ is the entire xy-plane.

Constructing Pedal Curves

This Interlude extends ideas developed in Section 15.2.

There are many ways to create new curves from old. This Interlude explores one of the classical ways of doing so.

Given a smooth curve C and a fixed point O, the **pedal curve** of C with respect to O is defined as follows. (We will always take O to be the origin, $(0, 0)$.) For a given point P on C, we draw the tangent line ℓ to C at P. The associated point Q is the point on ℓ such that $\overline{OQ}$ is perpendicular to ℓ. (If ℓ passes through O, then $Q = O$.) As P moves along C, the associated point Q traces out the pedal curve.

Figure 1(a), shows two points P and their associated points Q. Figure 1(b) shows a curve (the circle) and its pedal curve.

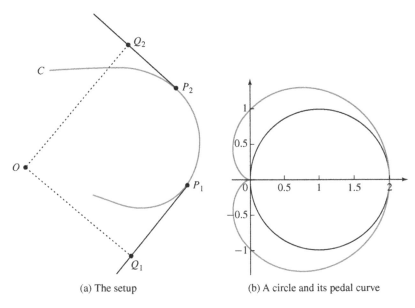

(a) The setup   (b) A circle and its pedal curve

FIGURE 1

Constructing a pedal curve

Finding a formula Suppose that the curve C is defined by the position vector $\mathbf{p}(t)$, with velocity vector $\mathbf{p}'(t)$. Then (some hints for deriving the formula are given below) the pedal curve is defined by the position vector

$$\mathbf{q}(t) = \mathbf{p}(t) - \frac{\mathbf{p}(t) \cdot \mathbf{p}'(t)}{|\mathbf{p}'(t)|^2}\mathbf{p}'(t). \tag{1}$$

PROBLEM 1 What is the pedal curve of a circle centered at the origin? First, give a geometric explanation. Then show how Equation 1 agrees with your answer.

PROBLEM 2 What is the pedal curve of a straight line? (Assume that the line misses the origin.) First give a geometric explanation for your answer. Then show how Equation 1 agrees with your answer.

PROBLEM 3 Study Figure 1(b); the pedal curve is the outside curve (in color). Can you guess the formula for this pedal curve? (HINT: Put your heart into it.) Then find a formula for the circle, and use Equation 1 to show that your guess is correct.
(HINT: Your guess should be a polar curve; you will need to change it to parametric form.)

PROBLEM 4 Study Figure 1(a); observe that the vector joining $\mathbf{P}(t)$ to $\mathbf{Q}(t)$ is the component of $P(t)$ in the direction of the velocity vector at $P(t)$. Using this fact, carefully derive Equation 1.

PROBLEM 5 For each curve following, find a vector formula for the pedal curve. Then use technology to plot both the curve and its pedal on the same axes.

1. The parabola $y = x^2$ (NOTE: In this case the pedal curve is a **cissoid of Diocles**, which is named for the ancient Greek mathematician who first recognized the special reflection properties of parabolic mirrors.)

2. The parabola $y = x^2 - 1/4$.

3. The ellipse $x = 2\cos t + \sqrt{3}$, $y = \sin t$, $0 \le t \le 2\pi$.

VECTOR CALCULUS

16.1 LINE INTEGRALS

A peek ahead In work so far we have met many variants on the basic single-variable definite integral $\int_a^b f(x)\,dx$. Now we meet yet another variant: the **line integral**

$$\int_\gamma \mathbf{f}(\mathbf{X}) \cdot d\mathbf{X},$$

About which, more soon.

where $\mathbf{f} : \mathbb{R}^2 \to \mathbb{R}^2$ is a vector-valued function, and γ is an oriented curve in $\mathbb{R}^2$. (Higher-dimensional versions exist as well.) As we will see a little later, line integrals have a close and far from obvious connection to double integrals. This important connection is known as Green's theorem. ◄

Line integrals are also essential tools in physics. They come with the territory whenever vector-valued functions are used, as when physicists model such phenomena as forces, fluid flow, electricity, and magnetism. The mathematical theory of line integrals was developed in the early 1800's partly to solve physical problems. Green's theorem, for example, can be understood as a quantitative property of fluid flow.

But we are getting far ahead of ourselves. Our first goal is to introduce line integrals themselves, starting from their basic ingredients.

Vector fields

For the moment we stick mainly to line integrals in $\mathbb{R}^2$.

The integrand in a line integral ◄ is a vector-valued function $\mathbf{f} : \mathbb{R}^2 \to \mathbb{R}^2$. The function $\mathbf{f}$ has the general form

$$\mathbf{f}(x, y) = \big(P(x, y), Q(x, y) \big),$$

where $P : \mathbb{R}^2 \to \mathbb{R}$ and $Q : \mathbb{R}^2 \to \mathbb{R}$ are real-valued functions of two variables. In effect, $\mathbf{f}$ is a *pair* of scalar functions.

That's four in all—too many to draw.

How might we visualize $\mathbf{f}$, which accepts two input variables and produces two output variables? ◄ The best approach for us will be to think of $\mathbf{f}$ as a **vector field** in $\mathbb{R}^2$: To any input point (x, y) in $\mathbb{R}^2$, $\mathbf{f}$ assigns the two-dimensional vector $\big(P(x, y), Q(x, y)\big)$, which we can represent as an arrow based at (x, y). By drawing (or having a machine draw) many of these arrows, we can get a sense of how $\mathbf{f}$ behaves.

E X A M P L E 1   Consider the vector-valued function $\mathbf{f}(x, y) = (x - y, x + y)$ as a vector field. Plot $\mathbf{f}$ on the input domain $[-3, 3] \times [-3, 3]$. Discuss the picture. Where are the arrows vertical and horizontal? Why?

Solution   Figure 1 shows a machine-drawn picture:

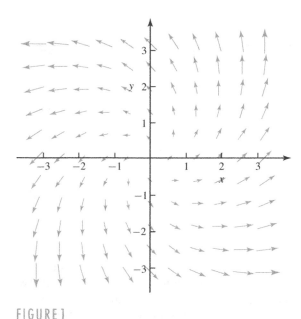

FIGURE 1
The vector field $\mathbf{f}(x, y) = (x - y,\ x + y)$

There is much to see:

- **Only a sample**   As with any plot, this one only samples values of $\mathbf{f}$. The picture shows 120 output arrows.

- **Arrows are scaled down**   By definition, $\mathbf{f}(1, 1) = (0, 2)$—a vertical arrow of length 2. In the picture, however, the arrow at $(1, 1)$ is much shorter. This scaling is usually done (by plotting programs and by humans) to prevent arrows from overlapping with their neighbors. Although some information is lost in this process, it seldom matters much. In any event, arrows that *look* longer *are* longer.

- **Horizontal and vertical arrows**   An arrow is vertical if its first coordinate (given here by $P(x, y) = x - y$) is zero. This occurs along the line $y = x$; the picture agrees. Arrows are horizontal where $Q(x, y) = x + y = 0$, that is, along the line $y = -x$.

- **Vanishing arrows**   Since $f(0, 0) = (0, 0)$, the arrow at the origin has zero length and so "vanishes." (The origin is the only such point.)

In general, the vector picture suggests something like a "reverse whirlpool" in which water wells slowly up at the center and then spins around counterclockwise, faster, and faster.

E X A M P L E 2   Plot $\mathbf{g}(x, y) = (y, x)$ as a vector field. Can you "see" a connection to the scalar function $h(x, y) = xy$?

Solution   Figure 2 shows the field, this time in the square $[-5, 5] \times [-5, 5]$:

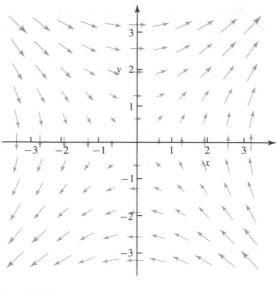

FIGURE 2
The vector field g(x, y) = (y, x)

Here, $P(x, y) = y$ vanishes along the x-axis; as the picture shows, arrows there are vertical. Similarly, $Q(x, y) = x$ is zero along the y-axis; there, arrows are horizontal. As in the preceding example, $\mathbf{g}(x, y) = (0, 0)$ *only* at the origin. Notice, too, that the vector field in Figure 2 is strongly reminiscent of the contour map of $h(x, y) = xy$. (A contour map appears in Figure 5, page 749.) This connection to $h(x, y)$ is no accident, of course. In fact,

$$\mathbf{g}(x, y) = (y, x) = \nabla h(x, y).$$

We will pursue this connection carefully in later sections (and briefly in this section's exercises).

Oriented curves

For curves, "orientation" is just math-speak for "direction." In other settings, "orientation" refers to something a little more complicated.

The other main ingredient in a line integral is the **oriented curve** over which the integral is taken. ← An oriented curve is, as the name suggests, a curve with a "direction of travel" specified from starting point to ending point. The curve may be either in the plane or in three-dimensional space. The upper half of the unit circle, for example, can be traversed counterclockwise from $(1, 0)$ to $(-1, 0)$; the "reversed" curve goes clockwise from $(-1, 0)$ to $(1, 0)$.

It's traditional to denote an oriented curve by γ (the Greek letter gamma) and the reversed curve by $-\gamma$. (Notice, however, that γ and $-\gamma$ are exactly the same set of points; only the direction is different.) This sign convention turns out to be a convenient aid to memory. We will see soon that, in line integrals, reversing a curve's orientation changes the *sign* of the integral.

We discussed parametrization of curves at length in Chapter 12.

To parametrize a curve γ in the xy-plane means to specify parametric equations and a parametrizing interval as follows: ←

$$x = x(t); \qquad y = y(t); \qquad a \le t \le b.$$

Notice that such a parametrization also automatically specifies an orientation: The curve starts at $(x(a), y(a))$ and ends at $(x(b), y(b))$. Therefore, if we want to parametrize a curve in a given direction, we may need to choose our parametrization to ensure that this comes about. (Example 3, below, reviews techniques for reversing a curve's direction.)

Smooth curves To avoid technical troubles in the calculations that follow it will help to assume that the curves we consider are **smooth**. This assumption means, geometrically, what the name suggests: smooth curves are free of gaps, sharp corners, and other features that might cause trouble for calculations that involve derivatives and integrals. In analytic language, a curve γ is **smooth** if it can be parametrized as just described with coordinate functions $x(t)$ and $y(t)$ such that the derivatives $x'(t)$ and $y'(t)$ exist, are continuous functions of t, and are not simultaneously zero except perhaps at the endpoints $t = a$ and $t = b$. (If both x' and y' are zero at the same input t, then the parametrization might "stop" or change direction suddenly.)

EXAMPLE 3 Let γ be the (smooth) curve with parametrization

$$x = \cos t; \qquad y = \sin t; \qquad 0 \le t \le \pi.$$

Describe γ and $-\gamma$. How would $-\gamma$ be parametrized?

Solution The curve γ is the upper half of the unit circle parametrized *counterclockwise*, and so $-\gamma$ is the same curve parametrized *clockwise*.

There are several ways to parametrize $-\gamma$. One approach is to use the parametrization

$$x = \cos(\pi - t); \qquad y = \sin(\pi - t); \qquad 0 \le t \le \pi,$$

which, in effect, reverses the direction of the parameter interval $[0, \pi]$. ➡

Another possibility is to think of $-\gamma$ as the graph of a *function*:

We described this trick in Section 12.2.

$$x = t; \qquad y = \sqrt{1 - t^2}; \qquad -1 \le t \le 1.$$

As t moves through the interval $[-1, 1]$, the point $\big(x(t), y(t)\big)$ moves from left to right along the semicircle.

Observe, finally, that plotting any of these recipes with technology would show the same semicircle but traversed in different directions. ∎

Calculating line integrals

With all ingredients at hand we can now calculate line integrals. We will start with a simple example. Watch both the calculations and the useful shorthands.

EXAMPLE 4 Let γ be the upper half-circle mentioned in Example 3, oriented counterclockwise, and let $\mathbf{f}(x, y) = (x - y, x + y)$ be the vector field shown in the first example. Evaluate the line integral

$$\int_\gamma \mathbf{f}(\mathbf{X}) \cdot d\mathbf{X}.$$

Solution To evaluate any line integral, the idea is to use a parametrization of γ to write everything in terms of one variable, say t. To parametrize the given curve γ, we will use the obvious choice:

$$\mathbf{X}(t) = \big(x(t), y(t)\big) = (\cos t, \sin t); \qquad 0 \le t \le \pi.$$

Along the curve γ—where the integral is taken—we can write

$$\mathbf{f}(\mathbf{X}) = \mathbf{f}\big(\mathbf{X}(t)\big) = (x - y, x + y) = (\cos t - \sin t, \cos t + \sin t);$$

this expresses the integrand as a (vector-valued) function of t alone.

Now we consider the $d\mathbf{X}$ factor. It follow from our parametrization that $dx/dt = -\sin t$ and $dy/dt = \cos t$; equivalently,

$$dx = -\sin t \, dt \qquad \text{and} \qquad dy = \cos t \, dt.$$

We can write the same information in vector form, as

$$d\mathbf{X} = (dx, dy) = (-\sin t \, dt, \cos t \, dt) = \mathbf{X}'(t) \, dt.$$

Now we have written both $\mathbf{f}(\mathbf{X})$ and $d\mathbf{X}$ as vector functions of t, and so the dot product in the integrand makes sense:

$$\mathbf{f}(\mathbf{X}) \cdot d\mathbf{X} = \mathbf{f}\big(\mathbf{X}(t)\big) \cdot \mathbf{X}'(t) \, dt$$
$$= (\cos t - \sin t, \cos t + \sin t) \cdot (-\sin t, \cos t) \, dt = 1 \, dt.$$

(The last equality comes from a trigonometric identity.) The result is a scalar function of t; all that remains is to integrate over the parameter interval. Our conclusion boils down to just one number:

$$\int_{\gamma} \mathbf{f}(\mathbf{X}) \cdot d\mathbf{X} = \int_{0}^{\pi} \mathbf{f}\big(\mathbf{X}(t)\big) \cdot \mathbf{X}'(t) \, dt = \int_{0}^{\pi} 1 \, dt = \pi.$$

The definition puts the technique just illustrated in general terms:

DEFINITION (Line integral) Let γ be a smooth, oriented curve in $\mathbb{R}^2$, and let $\mathbf{f} : \mathbb{R}^2 \to \mathbb{R}^2$ be a vector field defined on and near γ. Let $\mathbf{X}(t)$ be a differentiable parametrization of γ, with $a \leq t \leq b$. The line integral of $\mathbf{f}$ along γ, denoted by

$$\int_{\gamma} \mathbf{f}(\mathbf{X}) \cdot d\mathbf{X},$$

is defined by

$$\int_{a}^{b} \mathbf{f}\big(\mathbf{X}(t)\big) \cdot \mathbf{X}'(t) \, dt.$$

Here are some comments on this important definition:

For some reason, the letters P and Q are often used.

- **An equivalent notation** If we write $\mathbf{f}(x, y) = \big(P(x, y), Q(x, y)\big)$ and $\mathbf{X} = (x, y)$, then the line integral can be written in any of the alternative forms

$$\int_{\gamma} \mathbf{f}(\mathbf{X}) \cdot d\mathbf{X} = \int_{\gamma} \big(P(x, y), Q(x, y)\big) \cdot (dx, dy) = \int_{\gamma} P \, dx + Q \, dy.$$

(The last form is often used in print, perhaps for its typographical simplicity.) ◄

As we've seen, every curve has many different parametrizations.

- **Different parametrizations, same answer** For the definition to make good sense, the value of a line integral should depend on the vector field and the curve at hand—but not on the particular parametrization chosen for the curve. ◄

Otherwise this book would end right here.

In fact, different parametrizations of a smooth curve *do* turn out to produce the same value for a given line integral. ◄ We illustrate the idea in the following example, but a general proof takes some work, and we return to the matter in Section 16.2.

EXAMPLE 5 As in Example 4, consider the vector field $\mathbf{f}(x, y) = (x - y, x + y)$ and the curve $-\gamma$, the upper half-circle oriented *clockwise*. Evaluate the line integral

$$\int_{-\gamma} \mathbf{f}(\mathbf{X}) \cdot d\mathbf{X}$$

using each of the two different parametrizations of $-\gamma$ described in Example 3.

Solution   The calculations resemble those in Example 4. Parametrizing $-\gamma$ using

$$\mathbf{X}(t) = \big(\cos(\pi - t), \sin(\pi - t)\big); \qquad 0 \le t \le \pi$$

gives

$$\mathbf{f}\big(\mathbf{X}(t)\big) = (x - y, x + y) = \big(\cos(\pi - t) - \sin(\pi - t), \cos(\pi - t) + \sin(\pi - t)\big)$$

and

$$\mathbf{X}'(t) = \big(\sin(\pi - t), -\cos(\pi - t)\big).$$

Substituting the pieces into $\int_\gamma \mathbf{f}(\mathbf{X}) \cdot d\mathbf{X}$ gives

$$\int_0^\pi \big(\cos(\pi - t) - \sin(\pi - t), \cos(\pi - t) + \sin(\pi - t)\big) \cdot \big(\sin(\pi - t), -\cos(\pi - t)\big)\, dt$$

$$= \int_0^\pi -1\, dt = -\pi, \qquad \text{(the dot product collapses nicely)}$$

which is the opposite of what we found for the same curve in the opposite direction.
    The other parametrization for $-\gamma$ gives

$$x = t, \qquad y = \sqrt{1 - t^2}, \qquad dx = dt, \qquad \text{and} \qquad dy = \frac{-t}{\sqrt{1 - t^2}}\, dt; \qquad -1 \le t \le 1.$$

Substituting these data into the definition and simplifying symbolically gives the following integral: ➡

Notice the alternative notation; check details.

$$\int_\gamma \mathbf{f}(\mathbf{X}) \cdot d\mathbf{X} = \int_\gamma (x - y)\, dx + (x + y)\, dy = \int_{-1}^{1} \frac{-1}{\sqrt{1 - t^2}}\, dt = -\pi,$$

which is the same result as obtained above.   ■

Understanding line integrals; force and work

Let's rewrite the line integral formula in yet another way, as

$$\int_a^b \mathbf{f}\big(\mathbf{X}(t)\big) \cdot \frac{\mathbf{X}'(t)}{|\mathbf{X}'(t)|}\, |\mathbf{X}'(t)|\, dt.$$

This version leads to a useful and intuitively helpful physical interpretation of the line integral in terms of work.
    To see how, notice first that the vector to the right of the dot product is a unit vector. Let's think of $\mathbf{f}$ as describing a **force field**. ➡ Then, for each t, the dot product

Gravitational force is one possibility.

$$\mathbf{f}\big(\mathbf{X}(t)\big) \cdot \frac{\mathbf{X}'(t)}{|\mathbf{X}'(t)|}$$

is the scalar component of the force vector $\mathbf{f}\big(\mathbf{X}(t)\big)$ in the direction of $\mathbf{X}'(t)$, that is, in the direction tangent to the curve γ at the point $\mathbf{X}(t)$. Therefore, this dot product measures the *work* done per unit of distance by the force $\mathbf{f}\big(\mathbf{X}(t)\big)$. ➡ The last factor in the integral, $|\mathbf{X}'(t)|$, tells the *speed* of the curve at the point $\mathbf{X}(t)$, that is, the distance traveled per unit of time. Over a short time interval Δt, the force remains essentially constant. Over this interval, therefore, the work done by $\mathbf{f}$ along γ is approximately

We related work to the dot product in Chapter 12.

$$\mathbf{f}\big(\mathbf{X}(t)\big) \cdot \frac{\mathbf{X}'(t)}{|\mathbf{X}'(t)|}\, |\mathbf{X}'(t)|\, \Delta t.$$

(The factor $|\mathbf{X}'(t)|\,\Delta t$ approximates the distance traveled; the other factor measures force.) From this expression our conclusion follows:

> *The line integral $\int_\gamma \mathbf{f}(\mathbf{X}) \cdot d\mathbf{X}$ tells the work done by the force $\mathbf{f}$ along the oriented curve γ.*

For example, Figure 3 shows the vector field and the curve in Example 4:

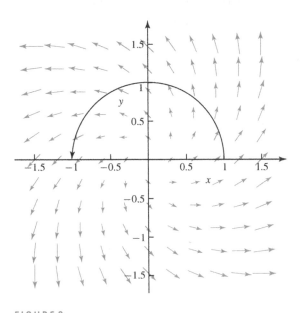

FIGURE 3
A path in a vector field

The curve runs counterclockwise—as does the force field. Therefore, the work done is positive, as we calculated.

Vector fields and line integrals in three dimensions   In this and the next two sections we focus on two-dimensional vector fields and line integrals. But the ideas, objects, and definitions of this section transfer almost unchanged from two to three dimensions. A **three-dimensional vector field $\mathbf{f}$** in $\mathbb{R}^3$, for example, is a function $\mathbf{f}: \mathbb{R}^3 \to \mathbb{R}^3$ of the general form

$$\mathbf{f}(x, y, z) = \big(P(x, y, z),\, Q(x, y, z),\, R(x, y, z) \big).$$

As in $\mathbb{R}^2$, a vector field in $\mathbb{R}^3$ might describe a force field or a fluid flow—but now in space rather than in the plane.

The line integral

$$\int_\gamma \mathbf{f}(\mathbf{X}) \cdot d\mathbf{X} = \int_\gamma P(x, y, z)\,dx + Q(x, y, z)\,dy + R(x, y, z)\,dz$$

of such a vector field over a curve γ in three-dimensional space is defined as one would expect from experience with the two-dimensional situation, and the same physical and geometric intuitions apply. We illustrate with a simple example.

EXAMPLE 6   Find the line integral $\int_\gamma \mathbf{f}(\mathbf{X}) \cdot d\mathbf{X}$, for the vector field

$\mathbf{f}(x, y, z) = (x, y, z)$ and the spiral curve γ parametrized by

$$x = \cos t, \qquad y = \sin t, \qquad z = t; \qquad 0 \le t \le 2\pi.$$

(The curve is shown in Figure 4, page 624.)

Solution   The parametrization given implies that

$$dx = -\sin t \, dt, \quad dy = \cos t \, dt, \qquad \text{and} \qquad dz = dt.$$

Substituting these data into the line integral gives

$$\int_\gamma x \, dx + y \, dy + z \, dz = \int_0^{2\pi} \left(\cos t, \sin t, t\right) \cdot \left(-\sin t, \cos t, 1\right) dt$$

$$= \int_0^{2\pi} t \, dt = 2\pi^2.$$

Notice the sign of the answer—it indicates (in the language of flow) a *positive* net flow along the spiral curve. ∎

BASIC EXERCISES

1. Let $\mathbf{f}(x, y) = (x, 0)$.
   (a) Draw the vector field $\mathbf{f}$ in the rectangle $[-2, 2] \times [-2, 2]$.
   (b) Evaluate $\int_\gamma \mathbf{f}(\mathbf{X}) \cdot d\mathbf{X}$, where γ is the curve $\mathbf{X}(t) = (\cos t, \sin t)$, $0 \le t \le 2\pi$. How does the vector field picture predict the sign of the answer?
   (c) Evaluate $\int_\gamma \mathbf{f}(\mathbf{X}) \cdot d\mathbf{X}$, where γ is the curve $\mathbf{X}(t) = (1 + \cos t, \sin t)$, $0 \le t \le 2\pi$. How does the vector field picture predict the sign of the answer?
   (d) Evaluate $\int_\gamma \mathbf{f}(\mathbf{X}) \cdot d\mathbf{X}$, where γ is the curve $\mathbf{X}(t) = (a + c \cos t, b + c \sin t)$, $0 \le t \le 2\pi$.

2. Repeat Exercise 1 with $\mathbf{f}(x, y) = (0, x)$.

Let $\mathbf{f}(x, y) = (x - y, x + y)$. (A picture is shown in Figure 1.) In Exercises 3–6, determine if the scalar component of the vector field $\mathbf{f}$ in the direction tangent to the curve γ at the point (x_0, y_0) is positive, negative, or zero.

3. γ is the line segment from $(0, 0)$ to $(6, 2)$; $(x_0, y_0) = (3, 1)$.

4. γ is the line segment from $(6, 2)$ to $(0, 0)$; $(x_0, y_0) = (3, 1)$.

5. γ is the curve $\mathbf{X}(t) = (t, -\sqrt{4 - t^2})$, $-2 \le t \le 2$; $(x_0, y_0) = (1, -\sqrt{3})$.

6. γ is the curve $\mathbf{X}(t) = (2 \sin t, 2 \cos t)$, $0 \le t \le \pi/2$; $(x_0, y_0) = (\sqrt{2}, \sqrt{2})$.

In Exercises 7–18, evaluate the line integral $\int_\gamma \mathbf{f}(\mathbf{X}) \cdot d\mathbf{X}$.

7. $\mathbf{f}(x, y) = (x, y)$; γ is the curve $x(t) = \cos t$, $y(t) = \sin t$, $0 \le t \le \pi$.

8. $\mathbf{f}(x, y) = (x, y)$; γ is the curve $x(t) = -t$, $y(t) = \sqrt{1 - t^2}$, $-1 \le t \le 1$.

9. $\mathbf{f}(x, y) = (x, y)$; γ is the curve $x(t) = t$, $y(t) = \sqrt{1 - t^2}$, $-1 \le t \le 1$.

10. $\mathbf{f}(x, y) = (x, y)$; γ is the curve $x(t) = t$, $y(t) = 0$, $-1 \le t \le 1$.

11. $\mathbf{f}(x, y) = (2x, -y)$; γ is the curve $x(t) = \cos t$, $y(t) = \sin t$, $0 \le t \le \pi$.

12. $\mathbf{f}(x, y) = (-y, x)$; γ is the curve $x(t) = \cos t$, $y(t) = \sin t$, $0 \le t \le 2\pi$.

13. $\mathbf{f}(x, y) = (-y, x)$; γ is the curve $x(t) = \cos t$, $y(t) = -\sin t$, $0 \le t \le 2\pi$.

14. $\mathbf{f}(x, y) = (x, 0)$; γ is the curve $x(t) = at$, $y(t) = bt$, $0 \le t \le 1$.

15. $\mathbf{f}(x, y) = (y, 0)$; γ is the curve $x(t) = at$, $y(t) = bt$, $0 \le t \le 1$.

16. $\mathbf{f}(x, y) = (x, y)$; γ is the curve $x(t) = a + t(c - a)$, $y(t) = b + t(d - b)$, $0 \le t \le 1$.

17. $\mathbf{f}(x, y, z) = (yz, xz, xy)$; γ is the curve $\mathbf{X}(t) = (t, t^2, t^3)$, $0 \le t \le 2$.

18. $\mathbf{f}(x, y, z) = (xy, e^z, z)$; γ is the curve $\mathbf{X}(t) = (t^2, -t, t)$, $0 \le t \le 1$.

In Exercises 19–22, evaluate the line integral $\int_\gamma P \, dx + Q \, dy$.

19. $(P, Q) = (x/(x^2 + y^2), y/(x^2 + y^2))$ and γ is the curve $x(t) = \cos t$, $y(t) = \sin t$, $0 \le t \le 2\pi$.

20. $(P, Q) = (-y/(x^2 + y^2), x/(x^2 + y^2))$ and γ is the curve $x(t) = \cos t$, $y(t) = \sin t$, $0 \le t \le 2\pi$.

21. $(P, Q) = (x, y)$; γ is the curve $x(t) = \cos t$, $y(t) = \sin t$, $0 \le t \le \pi$.

22. $(P, Q) = (3xy, x - y^2)$; γ is the line segment from $(0, 0)$ to $(2, 1)$.

23. Evaluate $\int_\gamma \mathbf{f}(\mathbf{X}) \cdot d\mathbf{X}$, where $\mathbf{f}(x, y) = (2, 0)$ and γ is the curve $\mathbf{X}(t) = (t^3, t^2)$, $-2 \le t \le 3$.

24. Evaluate $\int_\gamma \mathbf{f}(\mathbf{X}) \cdot d\mathbf{X}$, where $\mathbf{f}(x, y) = (x, 0)$ and γ is the curve $\mathbf{X}(t) = (1, e^t)$, $-2 \le t \le 3$.

25. Evaluate $\int_\gamma \mathbf{f}(\mathbf{X}) \cdot d\mathbf{X}$, where $\mathbf{f}(x, y) = (0, -3)$ and γ is the curve $\mathbf{X}(t) = (2t, 5 - t)$, $1 \le t \le 4$.

26. Evaluate $\int_\gamma \mathbf{f}(\mathbf{X}) \cdot d\mathbf{X}$, where $\mathbf{f}(x, y) = (3, e^{-x^2})$ and γ is the curve $\mathbf{X}(t) = (t^2, 3)$, $1 \le t \le 4$.

27. Let $\mathbf{f} = (x, y)$. Evaluate $\int_\gamma \mathbf{f}(\mathbf{X}) \cdot d\mathbf{X}$, where γ is

(a) the line segment from $(0, 0)$ to $(2, 4)$ parametrized by $\mathbf{X}(t) = (t, 2t)$, $0 \le t \le 2$.

(b) the line segment from $(0, 0)$ to $(2, 4)$ parametrized by $\mathbf{X}(t) = (t^2, 2t^2)$, $0 \le t \le \sqrt{2}$.

(c) the curve $y = x^2$ from $x = 0$ to $x = 2$ parametrized by $\mathbf{X}(t) = (t, t^2)$.

(d) the curve $y = x^2$ from $x = 0$ to $x = 2$ parametrized by $\mathbf{X}(t) = (t^2, t^4)$.

28. Let $\mathbf{f} = (y, x)$. Evaluate $\int_\gamma \mathbf{f}(\mathbf{X}) \cdot d\mathbf{X}$, where γ is

(a) the line segment from $(0, 0)$ to $(2, 4)$ parametrized by $\mathbf{X}(t) = (t, 2t)$, $0 \le t \le 2$.

(b) the line segment from $(0, 0)$ to $(2, 4)$ parametrized by $\mathbf{X}(t) = (t^2, 2t^2)$, $0 \le t \le \sqrt{2}$.

(c) the curve $y = x^2$ from $x = 0$ to $x = 2$ parametrized by $\mathbf{X}(t) = (t, t^2)$.

(d) the curve $y = x^2$ from $x = 0$ to $x = 2$ parametrized by $\mathbf{X}(t) = (t^2, t^4)$.

29. Let $h(x, y) = x + 2y$, and let $\mathbf{f}(x, y) = \nabla h$.

(a) Plot the vector field $\mathbf{f}$ in the rectangle $[0, 3] \times [0, 3]$.

(b) Add the level curves $h(x, y) = 1$, $h(x, y) = 2$, and $h(x, y) = y$ to the plot you created in part (a). What is the relationship between the level curves and the vector field?

(c) Let γ be the curve $x(t) = t$, $y(t) = (k - t)/2$, where k is a real number and $-\infty \le t \le \infty$. Show that $\mathbf{f}(\mathbf{X}(t)) \cdot \mathbf{X}'(t) = 0$. How is this result related to part (b)?

(d) Let γ be the curve $\mathbf{X}(t) = (t, 2 - t/2)$ for $0 \le t \le 3$. Evaluate the line integral $\int_\gamma \mathbf{f}(\mathbf{X}) \cdot d\mathbf{X}$.

30. Let $h(x, y) = y - x^2$ and let $\mathbf{f}(x, y) = \nabla h$.

(a) Plot the vector field $\mathbf{f}$ in the rectangle $[0, 3] \times [0, 3]$.

(b) Add the level curves $h(x, y) = 1$, $h(x, y) = 2$, and $h(x, y) = y$ to the plot you created in part (a). What is the relationship between the level curves and the vector field?

(c) Let γ be the curve $x(t) = t$, $y(t) = k + t^2$, where k is a real number and $-\infty \le t \le \infty$. Show that $\mathbf{f}(\mathbf{X}(t)) \cdot \mathbf{X}'(t) = 0$. How is this result related to part (b)?

(d) Let γ be the curve $\mathbf{X}(t) = (t, 2 + t^2)$ for $0 \le t \le 3$. Evaluate the line integral $\int_\gamma \mathbf{f}(\mathbf{X}) \cdot d\mathbf{X}$.

31. Evaluate $\int_\gamma \mathbf{f}(\mathbf{X}) \cdot d\mathbf{X}$, where $\mathbf{f}(x, y) = \left(x/|x|, \sqrt{1 + y^3}\right)$ and γ is the curve $\mathbf{X}(t) = (-t, 2)$, $0 \le t \le 4$.

32. Suppose that $\mathbf{f}(\mathbf{X}) = \nabla \mathbf{X}/|\nabla \mathbf{X}|$. Show that $\int_\gamma \mathbf{f}(\mathbf{X}) \cdot d\mathbf{X} = \text{length}(\gamma)$.

16.2 MORE ON LINE INTEGRALS; A FUNDAMENTAL THEOREM

We'll use both notations in this section.

In the preceding section we introduced the idea of the line integral $\int_\gamma \mathbf{f}(\mathbf{X}) \cdot d\mathbf{X}$, or, equivalently, $\int_\gamma P\, dx + Q\, dy$, ← where $\mathbf{f}(x, y) = \big(P(x, y), Q(x, y)\big)$ is a vector field in the plane and γ is an oriented curve in the domain of $\mathbf{f}$.

This section continues the story. We add some basic properties of line integrals, most of them closely analogous to properties of ordinary, single-variable integrals. The analogy culminates in a fundamental theorem for line integrals, which (like the "ordinary" fundamental theorem of elementary calculus) relates derivatives to integrals.

Building intuition: force and flow — but not area

Tempting as it is to think so.

Line integrals, unlike ordinary integrals, are not especially easy to visualize geometrically. In particular, the value of a line integral $\int_\gamma \mathbf{f}(\mathbf{X}) \cdot d\mathbf{X}$ is *not* in any helpful sense the "area under a graph." ← This is not to say that pictures are useless—just that they need to be interpreted quite differently from the standard pictures of elementary calculus.

Force   Line integrals were developed around problems in physics; not surprisingly, therefore, physical intuition is often best for understanding what line integrals say. We gave one such approach in the preceding section: If we interpret $\mathbf{f}$ (or (P, Q), in the other notation) as a force field in the plane, then the line integral becomes the work done by the force on an object that moves along the curve, from start to finish. This viewpoint helps explain, for instance, why reversing the orientation of a curve changes the sign of a line integral: The work done in moving an object along a curve in one direction is opposite to the work required in the other direction.

Fluid flow   Another physical way of thinking about line integrals is in terms of fluid flow. From this point of view, a vector field $\mathbf{f}$ describes, at each point, the velocity of fluid moving across the xy-plane. For an oriented curve γ, the line integral $\int_\gamma \mathbf{f} \cdot d\mathbf{X}$ is called a **flow integral**; it measures the net tendency of fluid to flow along the entire curve. If γ is a **closed curve,** ➡ with the same starting and ending point, then the line integral is called the **circulation** of $\mathbf{f}$ along γ.

That is, a loop.

E X A M P L E 1   Figure 1 shows two flow fields, each with a curve γ (the unit circle oriented counterclockwise) superimposed:

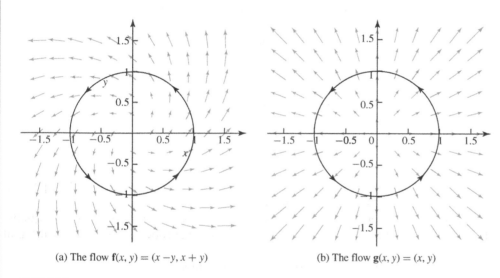

(a) The flow $\mathbf{f}(x, y) = (x - y, x + y)$ (b) The flow $\mathbf{g}(x, y) = (x, y)$

FIGURE 1
Two flow fields and a closed curve

What do the pictures suggest about the line integrals (i.e., flow integrals) $\int_\gamma \mathbf{f} \cdot d\mathbf{X}$ and $\int_\gamma \mathbf{g} \cdot d\mathbf{X}$? What are their values?

S o l u t i o n   In the left-hand picture the fluid appears to be swirling counterclockwise—in the same direction as γ—and so we would expect the flow integral to be *positive*. Indeed it is; a routine calculation ➡ shows that

Details are left to the exercises.

$$\int_\gamma \mathbf{f} \cdot d\mathbf{X} = \int_\gamma (x - y)\, dx + (x + y)\, dy = 2\pi.$$

In Figure 1(b) the flow is everywhere *perpendicular* to the curve γ, and so we would expect zero circulation. Again, the calculation agrees:

$$\int_\gamma \mathbf{g} \cdot d\mathbf{X} = \int_\gamma x\, dx + y\, dy = 0.$$ ∎

New line integrals from old

Line integrals have algebraic properties similar to those of ordinary integrals. The following theorem collects several useful properties.

THEOREM 1 (Algebra with line integrals)   Let $\mathbf{f} = (P, Q)$ and $\mathbf{g} = (R, S)$ be vector fields in $\mathbb{R}^2$, let c be a constant, and let γ be an oriented curve. Then

- $\displaystyle \int_\gamma (\mathbf{f} \pm \mathbf{g}) \cdot d\mathbf{X} = \int_\gamma \mathbf{f} \cdot d\mathbf{X} \pm \int_\gamma \mathbf{g} \cdot d\mathbf{X}.$

- $\displaystyle \int_\gamma (c\mathbf{f}) \cdot d\mathbf{X} = c \int_\gamma \mathbf{f} \cdot d\mathbf{X}.$

- Let $-\gamma$ be the same curve as γ but with orientation reversed. Then

$$\int_{-\gamma} \mathbf{f} \cdot d\mathbf{X} = -\int_\gamma \mathbf{f} \cdot d\mathbf{X}.$$

- Suppose that γ is the union of two oriented curves γ_1 and γ_2. Then

$$\int_\gamma \mathbf{f} \cdot d\mathbf{X} = \int_{\gamma_1} \mathbf{f} \cdot d\mathbf{X} + \int_{\gamma_2} \mathbf{f} \cdot d\mathbf{X}.$$

(The last part is mainly of interest when the two curves meet end to end. The upper and lower halves of the unit circle do so, for example.)

All parts of the theorem should seem reasonable; all are straightforward consequences of the definition of line integral and of the corresponding properties of ordinary integrals. We omit most of the proofs. ←

But one is an exercise in this section.

The theorem lets us combine old line integral results to find new ones.

EXAMPLE 2   Let $\mathbf{f}(x, y) = (x - y, x + y)$ and $\mathbf{g}(x, y) = (x, y)$. Use "known" results to calculate $\int_\gamma -y\, dx + x\, dy$, where γ is the unit circle oriented counterclockwise. ←

We've intentionally varied the notation.

Solution   Notice first that $\mathbf{f} - \mathbf{g} = (x - y, x + y) - (x, y) = (-y, x) = \mathbf{h}(x, y)$. Therefore,

$$\int_\gamma \mathbf{h} \cdot d\mathbf{X} = \int_\gamma -y\, dx + x\, dy$$

is the integral we seek. We have already discussed line integrals of both $\mathbf{f}$ and $\mathbf{g}$; we gave their values in Example 1. Combining these results produces our answer:

$$\int_\gamma \mathbf{h} \cdot d\mathbf{X} = \int_\gamma \mathbf{f} \cdot d\mathbf{X} - \int_\gamma \mathbf{g} \cdot d\mathbf{X} = 2\pi - 0 = 2\pi.$$

Figure 2 shows the vector field **h** near γ:

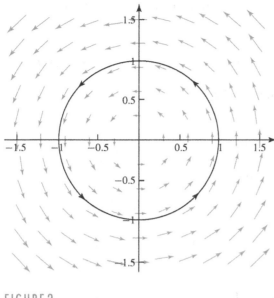

FIGURE 2
The flow h(x, y) = ($-y, x$)

The picture suggests why (as the calculation showed) the line integral is indeed a positive number: the flow points "along" the curve.

The curve matters, not the parametrization Another important property of line integrals—to which we alluded briefly in Section 5.1—is a little more subtle than those in preceding Examples. As the notation $\int_{\gamma} \mathbf{f} \cdot d\mathbf{X}$ suggests, a line integral certainly depends on both the vector field **f** and the curve γ. But any curve can be parametrized in many ways, and a line integral should not depend on the particular parametrization chosen for γ. This property, called **independence of parametrization**, guarantees that the definition of the line integral makes logical sense.

Fortunately, line integrals *do* enjoy this property. The next example illustrates why.

EXAMPLE 3 Let γ be the upper half of the unit circle, parametrized counterclockwise. Consider these two parametrizations of γ:

$$\mathbf{X}(s) = (\cos s, \sin s); \qquad 0 \le s \le \pi;$$
$$\mathbf{X}(t) = \left(\cos(t^2), \sin(t^2)\right); \qquad 0 \le t \le \sqrt{\pi}.$$

(Notice that the two parametrizations traverse the same curve, but at different speeds: $|\mathbf{X}'(s)| = 1$ while $|\mathbf{X}'(t)| = 2t$.) Show that, despite their differences, both parametrizations of the curve γ give the same value for the line integral $\int_{\gamma} -y \, dx + x \, dy$.

Solution The first parametrization gives

$$x = \cos s; \qquad dx = -\sin s \, ds; \qquad y = \sin s; \qquad dy = \cos s \, ds.$$

Therefore,

$$\int_{\gamma} -y \, dx + x \, dy = \int_{0}^{\pi} \left(\sin^2 s + \cos^2 s\right) ds = \pi.$$

The calculation is similar for the second parametrization:

$$x = \cos(t^2); \qquad dx = -2t\sin(t^2)\,dt; \qquad y = \sin(t^2); \qquad dy = 2t\cos(t^2)\,dt.$$

Check the last (easy) step.   This gives ←

$$\int_\gamma -y\,dx + x\,dy = \int_0^{\sqrt{\pi}} \left(\sin^2(t^2) + \cos^2(t^2)\right) 2t\,dt = \pi.$$

The results are the same, as they should be.   ∎

The general case   A general argument for independence of parametrization in line integrals resembles the specific calculation in Example 3. Suppose we are given any two differentiable parametrizations

$$\mathbf{X}_1(s) = \left(x_1(s),\, y_1(s)\right) \qquad \text{and} \qquad \mathbf{X}_2(t) = \left(x_2(t),\, y_2(t)\right)$$

for the curve γ, where $a \le s \le b$ and $c \le t \le d$. For the line integral $\int_\gamma P\,dx + Q\,dy$, these parametrizations lead, respectively, to two ugly integrals:

$$I_1: \qquad \int_a^b \left(P(\mathbf{X}_1(s)) \frac{dx_1}{ds} + Q(\mathbf{X}_1(s)) \frac{dy_1}{ds} \right) ds$$

$$I_2: \qquad \int_c^d \left(P(\mathbf{X}_2(t)) \frac{dx_2}{dt} + Q(\mathbf{X}_2(t)) \frac{dy_2}{dt} \right) dt.$$

These integrals look complicated, to be sure, and much notation is involved. But here we care only that I_1 and I_2 have the *same* value. To see this one shows first that s is a differentiable function of t such that $\mathbf{X}_2(t) = \mathbf{X}_1 \circ s$. In other words,

$$\mathbf{X}_2(t) = \left(x_2(t),\, y_2(t)\right) = \mathbf{X}_1(s(t)) = \left(x_1(s(t)),\, y_1(s(t))\right).$$

(Showing in general that $s(t)$ is differentiable is slightly delicate. In Example 3, we had simply $s = t^2$, which is clearly differentiable.) In particular, $s(c) = a$ and $s(d) = b$.

Assuming the foregoing facts, we can make the change of variable $s = s(t)$ in I_1, which (despite its complicated appearance) is an ordinary integral of the single-variable type. The happy result is that I_2 emerges:

$$
\begin{aligned}
I_1 &= \int_a^b \left(P(\mathbf{X}_1(s)) \frac{dx_1}{ds} + Q(\mathbf{X}_1(s)) \frac{dy_1}{ds} \right) ds \\
&= \int_c^d \left(P(\mathbf{X}_1(s(t))) \frac{dx_1}{ds} + Q(\mathbf{X}_1(s(t))) \frac{dy_1}{ds} \right) \frac{ds}{dt}\, dt \\
&= \int_c^d \left(P(\mathbf{X}_2(t)) \frac{dx_2}{dt} + Q(\mathbf{X}_2(t)) \frac{dy_2}{dt} \right) dt = I_2.
\end{aligned}
$$

(We used the ordinary chain rule in the last line, in the form

$$\frac{dx_1}{ds} \cdot \frac{ds}{dt} = \frac{d}{dt}\left(x_1(s(t))\right) = \frac{dx_2(t)}{dt}$$

for the first summand and similarly in y for the second.) We conclude that $I_1 = I_2$, just as desired.

The fundamental theorem for line integrals

One version of the fundamental theorem of elementary calculus relates the integral and the derivative: If a function f and its derivative f' are continuous on $[a, b]$, then

$$\int_a^b f'(x)\,dx = f(b) - f(a).$$

(There are other ways to state the fundamental theorem of calculus; this version is best for present purposes.)

A similar "fundamental theorem" holds for line integrals. To ensure that all the ingredients exist, we assume that the function $h(x, y)$ has continuous partial derivatives on and near the curve γ, and that γ is smooth, except perhaps at its endpoints.

THEOREM 2 (Fundamental theorem for line integrals) Let $h(x, y): \mathbb{R}^2 \to \mathbb{R}$ be a function. Let γ be a smooth, oriented curve starting at $\mathbf{X}_0 = (x_0, y_0)$ and ending at $\mathbf{X}_1 = (x_1, y_1)$. Then

$$\int_\gamma \nabla h \cdot d\mathbf{X} = h(\mathbf{X}_1) - h(\mathbf{X}_0).$$

If γ is a closed curve (i.e., $\mathbf{X}_1 = \mathbf{X}_0$), then

$$\int_\gamma \nabla h \cdot d\mathbf{X} = 0.$$

Proof   The proof is a pleasing and straightforward exercise with the chain rule. Let the curve γ be parametrized, as usual, by a function

$$\mathbf{X}(t) = \big(x(t), y(t)\big); \qquad a \le t \le b.$$

(Because γ is smooth, we can assume that x' and y' are continuous functions, and thus all the needed integrals exist.) Then the line integral has the form

$$I = \int_\gamma \nabla h \cdot d\mathbf{X} = \int_a^b \Big(h_x\big(\mathbf{X}(t)\big), h_y\big(\mathbf{X}(t)\big) \Big) \cdot \mathbf{X}'(t)\,dt.$$

Now the composite $h\big(\mathbf{X}(t)\big)$ is a new function of t, and, by the chain rule in several variables,

$$\frac{d}{dt}\Big(h\big(\mathbf{X}(t)\big) \Big) = \nabla h\big(\mathbf{X}(t)\big) \cdot \mathbf{X}'(t).$$

In other words, the integrand in the last integral above is the t-derivative of $h\big(\mathbf{X}(t)\big)$. Therefore, by the ordinary fundamental theorem of calculus,

$$I = \int_a^b \frac{d}{dt}\Big(h\big(\mathbf{X}(t)\big) \Big)\,dt = h\big(\mathbf{X}(b)\big) - h\big(\mathbf{X}(a)\big) = h(x_1, y_1) - h(x_0, y_0),$$

as claimed.

The gradient advantage   The fundamental theorem implies a remarkable property of any vector field that happens to be a gradient. For such fields, a line integral depends only on the endpoints of the curve γ—not on the curve itself. (All bets are off, of course, if the curve wanders outside the domain in which the gradient is defined.) Line integrals with this property are said to be **independent of path**. (Don't confuse this property with another sort of independence—*all* line integrals are independent of the particular parametrization chosen for γ.)

The fundamental theorem makes short work of evaluating line integrals when—but only when—we can find a function whose gradient is the vector field in the integrand. If $\nabla h = \mathbf{f}$, then h is called a **potential function** for the vector field $\mathbf{f}$.

Think like a mathematician!

EXAMPLE 4   Let $\mathbf{g}(x, y) = (y, x)$ and $\mathbf{f}(x, y) = (x - y, x + y)$. Let γ_1 be the line segment from $(0, 0)$ to $(2, 1)$, and let γ_2 be the part of the parabolic arch $y = x^2/4$ joining the same two points. Calculate the four line integrals

$$I_1 = \int_{\gamma_1} \mathbf{g} \cdot d\mathbf{X}, \qquad I_2 = \int_{\gamma_2} \mathbf{g} \cdot d\mathbf{X}, \qquad I_3 = \int_{\gamma_1} \mathbf{f} \cdot d\mathbf{X}, \qquad \text{and} \qquad I_4 = \int_{\gamma_2} \mathbf{f} \cdot d\mathbf{X}.$$

Do as little work as possible. ←

Solution   Notice first that if $h(x, y) = xy$, then

$$\nabla h(x, y) = (y, x) = \mathbf{g}(x, y).$$

(We found this potential function h purely by guessing; a more systematic approach is presented in a later example.) We could parametrize γ_1 and γ_2, but Theorem 2 makes that work unnecessary, at least for I_1 and I_2. By Theorem 2 we have

$$\int_{\gamma_1} \mathbf{g} \cdot d\mathbf{X} = \int_{\gamma_2} \mathbf{g} \cdot d\mathbf{X} = h(2, 1) - h(0, 0) = 2.$$

In particular, only the endpoints—not the curves themselves—determine the answer.

Now we consider I_3 and I_4. Guessing a potential function h for which

$$\nabla h = \mathbf{f}(x, y) = (x - y, x + y)$$

seems difficult. (In fact, it is impossible, and we will explain why in a moment.) Thus, Theorem 2 is no help, and so we'll have to parametrize γ_1 and γ_2 after all.

For these simple curves the work is easy. We can use

$$x = 2t; \qquad y = t; \qquad 0 \le t \le 1$$

for γ_1 and

$$x = t; \qquad y = \frac{t^2}{4}; \qquad 0 \le t \le 2$$

for γ_2. Substituting these data into our line integrals gives (after a little symbolic work)

$$I_3 = \int_0^1 5t\, dt = \frac{5}{2}; \qquad I_4 = \int_0^2 \left(t + \frac{t^2}{4} + \frac{t^3}{8} \right) dt = \frac{19}{6}.$$

We omitted some details of calculation because the result is the main point: For the vector field $\mathbf{f}$, the two line integrals are unequal even though the curves join the same points.   ∎

When is a vector field a gradient?   Why can't $\mathbf{f}(\mathbf{X})$ in the preceding example be a gradient? The reason is that if a potential function h existed, then we would have

$$h_x(x, y) = x - y \qquad \text{and} \qquad h_y(x, y) = x + y.$$

From this it would follow, in turn, that

$$h_{xy}(x, y) = -1 \quad \text{and} \quad h_{yx} = 1.$$

This, however, is impossible, because the "cross partial derivatives" of a well-behaved function must be equal. → This example illustrates a useful general fact—in effect, a test for a vector field *not* being a gradient. (The converse is false, as use will see.)

We discussed this earlier; See page 740.

FACT (Gradients and cross partials) A vector field $\mathbf{f}(x, y) = (P(x, y), Q(x, y))$ may or may not be a gradient in the vicinity of a curve γ. If $P_y \neq Q_x$ along γ, then $\mathbf{f}(x, y)$ is not a gradient field.

Finding potential functions

If a vector field passes the test just mentioned, a potential function can often be found by antidifferentiating in x or y separately. We illustrate by example.

EXAMPLE 5 Find a potential function for the vector field

$$(P, Q) = (y \cos(xy) + 1, x \cos(xy)).$$

Use it to calculate $\int_\gamma P \, dx + Q \, dy$, where γ is the upper half of the unit circle, oriented counterclockwise.

Solution A quick check shows that

$$P_y = \cos(xy) - xy \sin(xy) = Q_x,$$

and so it's worth continuing our search.

Note that if $h(x, y)$ is any potential function, then

$$h_x = y \cos(xy) + 1 \quad \text{and} \quad h_y = x \cos(xy).$$

Therefore, antidifferentiating in x (treating y as a constant) gives

$$h(x, y) = \int (y \cos(xy) + 1) \, dx = \int y \cos(xy) \, dx + \int 1 \, dx.$$

Let's make the u-substitution $u = xy$ (and $du = y \, dx$) in the first integral on the right. → This gives

Remember—we're treating y as a constant.

$$\int y \cos(xy) \, dx = \int \cos u \, du = \sin(xy) + C.$$

Putting the pieces together gives

$$h(x, y) = \int y \cos(xy) \, dx + \int 1 \, dx = \sin(xy) + x + C.$$

Now notice that C may legally depend on y, since our integration was only in the variable x. Therefore, we can write

$$h(x, y) = \int y \cos(xy) \, dx + \int 1 \, dx = \sin(xy) + x + C(y),$$

where $C(y)$ is some still-to-be-chosen function of y. To help choose $C(y)$, we differentiate our newly produced function $h(x, y)$ with respect to y:

$$h_y(x, y) = \frac{d}{dy}\big(\sin(xy) + x + C(y)\big) = x\cos(xy) + C'(y).$$

Now the requirement for h to be a potential function is that $h_y = x\cos(xy)$. Thus, $C'(y) = 0$, and so C can be taken as any real-number constant. We conclude (and it's easy to check) that, for any constant C, $h(x, y) = \sin(xy) + x + C$ is a potential function for (P, Q).

All our work in calculating h makes finding the line integral a snap. By the fundamental theorem, the answer depends only on the endpoints of γ:

$$\int_\gamma P\, dx + Q\, dy = h(-1, 0) - h(1, 0) = -2.$$

(Calculating the integral by parametrization would be unpleasant.)

The fundamental theorem reviewed

The fundamental theorem for line integrals and its associated ideas are important enough to deserve some review. The theorem says that, if a vector field **f** happens to be a gradient (that is, if $\mathbf{f} = \nabla h$ for some function h), then

$$\int_\gamma \mathbf{f} \cdot d\mathbf{X} = \int_\gamma \nabla h \cdot d\mathbf{X} = h(\mathbf{X_1}) - h(\mathbf{X_0}), \tag{1}$$

where $\mathbf{X_0}$ and $\mathbf{X_1}$ are the starting and ending points of γ. Several consequences of this identity will prove especially important:

- **Independence of path**   Equation 1 means that the line integral of a gradient field depends only on the endpoints $\mathbf{X_0}$ and $\mathbf{X_1}$—not on the particular curve joining them. If γ' is *another* curve joining $\mathbf{X_0}$ to $\mathbf{X_1}$, and γ' is also in the domain of h, then

$$\int_\gamma \mathbf{f} \cdot d\mathbf{X} = \int_{\gamma'} \mathbf{f} \cdot d\mathbf{X}.$$

- **Line integrals, closed curves, and gradient fields**   A curve is called **closed** if it starts and ends at the same point. In general, the line integral of a vector field around a closed curve may be postive, negative, or zero. The fundamental theorem guarantees, however, that integrating a gradient field around a closed curve *must* give zero: Because the starting and ending points are identical,

$$\int_\gamma \nabla h \cdot d\mathbf{X} = h(\mathbf{X_0}) - h(\mathbf{X_0}) = 0.$$

In physical terms, this result means that, if a force field **f** happens to be a gradient, then the force does zero total work in moving a particle around *any* closed curve. As we have also seen, a given vector field may or may not be the gradient field associated with a function.

- **Conservative vector fields**   Physicists use the adjective **conservative** to describe a force field with the property that every line integral taken around a closed curve in the domain of the vector field has the value zero. The term is appropriate because, in such fields, work and energy are "conserved" as one moves along any curve that returns to its initial position. The fundamental theorem says (among other things) that every gradient field is conservative.

EXAMPLE 6 Figure 3 shows several curves joining $(-1, 1)$ to $(1, 1)$; the vector field $\mathbf{f} = (P, Q) = (y^2, 2xy)$ is superimposed.

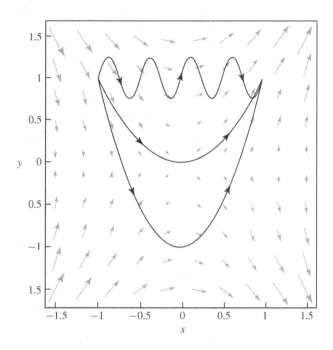

FIGURE 3
The vector field $(y^2, 2xy)$ and three curves

Explain why the line integral has the same value for all three curves. What *is* this value?

Solution It would be unpleasant to parametrize all three curves. Fortunately, we don't have to because the vector field $(y^2, 2xy)$ is a gradient. We can tell (either by guesswork or by the method outlined in the previous section) that $h(x, y) = xy^2$ is a suitable potential function, that is, $(y^2, 2xy) = \nabla(xy^2)$. Therefore, by the fundamental theorem, the line integral—for all three curves—is

$$\int_\gamma \nabla h \cdot d\mathbf{X} = h(1, 1) - h(-1, 1) = 2.$$

The answer looks reasonable, too; the curves are oriented generally *with* rather than *against* the flow, and so we expect a positive result. ➡

Take a close look at the picture to convince yourself.

BASIC EXERCISES

Let $\mathbf{f}(x, y) = (x, y)$. *(A picture is shown in Figure 1.) In Exercises 1–6, evaluate the line integral* $\int_\gamma \mathbf{f}(\mathbf{X}) \cdot d\mathbf{X}$ *for the given curve* γ *and explain, briefly, how the sign of the answer (positive, negative, or zero) could have been predicted from the picture.*

1. γ is the unit circle oriented counterclockwise.

2. γ is the upper half of the circle of radius 1 centered at $(1, 0)$ oriented counterclockwise.

3. γ is the lower half of the circle of radius 1 centered at $(1, 0)$ oriented counterclockwise.

4. γ is the line segment from $(0, 1)$ to $(1, 0)$.

5. γ is the line segment from $(0, 2)$ to $(1, 0)$.

6. γ is the line segment from $(0, 0)$ to $(1, 1)$.

Suppose that $\int_{\gamma} \mathbf{f}(\mathbf{X}) \cdot d\mathbf{X} = 17$ *if* γ *is the oriented curve* $\mathbf{X}(t) = (\cos t, \sin t)$, $0 \le t \le \pi/2$. *Use this to evaluate the line integrals in Exercises 7–10.*

7. $\int_{C} \mathbf{f}(\mathbf{X}) \cdot d\mathbf{X}$, where C is the oriented curve $\mathbf{X}(t) = (\sin t, \cos t)$, $0 \le t \le \pi/2$.

8. $\int_{C} \mathbf{f}(\mathbf{X}) \cdot d\mathbf{X}$, where C is the oriented curve $\mathbf{X}(t) = \left(t, \sqrt{1-t^2}\right)$, $0 \le t \le 1$.

9. $\int_{C} \mathbf{f}(\mathbf{X}) \cdot d\mathbf{X}$, where C is the oriented curve $\mathbf{X}(t) = \left(\sqrt{1-t^2}, t\right)$, $0 \le t \le 1$.

10. $\int_{C} \mathbf{f}(\mathbf{X}) \cdot d\mathbf{X}$, where C is the oriented curve $\mathbf{X}(t) = (\cos(\pi t), \sin(\pi t))$, $0 \le t \le 1/2$.

In Exercises 11–16, use the Fact on page 887 to test whether the given vector field could be a gradient. If the vector field passes the test, find a potential function h.

11. $(P, Q) = (x, y)$.

12. $(P, Q) = (y, x)$.

13. $(P, Q) = (-y, x)$.

14. $(P, Q) = (1, \sin x)$.

15. $(P, Q) = (\sin x, 1)$.

16. $(P, Q) = \left(x/(x^2 + y^2), y/(x^2 + y^2)\right)$.

17–22. For each vector field (P, Q) in Exercises 11–16, find the flow integral $\int_{\gamma} P\, dx + Q\, dy$, where γ is the line segment from $(1, 1)$ to $(2, 2)$.

In Exercises 23–26, use the fundamental theorem for line integrals to evaluate $\int_{\gamma} P\, dx + Q\, dy$.

23. $(P, Q) = (e^x \sin y, e^x \cos y)$; γ is the unit circle oriented counterclockwise.

24. $(P, Q) = (3x - 2y, 3y - 2x)$; γ is the curve $x = 4\cos t$, $y = 4\sin t$, $0 \le t \le \pi$ oriented counterclockwise.

25. $(P, Q) = (e^x \sin y, e^x \cos y)$; γ is the line segment from the origin to the point $(1, \pi/2)$.

26. $(P, Q) = (\sin y, x \cos y)$; γ is the line segment from the point $(1, \pi/2)$ to the origin.

27. Consider the curves and the vector field shown in Example 6. The vector field is $(P, Q) = (y^2, 2xy)$.

 (a) Explain why the line integral $\int_{\gamma} P\, dx + Q\, dy$ has the same value for any curve starting at $(0, 0)$ and ending at $(1, 1)$.

 (b) Let (a, b) be any point in the xy-plane, and let γ be any curve from $(0, 0)$ to (a, b). Find $\int_{\gamma} P\, dx + Q\, dy$.

28. Repeat Exercise 27 using the vector field

$$(P, Q) = \left(\frac{2x}{1 + x^2 + y^2}, \frac{2y}{1 + x^2 + y^2}\right).$$

Exercises 29 and 30 explore the fact that the force of gravity is conservative (in the sense discussed in this section).

29. Near the surface of the Earth, we can think of the force of gravity on an object as essentially constant and pointing straight down. We can model this situation in the xy-plane by letting $(P, Q) = (0, -k)$, where k is a positive constant.

 (a) Show that (P, Q) is a conservative force field by finding a potential function.

 (b) Suppose that an object moves once around a closed curve. How much work is done by the gravitational force?

30. For motion on a large scale (e.g., spacecraft flight), gravity is better modeled as a force whose direction is always toward the center of the Earth and whose magnitude is inversely proportional to the square of the distance to the center of the Earth. This can be modeled in the xy-plane by a force of the form

$$(P, Q) = k\left(\frac{-x}{(x^2 + y^2)^{3/2}}, \frac{-y}{(x^2 + y^2)^{3/2}}\right),$$

where k is any positive constant.

 (a) Show that (P, Q) is conservative by finding a potential function.

 (b) Find the work done by this force in moving an object from the point $(1, 1)$ to $(3, 4)$.

 (c) What does the sign of the answer to part (b) mean physically?

FURTHER EXERCISES

Suppose that $\mathbf{f}(x, y) = (ye^{xy}, xe^{xy})$. *In Exercises 31–34, indicate whether the statement must be true, cannot be true, or might be true. Justify your answers.*

31. $\mathbf{f}$ is a gradient field.

32. If γ_1 and γ_2 are smooth curves with the same starting and ending points, then $\int_{\gamma_1} \mathbf{f} \cdot d\mathbf{X} = \int_{\gamma_2} \mathbf{f} \cdot d\mathbf{X}$.

33. $\mathbf{f}$ is a conservative vector field.

34. If γ is a closed curve, then $\int_{\gamma} \mathbf{f} \cdot d\mathbf{X} = 0$.

Suppose that $\mathbf{f} = (P, Q)$ *is a conservative vector field. In Exercises 35–38, indicate whether the statement must be true, cannot be true, or might be true. Justify your answers.*

35. There is a function h such that $\nabla h = \mathbf{f}$.

36. $P_y - Q_x = 0$.

37. If γ is the path $(1 + 2\cos t, 3 + 4\sin t)$, $0 \le t \le 2\pi$, then $\int_\gamma P\,dx + Q\,dy = 0$.

38. Suppose that $\int_{\gamma_1} \mathbf{f} \cdot \mathbf{X} = 11$, where γ_1 is the line segment from $(-2, 0)$ to $(2, 0)$. If γ_2 is the path $(-t, 4 - t^2)$, $-2 \le t \le 2$, then $\int_{\gamma_2} \mathbf{f} \cdot \mathbf{X} = -11$.

Suppose that $\mathbf{f} = (P, Q)$ and that $P_y \ne Q_x$. In Exercises 39–42, indicate whether the statement must be true, cannot be true, or might be true. Justify your answers.

39. There is a function h such that $\nabla h = \mathbf{f}$.

40. $\mathbf{f}$ is a conservative vector field.

41. If γ is a closed curve, then $\int_\gamma \mathbf{f} \cdot d\mathbf{X} = 0$.

42. Suppose that $\int_{\gamma_1} \mathbf{f} \cdot \mathbf{X} = 11$, where γ_1 is the line segment from $(-2, 0)$ to $(2, 0)$. If γ_2 is the path $(-t, 4 - t^2)$, $-2 \le t \le 2$, then $\int_{\gamma_2} \mathbf{f} \cdot \mathbf{X} = -11$.

43. Let $\mathbf{f} = (P, Q) = \left(-y/(x^2 + y^2), x/(x^2 + y^2)\right)$.
  (a) Show that $P_y = Q_x$ for all $(x, y) \ne (0, 0)$.
  (b) Evaluate $\int_\gamma \mathbf{f} \cdot d\mathbf{X}$, where (i) γ is $(\cos t, \sin t)$ and (ii) γ is $(\cos t, -\sin t)$ for $0 \le t \le \pi$.

(c) Evaluate $\int_\gamma \mathbf{f} \cdot d\mathbf{X}$, where γ is the unit circle oriented counterclockwise.

(d) Is $\mathbf{f}$ a conservative vector field? Justify your answer.

44. Repeat Exercise 43 with $(P, Q) = \left(-x/(1 + x^2 + y^2), y/(1 + x^2 + y^2)\right)$.

45. Let $\mathbf{f} = (x + e^y, x(1 + e^y))$ and γ be a closed path. Explain why $\oint_\gamma \mathbf{f} \cdot d\mathbf{X} = \oint_\gamma x\,dx + x\,dy$.
  [HINT: $(x + e^y, x(1 + e^y)) = (e^y, xe^y) + (x, x)$.]

46. Let $\mathbf{f} = (xy + y\cos(xy), xy + x\cos(xy))$ and γ be a closed path. Explain why $\oint_\gamma \mathbf{f} \cdot d\mathbf{X} = \oint_\gamma xy\,dx + xy\,dy$.

47. Let $\mathbf{f}(x, y)$ be any vector field and γ any curve. Suppose that γ is parametrized by $\mathbf{X}(t)$, for $0 \le t \le 1$.
  (a) Explain why $-\gamma$ is parametrized by $\mathbf{X}(1 - t)$ for $0 \le t \le 1$.
  (b) Use the previous part to show that

$$\int_{-\gamma} \mathbf{f}(\mathbf{X}) \cdot d\mathbf{X} = -\int_\gamma \mathbf{f}(\mathbf{X}) \cdot d\mathbf{X}.$$

(In other words, turning the curve around changes the sign of the answer.)

16.3 GREEN'S THEOREM: RELATING LINE AND AREA INTEGRALS

Green's theorem, like other results we have seen, is in the spirit of the fundamental theorem of calculus. The fundamental theorem of line integrals (Theorem 2, page 885), for example, relates a line integral over a curve to values of a potential function at the *ends* of the curve. Green's theorem, by contrast, relates a double integral over a region in the plane to a line integral over the *boundary* of the region. ➤

Green's theorem says that, under appropriate assumptions that we will soon state carefully,

The endpoints of a curve are, in a sense, the curve's "boundary."

$$\oint_\gamma P\,dx + Q\,dy = \iint_R \left(\frac{\partial Q}{\partial x} - \frac{\partial P}{\partial y}\right) dA. \tag{1}$$

Let's take a careful look at the equation and its various parts. On the left, $(P(x, y), Q(x, y))$ is a vector field on $\mathbb{R}^2$, γ is a **simple closed curve,** ➤ and the special integral sign $(\oint)$ indicates that γ is the boundary of a region. (All curves are oriented counterclockwise unless stated otherwise.) On the right, the domain of integration, R, is assumed to be the region inside the curve γ. (In mathematical parlance, γ is the **boundary** of the region R.) Notice that the integrand, $Q_x(x, y) - P_y(x, y)$, is a *scalar*-valued function of two variables— the appropriate object to integrate over R.

We'll explain "simple" below.

The usual geometric relation between γ and R is as illustrated in Figure 1:

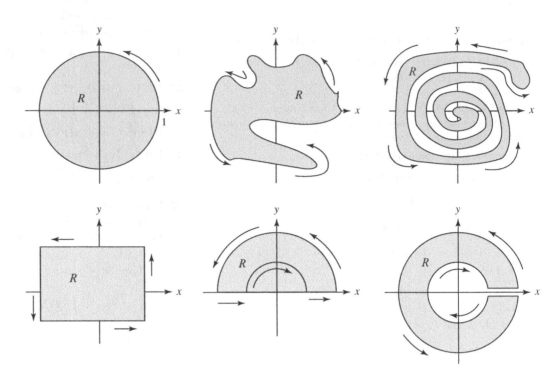

FIGURE 1
Six regions and their boundary curves

The upper left region is the **unit disk**; it is defined by the inequality $x^2 + y^2 \leq 1$, or, in polar coordinates, simply by $r \leq 1$.

Notice that, in each picture, the boundary curve γ is oriented so that the region R, shown shaded, remains on the *left* as one traverses γ. The boundary curves γ, moreover, are all **simple** (they have no self-intersections as a Figure-8 does), and they are either **smooth** or made up of a few smooth pieces. In full technical regalia, curves like those above, which bound reasonable areas, are called **piecewise-smooth Jordan curves**.

Jordan curves.   Curves in the plane—even closed curves that don't intersect themselves—can be very complicated. Though it seems reasonable that such a curve should divide the plane into two regions, one inside and one outside, it is surprisingly difficult to prove this rigorously. The first proof was given by the French mathematician Camille Jordan (1838–1922). In his honor, curves that bound area are known as Jordan curves.

Before stating the theorem carefully, we will explore what it asserts in some relatively simple cases.

EXAMPLE 1   Suppose the vector field (P, Q) is a gradient ∇h; that is,

$$\big(P(x, y), Q(x, y)\big) = \nabla h(x, y)$$

for some function h defined on and near a region R with boundary curve γ. Does Equation 1 hold?

Solution Yes. By the fundamental theorem of line integrals,

$$\oint_\gamma P\,dx + Q\,dy = \oint_\gamma \nabla h \cdot d\mathbf{X} = 0,$$

since γ is a closed curve. Now consider the double integral on the right side of Equation 1. By assumption, $(P, Q) = (h_x, h_y)$. Thus, for all (x, y),

$$Q_x - P_y = h_{yx} - h_{xy} = 0.$$

(The last equation holds because of the equality of cross-partial derivatives.) We have shown, therefore, that for any vector field (P, Q) that happens to be a gradient, both sides of Equation 1 are zero. ■

EXAMPLE 2 Let $(P, Q) = (x - y, x + y)$ and let R be the unit disk. What does Equation 1 say in this case? Is it true?

Solution The vector field and the boundary curve γ (the unit circle) are shown together in Example 1, page 881, and the line integral is calculated:

$$\oint_\gamma P\,dx + Q\,dy = 2\pi.$$

For the double integral in Equation 1 we have $Q_x - P_y = 2$. ➡ Therefore,   *Check it!*

$$\iint_R (Q_x - P_y)\,dA = \iint_R 2\,dA = 2\,\text{area}(R) = 2\pi.$$

Again, Equation 1 holds true. ■

A third example is a little more subtle.

EXAMPLE 3 Consider the vector field

$$(P, Q) = \left(\frac{-y}{x^2 + y^2}, \frac{x}{x^2 + y^2} \right)$$

on the unit disk. What does Equation 1 say now? Is it true?

Solution To find the line integral, we use the "usual" parametrization $\mathbf{X}(t) = (\cos t, \sin t)$ for $0 \le t \le 2\pi$. The result ➡ is   *We leave the important calculation to you; see the exercises.*

$$\oint_\gamma P\,dx + Q\,dy = \int_0^{2\pi} (\sin^2 t + \cos^2 t)\,dt = 2\pi.$$

For the double integral, we get ➡   *The calculation is worth a careful check.*

$$Q_x = P_y = \frac{y^2 - x^2}{(x^2 + y^2)^2},$$

and so $Q_x - P_y = 0$. Therefore—apparently—$\iint_R (Q_x - P_y)\,dA = 0$. In this case, it seems, Equation 1 fails. Is something wrong?

Actually, no; there's a simple explanation. The vector field (P, Q) is not defined at the point $(0, 0)$. Neither, therefore, are Q_x and P_y, and so there is no guarantee that the integral $\iint_R (Q_x - P_y)\, dA$ even exists. The moral, therefore, is that we need to take care with domains. We do so in the formal statement of the theorem. ∎

THEOREM 3 (Green's theorem) Let R be a region in $\mathbb{R}^2$ whose boundary is a piecewise-smooth Jordan curve γ. Let P and Q be scalar-valued functions, having continuous partial derivatives on and near R. Then

$$\oint_\gamma P\, dx + Q\, dy = \iint_R \left(\frac{\partial Q}{\partial x} - \frac{\partial P}{\partial y} \right) dA.$$

Older student makes good. George Green (1793–1841), a miller from Nottingham, England, spent only two years in elementary school. Despite this possible handicap, he published ten mathematical papers, including several on potential theory. He entered Cambridge University as an undergraduate at age 40, only 8 years before his death. Green's theorem itself, despite the name, probably predates George Green.

The idea of the proof A fully rigorous proof of Green's theorem is quite subtle. It would require, among other things, a rigorous definition of "counterclockwise," and a careful treatment of domains that can be quite irregular. However, the basic idea of the proof boils down to several careful applications of the ordinary fundamental theorem of calculus. To see how, suppose first that R happens to be a region of the special sort shown in Figure 2:

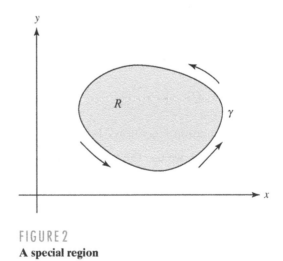

FIGURE 2
A special region

The region is special in that every horizontal line and every vertical line intersects γ at most twice. To prove the theorem, we will show two identities:

$$-\iint_R P_y\, dA = \oint_\gamma P\, dx \qquad \text{and} \qquad \iint_R Q_x\, dA = \oint_\gamma Q\, dy. \tag{2}$$

For the first identity, we will think of R as shown in Figure 3:

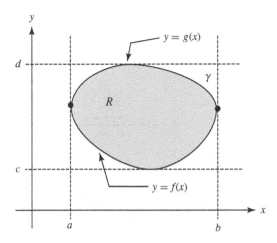

A region bounded by curves

The region is bounded by upper and lower boundary curves $y = g(x)$ and $y = f(x)$, respectively, with $a \le x \le b$. Then the double integral becomes

$$-\iint_R P_y(x, y)\, dA = -\int_{x=a}^{x=b} \left(\int_{y=f(x)}^{y=g(x)} P_y(x, y)\, dy \right) dx$$

$$= -\int_{x=a}^{x=b} \Big(P\big(x, g(x)\big) - P\big(x, f(x)\big) \Big) dx.$$

(We used the ordinary fundamental theorem, in the variable y, in the last step.) → *This is the key step.*

Now consider the line integral $\oint_\gamma P(x, y)\, dx$. Now γ consists of upper and lower boundary curves; we can parametrize both in the same way. For the lower curve, we use

$$x = t; \qquad y = f(t); \qquad a \le t \le b.$$

For the upper curve, we do almost the same thing:

$$x = t; \qquad y = g(t); \qquad a \le t \le b.$$

Notice that, in both cases, $dx = dt$. Also, since the upper curve is oriented from right to left, we attach a minus sign to the integral. Using these parametrizations, we obtain

$$\oint_\gamma P(x, y)\, dx = \int_{\gamma_{\text{bottom}}} P(x, y)\, dx - \int_{\gamma_{\text{top}}} P(x, y)\, dx$$

$$= \int_a^b P\big(t, f(t)\big)\, dt - \int_a^b P\big(t, g(t)\big)\, dt.$$

The last expression looks familiar: Except for the difference in variable names, this expression is exactly what we obtained for $\iint P_y\, dA$. Thus, the first identity in Equation 2 holds as claimed. The second identity is proved in a similar way. → *See the exercises.*

These arguments show that Green's theorem holds for regions of the special type shown above. To see that it holds on more general regions, the trick is to break R into several smaller regions, each of the special type just mentioned, as shown in Figure 4:

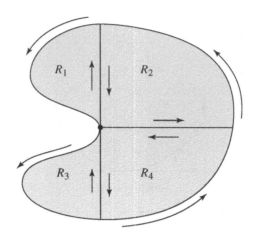

FIGURE 4
Breaking up a region

Notice that each "inner" boundary piece is traversed twice, once in each direction. We have shown that the theorem holds on each subregion R_i, with boundary γ_i, that is,

$$\iint_{R_i} (Q_x - P_y) \, dA = \oint_{\gamma_i} P \, dx + Q \, dy.$$

Adding these results together for $i = 1 \ldots 4$ gives $\iint_R (Q_x - P_y) \, dA$ on the left and $\oint_\gamma P \, dx + Q \, dy$ on the right (since the line integrals on inside boundary edges cancel each other out). This completes the proof of Green's theorem.   ■

Sometimes the trade goes the other way.

    The theorem sometimes saves trouble by trading a messy line integral for a simpler area integral. ◄

EXAMPLE 4   Find the work done by the force field $(P, Q) = (x - y, x + y)$ in moving an object around the square S with corners at $(-1, -1)$, $(1, -1)$, $(1, 1)$, and $(-1, 1)$.

Solution   Integrating around the square would require four separate parametrizations. Using Green's theorem to reduce to a double integral makes things much simpler. Here $Q_x - P_y = 2$, and so Green's theorem says

$$\oint_\gamma P \, dx + Q \, dy = \iint_{R_i} (Q_x - P_y) \, dA \qquad \text{(by Green's theorem)}$$

$$= \iint_S 2 \, dA = 2 \, \text{area}\,(S) = 8. \qquad \text{(S has area 4)}$$

■

Green's theorem on a region with holes

Green's theorem, as stated above, applies to regions with only one boundary curve. However, a simple trick shows that it applies, in slightly different form, to regions like the one in Figure 5:

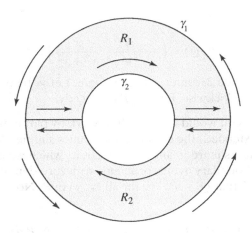

FIGURE 5
Cutting a donut

If R is the entire donut-shaped region, then we can imagine R as the union of the two simpler regions R_1 (the upper half) and R_2 (the lower half); we will call their boundary curves C_1 and C_2. (In the picture, γ_1 and γ_2 are the outer and inner circles, respectively. Each of C_1 and C_2 consists of two semicircles and two line segments.) Green's theorem *does* apply to R_1 and R_2 to give

$$\iint_{R_1} (Q_x - P_y)\, dA = \oint_{C_1} P\, dx + Q\, dy$$

and

$$\iint_{R_2} (Q_x - P_y)\, dA = \oint_{C_2} P\, dx + Q\, dy.$$

Adding these equations together (and keeping track of cancellations along the line segments, which are traversed twice in opposite directions) gives

$$\iint_{R} (Q_x - P_y)\, dA = \oint_{\gamma_1} P\, dx + Q\, dy - \oint_{\gamma_2} P\, dx + Q\, dy,$$

where, now, both line integrals are taken in the counterclockwise sense. ➡

The minus sign is there because the inner circle is traversed clockwise.

Here is the general principle:

FACT (Green's theorem for a region with holes)   Let the situation be as in Green's theorem, but assume that R has outer boundary curve γ_1 and inner boundary curve γ_2. Then

$$\iint_{R} (Q_x - P_y)\, dA = \oint_{\gamma_1} P\, dx + Q\, dy - \oint_{\gamma_2} P\, dx + Q\, dy.$$

EXAMPLE 5   Let

$$(P, Q) = \left(-\frac{y}{x^2 + y^2}, \frac{x}{x^2 + y^2} \right),$$

and let S be a square of side 2, centered at the origin. Let γ_1 be the boundary of the square, oriented counterclockwise. Find $\oint_{\gamma_1} P \, dx + Q \, dy$.

Solution   The line integral would be difficult to work by parametrization, but with Green's theorem (as extended) the calculation becomes simple. We will let R be the region between the outer square γ_1 and the unit circle, which we denote here by γ_2. (Using γ_2 as the inner boundary avoids problems with zero denominators at $(0, 0)$.) We calculated in Example 3 that $Q_x - P_y = 0$ for all (x, y) in R. Now the preceding Fact asserts that

$$0 = \iint_R (Q_x - P_y) \, dA = \oint_{\gamma_1} P \, dx + Q \, dy - \oint_{\gamma_2} P \, dx + Q \, dy,$$

which implies in turn that

$$\oint_{\gamma_1} P \, dx + Q \, dy = \oint_{\gamma_2} P \, dx + Q \, dy.$$

But we also showed above, in Example 3, that $\oint_{\gamma_2} P \, dx + Q \, dy = 2\pi$, and so this is also the value of the line integral around the less convenient "curve" γ_1.   ∎

BASIC EXERCISES

In Exercises 1–14, use Green's theorem to evaluate the line integral $\oint_{\gamma} P \, dx + Q \, dy$. (All curves are traversed counterclockwise.)

1. $(P, Q) = (y^2 + x, x + y)$; γ is the square with vertices at $(0, 0)$, $(1, 0)$, $(1, 1)$, and $(0, 1)$.

2. $(P, Q) = (x - y, x + y)$; γ is the square with vertices at $(0, 0)$, $(1, 0)$, $(1, 1)$, and $(0, 1)$.

3. $(P, Q) = (y^2 + x, x + y)$; γ is the circle of radius 1 with center $(0, 0)$.

4. $(P, Q) = (x - y, x + y)$; γ is the circle of radius 1 with center $(1, 0)$.

5. $(P, Q) = (y^2 + x, x + y)$; γ is the boundary of the region $0 \leq r \leq 1; 0 \leq \theta \leq \pi/2$.

6. $(P, Q) = (x - y, x + y)$; γ is the boundary of the region $0 \leq r \leq 1; 0 \leq \theta \leq \pi/2$.

7. $(P, Q) = (3xy, x - y^2)$; γ is the boundary of the rectangle with vertices $(0, 0)$, $(2, 0)$, $(2, 1)$, and $(0, 1)$.

8. $(P, Q) = (xy^2, x^2y)$; γ is the boundary of the rectangle with vertices $(0, 0)$, $(2, 0)$, $(2, 1)$, and $(0, 1)$.

9. $(P, Q) = (2y + e^{-x^2}, 3x - \sin(y^2))$; γ is the circle of radius 2 with center $(0, 0)$.

10. $\mathbf{f}(x, y) = (e^{x^2} - y, x + \cos(\sqrt{y}))$; γ is the circle of radius 3 with center $(1, 4)$.

11. $(P, Q) = (y^2, x^2)$; γ is the boundary of the rectangle $[1, 2] \times [3, 5]$.

12. $(P, Q) = (y^2, x^2)$; γ is the circle $x^2 + y^2 = 9$.

13. $(P, Q) = (\cos x, xy)$; γ is the path from $(0, 0)$ to $(3, 0)$ along the x-axis, then back to the origin along the curve $y = x^2 - 3x$, then back to the origin along the x axis.

14. $(P, Q) = (e^x, xy)$; γ is curve described by

$$\mathbf{X}(t) = \begin{cases} (t, t^2) & 0 \leq t \leq 1 \\ (2 - t, 2 - t) & 1 < t \leq 2. \end{cases}$$

FURTHER EXERCISES

15. Consider the vector field

$$(P, Q) = \left(-\frac{y}{x^2 + y^2}, \frac{x}{x^2 + y^2} \right);$$

let γ be the unit circle oriented counterclockwise.

(a) Evaluate $\oint_{\gamma} P \, dx + Q \, dy$.

(b) Is (P, Q) a gradient field on $\mathbb{R}^2$? Justify your answer.

(c) Find $\int_S P \, dx + Q \, dy$, where S is any closed path that surrounds the unit circle.   [HINT: Mimic Example 5.]

(d) Find $\int_C P \, dx + Q \, dy$, where C is any closed path that does not surround the origin.
[HINT: Does the ordinary version of Green's theorem apply now?]

16. Consider the vector field

$$(P, Q) = \left(\frac{x}{x^2 + y^2}, \frac{y}{x^2 + y^2}\right);$$

let γ be the unit circle oriented counterclockwise.

(a) Evaluate $\oint_\gamma P\, dx + Q\, dy$.

(b) Show that $Q_x(x, y) = P_y(x, y)$ for all $(x, y) \neq (0, 0)$.

(c) Does Green's theorem apply to (P, Q) on the unit disk? Justify your answer.

17. Evaluate $\oint_\gamma -y\, dx + x\, dy$, where γ is the cardioid with polar equation $r = 1 + \cos\theta$ oriented counterclockwise.

18. Evaluate $\oint_\gamma x\, dy$, where γ is the curve described by the polar equation $r = 2\sin(3\theta)$, $0 \leq \theta \leq \pi/3$ oriented counterclockwise.

19. Let R and γ be as in the statement of Green's theorem. Show that the area of R is equal to

(a) $\oint_\gamma x\, dy$.

(b) $-\oint_\gamma y\, dx$.

(c) $\frac{1}{2}\oint_\gamma -y\, dx + x\, dy$.

20. Use Green's theorem to find the area of the ellipse $x^2/a^2 + y^2/b^2 = 1$. [HINT: Use Exercise 19.]

21. Suppose that the boundary of the region R is described in polar coordinates as $r = f(\theta)$, $\alpha \leq \theta \leq \beta$. Use Green's theorem to show that the area of R is equal to $\frac{1}{2}\int_\alpha^\beta \left(f(\theta)\right)^2 d\theta$. [HINT: Use Exercise 19(c).]

22. Let (x_1, y_1), (x_2, y_2), and (x_3, y_3) be the vertices of a triangle in counterclockwise order. Use Green's theorem to show that the area of the triangle is

$$\text{area} = \frac{1}{2}\big((x_2 + x_1)(y_2 - y_1) + (x_3 + x_2)(y_3 - y_2) + (x_1 + x_3)(y_1 - y_3)\big).$$

23. Let (x_1, y_1), $(x_2, y_2), \dots$, (x_n, y_n), $(x_{n+1}, y_{n+1}) = (x_1, y_1)$ be points on a counterclockwise path around the boundary of a polygonal region in the xy-plane (i.e., the boundary of the region is made up of line segments connecting adjacent points on the list). Show that the area of the region is

$$\text{area} = \frac{1}{2}\sum_{k=1}^n \big((x_{k+1} + x_k)(y_{k+1} - y_k)\big).$$

24. Let $\mathbf{f}$ be the field $\mathbf{f}(x, y) = (2xy + x, xy - y)$ and γ be the perimeter of the square bounded by the lines $x = 0$, $x = 1$, $y = 0$, and $y = 1$. Use Green's theorem to evaluate the counterclockwise circulation of $\mathbf{f}$ around γ.

25. Let R, and γ be as in the statement of Green's theorem, and let A be the area of R. Show that the centroid of R is $(\bar{x}, \bar{y})$, where

$$\bar{x} = \frac{1}{2A}\oint_\gamma x^2\, dy \quad \text{and} \quad \bar{y} = -\frac{1}{2A}\oint_\gamma y^2\, dy.$$

26. We proved the first identity in Equation 2 by assuming that the region had a certain special shape. Mimic that proof to prove the remaining identity: $\iint_R Q_x\, dA = \oint_\gamma Q\, dy$.

[HINT: Because of its special shape, R can be assumed to have left boundary $x = h(y)$ and right boundary $x = k(y)$.]

27. Use the following steps to prove Green's theorem under the assumption that R is the rectangle $[a, b] \times [c, d]$.

(a) By parametrizing all four sides of the rectangle, show that

$$\oint_\gamma P\, dx + Q\, dy = \int_a^b P(t, c)\, dt - \int_a^b P(t, d)\, dt$$
$$+ \int_c^d Q(b, t)\, dt - \int_c^d Q(a, t)\, dt.$$

(b) Explain why $\iint_R (Q_x - P_y)\, dA = \iint_R Q_x\, dA - \iint_R P_y\, dA$.

(c) Show that

$$\iint_R Q_x\, dA = \int_c^d \int_a^b Q_x(x, y)\, dx\, dy$$
$$= \int_c^d \big(Q(b, y) - Q(a, y)\big)\, dy.$$

(d) Show that

$$-\iint_R P_y\, dA = \int_a^b \big(P(x, c) - P(x, d)\big)\, dx.$$

16.4 SURFACES AND THEIR PARAMETRIZATIONS

Calculating line integrals has offered us plenty of practice with parametrizing curves in the xy-plane. Parametrizations were key, in particular, to proving two fundamental theorems for line integrals.

The next few sections introduce *surface integrals*, in which the domain of integration is a two-dimensional surface in three-dimensional space, such as a plane, the surface of a sphere or ellipsoid, or the graph of a function $z = f(x, y)$. Surface integrals are higher-dimensional versions of line integrals; we will stress this connection whenever possible. In particular, surface integrals satisfy their own versions of the fundamental theorems we've studied for line integrals. Our goal in the rest of this chapter is to state and understand two such fundamental theorems relating various types of integrals and derivatives.

Essential to calculating such integrals, as one might expect, is the ability to parametrize surfaces in space. This brief section illustrates some useful ideas and techniques.

Curves, surfaces, and dimensions

The general setup for parametrizing a curve is as in Figure 1:

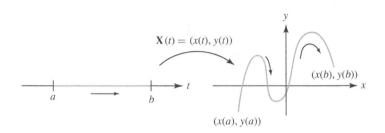

FIGURE 1
Parametrizing a curve in the plane

The curve is a one-dimensional object in two-dimensional space. It is the image of a one-dimensional t-interval $[a, b]$ mapped by a vector-valued function $\mathbf{X}(t)$. In effect, the function $\mathbf{X}$ "deforms" the one-dimensional t-interval into the one-dimensional curve.

Figure 2, by contrast, is the generic picture for parametrizing a *surface*:

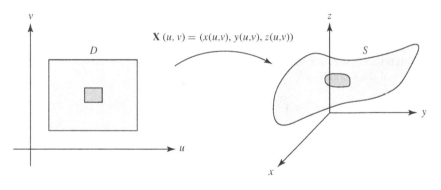

FIGURE 2
Parametrizing a surface in space

In this setting, the vector-valued function $\mathbf{X} : \mathbb{R}^2 \to \mathbb{R}^3$ maps a two-dimensional region D in the uv-plane onto the two-dimensional surface S in (x, y, z)-space. Roughly speaking, the function $\mathbf{X}$ deforms the two-dimensional flat region D into the two-dimensional surface S. (The small darker rectangle in the domain is mapped to a "near-rectangle" in the surface.) ◄

The region D need not be rectangular, but that case arises often, and may lead to simpler calculations.

It *can* happen, by the way, that a function $\mathbf{X} : \mathbb{R}^2 \to \mathbb{R}^3$ maps a two-dimensional set in the domain into a smaller-dimensional set in xyz-space. For example, the constant function defined by $\mathbf{X}(u, v) = (1, 2, 3)$ maps all of $\mathbb{R}^2$ into a single point. All of the functions we will use to parametrize surfaces, however, will "respect" dimensions, mapping two-dimensional sets to two-dimensional surfaces.

Surface parametrizations: a sampler

Any given surface in space can be parametrized in many ways (as can any curve in the plane) by using various domain sets. Some examples will give a sense of the possibilities.

Function graphs   The easiest surfaces to parametrize are graphs of functions $z = f(x, y)$— or, more often, *parts* of such graphs. Parametrizing a particular part of a graph may require fiddling with domains.

EXAMPLE 1   Parametrize S_1, the part of the graph of $z = x^2 + y^2$ that lies above the square $[0, 2] \times [0, 2]$ in the xy-plane.

Solution   Points on the surface satisfy the equation $z = x^2 + y^2$, and so we can set

$$\mathbf{X}(u, v) = (u, v, u^2 + v^2); \quad 0 \le u \le 2; \quad 0 \le v \le 2.$$

Plotting this surface (Figure 3) is an excellent way to check whether our parametrization is correct:

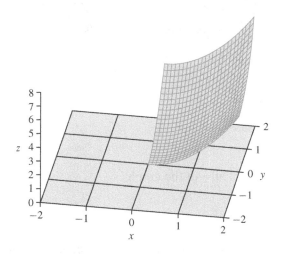

FIGURE 3
Part of the surface $z = x^2 + y^2$

The patch shown looks like what we intended.   ■

Another surface from the same graph   The surface patch shown in Figure 3 is only a small part of the full graph $z = x^2 + y^2$. Let's look at another surface carved from the same graph.

EXAMPLE 2   Parametrize S_2, the part of the graph of $z = x^2 + y^2$ that lies above the unit disk in the xy-plane.

Solution   Exactly the same parametrization works as in Example 1 except that the domain D is now the unit disk:

$$\mathbf{X}(u, v) = (u, v, u^2 + v^2), \quad \text{for } u \text{ and } v \text{ with } u^2 + v^2 \le 1.$$

If we prefer to parametrize S_2 using a *rectangle* as the domain set D, we can work in polar coordinates. With respect to r and θ, the unit disk is the rectangle described by $0 \le r \le 1$ and $0 \le \theta \le 2\pi$. In these coordinates, moreover, $x = r\cos\theta$, $y = r\sin\theta$, and $x^2 + y^2 = r^2$. For consistency of notation, we write $u = r$ and $v = \theta$. Now our desired parametrization becomes

$$\mathbf{X}(u, v) = (u\cos v, u\sin v, u^2); \qquad 0 \le u \le 1; \qquad 0 \le v \le 2\pi;$$

the domain set is a rectangle. Plotting these data (Figure 4) gives what we expect:

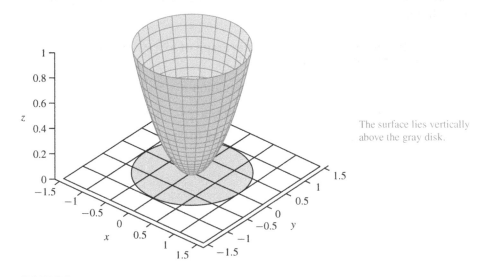

The surface lies vertically above the gray disk.

FIGURE 4
Another part of the surface $z = x^2 + y^2$

Again we see part—but not all—of the graph.

Area magnification

In working with parametrized surfaces and (in the next section) surface areas, it is important to know how *areas* in the uv-domain and on the surface itself are related. In Figure 4, for example, the unit circular disk represents the uv-domain for one possible parametrization of a paraboloid. As Figure 4 shows, the paraboloid has considerably larger surface area than the disk below. It also appears that the "stretching" required to map the disk onto the paraboloid is greater near the outer edge than near the center of the disk. In what follows we consider how areas in the uv-domain and on the surface are related, first for the simplest surfaces—planes—and then in the general case.

We have already raised the general question of area magnification in connection with changing variables in multiple integrals (see Example 10, page 831, for example). There, the Jacobian determinant turned out to describe the factor by which areas are magnified under a change of variable. We will see that, in the present setting, area magnification also involves a certain algebraic combination of derivatives.

Planes and area magnification   If a surface S is a plane, or part of a plane, then S can be parametrized with a *linear* parametrization function. The next example illustrates. Most important, the example shows how areas in the uv-domain and on the surface are related.

EXAMPLE 3   Let S be the part of the graph of $z = 2x + 3y$ that lies above the unit square $[0, 1] \times [0, 1]$. As Figure 5 shows, S is part of a plane:

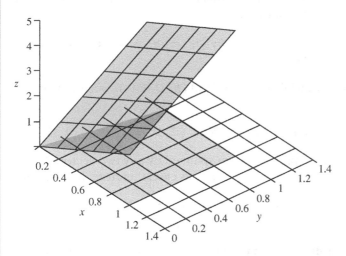

The surface S lies vertically above the gray rectangle.

FIGURE 5
A planar surface: part of the graph of $z = 2x + 3y$

How are the areas of the surface S and the base square $[0, 1] \times [0, 1]$ related?

Solution   The surface S is easily parametrized:

$$\mathbf{X}(u, v) = (u, v, 2u + 3v); \qquad 0 \le u \le 1; \qquad 0 \le v \le 1.$$

As Figure 5 illustrates, S is the parallelogram spanned by the two edges (thought of as vectors) that meet at the origin. The picture also shows that these spanning edges are simply the vectors

$$\mathbf{X}(1, 0) - \mathbf{X}(0, 0) = (1, 0, 2) \qquad \text{and} \qquad \mathbf{X}(0, 1) - \mathbf{X}(0, 0) = (0, 1, 3).$$

The area of the parallelogram spanned by any two vectors in space is given by the magnitude of their cross product. ➡ In the present case this gives

$$\text{area of } S = \left| (1, 0, 2) \times (0, 1, 3) \right| = \left| (-2, -3, 1) \right| = \sqrt{14}.$$

We discussed this property of the cross product in Section 12.9.

   Now the base square $[0, 1] \times [0, 1]$ has area 1, and so we see that the area of S represents a magnification by the factor $\sqrt{14} \approx 3.7$. ➡ We see, too, that because our parametrization is linear, area magnification is *constant* throughout the base square. ■

Figure 5 may suggest a lower magnification factor; this is due to the picture's compression in the z-direction.

   Next we consider a more general linear mapping—but come to a similar conclusion.

EXAMPLE 4   Describe the surface S parametrized by

$$\mathbf{X}(u, v) = (x_0 + Au + Bv, \, y_0 + Cu + Dv, \, z_0 + Eu + Fv);$$
$$0 \le u \le \Delta u ; \quad 0 \le v \le \Delta v,$$

where all letters except u and v denote constants. The uv-domain D has area $\Delta u \, \Delta v$. What is the area of the corresponding surface S?

Solution The parametrization function can be rewritten in the form

$$\mathbf{X}(u, v) = (x_0, y_0, z_0) + u(A, C, E) + v(B, D, F).$$

Therefore, the surface S consists of all points of this form, where u and v range from 0 to Δu and Δv, respectively. From this it follows:

The surface S is the parallelogram spanned by the vectors $\Delta u\,(A, C, E)$ and $\Delta v\,(B, D, F)$; one corner is at (x_0, y_0, z_0).

(Note that Δu and Δv are scalars, while (A, C, E) and (B, D, F) are vectors.) For brevity we will write ◄

$$\mathbf{X}_u = (A, C, E) \qquad \text{and} \qquad \mathbf{X}_v = (B, D, F).$$

Now recall that the area of the parallelogram spanned by any two vectors $\mathbf{X}_u$ and $\mathbf{X}_v$ is the absolute value of their cross product. ◄ In this case, therefore, the area of S is

$$\text{area } S = \left|(\Delta u\,\mathbf{X}_u) \times (\Delta v\,\mathbf{X}_v)\right| = \Delta u\,\Delta v\,|\mathbf{X}_u \times \mathbf{X}_v|. \qquad \blacksquare$$

We will explain the link to partial derivatives in a moment.

Section 12.9 explains this property of the cross product; see especially Figure 1, page 692.

Area magnification for any parametrization The notation in the preceding example is no accident: The vector $\mathbf{X}_u$ is the partial derivative with respect to u of the parametrization function:

$$\mathbf{X}_u = \frac{\partial}{\partial u}(x_0 + Au + Bv, y_0 + Cu + Dv, z_0 + Eu + Fv) = (A, C, E).$$

Similarly,

$$\mathbf{X}_v = \frac{\partial}{\partial v}(x_0 + Au + Bv, y_0 + Cu + Dv, z_0 + Eu + Fv) = (B, D, E).$$

The same notation is convenient for *any* parametrization function $\mathbf{X}(u, v) = (x(u, v), y(u, v), z(u, v))$. Thus, we write

$$\mathbf{X}_u = \left(\frac{\partial}{\partial u}x(u, v), \frac{\partial}{\partial u}y(u, v), \frac{\partial}{\partial u}z(u, v)\right),$$

and

$$\mathbf{X}_v = \left(\frac{\partial}{\partial v}x(u, v), \frac{\partial}{\partial v}y(u, v), \frac{\partial}{\partial v}z(u, v)\right).$$

We showed earlier that for a *linear* parametrization function, the cross product $|\mathbf{X}_u \times \mathbf{X}_v|$ describes the factor by which $\mathbf{X}$ increases areas.

A similar result holds for *any* differentiable parametrization function $\mathbf{X}$. Near a specific domain point (u_0, v_0), each coordinate function of $\mathbf{X}$ is closely approximated by its linear approximation function, which has the same derivatives $\mathbf{X}_u(u_0, v_0)$ and $\mathbf{X}_v(u_0, v_0)$ as does $\mathbf{X}$. An important result follows:

FACT *Near any point (u_0, v_0) in the domain, the parametrization $\mathbf{X}$ magnifies areas by a factor of approximately $|\mathbf{X}_u(u_0, v_0) \times \mathbf{X}_v(u_0, v_0)|$.*

EXAMPLE 5   Discuss area magnification for the parametrization

$$\mathbf{X}(u, v) = (u, v, u^2 + v^2)$$

for the surface in Example 2; see also Figure 4.

Solution   By the preceding Fact, the area magnification near a domain point (u_0, v_0) is $|\mathbf{X}_u(u_0, v_0) \times \mathbf{X}_v(u_0, v_0)|$. Here we have $\mathbf{X}_u(u_0, v_0) = (1, 0, 2u_0)$ and $\mathbf{X}_v(u_0, v_0) = (0, 1, 2v_0)$; this calculation gives

$$|\mathbf{X}_u(u_0, v_0) \times \mathbf{X}_v(u_0, v_0)| = |(1, 0, 2u_0) \times (0, 1, 2v_0)| = \sqrt{1 + 4u_0^2 + 4v_0^2}.$$

This formula helps explain what Figure 4 shows. At the origin, where $(u_0, v_0) = (0, 0)$, the magnification factor is 1, and *no* magnification occurs. Magnification increases as (u_0, v_0) approaches the edge of the unit circle; at the edge, $u_0^2 + v_0^2 = 1$, and areas are magnified by a factor of $\sqrt{5}$. ∎

We will use the preceding Fact and the idea of area magnification in the following section, where we calculate surface areas of nonplanar surfaces.

BASIC EXERCISES

In Exercises 1–6, find a parametrization of the given surface S over a rectangular domain D in the uv-plane. (If possible, use technology to plot the surface to see whether your parametrization is correct.)

1. S is the part of the plane $z = 5$ above the rectangle $[-1, 2] \times [0, 4]$.

2. S is the part of the plane $z = 2x + 3y$ above the rectangle $[0, 1] \times [4, 5]$.

3. S is the part of the cone $z = \sqrt{x^2 + y^2}$ that lies above the square $[-1, 1] \times [-1, 1]$ in the xy-plane.

4. S is the part of the plane $z = 2x + 3y + 4$ that lies above the square $[0, 1] \times [0, 1]$ in the xy-plane.

5. S is the part of the cone $z = \sqrt{x^2 + y^2}$ that lies above the unit disk in the xy-plane. [HINT: Use cylindrical coordinates.]

6. S is the part of the plane $z = 2x + 3y + 4$ that lies above the unit disk in the xy-plane.

In Exercises 7–10, a parametrization of a surface is given. Describe the surface and compute the area magnification $|\mathbf{X}_u(u_0, v_0) \times \mathbf{X}_v(u_0, v_0)|$ at the specified point (u_0, v_0).

7. $\mathbf{X}(u, v) = (u, v, 2u - 3v)$; $D = [-2, 1] \times [-3, 0]$; $(u_0, v_0) = (-1, -2)$.

8. $\mathbf{X}(u, v) = (u, v, v^2)$; $D = [1, 3] \times [2, 5]$; $(u_0, v_0) = (2, 4)$.

9. $\mathbf{X}(u, v) = (u \cos v, u \sin v, u^2)$; $D = [0, 2] \times [0, \pi]$; $(u_0, v_0) = (1, \pi/2)$.

10. $\mathbf{X}(u, v) = (u \cos v, u \sin v, u)$; $D = [0, 3] \times [\pi, 2\pi]$; $(u_0, v_0) = (2, 5\pi/4)$.

FURTHER EXERCISES

11. A surface S has the parametrization $\mathbf{X}(u, v) = (\sin v \cos u, \sin v \sin u, \cos v)$, with $0 \le u \le 2\pi$ and $0 \le v \le \pi$.
    (a) Use technology to plot the surface S. What is the surface?
    (b) Explain the link to spherical coordinates.

12. Let S be the upper half of the unit sphere $x^2 + y^2 + z^2 = 1$.
    (a) Parametrize S; let D be the unit disk $u^2 + v^2 \le 1$ in the uv-plane.
    (b) Parametrize S; let D be a rectangle in the uv-plane. [HINT: See the previous exercise.]

In Exercises 13–24, find a parametrization of the given surface S over a rectangular domain D in the uv-plane.

13. S is the part of the cylinder $z = x^2$ between the planes $y = -3$, $y = 2$, $z = 0$, and $z = 4$.

14. S is the part of the cylinder $x^2 + y^2 = 4$ that is between the planes $z = 0$ and $z = 1$.

15. S is the part of the paraboloid $z = x^2 + y^2$ that is below the plane $z = 9$.

16. S is the part of the paraboloid $z = x^2 + y^2$ that is between the planes $z = 4$ and $z = 16$.

17. S is the part of the plane $z = 5$ above the portion of the disk $x^2 + y^2 = 4$ that lies in the first quadrant

18. S is the part of the plane $z = x - y$ that is inside the cylinder $x^2 + y^2 = 1$.

19. S is the part of the sphere $x^2 + y^2 + z^2 = 1$ that is above the plane $z = 1/\sqrt{2}$. [HINT: Use spherical coordinates.]

20. S is the part of the sphere $x^2 + y^2 + z^2 = 1$ that is between the planes $z = 0$ and $z = 1/\sqrt{2}$.

21. S is the ellipsoid $x^2/a^2 + y^2/b^2 + z^2/c^2 = 1$, where a, b, and c are positive constants.

22. S is the elliptic paraboloid $x^2/a^2 + y^2/b^2 = z$ that is between the planes $z = 0$ and $z = 1$, where a and b are positive constants.

23. S is the elliptic cylinder $x^2/a^2 + y^2/b^2 = 1$, that is between the planes $z = 0$ and $z = 2$, where a and b are positive constants.

24. S is the elliptic cone $z = \sqrt{x^2/a^2 + y^2/b^2}$ that is between the planes $z = 0$ and $z = 2$, where a and b are positive constants.

16.5 SURFACE INTEGRALS

Surface integrals differ from line integrals in that, for the former, the domain of integration is a surface in space, not a curve in the plane. Line and surface integrals are similar, on the other hand, in that calculating both types of integrals begins with parametrizing the curve or surface in a convenient way. Once this is done, line and surface integrals reduce (albeit in somewhat different ways, as we will see) to "ordinary" integrals in one or two variables.

Line and surface integrals are also similar because both help answer natural physical questions about vector phenomena. If $\mathbf{f}$ is a vector field in the plane, representing the velocity of a flow near a closed oriented curve γ, then we have seen that the line integral $\int_\gamma \mathbf{f} \cdot d\mathbf{X}$ measures the circulation of the flow, that is, the tendency of the fluid to flow *around* γ with (or against) the direction of orientation. In a similar spirit, we will see that, if a vector field $\mathbf{f}$ represents a three-dimensional flow near a surface S, then we can use a special type of surface integral, called a **flux integral**, to measure the flow through the surface S. ◆ We will see flux integrals in the next section.

We think of the surface as permeable, like a fish net, so that fluid flows freely through it.

Defining the surface integral

Let S be a surface in $\mathbb{R}^3$, and let $f(x, y, z)$ be a scalar-valued function defined on S. How might we sensibly define

$$\iint_S f \, dS,$$

the surface integral of f on S? A "good" definition should, among other things, allow us to calculate the surface area of a surface as the integral $\iint_S 1 \, dS$. (The mnemonic symbol dS is analogous to dA for area integrals and dV for volume integrals.)

Parametrization is the key. Suppose that S is parametrized by a well-behaved function

$$\mathbf{X}(u, v) = \big(x(u, v), y(u, v), z(u, v) \big)$$

defined on a convenient domain D (a rectangle, say, or a disk) in the uv-plane. Composing f with $\mathbf{X}$ gives

$$f\big(\mathbf{X}(u, v)\big) = f\big(x(u, v), y(u, v), z(u, v) \big),$$

a function of u and v defined on D. It is tempting, perhaps, simply to integrate this function over D and call the result the surface integral. But that's a little too naive. After all, it is possible to parametrize a surface in various ways using uv-domains of various sizes and shapes. A "good" definition of surface integral, therefore, must somehow take account of *how* $\mathbf{X}$ maps D onto S.

Area magnification—the factor by which the parametrization $\mathbf{X}$ shrinks or stretches areas in mapping D onto S—turns out to be the important idea.

> EXAMPLE 1   In Figure 1, S is the part of the surface $z = 5 - x^2 - y^2$ that lies above the rectangle $[-1.5, 1.5] \times [-1.5, 1.5]$ in the xy-plane. (This rectangle is shown, too, as are some vectors we will discuss in a moment.)

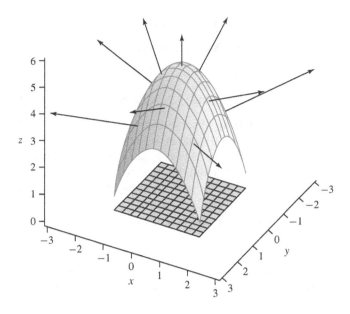

FIGURE 1
A surface and some normal vectors

The surface is parametrized by

$$\mathbf{X}(u, v) = (x, y, z) = (u, v, 5 - u^2 - v^2)$$

for (u, v) in the rectangle $[-1.5, 1.5] \times [-1.5, 1.5]$ in the uv-plane. (In this case, we can consider the uv-plane and the xy-plane as the same, but that's only because the present surface happens to be a graph.) Discuss how $\mathbf{X}$ magnifies areas.

S o l u t i o n   The picture shows both the domain and the range of the parametrization $\mathbf{X}$. The domain D is the flat rectangle in the xy-plane (or the uv-plane as we're thinking of it); the range S is the curved surface.

Notice especially the grids on both D and S. The rectangular grid on D is mapped by $\mathbf{X}$ to the grid of curves on S—each grid line in D is "lifted" straight up to a corresponding curve on S. In the same way, $\mathbf{X}$ lifts each small grid rectangle in D to a slightly curved, parallelogram-shaped grid element in S.

Now compare the relative areas of grid elements in D and S. As the picture shows, the degree of area magnification varies from place to place. The magnification is least at the vertex of S, where the surface is essentially horizontal; it is greatest at the "corners" of S, where the surface is steepest. *Everywhere* on this surface, however, the magnification factor appears to be greater than one.   ◼

Normal vectors   How can we calculate the magnification factor of a parametrizing function at various points (u, v) in the domain? We found the answer symbolically in the preceding section. According to the Fact on page 904, this factor is the magnitude

$$|\mathbf{X_u} \times \mathbf{X_v}|,$$

where

$$\mathbf{X}_u = \big(x_u(u, v), y_u(u, v), z_u(u, v) \big) \qquad \text{and} \qquad \mathbf{X}_v = \big(x_v(u, v), y_v(u, v), z_v(u, v) \big)$$

are the vectors found by partial differentiation of the parametrizing function.

Let's see how this works for the surface shown in Figure 1. First,

$$\mathbf{X}(u,v) = (u, v, 5 - u^2 - v^2) \implies \mathbf{X_u} = (1, 0, -2u) \quad \text{and} \quad \mathbf{X_v} = (0, 1, -2v).$$

Next, routine calculations show that

$$\mathbf{X_u} \times \mathbf{X_v} = (2u, 2v, 1) \quad \text{and} \quad |\mathbf{X_u} \times \mathbf{X_v}| = \sqrt{4u^2 + 4v^2 + 1}.$$

Find the point on the surface corresponding to $(u,v) = (1,1)$. Does it seem plausible that the area magnification is 3 there?

The last result agrees with what we observed earlier: The magnification factor is never less than one, and is *least* at $(u, v) = (0, 0)$, that is, at the vertex of the paraboloid. As u and v increase, so does the magnification factor. At $(u, v) = (1, 1)$, for instance, the magnification factor is 3—a plausible result, judging from the picture. ◄

The calculation also explains the vectors that appear in the picture. We calculated the perpendicular vector $\mathbf{X_u} \times \mathbf{X_v} = (2u, 2v, 1)$ at each (u, v) with integer coordinates; the vectors are shown based at $\mathbf{X}(u, v)$. As expected, each vector looks perpendicular to the surface at the given point.

The definition   There is still more to be seen in Figure 1. Given a grid of subdivisions in the uv-domain D, ◄ the mapping $\mathbf{X}$ produces a corresponding grid of subdivisions on S itself. ◄

The square grid at the bottom. The curved grid on the surface.

Let's use such a grid on S to define the surface integral. Given a function $f(x, y, z)$ defined on S, the surface integral is approximated by a sum of the form

$$\sum_{i=1}^{n} f(x_i, y_i, z_i) \cdot \text{area}(S_i),$$

where (x_i, y_i, z_i) is a point in the ith subdivision S_i. But thanks to the parametrization, $(x_i, y_i, z_i) = \mathbf{X}(u_i, v_i)$ for some point (u_i, v_i), and

$$\text{area}(S_i) \approx |\mathbf{X_u}(u_i, v_i) \times \mathbf{X_v}(u_i, v_i)| \cdot \text{area}(D_i),$$

where D_i is the subdivision of D that corresponds to S_i. Therefore,

$$\sum_{i=1}^{n} f(x_i, y_i, z_i) \cdot \text{area}(S_i) \approx \sum_{i=1}^{n} f(\mathbf{X}(u_i, v_i)) \, |\mathbf{X_u}(u_i, v_i) \times \mathbf{X_v}(u_i, v_i)| \cdot \text{area}(D_i).$$

Now the right side is an approximating sum for the integral

$$\iint_D f(\mathbf{X}(u, v)) \, |\mathbf{X_u} \times \mathbf{X_v}| \, du \, dv.$$

As the grid becomes finer and finer, the approximating sums tend to their respective integrals.

This analysis motivates the following definition:

DEFINITION (Surface integral of a function)   Let S be a surface in xyz-space parametrized by a function $\mathbf{X}(u, v)$ defined on a domain D in the uv-plane. Let $f(x, y, z)$ be a function defined on S. The surface integral of f on S is defined by

$$\iint_S f \, dS = \iint_D f(\mathbf{X}(u, v)) \, |\mathbf{X_u} \times \mathbf{X_v}| \, du \, dv,$$

if the integral exists.

An important special case occurs when the integrand is the constant function $f(x, y, z) = 1$:

> **DEFINITION (Surface area)**   Let S, D, and $\mathbf{X}$ be as in the previous definition. The **surface area** of S is defined as
> $$\iint_S 1\, dS = \iint_D |\mathbf{X_u} \times \mathbf{X_v}|\, du\, dv.$$

Surface area of a graph   If a surface has the form $z = f(x, y)$ for (x, y) in a domain D, then the preceding definition has an especially simple form. Here we can use the parametrization function

$$\mathbf{X}(u, v) = \big(u, v, f(u, v)\big),$$

and so $\mathbf{X_u} = (1, 0, f_u)$ and $\mathbf{X_v} = (0, 1, f_v)$. Taking the cross product gives ➡

Check the calculation.

$$\mathbf{X_u} \times \mathbf{X_v} = (-f_u, -f_v, 1) \quad \text{and} \quad |\mathbf{X_u} \times \mathbf{X_v}| = \sqrt{1 + f_u^2(u, v) + f_v^2(u, v)}.$$

Therefore, the area of S is given by the integral

$$\text{area} = \iint_D \sqrt{1 + f_u^2(u, v) + f_v^2(u, v)}\, du\, dv.$$

Notice the resemblance to the arclength formula for a one-dimensional curve $y = f(x)$ from $x = a$ to $x = b$:

$$\text{length} = \int_a^b \sqrt{1 + f'(x)^2}\, dx,$$

Notice, finally, that in the simplest case of all, in which $f(x, y)$ is constant, the surface S is parallel to D, and the surface area formula reduces simply to

$$\text{area}(S) = \iint_D \sqrt{1 + f_u^2(u, v) + f_v^2(u, v)}\, du\, dv = \iint_D 1\, du\, dv = \text{area}(D).$$

EXAMPLE 2   Find the area of the part of the paraboloid $z = x^2 + y^2$ that lies above the unit disk $x^2 + y^2 \le 1$.

Solution   By the formula just found, the area is $\iint_D \sqrt{1 + 4x^2 + 4x^2}\, dx\, dy$, where D is the unit disk. ➡ This integral is best handled in polar coordinates. With

We use x and y, not u and v, because still another set of coordinates is coming.

$$x = r\cos\theta, \quad y = r\sin\theta, \quad \text{and} \quad dx\, dy = r\, dr\, d\theta,$$

we obtain

$$\text{area} = \int_0^{2\pi} \int_0^1 \sqrt{1 + 4r^2}\, r\, dr\, d\theta = \frac{\pi\left(5\sqrt{5} - 1\right)}{6} \approx 5.33.$$
∎

Surface area for nongraphs   The same formula works—sometimes with a bit more mess—for surfaces that are given parametrically rather than as graphs of functions.

EXAMPLE 3   The sphere of radius a is not the graph of a function $z = f(x, y)$. Find its surface area anyway, using the spherical coordinate parametrization

$$\mathbf{X}(u, v) = (a\sin u\cos v, a\sin u\sin v, a\cos u),$$

for $0 \le u \le \pi$ and $0 \le v \le 2\pi$.

Solution   In this case we get

$$\mathbf{X}_u = a(\cos u \cos v, \cos u \sin v, -\sin u); \quad \mathbf{X}_v = a(-\sin u \sin v, \sin u \cos v, 0).$$

Straightforward but slightly messy calculations now show that

$$\mathbf{X_u} \times \mathbf{X_v} = a^2(\sin^2 u \cos v, \sin^2 u \sin v, \sin u \cos u)$$

and that

$$|\mathbf{X_u} \times \mathbf{X_v}| = a^2 \sin u.$$

Now the area formula gives

$$\text{area} = a^2 \int_{v=0}^{v=2\pi} \int_{u=0}^{u=\pi} \sin u \, du \, dv = 4\pi a^2,$$

as the classical formula says.   ∎

Nonconstant integrands; mass and center of mass   In surface area integrals the integrand function is always constant. In other integrals, of course, the integrand need not be constant.

An important physical application of nonconstant integrands concerns the **mass** and the **center of mass** of a surface. We have already calculated mass and center of mass of objects in the plane and in space; see, Example 1, page 798, for example. The formulas and methods in the context of surfaces are not much different.

Recall: Density is mass per unit area.

Suppose, then, that a surface S has variable density with density function $\rho(x, y, z)$ at different points (x, y, z). ◄ In this case, the **mass** of the surface is given by

$$\iint_S \rho(x, y, z) \, dS.$$

The surface's **center of mass** is the point $(\bar{x}, \bar{y}, \bar{z})$, with coordinates given by

$$\bar{x} = \frac{\iint_S x \, \rho(x, y, z) \, dS}{\text{mass}}; \qquad \bar{y} = \frac{\iint_S y \, \rho(x, y, z) \, dS}{\text{mass}}; \qquad \bar{z} = \frac{\iint_S z \, \rho(x, y, z) \, dS}{\text{mass}}.$$

(Each coordinate of the center of mass is, in effect, the "weighted average" of that coordinate over the surface.)

We're being informal about units; density would be measured in units of mass per unit of area.

EXAMPLE 4   Suppose that the paraboloid of Example 2 has constant density $\rho(x, y, z) = 1$. ◄ Find the mass and the center of mass.

Solution   Because the density is one, the mass is found from the same integral as in Example 2; we got

$$\frac{\pi \left(5\sqrt{5} - 1 \right)}{6} \approx 5.33.$$

Symmetry considerations suggest that the center of mass lies somewhere along the z-axis, and so we will search only for the z-coordinate. By definition,

$$\bar{z} = \frac{\iint_S z \, \rho(x, y, z) \, dS}{\text{mass}}.$$

Applying the definition to the integral in the numerator leads (by the same process as in Example 2) to the uv-integral

$$\iint_D (u^2 + v^2)\sqrt{1 + 4u^2 + 4v^2} \, du \, dv,$$

and from there, using polar coordinates, to

$$\int_0^{2\pi} \int_0^1 r^3 \sqrt{1+4r^2}\, dr\, d\theta.$$

This last integral can be calculated by standard symbolic methods, or even left to technology. The answer turns out to be

$$\int_0^{2\pi} \int_0^1 r^3 \sqrt{1+4r^2}\, dr\, d\theta = 2\pi \left(\frac{5\sqrt{5}}{24} + \frac{1}{120} \right) \approx 2.98.$$

Therefore, the z-coordinate of the center of mass is

$$\bar{z} = \frac{\iint_S z\, \rho(x,y,z)\, dS}{\text{mass}} \approx \frac{2.98}{5.33} \approx 0.56.$$

Thus, the center of mass is a little above the geometric center of the paraboloid, as we might expect given its shape. ■

BASIC EXERCISES

In Exercises 1–6, use the given parametrization **X** to find expressions for (a) the area magnification factor at the domain point (u, v) and (b) a vector normal to the surface at that point. (Answers will involve u and v.)

1. $\mathbf{X}(u, v) = (u, v, u^2 + v^2)$.

2. $\mathbf{X}(u, v) = (u, v, \sqrt{u^2 + v^2})$.

3. $\mathbf{X}(u, v) = (u\cos v, u\sin v, u^2)$.

4. $\mathbf{X}(u, v) = (u\cos v, u\sin v, u)$.

5. $\mathbf{X}(u, v) = (\sin u\cos v, \sin u\sin v, \cos u)$.

6. $\mathbf{X}(u, v) = ((R_1 + R_2\cos u)\cos v, (R_1 + R_2\cos u)\sin v, R_2\sin u)$, where R_1 and R_2 are constants such that $R_1 > R_2 > 0$.

In Exercises 7–12, use the vector $\mathbf{X}_u \times \mathbf{X}_v$ to find the tangent plane to the surface at the given point $\mathbf{X}(u_0, v_0)$. (If possible, use technology to plot both the surface and the tangent plane in an appropriate window.)

7. $\mathbf{X}(u, v) = (u, v, u^2 + v^2)$ at $(u_0, v_0) = (1, 1)$.

8. $\mathbf{X}(u, v) = (u, v, \sqrt{u^2 + v^2})$ at $(u_0, v_0) = (1, 1)$.

9. $\mathbf{X}(u, v) = (u\cos v, u\sin v, u^2)$ at $(u_0, v_0) = (\sqrt{2}, \pi/4)$.

10. $\mathbf{X}(u, v) = (u\cos v, u\sin v, u)$ at $(u_0, v_0) = (\sqrt{2}, \pi/4)$.

11. $\mathbf{X}(u, v) = (\sin u\cos v, \sin u\sin v, \cos u)$ at $(u_0, v_0) = (\pi/2, 0)$.

12. $\mathbf{X}(u, v) = (5\cos u\cos v, 5\cos u\sin v, 2\sin u)$ at $(u_0, v_0) = (\pi/4, 0)$.

In Exercises 13–18, use the integral formula to find the surface area both of the given surface and of the region it lies above (or below).

13. The part of the plane $z = 3$ that lies above the rectangle $[-1, 4] \times [2, 6]$.

14. The part of the plane $z = 3$ that lies above the unit disk $0 \le r \le 1$.

15. The part of the plane $z = 2x + 3y + 4$ that lies above the unit disk $0 \le r \le 1$.

16. The part of the plane $z = 2x + 3y + 4$ that lies above the unit square $[0, 1] \times [0, 1]$.

17. The part of the surface $z = x^2 - y^2$ that lies above (or below) the disk $x^2 + y^2 \le 1$.

18. The part of the surface $z = x^2 - y^2$ that lies above (or below) the disk $x^2 + y^2 \le a^2$.

19. Use cylindrical coordinates r, θ, and z to find the area of the part of the cone $z = r$ that lies between $z = 1$ and $z = 2$. [HINT: Parametrize the cone using $\mathbf{X}(u, v) = (u\cos v, u\sin v, u)$.]

20. Repeat Exercise 19 using the paraboloid $z = r^2$ rather than the cone $z = r$. [HINT: Parametrize the paraboloid using $\mathbf{X}(u, v) = (u\cos v, u\sin v, u^2)$.]

FURTHER EXERCISES

21. Use a surface integral to find the area of the portion of the plane $2x + 3y + 4z = 12$ that lies in the first octant.

22. Evaluate $\iint_S xy\, dS$, where S is the surface of the cylinder $x^2 + y^2 = 4$ between the planes $z = 0$ and $z = 3$.

23. Evaluate $\iint_S \sqrt{1 + x^2 + y^2}\, dS$, where S is the surface of the helicoid $(u\cos v, u\sin v, v)$, $0 \le u \le 2$, $0 \le v \le 2\pi$.

24. Evaluate $\iint_S yz\, dS$, where S is the hemisphere $x^2 + y^2 + z^2 = 1$, $z \ge 0$.

25. If S is a graph parametrized by $\mathbf{X}(u, v) = (u, v, f(u, v))$ for (u, v) in D, then it turns out that

$$\text{area}(S) = \iint_D \sec \alpha \, du \, dv,$$

where α is the angle between the normal vector $\mathbf{X_u} \times \mathbf{X_v}$ and the vertical vector $\mathbf{k}$. Show this fact. (This fact gives some insight into how and why area magnification changes with the "steepness" of the surface.)

26. Let S be the cone $z = \sqrt{x^2 + y^2}$ between $z = 0$ and $z = 4$. Find the center of mass of S if the density function is $\rho(x, y, z) = \sqrt{x^2 + y^2}$.

16.6 DERIVATIVES AND INTEGRALS OF VECTOR FIELDS

The next section presents our last two analogues of the fundamental theorem of calculus: the **divergence theorem** and **Stokes's theorem**. Whereas the elementary fundamental theorem relates derivatives and integrals of scalar-valued *functions*, these higher-dimensional theorems involve certain derivatives and integrals of *vector fields*.

The idea is exactly the same in $\mathbb{R}^3$.

We saw earlier in this chapter how to integrate a vector field $\mathbf{f}$ in $\mathbb{R}^2$ along an oriented curve γ, ← using the line integral $\int_\gamma \mathbf{f} \cdot d\mathbf{X}$. We interpreted the result physically either as work done along γ (if $\mathbf{f}$ is thought of as a force) or as circulation around γ (if $\mathbf{f}$ is thought of as the velocity field of a flow). In this section we will meet another type of vector integral. The **flux integral** of a vector field $\mathbf{f}$ over a surface S measures how much fluid flows across (i.e., perpendicular to) the surface in unit time. We will also see two ways of *differentiating* a vector field in space; each type of derivative has its own geometric and physical significance. In short, this section introduces the objects and operations needed to state our final theorems.

Flux integrals

Let $\mathbf{f}(x, y, z) = \big(P(x, y, z), Q(x, y, z), R(x, y, z)\big)$ be a vector field in $\mathbb{R}^3$; we'll think of $\mathbf{f}$ as the velocity field of a moving fluid. Let S be a surface in $\mathbb{R}^3$; imagine S as a permeable membrane suspended within the flow, like a fish net in a moving stream. Consider the problem of measuring the **flux** across S, that is, the rate of flow per unit of time across S. (In fish net terms, the question is: How much water flows through the net per unit of time?)

It is clear from physical intuition that the answer depends on the angle at which the surface S meets the flow. The flux will be greatest (in absolute value) if the surface is perpendicular to the flow and least if the surface is parallel to the flow. This means, in other words, that the flux at any point (x, y, z) on the surface is the component of the flow vector (P, Q, R) in the direction *perpendicular* to the surface. If $\mathbf{n} = \mathbf{n}(x, y, z)$ is a unit vector perpendicular to the surface at (x, y, z), then the dot product

$$\mathbf{n} \cdot \mathbf{f} = \mathbf{n} \cdot (P, Q, R)$$

Recall: Taking the dot product with a unit vector gives the component in that direction.

gives the component in question. ← Integrating this component over the surface gives the flux we are aiming for:

> **DEFINITION (Flux integral)**   Let $\mathbf{f}$ be a vector field and S a surface in $\mathbb{R}^3$, and let $\mathbf{n}(x, y, z)$ denote a unit vector normal to S at each point (x, y, z). The surface integral
>
> $$\iint_S \mathbf{f} \cdot \mathbf{n} \, dS,$$
>
> called the flux integral, measures the flow per unit time across S in the direction of $\mathbf{n}$.

Observe:

Not really different   The flux integral involves two vectors (**f** and **n**), but taking their dot product produces a scalar-valued function. Thus, the flux integral is really just a particular form of the surface integral we studied in the last section.

Two choices of normal vector   A two-dimensional surface in $\mathbb{R}^3$ has *two* normal directions at any point; they are opposite to each other. That is why the flux integral definition includes the proviso about flow in the direction of the normal vector. (In practice, the two directions are often easy to sort out.)

For some surfaces, however, there *is* no consistent choice of normal direction. (The Möbius strip is the simplest example.) Surfaces with this property are called **nonorientable**. We won't need to worry about this problem in this book. To be fully rigorous, however, the preceding definition would need to require that S be **orientable**.

Easy to calculate   Flux integrals are calculated exactly like any other surface integral. Recall that if the surface S is parametrized by a function $\mathbf{X}(u, v)$ defined on a domain D in uv-space, then the vector

$$\mathbf{X}_u \times \mathbf{X}_v$$

is normal to S at $\mathbf{X}(u, v)$. For a *unit* normal vector **n**, therefore, we may as well use

$$\mathbf{n} = \frac{\mathbf{X}_u \times \mathbf{X}_v}{|\mathbf{X}_u \times \mathbf{X}_v|}.$$

This may look formidable, but a pleasant surprise is in store. By our definition of the surface integral, ⇒

Recall it from the preceding section.

$$\iint_S \mathbf{f} \cdot \mathbf{n}\, dS = \iint_S \mathbf{f} \cdot \frac{\mathbf{X}_u \times \mathbf{X}_v}{|\mathbf{X}_u \times \mathbf{X}_v|}\, dS$$

$$= \iint_D \mathbf{f}\big(\mathbf{X}(u, v)\big) \cdot \frac{\mathbf{X}_u \times \mathbf{X}_v}{|\mathbf{X}_u \times \mathbf{X}_v|}\, |\mathbf{X}_u \times \mathbf{X}_v|\, du\, dv$$

$$= \iint_D \mathbf{f}\big(\mathbf{X}(u, v)\big) \cdot (\mathbf{X}_u \times \mathbf{X}_v)\, du\, dv.$$

The bottom line is that the flux integral may be even easier to calculate than a surface integral.

EXAMPLE 1   Let S be the part of the surface $z = x^2 + y^2$ above the unit disk, and let $\mathbf{f} = (x, y, z)$. Find the flux integral. In which direction does the normal vector **n** point?

Solution   We can parametrize S as a graph, using $\mathbf{X}(u, v) = (u, v, u^2 + v^2)$ with (u, v) in the unit disk D. We showed in the preceding section that, for this parametrization,

$$\mathbf{X}_u = (1, 0, 2u), \quad \mathbf{X}_v = (0, 1, 2v), \quad \text{and} \quad \mathbf{X}_u \times \mathbf{X}_v = (-2u, -2v, 1).$$

Therefore, the flux integral is

$$\iint_S \mathbf{f} \cdot \mathbf{n}\, dS = \iint_D (u, v, u^2 + v^2) \cdot (-2u, -2v, 1)\, du\, dv$$

$$= \iint_D (-u^2 - v^2)\, du\, dv = -\frac{\pi}{2}.$$

Convince yourself that the answer is right.

(The last integral is easily calculated in polar coordinates.) ⇒

Notice that our choice of normal vector **n** has positive z-coordinate, and so it points upward (or toward the inside of the paraboloid bowl). The sign of the answer means that, for the given flow field, more fluid flows "out of the bowl" than into it. ∎

Divergence and curl: derivatives of a vector field

A vector field $\mathbf{f}(x, y, z) = (P(x, y, z), Q(x, y, z), R(x, y, z))$ can be thought of as a function $\mathbf{f}: \mathbb{R}^3 \to \mathbb{R}^3$, with Jacobian matrix

$$\begin{pmatrix} P_x & P_y & P_z \\ Q_x & Q_y & Q_z \\ R_x & R_y & R_z \end{pmatrix}.$$

Many possible combinations of derivatives can be formed from all these data. The two following combinations turn out to have special physical interest:

> **DEFINITION (Divergence and curl)** Let $\mathbf{f} = (P, Q, R)$ be a vector field on $\mathbb{R}^3$. The **divergence** of **f** is the *scalar function* defined by
>
> $$\operatorname{div} \mathbf{f} = P_x + Q_y + R_z.$$
>
> The **curl** of **f** is the *vector field* defined by
>
> $$\operatorname{curl} \mathbf{f} = (R_y - Q_z, P_z - R_x, Q_x - P_y).$$

Notice:

From the Jacobian matrix Both the divergence and the curl are formed in regular ways from the entries of the Jacobian matrix above: The divergence is the sum of the diagonal entries (called the **trace** of the matrix). Each component of the curl field is the difference of two Jacobian entries that are symmetric with respect to the diagonal.

Easy to calculate Calculating the divergence and curl of a vector field is a mechanical matter—indeed, *Maple* and other such programs have commands that do so. It is more interesting to see what the answers mean.

What div means If we think of **f** as a flow, then at any given point (x, y, z) in $\mathbb{R}^3$, the divergence $P_x + Q_y + R_z$ measures the total tendency of fluid to flow *away* from the point.

To get some feeling for why this is so, notice first that $P(x, y, z)$ describes the flow's velocity in the x-direction. Thus, P_x is the acceleration in the x-direction: If $P_x(x, y, z) > 0$, the fluid is speeding up in the x-direction at (x, y, z) and so tends to "diverge" from (x, y, z). If $P_x(x, y, z) < 0$, the fluid is "slowing down" and so tends to "converge," or pile up, at (x, y, z). Adding up the similar contributions in the y- and z-directions gives the bottom line on whether the fluid diverges or converges at (x, y, z).

Curl and grad If **f** happens to be a gradient field, that is, if $\mathbf{f} = \nabla h = (h_x, h_y, h_z)$ for some function $h(x, y, z)$, then a simple but important calculation (left to the exercises) shows that

$$\operatorname{curl} \nabla h = (0, 0, 0).$$

In words:

> *Every gradient field has curl* $(0, 0, 0)$.

Thus, the curl of a vector field **f** measures, in some sense, the extent to which **f** *differs* from being a gradient field. As we have already seen, vector fields that are *not* gradients do appear to "curl" around certain points (compare Figures 1(a) and 1(b), for instance).

Picturing divergence and curl We defined divergence and curl for vector fields in $\mathbb{R}^3$. Unfortunately, three-dimensional vector fields are quite difficult to draw accurately on a flat page—too much detail is lost in projecting three dimensions onto two. In the following pictures, therefore, we will think mainly of the two-dimensional versions of divergence and curl. The idea is that a two-dimensional vector field can, when convenient, be thought of as a three-dimensional field that happens to be independent of z. The vector field $\mathbf{f}(x, y) = (x - y, x + y)$, for instance, can be thought of as the "slice" at $z = 0$ of the three-dimensional field $\mathbf{f}(x, y, z) = (x - y, x + y, 0)$. It's natural, therefore, to define the divergence and curl of a two-dimensional field $\mathbf{f}(x, y) = \big(P(x, y), Q(x, y) \big)$ as follows: ➡

$$\operatorname{div} \mathbf{f} = P_x + Q_y; \quad \operatorname{curl} \mathbf{f} = \big(0, 0, Q_x - P_y \big).$$

Check that this definition of curl is consistent with the earlier one.

Notice a property of the curl vector: It points in the z-direction, *perpendicular* to the xy-plane. The curl vector is in effect a perpendicular axis around which the flow "curls."

EXAMPLE 2 Discuss the divergence and curl of the vector fields

$$\mathbf{f}(x, y) = (x - y, x) \qquad \text{and} \qquad \mathbf{g}(x, y) = (x^2, 2y)$$

shown in Figure 1.

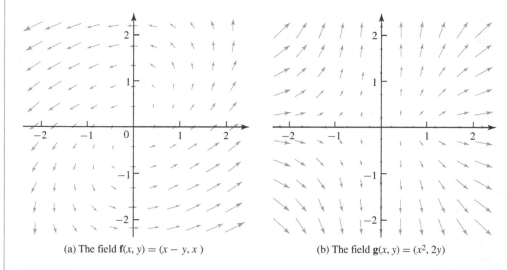

(a) The field $\mathbf{f}(x, y) = (x - y, x)$ (b) The field $\mathbf{g}(x, y) = (x^2, 2y)$

FIGURE 1
Two vector fields

Solution All the symbolic calculations are easy. ➡ Let us start with divergence; calculating $P_x + Q_y$ gives the scalar-valued functions

But check details for yourself.

$$\operatorname{div} \mathbf{f} = 1; \quad \operatorname{div} \mathbf{g} = 2x + 2.$$

Both results can be seen, at least qualitatively, in the pictures. Look first at **f**. At *every* point (x, y), incoming arrows are shorter than outgoing arrows. Thus, the divergence—which measures outflow—is everywhere positive for **f**. For the field **g**, the formula $\operatorname{div} \mathbf{g} = 2x + 2$ means that the divergence changes sign at $x = -1$. This can be seen in the picture, too: Where $x < -1$, comparing lengths of incoming and outgoing arrows reveals a net *inflow*, or negative divergence. Where $x > -1$, the same feature shows a net *outflow*, or positive divergence.

Calculating the curl is easy, too; here are the results:

$$\text{curl } \mathbf{f} = (0, 0, 2); \quad \text{curl } \mathbf{g} = (0, 0, 0).$$

These results also appear (qualitatively) in the pictures. The field $\mathbf{f}$ *does* appear to curl counterclockwise around a vertical axis. (A clockwise curl would produce a negative z-coordinate.) The field $\mathbf{g}$, by contrast, does not seem to curl around on itself, and so the curl vector appears to be zero. There's a good reason for this: $\mathbf{g}$ is a gradient field. Specifically,

$$\mathbf{g} = (x^2, 2y) = \nabla\left(\frac{x^3}{3} + y^2\right).$$

As we observed earlier, *every* gradient field has zero curl. ◼

BASIC EXERCISES

In Exercises 1–4, let S be the part of the surface $z = x^2 + y^2$ that lies above the unit disk, as in Example 1. (Use the same parametrization as was used there.)

1. Let $\mathbf{f}(x, y, z) = (x, 0, 0)$. Find the flux across S; discuss the sign of the answer (as in Example 1).

2. Let $\mathbf{f}(x, y, z) = (0, 1, 0)$. Find the flux across S; discuss the sign of the answer.

3. Let $\mathbf{f}(x, y, z) = (0, 0, z)$. Find the flux across S.

4. Let $\mathbf{f}(x, y, z) = (0, y, 0)$. Find the flux across S.

5–8. Repeat Exercises 1–4 using the cylindrical coordinate parametrization $\mathbf{X}(u, v) = (u \cos v, u \sin v, u^2)$.

9. Let the surface S be the triangle with corners at $(1, 0, 0)$, $(0, 1, 0)$, and $(0, 0, 1)$.

  (a) Use an appropriate cross product to find the area of S without integration.

  (b) Parametrize S as the graph of a function $z = f(x, y)$ for (x, y) in an appropriate region D.

  (c) Use the parametrization of part (b) to find the surface area of S.

  (d) Let $\mathbf{f}(x, y, z) = (x, y, z)$. Find the flux across S.

  (e) Let $\mathbf{f}(x, y, z) = (a, b, c)$; a, b, and c are all constants. Find the flux across S.

  (f) Under what conditions on a, b, and c is the flux in part (e) zero?

10. Let S be the part of the cylinder $x^2 + y^2 = 1$ from $z = 0$ to $z = 1$.

  (a) Find the surface area of S by elementary means. [HINT: Imagine cutting the cylinder and unrolling it.]

  (b) Parametrize S using cylindrical coordinates. Use the result to evaluate the surface area of S by integration.

  (c) Let $\mathbf{f}(x, y, z) = (x, 0, 0)$. Find the flux across S.

  (d) Let $\mathbf{f}(x, y, z) = (0, 0, R(x, y, z))$. Show that the flux across S is zero regardless of the function $R(x, y, z)$.

In Exercises 11–14, find the divergence and the curl of the vector field.

11. $\mathbf{f} = (x, y, z)$.

12. $\mathbf{f} = (y, z, x)$.

13. $\mathbf{f} = (-y, x, z)$.

14. $\mathbf{f} = \left(-y/(x^2 + y^2), x/(x^2 + y^2), 0\right)$.

In Exercises 15 and 16, $\mathbf{f}$ is the vector field $\mathbf{f}(x, y) = (\sin(xy), \cos(x))$. A picture of $\mathbf{f}$ is shown below.

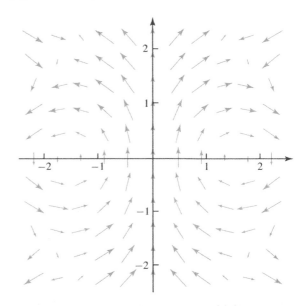

15. (a) Find a formula for the divergence of $\mathbf{f}$.

  (b) Show that the divergence of $\mathbf{f}$ is zero everywhere along the x-axis. How does this appear in the picture?

(c) Find the divergence of **f** at the points $(1, 1)$, $(-1, 1)$, $(-1, -1)$, and $(1, -1)$. How do the signs of the answers appear in the picture?

16. (a) Find a formula for the curl of **f**.

   (b) Find the curl of **f** at the points $(\pi/2, 0)$ and $(-\pi/2, 0)$. Relate the sign difference to the direction of curl.

In Exercises 17–20, calculate the divergence and the curl of the vector field and then plot the given vector field in the rectangle

$[-2, 2] \times [-2, 2]$ *in the xy-plane. (When plotting each field, simply ignore the z-coordinate.) In each case, try to "see" where the divergence is positive and negative and look for the presence or absence of curl.*

17. $\mathbf{f} = (1, 0, 0)$.

18. $\mathbf{f} = (x, 0, 0)$.

19. $\mathbf{f} = (x^2, 0, 0)$.

20. $\mathbf{f} = (x - y, x + y, 0)$.

FURTHER EXERCISES

*In Exercises 21–28, show that the given identity is true for any differentiable vector fields **f** and **g**, any real numbers a and b, and any functions h and j from $\mathbb{R}^3$ to $\mathbb{R}$.*

21. $\operatorname{div}(a\mathbf{f} + b\mathbf{g}) = a \operatorname{div}\mathbf{f} + b \operatorname{div}\mathbf{g}$.

22. $\operatorname{curl}(a\mathbf{f} + b\mathbf{g}) = a \operatorname{curl}\mathbf{f} + b \operatorname{curl}\mathbf{g}$.

23. $\operatorname{div}(h \cdot \mathbf{f}) = h \cdot (\operatorname{div}\mathbf{f}) + (\nabla h) \cdot \mathbf{f}$.

24. $\operatorname{curl}(h \cdot \mathbf{f}) = h \cdot (\operatorname{curl}\mathbf{f}) + (\nabla h) \times \mathbf{f}$.

25. $\operatorname{div}(\operatorname{curl}\mathbf{f}) = 0$.

26. $\operatorname{curl}(\nabla h) = \mathbf{0}$.

27. $\operatorname{div}(\nabla h \times \nabla j) = 0$.

28. $\operatorname{div}(\mathbf{f} \times \mathbf{g}) = (\operatorname{curl}\mathbf{f}) \cdot \mathbf{g} - \mathbf{f} \cdot (\operatorname{curl}\mathbf{g})$.

29–36. The notations $\nabla\cdot$ and $\nabla\times$ are sometimes used to denote div and curl, respectively. Rewrite the identities in Exercises 21–28 using this alternate notation.

37. Let **f** be a differentiable vector field. Show that
$$\operatorname{div}(\nabla\mathbf{f}) = \frac{\partial^2 f}{\partial x^2} + \frac{\partial^2 f}{\partial y^2} + \frac{\partial^2 f}{\partial z^2}.$$

38. Let **f** be the field $\mathbf{f}(x, y) = (2xy + x, xy - y)$ and γ be the perimeter of the square bounded by the lines $x = 0$, $x = 1$, $y = 0$, and $y = 1$. Use Green's theorem to evaluate the outward flux of **f** through γ.

16.7 BACK TO FUNDAMENTALS: STOKES'S THEOREM AND THE DIVERGENCE THEOREM

A calculus course traditionally ends with a "fundamental theorem," one that links the main concepts of differentiation and integration. We will adhere to that tradition. In fact, we'll raise the ante by stating no fewer than *five* fundamental theorems. We have developed all the necessary objects and processes, and it is time to assemble the pieces.

Five fundamental theorems

We will collect all five theorems for comparison. Notation is as follows (bold symbols indicate vector quantities):

   γ: an oriented smooth curve in $\mathbb{R}^2$ or $\mathbb{R}^3$,

   D: a region in $\mathbb{R}^2$,

   S: a two-dimensional surface in $\mathbb{R}^3$,

   V: a three-dimensional solid in $\mathbb{R}^3$,

   n: a unit vector field, normal to a surface S at each point of S,

   f: a vector field in $\mathbb{R}^2$ or $\mathbb{R}^3$,

   f: a scalar-valued function of one or more variables,

   a, b: fixed points in $\mathbb{R}$,

   a, b: fixed points in $\mathbb{R}^2$ or $\mathbb{R}^3$.

Technical hypotheses To avoid sidetracks, we make several technical assumptions. We state these assumptions mainly for the record—all are satisfied in typical simple examples.

We discussed orientability briefly in the preceding section; see page 912.

We should add, however, that the assumptions are genuinely important; in their absence there is no guarantee that the objects in question even exist. For a nonorientable surface, for instance, there *is* no suitable choice of normal vector, and so the surface integrals in question don't make sense. ←

We will assume, then, that all functions and derivatives mentioned in the following theorems exist and are continuous; this ensures, in turn, that the integrals exist. Curves are assumed to be either smooth or piecewise-smooth (i.e., the union of several smooth curves or surfaces, with "kinks" only where the pieces join). Surfaces are orientable, and smooth except perhaps along edges where smooth pieces join.

Five theorems   With these provisos, here are the theorems; the last two are new.

THEOREM 4 (Fundamental theorem of calculus)

$$\int_a^b f'(x)\, dx = f(b) - f(a).$$

THEOREM 5 (Fundamental theorem for line integrals)   If γ starts at **a** and ends at **b**, then

$$\int_\gamma \nabla f \cdot d\mathbf{X} = f(\mathbf{b}) - f(\mathbf{a}).$$

THEOREM 6 (Green's theorem)   Let $\mathbf{f} = (P, Q)$ be a vector field in $\mathbb{R}^2$, γ a closed curve (oriented counterclockwise), and D the region inside γ. Then

$$\iint_D (Q_x - P_y)\, dA = \int_\gamma P\, dx + Q\, dy.$$

THEOREM 7 (Stokes's theorem)   Let $\mathbf{f} = (P, Q, R)$ be a vector field in $\mathbb{R}^3$. Let S be a surface in $\mathbb{R}^3$, bounded by a closed curve γ, with unit normal **n**. Then

$$\iint_S (\text{curl } \mathbf{f}) \cdot \mathbf{n}\, dS = \pm \oint_\gamma \mathbf{f} \cdot d\mathbf{X}.$$

(The sign depends on the direction of **n**.)

THEOREM 8 (Divergence theorem)   Let $\mathbf{f} = (P, Q, R)$ be a vector field in $\mathbb{R}^3$. Let V be a solid region in $\mathbb{R}^3$, bounded by a surface S, with outward unit normal **n**. Then

$$\iiint_V \text{div } \mathbf{f}\, dV = \iint_S \mathbf{f} \cdot \mathbf{n}\, dS.$$

All five theorems have the same theme: A function or vector field **f** is given. On the left side of each equation, some sort of derivative of **f** is integrated over some domain in $\mathbb{R}$, $\mathbb{R}^2$, or $\mathbb{R}^3$. On the right side of each equation, the expression involves **f** itself, evaluated on a lower-dimensional set—the *boundary* of the original domain.

More on Stokes's theorem

Let's see first, by example, what the theorem says.

EXAMPLE 1 Let S be the part of the surface $z = x^2 + y^2$ above the unit disk, and let $\mathbf{f} = (P, Q, R) = (-y, x, z)$. What does Stokes's theorem say in this case? Is it true?

Solution We will calculate the integrals on both sides of Stokes's theorem and see that they are equal.

Notice first that the boundary of S is a circle of radius 1 in $\mathbb{R}^3$—the intersection of the paraboloid and the plane $z = 1$. We can parametrize this boundary using

$$\mathbf{X}(t) = (\cos t, \sin t, 1); \qquad 0 \le t \le 2\pi.$$

The line integral $\int_\gamma \mathbf{f} \cdot d\mathbf{X}$ is now easy to calculate: →   *But check details.*

$$\int_\gamma P \, dx + Q \, dy + R \, dz = \int_0^{2\pi} \left(\sin^2 t + \cos^2 t \right) dt = 2\pi.$$

To handle the surface integral, recall that we used the same surface in Example 1, page 913. There we used the parametrization $\mathbf{X}(u, v) = (u, v, u^2 + v^2)$ with (u, v) in the unit disk D; we found the normal vector $\mathbf{X}_u \times \mathbf{X}_v = (-2u, -2v, 1)$.

For the vector field $\mathbf{f} = (-y, x, z)$, calculation gives → $\operatorname{curl} \mathbf{f} = (0, 0, 2)$. Therefore,   *Check it!*
the flux integral is

$$\iint_S (\operatorname{curl} \mathbf{f}) \cdot \mathbf{n} \, dS = \iint_D (0, 0, 2) \cdot (-2u, -2v, 1) \, du \, dv = \iint_D 2 \, du \, dv.$$

The last integral is twice the area of the unit disk, or 2π.

Thus, the line and surface integrals are the same in this case; Stokes's theorem holds. (With the opposite choice of normal vector the flux integral would change sign, but Stokes's theorem permits that.) ∎

EXAMPLE 2 Let S be the part of the surface $z = x^2 + y^2$ above the unit disk (as in Example 1), let $h(x, y, z)$ be a smooth function, and let $\mathbf{f} = \nabla h = (h_x, h_y, h_z)$. What does Stokes's theorem say in this case? Is it true?

Solution The curl of every gradient field is the zero vector field. → That is,   *It is easy (and amusing) to check this for oneself.*

$$\operatorname{curl} \nabla h = (0, 0, 0)$$

regardless of the function h. Thus, the *left* side of Stokes's theorem is trivially zero.

Now the *right* side of Stokes's theorem involves the line integral $\oint_\gamma \nabla h \cdot d\mathbf{X}$, where γ is the upper boundary curve of our surface S. The fundamental theorem of line integrals guarantees that

$$\oint_\gamma \nabla h \cdot d\mathbf{X} = h(\mathbf{b}) - h(\mathbf{a}),$$

where $\mathbf{a}$ and $\mathbf{b}$ are the beginning and ending points of γ. But γ is a *closed* curve, and so $\mathbf{b} = \mathbf{a}$. Thus, the right side of Stokes's theorem vanishes, as does the left, and so the theorem's claim holds true. ∎

From Stokes to Green Stokes's theorem is, in a natural sense, an extension to $\mathbb{R}^3$ of Green's theorem, which "lives" in $\mathbb{R}^2$. (Indeed, Stokes's theorem is sometimes known as "Green's theorem for surfaces.") We will not prove Stokes's theorem since its proof involves the same main ideas (appropriately translated) as appear in the proof of Green's theorem.

Instead of proving Stokes's theorem, let's see exactly how Green's theorem is a special case. First, recall that a vector field $\mathbf{f} = (P(x, y), Q(x, y))$ in $\mathbb{R}^2$ can be thought of, when convenient, as a special three-dimensional vector field $(P, Q, 0)$, which depends only on x and y and has zero for the last coordinate. ← For this new vector field, an easy calculation shows that

We discussed this viewpoint in the preceding section; see page 915.

$$\text{curl}(P, Q, R) = (R_y - Q_z, P_z - R_x, Q_x - P_y) = (0, 0, Q_x - P_y).$$

Thus, the third coordinate of curl $\mathbf{f}$ turns out to be the area integrand in Green's theorem.

The second step is to think of the domain D in Green's theorem as a very simple surface in $\mathbb{R}^3$. Since D lies flat in the xy-plane, the upward-pointing vector $\mathbf{k}$ can be taken as a unit normal to D. (The downward-pointing vector $-\mathbf{k}$ could also have been chosen; doing so would change the sign of the surface integral.)

Taken together, these remarks mean that, in the present situation,

$$\iint_D (\text{curl } \mathbf{f}) \cdot \mathbf{n} \, dS = \iint_D (0, 0, Q_x - P_y) \cdot \mathbf{k} \, dS = \iint_D (Q_x - P_y) \, dA.$$

In other words, the flux integral in Stokes's theorem boils down, in this special case, to the area integral of Green's theorem.

In a similar way, the boundary curve γ—which lies in the plane $z = 0$—can be parametrized either as a curve in $\mathbb{R}^2$ or as a curve in $\mathbb{R}^3$. If the function

$$\mathbf{X}(t) = (x(t), y(t)); \qquad a \le t \le b$$

parametrizes γ in $\mathbb{R}^2$, then

$$\mathbf{X}(t) = (x(t), y(t), 0); \qquad a \le t \le b$$

parametrizes γ in $\mathbb{R}^3$. In the latter case, the line integral becomes

$$\int_\gamma P \, dx + Q \, dy + R \, dz = \int_\gamma P \, dx + Q \, dy$$

because $dz = 0$. Thus, the line integrals in Stokes's theorem and Green's theorem turn out, in this special case, to be identical.

We've shown what we wanted to show: For domains in the plane, each side of Stokes's identity reduces to the corresponding side of Green's identity.

Curl, flux, and circulation: a physical interpretation   The mathematical theory of line and surface integrals grew up around physical problems, and the terms "curl," "flux," and "circulation" are all borrowed from physics. Naturally enough, therefore, Stokes's theorem can be interpreted, at least intuitively, in the physical language of flow.

We start by imagining a three-dimensional fluid flow $\mathbf{f}$ defined on and near a surface S. At each point (x, y, z) on the surface S, the derivative curl $\mathbf{f}$ is a new vector, which measures the tendency of the flow to "curl" at (x, y, z). (The vector curl $\mathbf{f}$ acts as an axis around which the curling occurs.) It follows that if $\mathbf{n}$ is a unit normal to S at (x, y, z), then the dot product $(\text{curl } \mathbf{f}) \cdot \mathbf{n}$—the left-hand integrand in Stokes' theorem—tells how much of the fluid's rotation occurs along the surface (rather than, say, perpendicular to it). Therefore, the surface integral

The reasoning here is rough and ready, but it can be made mathematically precise.

$$\iint_S (\text{curl } \mathbf{f}) \cdot \mathbf{n} \, dS$$

measures, in some sense, the *total* rotation of the flow along the surface. ←

The line integral $\pm \oint_\gamma \mathbf{f} \cdot d\mathbf{X}$ is easier to interpret: It measures the fluid's circulation around the boundary of S. Stokes's theorem asserts, then, that two quantities are equal: (i) the fluid's circulation around the boundary of S, and (ii) the total rotation of fluid along S itself. That (i) and (ii) are equal seems physically believable, since each phenomenon can be thought of as causing the other.

More on the divergence theorem

The divergence theorem asserts that, under appropriate hypotheses,

$$\iiint_V \text{div}\,\mathbf{f}\,dV = \iint_S \mathbf{f} \cdot \mathbf{n}\,dS. \tag{1}$$

Let's pick the equation apart, thinking in physical terms.

On the left: a triple integral The left side of Equation 1 is a triple integral over the solid region V; the integrand is the scalar-valued function div $\mathbf{f}$. If $\mathbf{f}$ represents a flow, then we have seen that, at any domain point (x, y, z), the function div $\mathbf{f}$ measures the net flow *away* from (x, y, z) per unit time. Integrating div $\mathbf{f}$ over V, therefore, gives the total net flow out of V per unit time, that is, the net rate at which fluid "leaves" V.

On the right: a flux integral The right side of Equation 1 is a flux integral of the type described in the previous section; it measures the rate at which fluid crosses the boundary surface S. Because the unit normal $\mathbf{n}$ is chosen to point outward from V, the flux integral measures the flow across S in the outward direction.

Divergence and flux: why they are equal The last two paragraphs say, in effect, that the integrals on both sides of Equation 1 measure the same thing: the flow out of V. From this point of view, the divergence theorem should sound physically reasonable: Two integrals that measure the same quantity should have the same value.

EXAMPLE 3 Let V be the solid in $\mathbb{R}^3$ bounded above by the plane $z = 1$ and below by the paraboloid $z = x^2 + y^2$. Let $\mathbf{f}$ be the vector field $(P, Q, R) = (x, y, z)$. Figure 1 shows the solid and the field:

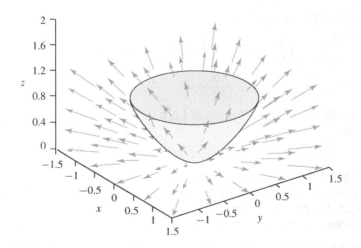

FIGURE 1

A vector field and a solid

What does the divergence theorem say in this case? Is it true?

Solution   In this case the boundary surface S has two parts: (i) S_1, the part of the paraboloid $z = x^2 + y^2$ below the plane $z = 1$ (the "drum"), and (ii) S_2, the part of the plane $z = 1$ for which $x^2 + y^2 \leq 1$ (the "drumhead").

Let's calculate both sides of Equation 1, starting on the left. It is easy to see, first, that

$$\text{div } \mathbf{f} = \text{div}(x, y, z) = 3.$$

(The picture suggests, too, that the divergence is everywhere positive.) Thus, the volume integral is

$$\iiint_V 3 \, dV = 3 \times \text{volume of } V.$$

Some details are left to the exercises.

This integral is simplest in cylindrical coordinates; we get ◄

$$\iiint_V 3 \, dV = \int_{\theta=0}^{\theta=2\pi} \int_{r=0}^{r=1} \int_{z=r^2}^{z=1} 3r \, dz \, dr \, d\theta = \frac{3\pi}{2}.$$

Now for the right side of Equation 1. The flux integral has two parts, one for each part of the surface. We have already handled the surface S_1 several times—most recently in Example 1. There, we parametrized S_1 by $\mathbf{X}(u, v) = (u, v, u^2 + v^2)$ with (u, v) in the unit disk D. Then the vector $\mathbf{X}_u \times \mathbf{X}_v = (-2u, -2v, 1)$ is normal to S_1. Notice,

Look closely at the picture to convince yourself.

however, that this vector has a positive z-coordinate, and so it points *into* V. ◄ For an *outward* unit normal, therefore, we can take the reversed vector

$$\mathbf{n} = \frac{(2u, 2v, -1)}{\sqrt{1 + 4u^2 + 4v^2}}.$$

Some details are left to the exercises.

With this parametrization, therefore, the flux integral over S_1 becomes ◄

$$\iint_{S_1} \mathbf{f} \cdot \mathbf{n} \, dS = \iint_D (u, v, u^2 + v^2) \cdot (2u, 2v, -1) \, du \, dv = \frac{\pi}{2}.$$

The surface S_2 is even easier to parametrize by using $\mathbf{X}(u, v) = (u, v, 1)$, with (u, v) in the unit disk D. Because S_2 is parallel to the xy-plane, moreover, the upward-pointing vector $\mathbf{k}$ is a suitable outward unit normal. Thus, the flux integral over S_2 becomes

$$\iint_{S_2} \mathbf{f} \cdot \mathbf{n} \, dS = \iint_D (u, v, 1) \cdot (0, 0, 1) \, du \, dv = \iint_D 1 \, du \, dv = \pi.$$

Therefore, the total flux integral is

$$\iint_S \mathbf{f} \cdot \mathbf{n} \, dS = \iint_{S_1} \mathbf{f} \cdot \mathbf{n} \, dS + \iint_{S_2} \mathbf{f} \cdot \mathbf{n} \, dS$$

$$= \frac{\pi}{2} + \pi = \frac{3\pi}{2}.$$

Here, therefore, the two sides of Equation 1 are indeed equal, as the divergence theorem asserts. ∎

EXAMPLE 4   Let V be the *solid* sphere of radius a, centered at the origin in xyz-space. The boundary surface S is then the *hollow* sphere of radius a with equation $x^2 + y^2 + z^2 = a^2$. Let $\mathbf{f}$ be the vector field $(P, Q, R) = (x, y, z)$, which is shown in Figure 1 (along with a different surface). What does the divergence theorem say now?

Solution   As in Example 3, we have div $\mathbf{f} = 3$, and so the left side of Equation 1 is simply

$$\iiint_V 3 \, dV = 3 \times \text{volume of } V = 4\pi a^3.$$

To evaluate the *right* side of Equation 1 we notice that the vector field $\mathbf{f} = (x, y, z)$ is itself normal to S (and outward-pointing) at each point of S and has length $\sqrt{x^2 + y^2 + z^2} = \sqrt{a^2} = a$ at each such point. Thus, we can use

$$\mathbf{n} = \frac{1}{a}(x, y, z)$$

as the needed outward *unit* normal, and so the right side of Equation 1 becomes

$$\iint_S \mathbf{f} \cdot \mathbf{n} \, dS = \iint_S \frac{1}{a}(x, y, z) \cdot (x, y, z) \, dS = \iint_S \frac{1}{a} \left(x^2 + y^2 + z^2 \right) dS.$$

But $\left(x^2 + y^2 + z^2 \right) = a^2$ for (x, y, z) on S, and so

$$\iint_S \mathbf{f} \cdot \mathbf{n} \, dS = \iint_S \frac{a^2}{a} \, dS = a \times \text{surface area of } S.$$

We calculated the surface area of S to be $4\pi a^2$ in Example 3, page 909. Thus, both sides of Equation 1 have the value $4\pi a^3$. ◼

Proving the divergence theorem: the idea   The proof of the divergence theorem is similar to that for Green's theorem. Notice first that, if $\mathbf{f} = (P, Q, R)$ and the outward unit normal to S has the form $\mathbf{n} = (n_1, n_2, n_3)$, then the divergence theorem says that

$$\iiint_V (P_x + Q_y + R_z) \, dV = \iint_S (P\, n_1 + Q\, n_2 + R\, n_3) \, dS.$$

Thus, it is enough to prove, separately, three simpler identities:

$$\iiint_V P_x \, dV = \iint_S P\, n_1 \, dS; \qquad \iiint_V Q_y \, dV = \iint_S Q\, n_2 \, dS;$$

$$\iiint_V R_z \, dV = \iint_S R\, n_3 \, dS.$$

We will prove the third identity, assuming that the boundary S of V has two parts: a lower surface S_1 and an upper surface S_2 (as in the preceding example) and that both S_1 and S_2 are graphs of functions $z = g(x, y)$ and $z = h(x, y)$, respectively, for (x, y) in a region D in the xy-plane. Then, as we've seen, the vectors $(-g_x, -g_y, -1)$ and $(h_x, h_y, 1)$ are normal to the surfaces S_1 and S_2. (We used the downward-pointing normal for the lower surface S_1.)

Finally, we write out both sides of the third identity and compare results. On the right, the preceding parametrizations imply that

$$\iint_{S_1} R\, n_3 \, dS = -\iint_D R(x, y, g(x, y)) \, dA$$

and

$$\iint_{S_2} R\, n_3 \, dS = \iint_D R(x, y, h(x, y)) \, dA.$$

(The two integrals have opposite signs because of their different normal directions.) Adding these results gives the total surface integral:

$$\iint_S R\, n_3 \, dS = \iint_D \Big(R(x, y, h(x, y)) - R(x, y, g(x, y)) \Big) \, dA.$$

On the left, the triple integral can be written in iterated form, as follows:

$$\iiint_V R_z \, dV = \iint_D \left(\int_{z=g(x,y)}^{z=h(x,y)} R_z \, dz \right) dA.$$

Applying the ordinary fundamental theorem to the inner integral (i.e., antidifferentiating in z) gives

$$\iiint_V R_z \, dV = \iint_D \Big(R(x, y, h(x, y)) - R(x, y, g(x, y)) \Big) \, dA.$$

Thus, the surface and volume integrals are equal. Similar arguments apply to the integrals involving P and Q; they complete the idea of proof.

BASIC EXERCISES

In Exercises 1–4, use Stokes's theorem to find the value of the surface integral

$$\iint_S (\text{curl } \mathbf{f}) \cdot \mathbf{n} \, dS.$$

Do so by transforming the surface integral into an equivalent line integral and then calculating the line integral.

1. Let S be the upper half of the sphere $x^2 + y^2 + z^2 = 1$, and let $\mathbf{f} = (x, y, z)$.

2. Let S be the upper half of the sphere $x^2 + y^2 + z^2 = 1$, and let $\mathbf{f} = (-y, x, z)$.

3. Let S be the part of the paraboloid $z = x^2 + y^2$ with $z \leq 1$, and let $\mathbf{f} = (y, z, x)$.

4. Let S be the part of the plane $z = y + 1$ above the disk $x^2 + y^2 \leq 1$, and let $\mathbf{f} = (2z, -x, x)$.

5. Carefully work out each of the volume and surface integrals in Example 3.

6. Consider the situation in Example 3, but use the vector field $\mathbf{f} = (x, 0, 0)$.

(a) Calculate the volume integral $\iiint_V \text{div } \mathbf{f} \, dV$.

(b) Calculate the surface integral $\iint_S \mathbf{f} \cdot \mathbf{n} \, dS$.

7. Let V be the cube in $\mathbb{R}^3$ defined by $0 \leq x \leq 1$, $0 \leq y \leq 1$, $0 \leq z \leq 1$. Let $\mathbf{f}$ be the vector field $\mathbf{f} = (x, y, z)$. Let S be the boundary of V; note that S has six faces.

(a) What does the divergence theorem say in this case?

(b) Calculate the triple integral $\iiint_V \text{div } \mathbf{f} \, dV$. [HINT: Very little calculation is involved!]

(c) Calculate the flux integral $\iint_S \mathbf{f} \cdot \mathbf{n} \, dS$. [HINTS: There are six parts, but all are quite simple. Notice that for each face of S, $\mathbf{i}, \mathbf{j}$, or $\mathbf{k}$ can be used as a unit normal. Be careful with signs!]

8. Let V be a solid in $\mathbb{R}^3$ with smooth boundary S. Let $\mathbf{f}$ be a smooth vector field (P, Q, R) in $\mathbb{R}^3$. Use the divergence theorem to show that

$$\iint_S (\text{curl } \mathbf{f}) \cdot \mathbf{n} \, dS = 0.$$

In Exercises 9–12, use the divergence theorem to evaluate the flux integral $\iint_S \mathbf{f} \cdot \mathbf{n} \, dS$, where S is the unit sphere $x^2 + y^2 + z^2 = 1$ with outward unit normal.

9. $\mathbf{f} = (x, 2y, 3z)$.

10. $\mathbf{f} = (x, y^2, 0)$.

11. $\mathbf{f} = (0, y^2, 0)$.

12. $\mathbf{f} = (-e^x \cos y, e^x \sin y, 1)$.

13. Let γ be the unit circle, oriented counterclockwise, in the xy-plane, and consider the vector field $\mathbf{f} = (-y, x, z)$. (Think of γ as a curve in 3-space.)

(a) Calculate the line integral $\int_\gamma \mathbf{f} \cdot d\mathbf{X}$.

(b) Let the surface S be the upper half of the unit sphere. Then γ is the boundary of S. What does Stokes's theorem say in this case? Is it true? [HINT: No calculation is needed; use the fact that the unit sphere has surface area 4π.]

(c) Now let the surface S be the part of the paraboloid $z = 1 - x^2 - y^2$ that lies above the xy-plane. Again, γ is the boundary of S. What does Stokes's theorem say now? Is it true? [HINT: This time a surface integral calculation is needed, but it is relatively easy.]

14. Let $\mathbf{f}(x, y, z) = (z - x, x - y, y - z)$, and let T be the sphere $x^2 + y^2 + z^2 = 4$.

(a) Evaluate curl $\mathbf{f}$.

(b) Evaluate div $\mathbf{f}$.

(c) Is $\iint_T \mathbf{f} \cdot \mathbf{n} \, dS$ positive, negative, or zero? Justify your answer.

15. Let S be the portion of the surface $z = x^2 + y^2$ that lies above the region R: $1 \leq x \leq 2$, $2 \leq y \leq 3$.

(a) Find a vector normal to S at the point $(1, 2, 5)$.

(b) Find an equation for the plane tangent to S at the point $(1, 2, 5)$.

(c) Find the area of S.

(d) Find the area of the portion of the tangent plane in part (b) that lies above R.

16. Let $\mathbf{f}$, S, γ, and $\mathbf{n}$ be as in the statement of Stokes's theorem, and assume that $\mathbf{f} = \nabla h$ for some smooth function h defined in $\mathbb{R}^3$.

(a) Show that curl $\mathbf{f} = (0, 0, 0)$.

(b) Use Stokes's theorem to show that $\oint_\gamma \mathbf{f} \cdot d\mathbf{X} = 0$. (The direction of γ doesn't matter.)

(c) Use the fundamental theorem for line integrals—not Stokes's theorem—to show that $\oint_\gamma \mathbf{f} \cdot d\mathbf{X} = 0$.

17. Let $\mathbf{f}$ be a vector field in $\mathbb{R}^3$ such that, on the surface of the unit sphere, $\mathbf{f}(x, y, z) = (0, 0, z^3)$. If div $\mathbf{f} = a$ is a constant in all of $\mathbb{R}^3$, what is the value of a?

18. Suppose that $\mathbf{f}$ is a vector field such that curl $\mathbf{f} = (1, 2, 5)$ at every point in $\mathbb{R}^3$. Find an equation of a plane through the origin with the property that $\oint_\gamma \mathbf{f} \cdot d\mathbf{X} = 0$ for any closed curve γ lying in the plane.

Matrices and Matrix Algebra: A Crash Course

This appendix assumes basic familiarity with vectors, including vector addition, scalar multiplication, and the dot product.

What is a matrix?

A **matrix** is a rectangular array of real numbers or symbols. ➤ Here are some examples:

The symbols stand for real numbers.

$$A = \begin{bmatrix} 1 & 2 & 3 \end{bmatrix} \quad X = \begin{bmatrix} x \\ y \\ z \end{bmatrix} \quad B = \begin{bmatrix} a & b & c \\ d & e & f \\ g & h & i \end{bmatrix} \quad C = \begin{bmatrix} 1 & 2 & 3 \\ 4 & 5 & 6 \end{bmatrix} \quad I_2 = \begin{bmatrix} 1 & 0 \\ 0 & 1 \end{bmatrix}.$$

As the examples suggest, matrices may have any number of rows and any number of columns. A little vocabulary and notation will help us navigate among the possibilities:

Dimensions An $m \times n$ matrix has m rows and n columns. Among the matrices above, A is a 1×3 matrix, X is 3×1, and C is 2×3. An $m \times m$ matrix, such as B and I_2 above, is called **square**.

Rows, columns, and indices Matrix notation uses *two* index variables, ➤ one for rows and one for columns. Keeping clear which is which is essential. Fortunately, there is an unbreakable rule:

The letters i, j, and k are popular—but not sacred—choices.

First rows, then columns.

For instance, M_{23} is the entry in row 2, column 3, and a 4×7 matrix has 4 rows and 7 columns—not the other way around.

Matrices and vectors Matrices and vectors are closely related. An $m \times 1$ matrix is sometimes called a **column vector**; a $1 \times m$ matrix is a **row vector**. (See the matrices X and A above.) In fact, we can think of any matrix—regardless of its dimensions—as being built up either from column vectors or from row vectors. Which point of view is more useful depends on the circumstances; we'll find uses for both.

Entries The numbers or symbols in any matrix M are called **entries**; the i,jth entry, denoted M_{ij}, is the one in row i and column j. Above, for instance, $A_{12} = 2$, $X_{31} = z$, and

Matrices B and C illustrate this possibility.

$B_{23} = f$, while A_{21}, X_{13}, and B_{45} are not defined. In general, of course, M_{ij} and M_{ji} need not be equal, even if both quantities are defined. ◄

Diagonals and identity matrices   The **main diagonal** of a matrix M consists of all entries of the form M_{ii}. In an $m \times m$ (square) matrix, the main diagonal starts at the upper left corner and ends at the lower right corner. ◄ The $m \times m$ matrix that has ones along the main diagonal and zeros everywhere else is called the $m \times m$ **identity matrix**; it is denoted I_m (I_2 appears above). (The name is appropriate because I_m behaves, as we will see in a moment, as an "identity" for matrix multiplication.)

In a nonsquare matrix, the main diagonal does not end at the lower right corner.

What are matrices for?

Like tables, arrays, and spreadsheets, matrices are used to store information concisely and efficiently—especially when the information has, like the matrix itself, a natural "two-dimensional" structure. For instance, the following table gives road mileage between several pairs of Texas cities:

	Austin	Dallas	Houston	San Antonio
Austin	0	195	161	78
Dallas	195	0	240	272
Houston	161	240	0	199
San Antonio	78	272	199	0

"T" is For Texas—the cities are now understood, not explicitly stated.

All the numerical information fits naturally into a 4×4 matrix, say T: ◄

$$T = \begin{bmatrix} 0 & 195 & 161 & 78 \\ 195 & 0 & 240 & 272 \\ 161 & 240 & 0 & 199 \\ 78 & 272 & 199 & 0 \end{bmatrix}.$$

How much does it cost, in dollars, to drive from one Texas city to another? If we multiply every entry in T by 0.31 (some companies reimburse employees for auto expenses at \$0.31 per mile), we get a new matrix

$$E = \begin{bmatrix} 0 & 60.45 & 49.91 & 24.18 \\ 60.45 & 0 & 74.40 & 84.32 \\ 49.91 & 74.40 & 0 & 61.69 \\ 24.18 & 84.32 & 61.69 & 0 \end{bmatrix};$$

More below on other matrix operations.

each entry of E tells the mileage reimbursement associated with the given trip. Under these circumstances, it is reasonable to write, simply, $E = 0.31\,T$. ◄

The matrices T and E have some special properties, such as having all zeros along the main diagonal. (The distance from any city to itself is zero!) Notice, too, that each entry has an identical twin in the "mirror" position across the main diagonal. This property has a formal name:

> **DEFINITION** An $n \times n$ matrix M is **symmetric** if, for all i and j from 1 to n,
>
> $$M_{ij} = M_{ji}.$$

Not every useful matrix is symmetric, but enough are to have spawned a considerable mathematical theory. In multivariable calculus, symmetric matrices occur in the context of second derivatives of functions of several variables.

Matrix algebra

Addition, subtraction, and scalar multiplication Matrices can be added, subtracted, and multiplied by scalars in *exactly* the same way as vectors, their close cousins. The following simple examples illustrate what this means; all the symbols stand for real numbers:

$$\begin{bmatrix} a & b \\ c & d \end{bmatrix} + \begin{bmatrix} e & f \\ g & h \end{bmatrix} = \begin{bmatrix} a+e & b+f \\ c+g & d+h \end{bmatrix}; \qquad r \begin{bmatrix} a & b \\ c & d \end{bmatrix} = \begin{bmatrix} ra & rb \\ rc & rd \end{bmatrix}.$$

(The Texas mileage expense matrix is another example of scalar multiplication.) The examples show that matrices—just like vectors—can be added, subtracted, and multiplied by scalars in the simplest possible way, "entry-by-entry." ⇒ (In particular, $A \pm B$ can make sense only if A and B have the same dimensions.) The formal definitions say the same thing, but in full symbolic regalia:

Vectors are special types of matrices, and so the similarity in behavior is no surprise.

> **DEFINITION** Let A and B be $m \times n$ matrices and let r be a scalar. Then $A + B$, $A - B$, and rA are new $m \times n$ matrices defined respectively by
>
> $$(A+B)_{ij} = A_{ij} + B_{ij}; \qquad (A-B)_{ij} = A_{ij} - B_{ij}; \qquad (rA)_{ij} = r A_{ij}.$$

Like their vector counterparts (and for the same reasons), these matrix operations enjoy pleasant algebraic properties, such as **commutativity** of addition (i.e., $A + B = B + A$) and **distributivity** (i.e., $r(A + B) = rA + rB$).

EXAMPLE 1 In multivariable calculus, matrices are often associated with linear equations and linear functions. Consider the following system of three linear equations in three unknowns:

$$1x + 2y + 3z = 7$$
$$2x + 3y + 1z = 8$$
$$3x + 2y + 1z = 9.$$

Rewrite this system using matrices.

Solution The system suggests three different matrices:

$$D = \begin{bmatrix} 1 & 2 & 3 \\ 2 & 3 & 1 \\ 3 & 2 & 1 \end{bmatrix}, \qquad X = \begin{bmatrix} x \\ y \\ z \end{bmatrix}, \quad \text{and} \quad E = \begin{bmatrix} 7 \\ 8 \\ 9 \end{bmatrix}.$$

Here D is called the **coefficient matrix** of the system, B contains the "right side" of the system of equations, and X stores the variable names. (We will see shortly why it's convenient for X to have the shape of a column, not a row.)

Solving linear systems of equations (including very large ones) is an important mathematical problem, but it can be tedious and error prone. Storing information in matrices as one goes along helps avoid errors and unnecessary duplication. The system in this example can be written entirely in matrix notation as

$$\begin{bmatrix} 1 & 2 & 3 \\ 2 & 3 & 1 \\ 3 & 2 & 1 \end{bmatrix} \begin{bmatrix} x \\ y \\ z \end{bmatrix} = \begin{bmatrix} 7 \\ 8 \\ 9 \end{bmatrix};$$

written entirely in symbols, the equation is simply $DX = E$.

Notice especially the left side of the preceding equation: DX is a matrix product—an idea to which we now turn.

Matrix multiplication   How can two matrices A and B be multiplied to form a third matrix, AB, that deserves to be called the matrix product? One reasonable guess, following the pattern for addition, is simply to multiply element-by-element. It turns out, however, that another definition of matrix multiplication is much more useful. We will first state the definition formally and then explore what it means.

DEFINITION (Matrix multiplication)   Let A be an $m \times p$ matrix and B a $p \times n$ matrix. The product AB is an $m \times n$ matrix with entries given by

$$(AB)_{ij} = \sum_{k=1}^{p} A_{ik} B_{kj} = A_{i1} B_{1j} + A_{i2} B_{2j} + \cdots + A_{ip} B_{pj}.$$

This somewhat forbidding-looking definition is best unpacked through simple examples. Consider these carefully:

$$\begin{bmatrix} a & b \\ c & d \end{bmatrix} \begin{bmatrix} x \\ y \end{bmatrix} = \begin{bmatrix} ax + by \\ cx + dy \end{bmatrix} \qquad \begin{bmatrix} 2 & 3 \end{bmatrix} \begin{bmatrix} x \\ y \end{bmatrix} = \begin{bmatrix} 2x + 3y \end{bmatrix}$$

$$\begin{bmatrix} 1 & 2 & 3 \\ 2 & 3 & 1 \\ 3 & 2 & 1 \end{bmatrix} \begin{bmatrix} x \\ y \\ z \end{bmatrix} = \begin{bmatrix} 1x + 2y + 3z \\ 2x + 3y + 1z \\ 3x + 2y + 1z \end{bmatrix} \qquad \begin{bmatrix} a & b \\ c & d \end{bmatrix} \begin{bmatrix} x & z \\ y & w \end{bmatrix} = \begin{bmatrix} ax + by & az + bw \\ cx + dy & cz + dw \end{bmatrix}.$$

Here are some lessons the definition and examples teach:

Possible shapes   The product AB makes sense if (but only if) A has the same number of columns as B has rows. Equivalently, the rows of A must be exactly as long as the columns of B. In particular, A, B, and AB may *all* have different shapes. At the other extreme, if both A and B are (square) $n \times n$ matrices, then so is AB.

Order matters   Matrix multiplication is not commutative. More often than not, $AB \neq BA$—even if both products happen to make sense. ← For example, reversing the factors in the last example above gives

Of which there is no guarantee.

$$\begin{bmatrix} x & z \\ y & w \end{bmatrix} \begin{bmatrix} a & b \\ c & d \end{bmatrix} = \begin{bmatrix} ax + cz & bx + dz \\ ay + cw & by + dw \end{bmatrix}.$$

Matrix multiplication and dot products   Matrix multiplication is closely linked to the dot product. Indeed, if A is a $1 \times n$ row vector and B is an $n \times 1$ column vector, then the matrix product AB has just one entry: the dot product of A and B, thought of as vectors. In fact, *every* matrix product AB is found by taking appropriate dot products. More precisely: ➤

Read this carefully—it's a nice way to remember the recipe for matrix multiplication.

The ijth entry of AB is the dot product of the ith row of A and the jth column of B.

Zero and identity matrices   The $n \times n$ matrix O with all entries zero is called the **zero matrix**. It behaves as any self-respecting zero should: For every $n \times n$ matrix A, $AO = O = OA$ and $A + O = A = O + A$. In a similar vein, the $n \times n$ **identity matrix** I_n (with ones on the main diagonal and zeros elsewhere) ➤ is a multiplicative identity: For every $n \times n$ matrix A, $AI_n = A = I_n A$.

See I_2 at the beginning of this appendix.

Matrix algebraic expressions   With various matrix operations understood, we can make sense of algebraic expressions and equations that involve matrices. In Example 1, for example, we expressed the system of linear equations as the *matrix* equation $DX = E$, where all the symbols represent matrices of appropriate shapes and sizes. Similarly, we can now understand more complicated matrix equations such as

$$PAQ = B, \qquad A(3B + C) = 3AB + AC, \qquad \text{and} \qquad AB = I_3.$$

For these expressions to make good sense requires, of course, that the various matrices have compatible dimensions.

Multiplicative inverses   Two real numbers a and b are called multiplicative inverses if $ab = 1$, that is, if their product is the multiplicative identity for real numbers. Thus, 3 and $1/3$ are inverses, as are $-17/12$ and $-12/17$; in fact, the real numbers a and $1/a$ are inverses for any $a \neq 0$.

Similar ideas hold for square matrices. ➤ Two $n \times n$ matrices A and B are called **inverses** if $AB = I_n$, that is, if their product is an identity matrix. In this case, we write $A = B^{-1}$ and $B = A^{-1}$. For instance, it is easy to check that if

Square matrices work best since they can be multiplied in either order.

$$A = \begin{bmatrix} 3 & 2 \\ 2 & 1 \end{bmatrix} \quad \text{and} \quad B = \begin{bmatrix} -1 & 2 \\ 2 & -3 \end{bmatrix}, \quad \text{and} \quad I_2 = \begin{bmatrix} 1 & 0 \\ 0 & 1 \end{bmatrix},$$

then $AB = BA = I_2$, and so $B = A^{-1}$ and $A = B^{-1}$.

(To be fully rigorous about inverses, we should require *both* $AB = I_n$ and $BA = I_n$. It turns out, however, that for square matrices the two conditions are equivalent. This can be shown using techniques from linear algebra.)

Not every square matrix *has* an inverse. For example, the $n \times n$ zero matrix O certainly does not since, for every $n \times n$ matrix A, $AO = O \neq I_n$. But O is not the only square matrix without an inverse. The next example—which applies to *any* 2×2 matrix—suggests why some square matrices have inverses and some do not.

EXAMPLE 2   Consider the 2×2 matrices

$$A = \begin{bmatrix} a & b \\ c & d \end{bmatrix} \quad \text{and} \quad B = \begin{bmatrix} \frac{d}{ad-bc} & -\frac{b}{ad-bc} \\ -\frac{c}{ad-bc} & \frac{a}{ad-bc} \end{bmatrix}.$$

How are A and B related?

Solution   For B to make any sense, the common denominator $ad - bc$ must be nonzero. But if the denominator isn't zero, then straightforward calculation ➤ shows that $AB = I_2$, and so A and B are inverses.   ∎

Try it—the result is satisfying!

More to the story ...   The general theory of matrices and their inverses and how (if possible) to find one from the other is studied in much more detail in linear algebra courses. For multivariable calculus we need only basic ideas and definitions.

Matrices through the ages.   Matrices have a long history. Ordered tables of numbers and recipes for manipulating them go back as far as the ancient Babylonian and Chinese mathematicians, who may have used matrices in solving practical problems that led to linear equations. The modern abstract theory of matrices (and the word "matrix") dates back to the mid-1800's. The English lawyer and mathematician Arthur Cayley first defined the matrix operations of addition, scalar multiplication, and inversion.

Determinants

The quantity $ad - bc$ in Example 2 is called the **determinant** of the matrix $A = \begin{bmatrix} a & b \\ c & d \end{bmatrix}$; it is denoted by $\det A$. The determinant turns out to . . . well . . . determine whether or not A has an inverse.

Defining determinants   Determinants can be defined for square matrices M of any size. In all cases, $\det M$ is a number (not a vector or a matrix) calculated from the entries of M; it tells (among other things) whether M has an inverse. The general definition of $\det M$ is a bit complicated; fortunately, we will need the idea only in dimensions two and three.

DEFINITION (Determinant)   For a 2×2 matrix, the determinant is

$$\det \begin{bmatrix} a & b \\ c & d \end{bmatrix} = ad - bc.$$

For a 3×3 matrix, the determinant is

$$\det \begin{bmatrix} a & b & c \\ d & e & f \\ g & h & i \end{bmatrix} = aei - afh + bfg - bdi + cdh - ceg.$$

This is easy to see for the 2×2 case; the 3×3 case needs a closer look.

Observe that, in both cases, the determinant involves several summands, having alternating positive and negative signs. Each summand is the product of one factor from each row and each column. ← There are six such summands for a 3×3 matrix. One way to organize the summands and keep track of signs in the 3×3 case is to write the matrix in a "double" array:

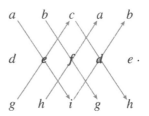

Try it.

Then the determinant is the sum of six threefold diagonal products with "southeast-pointing" diagonals counted positive and "northeast-pointing" diagonals negative. ←

For larger matrices, determinants are normally calculated (if at all) with help from technology. However they are calculated, determinants have a useful relationship to inverses.

> **FACT** An $n \times n$ matrix M has an inverse if and only if $\det M \neq 0$.

Example 2 showed what the Fact means (and why it's true) for 2×2 matrices. ⟶ More general discussion can be found in any linear algebra text.

The exercises pursue the matter a little further, too.

Determinants geometrically: area and volume What else can determinants tell us about matrices? For multivariable calculus purposes there are good geometric answers.

For a 2×2 matrix $M = \begin{bmatrix} a & b \\ c & d \end{bmatrix}$, the determinant measures the *area* of a certain parallelogram. To see why this is so, think of the rows of M as vectors (a, b) and (c, d) in the xy-plane; the parallelogram in question is "spanned" by these two vectors, as shown in Figure 1:

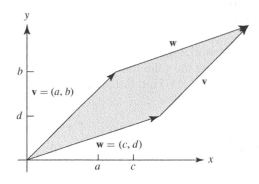

FIGURE 1
Two vectors span a parallelogram

Similarly, for a 3×3 matrix we can regard the rows as vectors in xyz-space and consider the three-dimensional solid (called a **parallelepiped**) spanned by these three vectors, as shown in Figure 2:

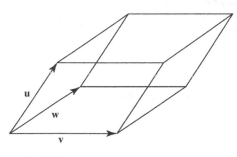

FIGURE 2
The solid spanned by three vectors

This time, the determinant measures the volume. Here are the precise statements:

> **FACT (What the determinant tells)**   For a 2×2 matrix M,
>
> $$|\det M| = \text{area of parallelogram spanned by rows of } M.$$
>
> For a 3×3 matrix M,
>
> $$|\det M| = \text{volume of parallelepiped spanned by rows of } M.$$

We will not pause to prove this Fact here; further discussion of its two claims appear in Section 12.9 in connection with the cross product. Observe:

Determinant zero If $\det M = 0$, then the rows of M span zero area or zero volume. In the 2×2 case, this means that the two row vectors of M are collinear—one is simply a multiple of the other. In the 3×3 case, spanning zero *volume* means that the three row vectors of M are coplanar. (In linear algebra jargon, these conditions say that the rows of M are "linearly dependent.")

Positive or negative? The Fact refers only to the magnitude $|\det M|$; determinants may be either positive or negative. For our purposes, the sign of the determinant is usually less important than its magnitude.

EXAMPLE 3 Let P be the parallelogram with adjacent edges determined by the vectors $(1, 1)$ and $(2, 0)$. Let S be the solid parallelepiped with edges determined by the vectors $(1, 0, 0)$, $(0, 1, 0)$, and $(1, 1, 1)$. Find the area of P and the volume of S.

Solution Consider the matrices

$$M = \begin{bmatrix} 1 & 1 \\ 2 & 0 \end{bmatrix} \quad \text{and} \quad N = \begin{bmatrix} 1 & 0 & 0 \\ 0 & 1 & 0 \\ 1 & 1 & 1 \end{bmatrix}.$$

Remember, it's the absolute value that counts.

Easy calculations show that $\det M = -2$ and $\det N = 1$. Therefore, P has area 2 and Q has volume 1.

Determinants—how important? Determinants were especially popular in the 18th and 19th centuries, when their theory was developed by such famous mathematicians as Leibniz, Maclaurin, Gauss, Lagrange, and Cauchy. But mathematical fashions change. More recently, the importance of determinants (as opposed to other mathematical tools for addressing similar questions) has been questioned. As evidence, consider the title of a recent journal article: "Down with determinants!" by S. Axler, *American Mathematical Monthly*, February 1995, pp. 139–154.

EXERCISES

NOTES: Several problems below use the following matrices:

$$A = \begin{bmatrix} 1 & 2 \\ 3 & 5 \end{bmatrix}; \quad B = \begin{bmatrix} -5 & 2 \\ 3 & -1 \end{bmatrix}; \quad C = \begin{bmatrix} 0 & 1 & 2 \\ 0 & 0 & 1 \\ 1 & 0 & 2 \end{bmatrix};$$

$$D = \begin{bmatrix} 0 & 1 & 4 \\ 2 & 0 & 1 \end{bmatrix}; \quad E = \begin{bmatrix} x & u \\ y & v \\ z & w \end{bmatrix}; \quad X = \begin{bmatrix} x \\ y \end{bmatrix}; \quad U = \begin{bmatrix} u & v \end{bmatrix};$$

1. (a) Calculate AB, AD, AX, UA, and UX.
   (b) For each product MN in the preceding part, consider the reversed product NM. Compute those that make sense.
   (c) Are there any pairs of inverses among the matrices above?

2. (a) Calculate CE, C^2, DC, UD, EX.
   (b) For each product MN in the preceding part, consider the reversed product NM. Compute those that make sense.

3. **(a)** Find the determinants of A, B, AB, C, and C^2.

   (b) How is the determinant of AB related to the determinants of A and B?

   (c) How is the determinant of C^2 related to the determinant of C?

4. For real numbers a and b, $ab = 0$ implies that at least one of a and b is zero. Matrices do not have this property. To illustrate this, find two 2×2 matrices A and B with all nonzero entries such that $AB = \left[\begin{smallmatrix} 0 & 0 \\ 0 & 0 \end{smallmatrix}\right]$.

5. Consider the 2×2 matrices $M = \left[\begin{smallmatrix} a & b \\ c & d \end{smallmatrix}\right]$ and $N = \left[\begin{smallmatrix} e & f \\ g & h \end{smallmatrix}\right]$. Show that $\det(MN) = (\det M)(\det N)$. (Use "brute force"—i.e., just crank out both sides of the equation.)

6. This exercise is about Example 3, page A-8.

   (a) Draw the parallelogram P. Convince yourself by elementary methods (e.g., areas of triangles) that P has area 2, as the determinant formula says.

   (b) Consider the matrix M in Example 3. Let M' be the same as M but with the first and second rows reversed. Show that $\det M' = -\det M$. What does $\det M'$ mean geometrically?

   (c) Draw the parallelepiped Q. (Do this by hand.) Can you convince yourself by elementary methods that Q has volume 1 as the determinant formula says?

7. Consider the Texas mileage matrix T on page A-2; we observed that T is symmetric. What property of the real world guarantees this symmetry?

8. Recall that a square matrix M is symmetric (see page A-3) if $M_{ij} = M_{ji}$ for all index values i and j. Similarly, a square matrix M is called **skew-symmetric** if $M_{ij} = -M_{ji}$ for all i and j. In each of the following parts, write out the 3×3 matrix M determined by the given rule and decide whether the matrix is symmetric, skew-symmetric, or neither.

   (a) $M_{ij} = i + j$.

   (b) $M_{ij} = i - j$.

   (c) $M_{ij} = i^2 + j^2$.

   (d) $M_{ij} = i^2 - j$.

9. Let M be a 3×3 matrix. (See the previous problem for the definition of a skew-symmetric matrix.)

   (a) Show that if M is skew-symmetric, then all the diagonal entries of M must be zeros.

   (b) Can M be both symmetric and skew-symmetric? If so, give an example. If not, why not?

10. Let A and B be matrices, both of dimension $m \times n$. We will denote by $A \star B$ the new $m \times n$ matrix formed by multiplying element-by-element, that is, $(A \star B)_{ij} = A_{ij}B_{ij}$. (The matrix $A \star B$ is called the **Hadamard** product of A and B after the French mathematician Jacques Hadamard.)

   (a) Calculate the Hadamard products

   $$\begin{bmatrix} 1 & 2 & 3 \\ 4 & 5 & 6 \end{bmatrix} \star \begin{bmatrix} 7 & 8 & 9 \\ 10 & 11 & 12 \end{bmatrix} \quad \text{and} \quad \begin{bmatrix} a & b \\ c & d \end{bmatrix} \star \begin{bmatrix} e & f \\ g & h \end{bmatrix}.$$

   (b) Ordinary matrix multiplication is not commutative. Is Hadamard multiplication commutative? Why or why not?

   (c) Among all 2×2 matrices, which matrix I deserves to be called an identity for Hadamard multiplication? Why?

   (d) Let A and B be 2×2 matrices. We will say that A and B are **Hadamard inverses** if $A \star B = I$, where I is the matrix of the previous part. Which 2×2 matrices have Hadamard inverses? If A has a Hadamard inverse, how is it calculated?

11. Let M be a 2×2 matrix and suppose that $MN = NM$ for all 2×2 matrices N. Show that M is a scalar multiple of the identity matrix.

THEORY OF MULTIVARIABLE CALCULUS: BRIEF GLIMPSES

This appendix offers a brief sampler of definitions and proofs in the theory of multivariable calculus. These samples are intended to be read "as needed." But some readers may enjoy the material for its own sake as an introduction to some formal ideas and methods of analytic mathematics.

The limit of a function: a formal definition

The rigorous theory of calculus—in any number of variables—is based firmly on a clear and unambiguous notion of limit. Continuous functions, derivatives, integrals, and other standard objects of calculus are all defined in terms of limits. Recall the formal definition of limit for functions of one variable:

DEFINITION (Limit of a function of one variable) Let $f(x)$ be defined for x in an interval containing a except perhaps *at* $x = a$. Suppose that, for every positive number ϵ (epsilon), no matter how small, there is a corresponding positive number δ (delta) so that

$$|f(x) - L| < \epsilon \quad \text{whenever} \quad 0 < |x - a| < \delta.$$

Then, $\lim_{x \to a} f(x) = L$.

The corresponding definition for functions of two (or more) variables is as follows:

DEFINITION (Limit of a function of two variable) Consider a function $f : \mathbb{R}^2 \to \mathbb{R}$ defined for (x, y) in an interval containing (a, b) except perhaps at $(x, y) = (a, b)$. Suppose that for every positive number ϵ, no matter how small, there is a corresponding positive number δ such that

$$|f(x, y) - L| < \epsilon \quad \text{whenever} \quad 0 < \sqrt{(x - a)^2 + (y - b)^2} < \delta.$$

Then $\lim_{(x,y) \to (a,b)} f(x) = L$.

Observe some similarities and differences between the two definitions:

- **A missing value?** A close look reveals that neither $f(a)$ (in the first case) nor $f(a, b)$ (in the second case) plays any role in the definitions. ➤ This is no accident—indeed, the function f need not even be defined at the "target" point $x = a$ or $(x, y) = (a, b)$ to have a limit there.

- **What they say** Both versions of the definition say that outputs from f are "near" the number L whenever inputs are "near" the domain point ($x = a$ or $(x, y) = (a, b)$) at which the limit is taken. The two definitions differ mainly in how "nearness" is measured.

- **Measuring distance** In the first definition, distance is measured using *absolute values*, and so $|f(x) - L|$ is the distance between $f(x)$ and L, while $|x - a|$ is the distance between x and a. In the second definition, $f(x, y)$ and L are ordinary numbers, and so $|f(x, y) - L|$ measures distance exactly as in the first definition. But inputs to $f(x, y)$ are two-dimensional points, and so distance is measured using the two-dimensional distance formula $\sqrt{(x - a)^2 + (y - b)^2}$, which involves a square root.

The inequality $0 < |x - a|$ specifically excludes $x = a$.

Using the definition Next we apply the definition directly to verify a simple two-variable limit. The claim itself is hardly surprising; the point is to see the definition in action. There is a moral to draw as well: Even the simplest multivariate limits may take some effort to verify rigorously.

EXAMPLE 1 Let $f(x, y) = x + y$. Show that $\lim\limits_{(x,y) \to (0,0)} f(x, y) = 0$.

Solution In this case we have

$$|f(x, y) - L| = |x + y| \qquad \text{and} \qquad \sqrt{(x - a)^2 + (y - b)^2} = \sqrt{x^2 + y^2}.$$

Thus, the definition requires that, for a given $\epsilon > 0$, we need to find some $\delta > 0$ so that

$$|x + y| < \epsilon \qquad \text{whenever} \qquad \sqrt{x^2 + y^2} < \delta.$$

To this end, we notice that

$$|x + y| \le |x| + |y| \le \sqrt{x^2 + y^2} + \sqrt{x^2 + y^2} = 2\sqrt{x^2 + y^2}.$$

This chain of inequalities implies that $\delta = \epsilon/2$ "works" in the sense of the definition. In other words,

$$\sqrt{x^2 + y^2} < \delta \qquad \text{implies that} \qquad |x + y| \le 2\sqrt{x^2 + y^2} < 2\frac{\epsilon}{2} = \epsilon,$$

as desired. ∎

New limits from old To avoid the work of using the limit definition directly, we usually find new limits by combining a few basic "known" limits in allowable algebraic ways. The expected rules for combining limits do indeed hold (but must themselves be proved using the definition!) for multivariable functions. ➤ For instance, it can be shown by basic arguments like the one above that

The limit of a sum is the sum of the respective limits, for instance.

$$\lim_{(x,y) \to (2,1)} x = 2 \qquad \text{and} \qquad \lim_{(x,y) \to (2,1)} y = 1.$$

Combining these limits tells us, for example, that

$$\lim_{(x,y)\to(2,1)} \frac{x^2+2y}{3x^2+4y} = \frac{2^2+2\cdot1}{3\cdot2^2+4\cdot1} = \frac{6}{16}.$$

Continuity

Continuity can be understood informally, but the formal definition involves a limit: A function $f : \mathbb{R}^2 \to \mathbb{R}$ is **continuous** at a domain point (a, b) if

$$\lim_{(x,y)\to(a,b)} f(x, y) = f(a, b).$$

The function f is **discontinuous** at (a, b) if the limit above either fails to exist or has some other value than $f(a, b)$.

For example, the function $f(x, y) = x + y$ is continuous at $(0, 0)$ because

$$\lim_{(x,y)\to(0,0)} (x + y) = f(a, b) = 0,$$

as we proved above in Example 1. We also showed in Example 1 that the function $f(x, y) = (x^2 + 2y)/(3x^2 + 4y)$ is continuous at $(2, 1)$ because its limit and its value coincide at that point.

EXAMPLE 2   Suppose that $f(x, y) = \dfrac{xy}{x^2 + y^2}$ for $(x, y) \neq (0, 0)$ and $f(0, 0) = 0$. Is f continuous at $(0, 0)$? (Further discussion of this function, and a useful picture, appear in Example 1, page 765.)

Solution   The question boils down to whether or not

$$\lim_{(x,y)\to(0,0)} \frac{xy}{x^2 + y^2} = 0.$$

The answer turns out to be no because the definition of limit is not satisfied in this case. To see why, suppose we choose, say, $\epsilon = 0.1$. If the limit were indeed 0, there would be some $\delta > 0$ such that

$$|f(x, y) - 0| = \frac{xy}{x^2 + y^2} < 0.1$$

for all (x, y) inside a circle of radius δ about $(0, 0)$.

But a look at the formula for f shows that, for every input (x, y) with $x = y$, we have

$$f(x, y) = f(x, x) = \frac{x^2}{x^2 + x^2} = 0.5.$$

(In other words, $f(x, y) = 0.5$ for all (x, y) on the line $y = x$.) Because part of this line lies inside every circle centered at $(0, 0)$, no matter how small, it follows that *no* suitable positive δ can be chosen.

We conclude that the limit at $(0, 0)$ is not 0, and so setting $f(0, 0) = (0, 0)$ makes f discontinuous at $(0, 0)$. Indeed, our argument can be extended to show a little more: the function f has *no* limit at the origin. Thus, there is no way to "repair" the discontinuity of f at $(0, 0)$ by choosing $f(0, 0)$ cleverly.   ∎

Differentiability and the total derivative

Recall the definition (see page 766):

DEFINITION Let $f(x, y)$ be a function and $\mathbf{X_0} = (x_0, y_0)$ a point of its domain. Let

$$L(x, y) = f(x_0, y_0) + f_x(x_0, y_0)(x - x_0) + f_y(x_0, y_0)(y - y_0)$$
$$= f(\mathbf{X_0}) + \nabla f(\mathbf{X_0}) \cdot (\mathbf{X} - \mathbf{X_0})$$

be the linear approximation to f at (x_0, y_0). If

$$\lim_{\mathbf{X} \to \mathbf{X_0}} \frac{f(\mathbf{X}) - L(\mathbf{X})}{|\mathbf{X} - \mathbf{X_0}|} = 0,$$

then f is **differentiable** at $\mathbf{X_0}$, and the vector $\nabla f(\mathbf{X_0})$ is the **total derivative** of f at $\mathbf{X_0}$.

Showing rigorously that a function is differentiable in this sense depends—like so much else—on careful work with limits.

EXAMPLE 3 Show carefully that the function $f(x, y) = x^2 + y^2$ is differentiable at $(2, 1)$, and find the total derivative.

Solution In Example 1, page 724, we calculated

$$\nabla f(2, 1) = (4, 2) \quad \text{and} \quad L(x, y) = 4(x - 2) + 2(y - 1) + 5.$$

Thus, it remains only to prove that

$$\lim_{(x,y) \to (2,1)} \frac{f(x, y) - L(x, y)}{\sqrt{(x - 2)^2 + (y - 1)^2}} = 0$$

to show that $(4, 2)$ is the total derivative at $(2, 1)$.

To prove the desired limit, we manipulate the expression inside the limit:

$$\frac{f(x, y) - L(x, y)}{\sqrt{(x - 2)^2 + (y - 1)^2}} = \frac{x^2 + y^2 - (4(x - 2) + 2(y - 1) + 5)}{\sqrt{(x - 2)^2 + (y - 1)^2}} \qquad \text{substituting}$$

$$= \frac{x^2 - 4x + 4 + y^2 - 2y + 1}{\sqrt{(x - 2)^2 + (y - 1)^2}} \qquad \text{rearranging terms}$$

$$= \frac{(x - 2)^2 + (y - 1)^2}{\sqrt{(x - 2)^2 + (y - 1)^2}} \qquad \text{completing the square; then canceling}$$

$$= \sqrt{(x - 2)^2 + (y - 1)^2}.$$

The result is now clear: As $(x, y) \to (2, 1)$, the last quantity clearly tends to zero, as desired.

Equality of mixed partial derivatives

We met the following theorem in Section 13.5:

> **THEOREM 1 (Equality of mixed partial derivatives)** Let $f(x, y)$ be a function; assume that the second partial derivatives f_{xy} and f_{yx} are defined and continuous on the domain of f. Then, for all (x, y),
>
> $$f_{xy}(x, y) = f_{yx}(x, y).$$

Perhaps surprisingly, a relatively simple proof of the theorem can be given using an iterated double integral.

The idea of the proof We will explain why $f_{xy}(0, 0) = f_{yx}(0, 0)$. (That will suffice because there is nothing special about $(0, 0)$.) To do so, we show that, for any small square $R = [0, h] \times [0, h]$,

$$\iint_R f_{xy}(x, y)\, dA = \iint_R f_{yx}(x, y)\, dA. \tag{1}$$

Before proving Equation 1, let's see why it helps our cause. Suppose, for instance, that $f_{xy}(0, 0) > f_{yx}(0, 0)$. Then, since f_{xy} and f_{yx} are continuous functions (by our technical assumption), we would have $f_{xy}(x, y) > f_{yx}(x, y)$ for all (x, y) sufficiently near $(0, 0)$. In particular, we would have $f_{xy}(x, y) > f_{yx}(x, y)$ for all (x, y) on some small rectangle $R = [0, h] \times [0, h]$, and in this case Equation 1 could not possibly hold. It suffices, therefore, to convince ourselves of Equation 1.

To do this we calculate both sides of Equation 1 as iterated integrals. The left side (LHS) is

$$\text{LHS} = \iint_R f_{xy}(x, y)\, dA = \int_0^h \left(\int_0^h f_{xy}(x, y)\, dy \right) dx.$$

Now notice that f_{xy} is (by definition!) the y-derivative of f_x, and f_x is the x-derivative of f. This means that we can calculate our integrals by antidifferentiation:

$$\text{LHS} = \int_0^h \left(\int_0^h f_{xy}(x, y)\, dy \right) dx \qquad \text{integrate first in } y, \text{ then in } x$$

$$= \int_0^h \left(f_x(x, y) \Big]_0^h \right) dx = \int_0^h \left(f_x(x, h) - f_x(x, 0) \right) dx \qquad \text{antidifferentiate in } y$$

$$= \left(f(x, h) - f(x, 0) \right) \Big]_0^h \qquad\qquad \text{antidifferentiate both terms in } x$$

$$= f(h, h) - f(0, h) - f(h, 0) + f(0, 0).$$

A similar calculation—but now integrating first in x and then in y—shows that the right-hand integral in Equation 1 has the same value.

TABLE OF DERIVATIVES AND INTEGRALS

Basic Forms

1. $\dfrac{d}{dx}c = (c)' = 0$

2. $\dfrac{d}{dx}x = (x)' = 1$

3. $\dfrac{d}{dx}x^n = (x^n)' = nx^{n-1}$

4. $\dfrac{d}{dx}e^x = (e^x)' = e^x$

5. $\dfrac{d}{dx}\ln x = (\ln x)' = \dfrac{1}{x}$

6. $\dfrac{d}{dx}\sin x = (\sin x)' = \cos x$

7. $\dfrac{d}{dx}\cos x = (\cos x)' = -\sin x$

8. $\dfrac{d}{dx}\tan x = (\tan x)' = \sec^2 x$

9. $\dfrac{d}{dx}\sec x = (\sec x)' = \sec x \tan x$

10. $\dfrac{d}{dx}\arcsin x = (\arcsin x)' = \dfrac{1}{\sqrt{1-x^2}}$

11. $\dfrac{d}{dx}\arctan x = (\arctan x)' = \dfrac{1}{1+x^2}$

12. $\displaystyle\int x^n \, dx = \dfrac{x^{n+1}}{n+1}, \quad n \neq -1$

13. $\displaystyle\int \dfrac{dx}{x} = \ln |x|$

14. $\displaystyle\int e^x \, dx = e^x$

15. $\displaystyle\int b^x \, dx = \dfrac{1}{\ln b}b^x$

16. $\displaystyle\int \sin x \, dx = -\cos x$

17. $\displaystyle\int \cos x \, dx = \sin x$

18. $\displaystyle\int \tan x \, dx = \ln |\sec x| = -\ln |\cos x|$

19. $\displaystyle\int \cot x \, dx = \ln |\sin x| = -\ln |\csc x|$

20. $\displaystyle\int \sec x \, dx = \ln |\sec x + \tan x| = \ln \left| \tan \left(\dfrac{x}{2} + \dfrac{\pi}{4} \right) \right|$

21. $\displaystyle\int \csc x \, dx = \ln |\csc x - \cot x| = \ln \left| \tan \left(\dfrac{x}{2} \right) \right|$

22. $\displaystyle\int \sec^2 x \, dx = \tan x$

23. $\displaystyle\int \csc^2 x \, dx = -\cot x$

24. $\displaystyle\int \sec x \tan x \, dx = \sec x$

25. $\displaystyle\int \csc x \cot x \, dx = -\csc x$

26. $\displaystyle\int \dfrac{dx}{x^2 + a^2} = \dfrac{1}{a}\arctan \left(\dfrac{x}{a} \right), \quad a \neq 0$

27. $\displaystyle\int \dfrac{dx}{x^2 - a^2} = \dfrac{1}{2a}\ln \left| \dfrac{x-a}{x+a} \right|$

28. $\displaystyle\int \dfrac{dx}{\sqrt{a^2 - x^2}} = \arcsin \left(\dfrac{x}{a} \right), \quad a > 0$

29. $\displaystyle\int \ln x \, dx = x(\ln x - 1)$

Expressions Containing $ax+b$

30. $\displaystyle\int (ax+b)^n\, dx = \frac{(ax+b)^{n+1}}{a(n+1)}, \quad n \neq -1$

31. $\displaystyle\int \frac{dx}{ax+b} = \frac{1}{a}\ln|ax+b|$

32. $\displaystyle\int \frac{x}{ax+b}\, dx = \frac{x}{a} - \frac{b}{a^2}\ln|ax+b|$

33. $\displaystyle\int \frac{x}{(ax+b)^2}\, dx = \frac{b}{a^2(ax+b)} + \frac{1}{a^2}\ln|ax+b|$

34. $\displaystyle\int \frac{dx}{x(ax+b)} = \frac{1}{b}\ln\left|\frac{x}{ax+b}\right|$

35. $\displaystyle\int \frac{dx}{x^2(ax+b)} = -\frac{1}{bx} + \frac{a}{b^2}\ln\left|\frac{ax+b}{x}\right|$

36. $\displaystyle\int \sqrt{ax+b}\, dx = \frac{2}{3a}\sqrt{(ax+b)^3}$

37. $\displaystyle\int x\sqrt{ax+b}\, dx = \frac{2(3ax-2b)}{15a^2}\sqrt{(ax+b)^3}$

38. $\displaystyle\int \frac{dx}{\sqrt{ax+b}} = \frac{2\sqrt{ax+b}}{a}$

39. $\displaystyle\int \frac{dx}{x\sqrt{ax+b}} = \frac{1}{\sqrt{b}}\ln\left|\frac{\sqrt{ax+b}-\sqrt{b}}{\sqrt{ax+b}+\sqrt{b}}\right|, \quad b > 0$

40. $\displaystyle\int \frac{dx}{x\sqrt{ax-b}} = \frac{2}{\sqrt{b}}\arctan\sqrt{\frac{ax-b}{b}}, \quad b > 0$

41. $\displaystyle\int x^n\sqrt{ax+b}\, dx = \frac{2}{a(2n+3)}\left(x^n\sqrt{(ax+b)^3} - nb\int x^{n-1}\sqrt{ax+b}\, dx\right)$

42. $\displaystyle\int \frac{dx}{x^n\sqrt{ax+b}} = -\frac{\sqrt{ax+b}}{(n-1)bx^{n-1}} - \frac{(2n-3)a}{(2n-2)b}\int \frac{dx}{x^{n-1}\sqrt{ax+b}}$

Expressions Containing $ax^2 + c$, $x^2 \pm p^2$, and $p^2 - x^2$, $p > 0$

43. $\displaystyle\int \frac{dx}{p^2-x^2} = \frac{1}{2p}\ln\left|\frac{p+x}{p-x}\right|$

44. $\displaystyle\int \frac{dx}{ax^2+c} = \frac{1}{\sqrt{ac}}\arctan\left(x\sqrt{\frac{a}{c}}\right), \quad a > 0,\ c > 0$

45. $\displaystyle\int \frac{dx}{ax^2-c} = \frac{1}{2\sqrt{ac}}\ln\left|\frac{x\sqrt{a}-\sqrt{c}}{x\sqrt{a}+\sqrt{c}}\right|, \quad a > 0,\ c > 0$

46. $\displaystyle\int \frac{dx}{(ax^2+c)^n} = \frac{1}{2(n-1)c}\frac{x}{(ax^2+c)^{n-1}} + \frac{2n-3}{2(n-1)c}\int \frac{dx}{(ax^2+c)^{n-1}}, \quad n > 1$

47. $\displaystyle\int x\left(ax^2+c\right)^n\, dx = \frac{1}{2a}\frac{(ax^2+c)^{n+1}}{n+1}, \quad n \neq -1$

48. $\displaystyle\int \frac{x}{ax^2+c}\, dx = \frac{1}{2a}\ln|ax^2+c|$

49. $\displaystyle\int \sqrt{x^2 \pm p^2}\, dx = \frac{1}{2}\left(x\sqrt{x^2 \pm p^2} \pm p^2\ln\left|x+\sqrt{x^2 \pm p^2}\right|\right)$

50. $\displaystyle\int \sqrt{p^2 - x^2}\, dx = \frac{1}{2}\left(x\sqrt{p^2-x^2} + p^2\arcsin\left(\frac{x}{p}\right)\right), \quad p > 0$

51. $\displaystyle\int \frac{dx}{\sqrt{x^2 \pm p^2}} = \ln\left|x + \sqrt{x^2 \pm p^2}\right|$

Expressions Containing Trigonometric Functions

52. $\displaystyle\int \sin^2(ax)\, dx = \frac{x}{2} - \frac{\sin(2ax)}{4a}$

53. $\displaystyle\int \sin^3(ax)\, dx = -\frac{1}{a}\cos(ax) + \frac{1}{3a}\cos^3(ax)$

54. $\displaystyle\int \sin^n(ax)\, dx = -\frac{\sin^{n-1}(ax)\cos(ax)}{na} + \frac{n-1}{n}\int \sin^{n-2}(ax)\, dx,\quad n > 0$

55. $\displaystyle\int \cos^2(ax)\, dx = \frac{x}{2} + \frac{\sin(2ax)}{4a}$

56. $\displaystyle\int \cos^3(ax)\, dx = \frac{1}{a}\sin(ax) - \frac{1}{3a}\sin^3(ax)$

57. $\displaystyle\int \cos^n(ax)\, dx = \frac{\cos^{n-1}(ax)\sin(ax)}{na} + \frac{n-1}{n}\int \cos^{n-2}(ax)\, dx$

58. $\displaystyle\int \sin(ax)\cos(bx)\, dx = -\frac{\cos((a-b)x)}{2(a-b)} - \frac{\cos((a+b)x)}{2(a+b)},\quad a^2 \neq b^2$

59. $\displaystyle\int \sin(ax)\sin(bx)\, dx = \frac{\sin((a-b)x)}{2(a-b)} - \frac{\sin((a+b)x)}{2(a+b)},\quad a^2 \neq b^2$

60. $\displaystyle\int \cos(ax)\cos(bx)\, dx = \frac{\sin((a-b)x)}{2(a-b)} + \frac{\sin((a+b)x)}{2(a+b)},\quad a^2 \neq b^2$

61. $\displaystyle\int x\sin(ax)\, dx = \frac{1}{a^2}\sin(ax) - \frac{x}{a}\cos(ax)$

62. $\displaystyle\int x\cos(ax)\, dx = \frac{1}{a^2}\cos(ax) + \frac{x}{a}\sin(ax)$

63. $\displaystyle\int x^n \sin(ax)\, dx = -\frac{x^n}{a}\cos(ax) + \frac{n}{a}\int x^{n-1}\cos(ax)\, dx,\quad n > 0$

64. $\displaystyle\int x^n \cos(ax)\, dx = \frac{x^n}{a}\sin(ax) - \frac{n}{a}\int x^{n-1}\sin(ax)\, dx,\quad n > 0$

65. $\displaystyle\int \tan^n(ax)\, dx = \frac{\tan^{n-1}(ax)}{a(n-1)} - \int \tan^{n-2}(ax)\, dx,\quad n \neq 1$

66. $\displaystyle\int \sec^n(ax)\, dx = \frac{\sec^{n-2}(ax)\tan(ax)}{a(n-1)} + \frac{n-2}{n-1}\int \sec^{n-2}(ax)\, dx,\quad n \neq 1$

Expressions Containing Exponential and Logarithm Functions

67. $\displaystyle\int xe^{ax}\, dx = \frac{e^{ax}}{a^2}(ax - 1)$

68. $\displaystyle\int x^n e^{ax}\, dx = \frac{1}{a}x^n e^{ax} - \frac{n}{a}\int x^{n-1}e^{ax}\, dx,\quad n > 0$

69. $\displaystyle\int e^{ax}\sin(bx)\, dx = \frac{e^{ax}}{a^2+b^2}\left(a\sin(bx) - b\cos(bx)\right)$

70. $\displaystyle\int e^{ax}\cos(bx)\, dx = \frac{e^{ax}}{a^2+b^2}\left(a\cos(bx) + b\sin(bx)\right)$

71. $\displaystyle\int x^n \ln(ax)\, dx = x^{n+1}\left(\frac{\ln(ax)}{n+1} - \frac{1}{(n+1)^2}\right),\quad n \neq -1$

72. $\int (\ln x)^n \, dx = x(\ln x)^n - n \int (\ln x)^{n-1} \, dx$

73. $\int \dfrac{dx}{a + be^{px}} = \dfrac{x}{a} - \dfrac{1}{ap} \ln |a + be^{px}|$

Expressions Containing Inverse Trigonometric Functions

74. $\int \arcsin(ax) \, dx = x \arcsin(ax) + \dfrac{1}{a}\sqrt{1 - a^2x^2}$

75. $\int \arccos(ax) \, dx = x \arccos(ax) - \dfrac{1}{a}\sqrt{1 - a^2x^2}$

76. $\int \text{arccsc}(ax) \, dx = x \, \text{arccsc}(ax) + \dfrac{1}{a} \ln \left| ax + \sqrt{a^2x^2 - 1} \right|$

77. $\int \text{arcsec}(ax) \, dx = x \, \text{arcsec}(ax) - \dfrac{1}{a} \ln \left| ax + \sqrt{a^2x^2 - 1} \right|$

78. $\int \arctan(ax) \, dx = x \arctan(ax) - \dfrac{1}{2a} \ln(1 + a^2x^2)$

79. $\int \text{arccot}(ax) \, dx = x \, \text{arccot}(ax) + \dfrac{1}{2a} \ln(1 + a^2x^2)$

ANSWERS TO SELECTED EXERCISES

Section 11.1

1. no limit   **3.** ∞

5. 0   **7.** $\pi/2$

9. 0   **11.** 1

13. 0   **15.** 0

17. 1   **19.** $a_k = (-1)^k/k$

21. $a_k = k$ **23.** $a_k = e^{-k}$

27. 1   **29.** 1

31. 0   **33.** $x \le 0$

35. $-\sin 1 < x \le \sin 1$ **39.** $e^{-1/2}$

41. (c) converges   **47.** no

49. yes   **53.** yes

55. $x \ge 0$

Section 11.2

1. (a) 1/5; 1/25; 1/3125; 1/9,765,625; 6/5; 31/25; 3906/3125; 12,207,031/9,765,625

   (d) 5/4

   (e) 1/20; 1/100; 1/12,500; 1/39,062,500

   (g) 0

7. $\pi^4/90$ **9.** (a) $S_n = (n+1)a$

11. 1/8   **13.** $e/(e-1)$

15. $\pi^2/(16 - 4\pi)$ **17.** $-1/48$

21. $S_n = \arctan(n+1)$; $S = \pi/2$

23. $S_n = 1 + 1/\sqrt{2} - 1/\sqrt{n+1} - 1/\sqrt{n+2}$; $S = 1 + 1/\sqrt{2}$

25. (a) 4.97   (b) 5

   (c) 0

27. $\pi^2/24$ **29.** $\pi^2/12$

31. $-1 < x < 1$; $1/(1-x)$

33. $-1 < x < 1$; $x^{10}/(1 - x^2)$

35. $-2 < x < 0$; $-(1+x)^3/x$

37. 3/2   **39.** diverges

41. 3/2   **43.** diverges

45. 9/16   **47.** diverges

49. 1/5   **51.** diverges

53. 20 feet

59. (a) yes; $S = 3$ (b) $\lim\limits_{k \to \infty} a_k = 0$

61. (b) $a_k = (-1)^k$

Section 11.3

1. (c) $S_{10} \approx 1.6963$ (d) no

3. $\sum\limits_{k=2}^{n} a_k < \int_{1}^{n} a(x)\,dx < \sum\limits_{k=1}^{n-1} a_k$

9. $1/2 < S < 3/2$

11. $2e^{-1} \le S \le 3e^{-1}$ **13.** (a) no

17. $1/2 < S < 2$ **19.** $1/\sqrt{2} < S < 2$

25. (c) no

27. (a) nothing

   (b) converges

31. converges; $N \ge 1000$

33. diverges; $N \ge 2999$

35. converges; $N \ge 4$ **41.** yes

45. converges; $\pi/8 < S < \pi/8 + 3\pi^2/32$

47. diverges **49.** $1/3 < S < 1/2$

51. converges; $1/2 < S < 5/2$

55. $N \geq 6$

Section 11.4

1. conditionally

3. $14.902 < S < 14.918$

5. (a) $S_{50} \approx 0.23794$
   (c) $0.23774 < S < 0.23814$

7. no **9.** $S \approx -0.94985$

11. $S \approx -1.625$ **13.** no

15. $p > 1$ **17.** $p > 1$

19. converges absolutely; $3/4 < S < 1$

21. diverges **23.** diverges

25. converges absolutely; $0 < S < 1/12$

31. $a_k = (-1)^k/\sqrt{k}$

Section 11.5

3. 2 **5.** 1

7. 1; $(1,3)$ **9.** 1; $[-6,-4]$

15. $\sum_{k=1}^{\infty} \dfrac{x^k}{k4^k}$ **17.** $\sum_{k=1}^{\infty} \dfrac{(x-2)^k}{3^k K^2}$

19. $\sum_{k=1}^{\infty} \dfrac{(12-x)^k}{k4^k}$

25. (a) $R = 14$ (b) $b = 3$

27. cannot **29.** may

31. may **33.** may

35. may **37.** may

41. 0.820 **43.** $(-\infty, \infty)$

45. 0.581 **47.** $[-9,1]$

49. 1.198

Section 11.6

1. 2 **3.** 2

5. $\sum_{k=0}^{\infty} (-1)^k x^{k+2}$ **7.** $\sum_{k=1}^{\infty} k(-x)^{k-1}$

9. $\sum_{k=0}^{\infty} (-1)^k \dfrac{(2x)^{2k+1}}{2k+1}$; $R = 1/2$

11. $\sum_{k=0}^{\infty} (-1)^k \dfrac{x^{2k+3}}{(2k+1)!}$; $R = \infty$

13. 29/48

17. $\dfrac{1}{2} \sum_{k=0}^{\infty} (-1)^k \left(\dfrac{1}{2}\right)^k$; $R = 2$

19. $\sum_{k=0}^{\infty} (-1)^k \left(\dfrac{x^{2k}}{(2k)!} + \dfrac{x^{2k+1}}{(2k+1)!} \right)$; $R = \infty$

21. $\sum_{k=0}^{\infty} (-1)^{k+1} \dfrac{(4k^2+2k+1)x^{2k+1}}{(2k+1)!}$; $R = \infty$

29. 1 **31.** 1/2

33. $-1/2$ **35.** 2

37. (b) $[-1,1)$

41. (a) $\sum_{k=0}^{\infty} \dfrac{(-1)^k}{(2k+1)\cdot k!} x^{2k+1}$ (b) 26/35

43. $2557/7020 \approx 0.364$

45. $x^3 + 2x^4 + 2x^5 + \frac{5}{6}x^6$

47. $1 + 2x + \frac{5}{2}x^2 + \frac{8}{3}x^3$

49. $1 + x + x^2/2 - x^4/8$

51. $(1-x)^{-2}$ **53.** $x/(1+x)$

Section 11.7

1. (b) 9.1×10^{-5} **3.** $2^{100}/100!$

5. (a) $\sum_{k=0}^{\infty} \dfrac{(-1)^k x^k}{2^{k+1}}$ (b) $-259!/2^{260}$

9. (b) yes

11. (a) $\sum_{k=0}^{\infty} \dfrac{(-1)^k x^{2k}}{(2k+1)!}$ (b) $(-\infty, \infty)$

   (c) $f'''(1) \approx \dfrac{37}{210}$

Section 11.8

1. ∞ **3.** ∞

5. 0 **7.** 0

9. converges absolutely; $S < e$

11. diverges

13. converges absolutely; $S < 1/3$

15. converges absolutely; $S < 1 + \sqrt{\pi}/2$

17. diverges

19. converges absolutely; $S < 3$

21. converges absolutely; $S < 1$

23. converges conditionally; $S < -1/2$

25. diverges

27. converges absolutely; $S = 583/120$

29. diverges **31.** diverges

33. diverges

35. (a) $n^n e^{1-n} < n!$ (b) $N > be$

37. $[-1, 1]$ **39.** $(-1/3, 1/3)$

41. $[-1/3, 1/3)$ **43.** $(1, 5)$

45. $[-3, 5)$ **47.** $[-6, -4]$

49. cannot   **51.** must

53. cannot   **55.** 0

57. 1/3

59. $\displaystyle\sum_{k=0}^{\infty} \frac{(x \ln 2)^k}{k!}; R = \infty$

61. $\displaystyle -\sum_{k=0}^{\infty} \left(\frac{2^{k+2} + (-1)^k}{2^{k+1}} \right) x^k; R = 1$

63. $a_k = (-1)^{k+1}$

65. (a)   no
     (b)   yes

69. $f(x) = 1 + 3x^4 + 3x^8 + x^{12}$

71. $g(x) \approx 1 - 3x^2/2 + 15x^4/8 - 35x^6/16$

73. 252/625

CHAPTER 12

Section 12.1

1. $x^2 + y^2 + z^2 = 4$ **3.** $x^2 + z^2 = 1$

5. $z = \sin x$ **7.** (b) Yes, y-direction.

11. $y = -x^2$, parabola   **13.** $z = y$, line

15. $z = x^2$, parabola

17. (a) $x^2 + y^2 = 1$ (b) $x^2 + y^2 = 3/4$
     (c) $(0, 0)$ (d) empty set

19. $2y + 3z = 3$ **21.** $y = -x/2$

23. $y = -z^3$ **25.** x-axis

27. parabolic tent   **29.** $(0, 3, 2); r = \sqrt{13}$

31. (a) $-A/B$ (b) $B = 0$

33. (a) $x = C/A$ (b) $A = 0$

35. (a) $x = D/A$ (b) $y + z = 1$

37. $x = 3$

39. $x = 1; y = 1/2$

43. (a) $\sqrt{50}$
     (b) $(x - 1)^2 + (y - 2)^2 + (z - 3)^2 = 50$

45. (b) $\sqrt{5}$
     (c) $\sqrt{14}$
     (d) $\sqrt{14}$

47. $z = b^2/9 - x^2/16$

Section 12.2

1. The curve is the upper half of the unit circle.

3. The curve is the right half of the unit circle.

5. The curve is the unit circle.

7. $x = t, y = 2t, 0 \le t \le 1$

9. $x = \cos t, y = -\sin t, 0 \le t \le 2\pi$

11. $x = \cos(2\pi t), y = \sin(2\pi t), 0 \le t \le 1$

13. $(2, 1), (5, 3), y = 2(x - 2)/3 + 1$

15. $(0, b), (1, m + b), y = mx + b$

17. $(x_0, y_0), (x_1, y_1), y = y_0 + (y_1 - y_0)(x - x_0)/(x_1 - x_0)$

19. (a) quickly: $t = 3, t = 4, t = 9$, and $t = 10$; slowly: $t = 0, t = 1$,
     $t = 6$, and $t = 7$
     (b) ≈ 3 units per second
     (c) ≈ 1 unit per second

21. (c) $x = 2 + \sqrt{13} \cos t, y = 3 + \sqrt{13} \sin t, 0 \le t \le 2\pi$

23. $(t, \sqrt{1 - t^2/4}), -2 \le t \le 2; (2 \cos t, \sin t), 0 \le t \le \pi$

27. $(2 + 4 \cos t, -3 - 4 \sin t), 0 \le t \le \pi$

29. $\left(R \cos(2\pi t), R \sin(2\pi t), Ht/N \right), 0 \le t \le N.$

Section 12.3

1. $(\pi, 0); (-\pi, \pi); (\pi, 2\pi)$

3. $(\sqrt{2}, \pi/4); (\sqrt{2}, -7\pi/4); (-\sqrt{2}, -3\pi/4); (-\sqrt{2}, 5\pi/4)$

5. $(\sqrt{5}, 1.1071); (\sqrt{5}, -5.1760); (-\sqrt{5}, 4.2487)$

7. $(\sqrt{17}, 1.3258); (\sqrt{17}, -4.9574); (-\sqrt{17}, 4.4674)$

9. $(\sqrt{2}, \sqrt{2})$ **11.** $(\sqrt{3}/2, 1/2)$

13. $(0.5403, 0.8415)$ **15.** $(\sqrt{2}, \sqrt{2})$

17. $x = 2$ **19.** $y = \sqrt{3}x$

21. $r = 3$ **23.** $\tan \theta = 2$

27. (a) 2; 1.866; 1.5; 1; 0.5; 0.134; 0; 0.134; 0.5; 1; 1.5; 1.866; 2
     (c) x-axis
     (d) symmetric about $\theta = \pi$

29. (a) $x = \sqrt{2}t/2, y = \sqrt{2}t/2$
     (b) $x = \cos t, y = \sin t$
     (c) $x = t \cos t, y = t \sin t$

31. (b) $t = 0, \pi, 2\pi, \ldots, 10\pi$
     (c) $t = \pi/2, 5\pi/2, 9\pi/2, \ldots; t = 3\pi/2, 7\pi/2, 11\pi/2, \ldots$

33. $x = 2 \cos \theta, y = 2 \sin \theta, 0 \le \theta \le 2\pi$

55. $\sqrt{a^2 + b^2}/2, (a/2, b/2)$

57. (b) $(x^2 + y^2)^3 = x^4$

59. $\sqrt{r_1^2 + r_2^2 - 2r_1 r_2 \cos(\theta_1 - \theta_2)}$

Section 12.4

1. $(3, 5)$

3. $(-4, -5)$

5. $(1/\sqrt{5}, 2/\sqrt{5})$

7. $(-2/\sqrt{5}, 1/\sqrt{5})$

9. $(-21/5, -28/5)$

11. (a) $|\mathbf{v}| = \sqrt{14}$
(b) $\mathbf{u} = (1, -2, 3)/\sqrt{14}$

23. triangle inequality

Section 12.5

1. (a) $\mathbf{L}(0) = (1, 2)$; $\mathbf{L}(1) = (3, 5)$; $\mathbf{L}(2) = (5, 8)$; $\mathbf{L}(-1) = (-1, -1)$
(b) ray from $P = (1, 2)$ in same direction as the vector $(2, 3)$
(c) line segment from $(-1, -1)$ to $(3, 5)$

3. $\sqrt{65}$

5. $\mathbf{S}(t) = (-2, 1) + t(5, 1)$

7. (a) an ellipse
(c) $\boldsymbol{\ell}(s) = (\sqrt{3}, 1/2) + s(-1, \sqrt{3}/2)$

9. $\boldsymbol{\ell}(t) = (1, 2, 1) + t(2, 2, 3)$; $|\mathbf{f}'(1)| = \sqrt{17}$

13. $\sqrt{10}$

17. ≈ 80.82

15. 2π

19. $\mathbf{l}(t) = (\sqrt{2}/2, \sqrt{2}/2) + t(-\sqrt{2}/2, \sqrt{2}/2)$

21. $\mathbf{l}(t) = (0, -\pi) + t(-\pi, -1)$

23. no

25. a line

27. (b) $\mathbf{g}(t) = \sqrt{2}\mathbf{f}(t)/2$

29. (b) $\boldsymbol{\ell}(t) = (0, 1, \pi) + t(-1/2, 0, 1)$
(d) $2\sqrt{5}\pi$

31. ≈ 31.312

Section 12.6

3. $\mathbf{v}(t) = 1$, $\mathbf{p}(t) = t$; $\mathbf{v}(10) = 1$, $\mathbf{p}(10) = 10$

5. $\mathbf{v}(t) = t^2/2$, $\mathbf{p}(t) = t^3/6$; $\mathbf{v}(10) = 50$, $\mathbf{p}(10) = 1000/6$

7. (b) $y = -x$ (c) $5000\sqrt{2}$

9. (a) $\mathbf{v}(t) = (50, 50\sqrt{3} - gt)$; $\mathbf{p}(t) = t(50, 50\sqrt{3}) + t^2(0, -g/2)$
(b) $t_0 = 100\sqrt{3}/g$
(d) $50\sqrt{3}/g$; $7500/2g$
(e) 50; $(50, 0)$

11. (a) yes; $t = 1$; $(-2, -1)$
(b) yes; $t = -1$; $(2, 3)$
(c) no

13. (a) $y = -\frac{1}{200}x^2 g \sec^2\alpha + x\tan\alpha$
(b) $10,000/g$ meters

15. $(-3\pi/2, 3, \pi)$

17. (a) $\mathbf{v}(t) = (1/5, -2/5, 2/5)$
(b) $\mathbf{a}(t) = (0, 0, 0)$

19. $\mathbf{p}(t) = (4 - 3\cos(2t/3), 5 + 3\sin(2t/3))$, $0 \le t \le 3\pi$

Section 12.7

1. $(1, 0)$

3. $(\sqrt{2}/2, -\sqrt{2}/2)$

5. $(1, 0)$

7. $r = 2, \theta = 0$

9. $r = 2, \theta = \pi/6$

13. (a) $(0, 2)$ (b) $(3, 0)$
(c) 0 (d) $(3, 2)$

15. (a) $50\sqrt{2}$ foot-pounds
(b) $5\sqrt{2}$ feet
(c) same

23. $(6\sqrt{5}/5, -3\sqrt{5}/5)$; two

25. yes

27. (a) $\sqrt{14}$
(b) $1/\sqrt{14}$; $2/\sqrt{14}$; $3/\sqrt{14}$

29. 3

31. $3\sqrt{2}/2$

41. $\cos\theta = \sqrt{3}/3$

47. (a) $\mathbf{f}(0) = (0, 0)$, $\mathbf{f}'(0) = (1, 1)$;
$\mathbf{f}(\pi/4) = (\pi/4, \sqrt{2}/2)$, $\mathbf{f}'(\pi/4) = (1, \sqrt{2}/2)$;
$\mathbf{f}(\pi/2) = (\pi/2, 1)$, $\mathbf{f}'(\pi/2) = (1, 0)$;
$\mathbf{f}(\pi) = (\pi, 0)$, $\mathbf{f}'(\pi) = (1, -1)$;

49. (a) $8\sqrt{65}/65$; acute
(b) $(16/13, 24/13, 0)$
(c) $(0, 0, 1)$

59. $\sqrt{13}$

61. $\sqrt{13}$

Section 12.8

1. $\mathbf{X}(t) = (0, t, 0)$

3. $\mathbf{X}(t) = (1, 2, 3) + t(2, 3, 4)$

5. $(4, 5, 6)$

7. no intersection

9. $3x + 4y + 5z = 26$; $(3, 4, 5) \cdot (\mathbf{X} - (1, 2, 3)) = 0$

11. $3x + 2y + z = 10$; $(\mathbf{X} - (1, 2, 3)) \cdot (3, 2, 1) = 0$

13. $2x + 3y - z = -5$; $(2, 3, -1) \cdot (\mathbf{X} - (0, 0, 5)) = 0$

15. $x + 2y + 3z = 6$

17. $\mathbf{X}(s, t) = s\mathbf{i} + t\mathbf{j}$

19. $\mathbf{X}(s, t) = (1, 2, 3) + s(1, 0, 0) + t(1, 1, 0)$

21. $\mathbf{X}(s, t) = s\mathbf{i} + t\mathbf{j}$, $0 \le s \le 1$, $0 \le t \le 2$

23. no

25. $x(t) = t$, $y(t) = 3 - 2t$, $z(t) = t$

27. yes

29. no

31. $2x - y + z = 6$

33. $x - 3y + 4z = 7$

35. no

39. $10x - 3y + 4z = 11$

41. (b) $|d_1 - d_2| / \sqrt{a^2 + b^2 + c^2}$

43. $\sqrt{3561/77}$

Section 12.9

1. $(-3, 6, -3)$

3. $(-30, 60, -30)$

5. $(1, -1, 1)$

7. $(-v_3, 0, v_1)$

9. $x - 5y + 3z = 0$.

11. $5x + y - 13z = 4$

13. $(-\sqrt{6}/6, \sqrt{6}/3, -\sqrt{6}/6)$

15. $(\sqrt{6}/6, -\sqrt{6}/3, \sqrt{6}/6)$

17. (a) ≈ 0.94 newton-meters; upward

(b) 0

(c) 0

19. $\pi < \theta < 2\pi$

23. no, $\mathbf{i} \times (\mathbf{i} \times \mathbf{j}) \neq (\mathbf{i} \times \mathbf{i}) \times \mathbf{j}$

27. (a) 1, yes

(b) abc, yes

(c) 0

(d) $|\mathbf{v} \times \mathbf{w}|$

29. $(65/98, 43/98, 54/49)$

31. $x + y + z = 24$

43. $(0, -6\sqrt{3}, 0)$

Section 12.10

1. $2x + 3y + 4z = 20$.

3. yes

5. $x(t) = 1 - t$, $y(t) = t$, $0 \le t \le 1$; $x(t) = \cos t$, $y(t) = \sin t$, $0 \le t \le \pi/2$

7. $x = 1 + 3\cos t$, $y = 2 + 3\sin t$, $\pi/2 \le t \le 3\pi/2$

9. $(-\sqrt{6}/6, \sqrt{6}/3, -\sqrt{6}/6)$

11. (a) $\ell(t) = (1, 2, 3) + t(2, 3, 4)$

(b) Yes, $\ell(25) = (51, 77, 103)$.

13. (a) $x - 2y + z = 0$

(b) yes

15. (a) $\mathbf{v}(t) = (2t, 3t^2)$; $\mathbf{p}(t) = (t, t^3)$

(b) $\int_0^{10} \sqrt{4t^2 + 9t^4}\, dt$

17. $x = 12 + 13\cos t$, $y = 5 + 13\sin t$, $0 \le t \le 2\pi$

19. $(\sqrt{14}/14, \sqrt{14}/7, 3\sqrt{14}/14)$

21. $x + y + z = 6$

23. $34\sqrt{83}/83$

25. $(3/7, 6/7, 9/7)$

27. $\sqrt{6}$

29. $\sqrt{6}/3$

31. $(1, 2, 3) \cdot (\mathbf{X} - \mathbf{P}) = 0$

33. $\sqrt{13}\pi$

35. $(4 + t, 5 + 2t, 6 + 3t)$

37. $y + z = 11$.

39. $\ell_1(t) = (1, 2, 3) + t(4, 5, 6)$; $\ell_2(s) = (1, 0, 0) + s(4, 5, 6)$

43. $(x - 1)^2 + (y - 2)^2 + (z - 3)^2 = 50$

45. $(x - 2)^2 + (y - 1)^2 = 20$

47. $x = 2 + \sqrt{20}\cos t$, $y = 1 + \sqrt{20}\sin t$, $0 \le t \le 2\pi$

49. (a) $\mathbf{f}(2) = (5, 11)$; $\mathbf{v}(2) = (4, 3)$; speed $= 5$; $\mathbf{a}(2) = (2, 0)$

(b) $\ell(t) = (5, 11) + t(4, 3)$

51. $\sqrt{2}/2$

55. $3\sqrt{5}/5$

57. no

59. yes

61. $\mathbf{X}(t) = (0, 0, 1) + t(1, -2, 1)$

63. no

65. $\mathbf{X}(t) = (4, 5) + t(3, 2)$; $7\sqrt{13}/13$

67. yes, at $(3, 5, 8)$

69. yes, at $(2, 3, 5)$

71. $x + y + z = 6$

73. $3x + 4y + 5z = 26$

75. $z = 0$

77. $\mathbf{X}(t) = (1, 2, 3) + t(-1, 2, -1)$

79. $x = t$, $y = \sin t$, $0 \le t \le 2\pi$

81. (a) $\mathbf{v}(t) = (t, -2t + 10)$; $\mathbf{p}(t) = (t^2/2, -t^2 + 10t)$

(b) $y = -2x + 10\sqrt{2x}$

(c) $t_0 = 10$; $\mathbf{a}(10) = (1, -2)$; $\mathbf{v}(10) = (10, -10)$; $s(10) = 10\sqrt{2}$; $\mathbf{p}(10) = (50, 0)$

83. (a) $\mathbf{v}(t) = (-e^{-t} + k_1, -t + k_2)$; $\mathbf{p}(t) = (t + e^{-t} - 1, -t^2/2)$

(b) $\mathbf{p}(5) = (e^{-5} + 4, -25/2)$; $\mathbf{a}(5) = (e^{-5}, -1)$; $\mathbf{v}(5) = (1 - e^{-5}, -5)$; $s(5) = \sqrt{1 - 2e^{-5} + e^{-10} + 25}$

(c) $\mathbf{p}(t) = (t - te^{-5} + 6e^{-5} - 1, -5t + 25/2)$

CHAPTER 13

Section 13.1

1. $\mathbb{R}^2$; $[0, \infty)$

3. $\mathbb{R}^2$ except $(0, 0)$; $(0, \infty)$

5. surface and interior of the unit sphere; $[0, 1]$

7. (a) $(0, 0)$; $x^2 + y^2 = 1$; $x^2 + y^2 = 2$

(c) circles; $(0, 0)$; $x^2 + y^2 = ya$; $x^2 + y^2 = 4a$

(d) parabolas; $z = ax^2$; $z = ax^2 + a$; $z = ax^2 + 4a$

9. (a) circles

(b) circles

11. (a) parabolas

(b) parabolas

13. (a) $\mathbb{R}^2$ except $(0, 0)$

(b) $\mathbb{R}$

(c) ellipses

15. (a) lines with slope 2/3

(b) lines with slope 2/3

17. (a) a plane

(b) $\ell(t) = (0, 0, 4) + t(2, 3, -1)$

(d) parallel lines

19. sphere of radius $\sqrt{c}$ centered at the origin

21. (a) $\mathbb{R}^2$; $[0, \infty)$

(b) a cone

(c) circle of radius 5, center at $(0, 0)$

(d) circles with center at $(0, 0)$

23. $g(x, y) = g(y)$

25. might

27. cannot

29. $x = 0$; $x = 3$

31. $y = -x$; $y = 3 - x$

33. $x = 0$; $x = 3$

35. $x + y + z = 0$; $x + y + z = 3$

Section 13.2

1. $f_x(x, y) = 2x$; $f_y(x, y) = -2y$

3. $f_x(x, y) = 2x/y^2$; $f_y(x, y) = -2x^2/y^3$

5. $f_x(x, y) = -\sin x \cos y$; $f_y(x, y) = -\cos x \sin y$

7. $f_x(x, y, z) = y^2z^3$; $f_y(x, y, z) = 2xyz^3$; $f_z(x, y, z) = 3xy^2z^2$

9. $f_x(x, y, z) = (1 - y^2 z)/(1 + xyz)^2$; $f_y(x, y, z) = (1 - x^2 z)/(1 + xyz)^2$; $f_z(x, y, z) = -(x + y)xy/(1 + xyz)^2$

11. $-2\pi r/T^2$

13. $(3x^4 + 10x^3 y^4)\cos(3x^4 y + 2x^3 y^5)$

15. $-y/\left(x^2 + y^2 + z^2\right)^{3/2}$ **17.** $-2.01; 3.01$

19. $-2; 1$ **21.** $-4; 4$

23. $4; -6$ **25.** (b) $2; -3$

27. (b) $\approx 1; 0$ (c) $\approx 0; 0$

29. (a) negative   (b) $\approx 4.25; \approx -7.5$

31. (b) $f_x(x, 1) = 2x - 3$ (c) $1, -3, -5$

     (d) $z = 10 - 6y$ (e) $f_y(2, y) = -6$

     (f) $-6; -6; -6$

33. (b) local maximum   **37.** (b) 5

39. 6 **41.** no such function

43. no such function   **45.** $g(x, y) = 3(2x - y)/2$

47. $g(x, y) = 2x - y$ **51.** $-96; -16$

53. $x(t) = t,\ y(t) = 1,\ z(t) = \sqrt{11} - 2\sqrt{11}(t - 2)/11$

55. $g(y) = e^{-cy}$

Section 13.3

1. (b) $-1.5; -4.5$

     (c) $L(x, y) = -1.5x - 4.5y + 10.5$

3. (a) $L(x, y) = x + 2y - 2$

5. (a) $f_x(x, y) = \cos x + y$; $f_y(x, y) = 2 + x$

     (b) $L(x, y) = x + 2y$

7. $L(x, y) = 4x + 2y - 5$

9. $L(x, y) = x + y$

11. $L(x, y) = 12x - 20y + 4$

13. (a) $L(x, y) = 3 + 2y$

     (b) $(\mathbf{X} - (0, 0, 3)) \cdot (0, 2, -1) = 0$

15. $4x + 2y - z = 5$ **17.** $z = 1$

19. $\mathbf{X}(t) = (2, 1, 5) + t(-4, -2, 1)$

21. $4; -2; 3$

23. (b) $f_x(x, y) = 0$; $f_y(x, y) = \cos y$

     (c) $L(x, y) = y + 2$

25. no

27. (a) $L(x, y) = 0.6x + 0.8y$

     (b) $5.02; 6.4$

     (c) no

29. $-1; 4; 5; -3$ **31.** $|R_2 - 100| \le 2.$

Section 13.4

5. $(3/5, 4/5)$ **7.** $(1, 1/2, 1/3)$

9. $\nabla f = (1, 0)$; $x = 2$

11. $\nabla f(2, 1) = (4, -1)$; $y = x^2 - 3$

13. $4x + 2y - z = 5$ **15.** $z = 1$

17. $x + y - z = -1$ **19.** $(\pm\sqrt{6}/2, 0)$; $(0, y)$

21. $(\pm 2, 0)$; $(1, \pm 3)$ **23.** $7\sqrt{2}/10$

25. 0 **27.** $f(x, y) = x^2 + xy$

29. $f(x, y, z) = x^2 + 2y^2 + 3z^2$

31. (a) $4; 4\sqrt{2}/2; 0; -4\sqrt{2}/2; -4; -4\sqrt{2}/2; 0; 4\sqrt{2}/2$

     (c) $(3/4, \pm\sqrt{7}/4)$

33. (a) $ax + by = ax_0 + by_0$

37. $-\sqrt{9/2 - 2\sqrt{2} + \pi^2/32}$; $(\sqrt{2}/2 - 2, \sqrt{2}\pi/8)$

39. $\sqrt{84}/4$

41. (a) $(2, -2)$ (b) $-2\sqrt{13}/13$

43. $17/2$ **45.** $(0, -1)$ or $(4/5, -3/5)$

Section 13.5

1. $f_{xx} = 6xy^4$; $f_{yy} = 12x^3 y^2$

3. $f_{uu} = (4u^2 - 2)e^{-(u^2 + v^2)}$; $f_{vv} = (4v^2 - 2)e^{-(u^2 + v^2)}$

9. $\begin{pmatrix} -y^2 \sin(xy) & \cos(xy) - xy\sin(xy) \\ \cos(xy) - xy\sin 6xy & -x^2 \sin(xy) \end{pmatrix}$

11. $\begin{pmatrix} -\sin x & 0 \\ 0 & -4\cos(2y) \end{pmatrix}$

13. $\begin{pmatrix} 2 & 0 \\ 0 & -2 \end{pmatrix}$

15. $\begin{pmatrix} -2\sin(x+y) - x\cos(x+y) & -\sin(x+y) - x\cos(x+y) \\ -\sin(x+y) - x\cos(x+y) & -x\cos(x+y) \end{pmatrix}$

17. $\begin{pmatrix} 2A & C \\ C & 2B \end{pmatrix}$

19. zero matrix

21. $Q(x, y) = xy$ **23.** $Q(x, y) = xy$

25. $Q(x, y) = -1 - 2(y - \pi/4) + (x + \pi/2)^2/2$

27. $Q(x, y) = x^2 - y^2$

29. $Q(x, y, z) = 3 + 2(x - 2)/3 + 2(y - 2)/3 + (z - 1)/3 + 5(x - 2)^2/54 + 5(y - 2)^2/54 + 4(z - 1)^2/27 - 4(x - 2)(y - 2)/27 - 2(x - 2)(z - 1)/27 - 2(y - 2)(z - 1)/27$

41. $Q(x, y, z) = f(0, 0, 0) + f_x x + f_y y + f_z z + (f_{xx} x^2 + f_{yy} y^2 + f_{zz} z^2 + 2f_{xy} xy + 2f_{xz} xz + 2f_{yz} yz)/2$; 10

Section 13.6

1. (a) remains constant

     (b) rises all the way; $(0.5, 1)$; 0.5

3. (a) $(1, \pi/2)$ is a local minimum; $(1, -\pi/2)$ is a local maximum

(b) $(0, 0)$, $(2, 0)$, $(1, \pi/2)$, and $(1, -\pi/2)$

(c) $15; -15$

5. (b) saddle point   **7.** $(0, 0)$; $(4/3, 4/3)$

9. $(0, 0)$; local maximum   **11.** $(0, 0)$; local minimum

13. $(0, 0)$ is a local minimum point; $(\pm 2\sqrt{2}, 2)$ are saddle points

15. $(0, 0)$ and $(-2, 2)$ are saddle points; $(0, 2)$ is a local minimum point; $(-2, 0)$ is a local maximum point

19. $g(x, y) = y^2$

21. $k(x, y) = (x - 3)^2 + (y - 4)^2$

25. (c) global minimum

Section 13.7

1. $\begin{pmatrix} 1 & 2 \\ 4 & 5 \end{pmatrix}$ **3.** $\begin{pmatrix} 0 \\ 1 \\ 1 \end{pmatrix}$

5. $\begin{pmatrix} 32 & 43 \\ 37 & 50 \\ 42 & 57 \end{pmatrix}$ **7.** $\begin{pmatrix} 11 & 17 \\ 10 & 16 \\ 9 & 15 \end{pmatrix}$

9. $\begin{pmatrix} 2 \\ 1 \\ 1 \end{pmatrix}$

11. (a) $L_g(x, y) = 4x + 2y - 5$

(b) $L_f(t) = t - 5$

(c) $(f \circ g)(x, y) = x^4 + 2x^2y^2 + y^4 - 9x^2 - 9y^2 + 20$; $(L_f \circ L_g)(x, y) = 4x + 2y - 10$

13. $\left(6x_0 \left(x_0^2 + y_0^2 \right)^2, 6y_0 \left(x_0^2 + y_0^2 \right)^2 \right)$

15. $\left(44 + 58x_0 + 72y_0, 54 + 72x_0 + 90y_0 \right)$

17. (a) $\begin{pmatrix} -3t^2 \sin(t^3) \\ 3t^2 \cos(t^3) \\ 3t^2 \end{pmatrix}$

19. $\begin{pmatrix} 32 & -4 \\ 8 & 0 \end{pmatrix}$

21. $\dfrac{\partial u}{\partial r} = \dfrac{\partial u}{\partial x}\dfrac{\partial x}{\partial r} + \dfrac{\partial u}{\partial y}\dfrac{\partial y}{\partial r} + \dfrac{\partial u}{\partial z}\dfrac{\partial z}{\partial r}$; $\dfrac{\partial u}{\partial s} = \dfrac{\partial u}{\partial x}\dfrac{\partial x}{\partial s} + \dfrac{\partial u}{\partial y}\dfrac{\partial y}{\partial s} + \dfrac{\partial u}{\partial z}\dfrac{\partial z}{\partial s}$

23. (a) $\mathbf{k}(x, y) = x^2 + y^2$ (b) $\mathbf{k}(t) = (t, t^2)$

25. -0.000288 **29.** 0

31. $(16 + 5\cos(1)\sin(1))/\sqrt{25 - 5\cos^2(1)}$

37. $4\,°C; 1\,°C$ **39.** $(16/25, 4/25)$

Section 13.8

1. (a) $f(x, mx) = m/(1 + m^2)$

(c) no

(d) only $\mathbf{u} = \mathbf{i}$ or $\mathbf{u} = \mathbf{j}$

3. $L(x, y) = x + y$ **5.** $L(x, y) = -1 + 2x + y$

7. $(6/25, -3/25)$ **9.** no

13. (b) 0 **15.** yes; $\nabla f(0, 0) = (0, 0)$

Section 13.9

3. (b) $f(x, y) = x^2 - y^2$

(c) $g(x, y) = (x - 2)^2 - (y - 1)^2$

7. (a) $(1, 2)$ (b) saddle point

(c) $L(x, y) = 3$; $Q(x, y) = 3 + (x - 1)^2 + 2(x - 1)(y - 2)$

9. (a) $(1, 2)$ (b) saddle point

(c) $L(x, y) = 0$; $Q(x, y) = (x - 1)(y - 2)$

11. $8\sqrt{2}$ **13.** $3\sqrt{2}/2$

15. $f(x, y) = -(x^2 + y^2)$

17. $Q(x, y) = 2 + x + xy + x^2/2$

19. (a) $(4, 1)$ (b) $3\sqrt{2}/2$

29. $f(x, y) = y/x$; $\mathbb{R}^2$ except the line $x = 0$

31. $f(x, y) = \left| \sqrt{x^2 + y^2} - 1 \right|$; $\mathbb{R}^2$

33. (a) $\kappa = \left| f''(t) \right| / \left(1 + \left(f'(t) \right)^2 \right)^{3/2}$

(b) $\kappa = 2$

(c) $\kappa = 1/r$

37. (a) $\sqrt{17}$

(b) $\ell(t) = (1, 2) + t(1, -4)$

(c) $4x + y - z = 3$

39. (a) $\left(y\cos(x + y), \sin(x + y) + y\cos(x + y) \right)$

(b) $-7\pi/5$

41. yes; $f(x, y) = \sin x + \sin(x + y) + C$

43. (a) -1 (b) -0.90827

47. $5; -3$ **51.** no

55. $5x + 2y - z = 6$

57. (a) $\left(5\sqrt{29}/29, 2\sqrt{29}/29 \right)$ (b) $\sqrt{29}$

59. $(4\cos 5, \cos 5)$

63. $(2, 1)$ is a local minimum point; $(-2, 1)$ is a saddle point

65. no

67. $(1, 1/2)$ is a local minimum point; $(-1, 1/2)$ is a saddle point

69. (a) $L(x, y) = 2 + y$

(b) $Q(x, y) = 2 + y - x^2/2 + xy$

71. $(0, 0)$ is a saddle point; $(1, 1)$ is a local minimum point

73. (a) $2ax + 2by - z = a^2 + b^2$

(b) $x = (a^2 + b^2)/2a$; $y = (a^2 + b^2)/2b$; $z = -a^2 - b^2$

75. (a) $(1, -8)$ (b) $x = 2y^2 - 7$

77. $(0, 0)$ and $(-4, 3)$ are saddle points; $(0, 3)$ is a local minimum point; $(-4, 0)$ is a local maximum point

79. (a) $\left(ye^{xy}, xe^{xy} + 3y^2 \right)$ (b) $x + 3y - z = 1$

CHAPTER 14

Section 14.1

1. (a) 0.2133

(b) 0.211498

5. $f(x, y) = 1.25$; 1875

7. 1562.5

9. (a) 144

(b) 96

(c) 120

(d) 120

11. 60; 20/3

13. 32; 32/9

15. 12

17. 8

19. 81

21. $32\pi/3$

25. (a) 5/4

(b) $I = 8/3$

27. (c) 21/8

31. $[-1, 0] \times [0, 1]$

37. 1

39. 16

41. 48

43. 69

45. 325π

47. $325\pi - 104$

49. -14

Section 14.2

1. 1

3. 512/3

5. $12\sqrt{3}/5 - 32\sqrt{2}/15 + 4/15$

7. $5\ln 5 - 16\ln 2 + 3\ln 3$

9. $e^5(e^3 - e^2 - e + 1)/2$

11. 1274/5

13. π

15. 3

17. 9/2

19. 28

21. 1073

23. $29(d^2 - c^2)/2 + 37(b^2 - a^2)/2$

25. $87(d - c) - 111(b - a)$

Section 14.3

3. $\bar{x} = 0$; $\bar{y} = 3/5$.

5. (a) 108

(b) 108

(c) 108

7. 215,000 cubic meters

9. 0

11. 3/20

13. 0

15. 3/20

17. (a) 1

(b) 1

19. 6/35

21. (a) $\displaystyle\int_a^b f(x)\, dx$

23. (a) $\displaystyle\int_1^3 (2 + \cos x)\, dx$

(b) $\displaystyle\int_1^3 \int_0^{2+\cos x} 1\, dy\, dx$

25. $(0, 4/3)$

27. $(0, 4/3\pi)$

29. $\displaystyle\int_0^1 \int_0^{\sqrt{y}} f(x, y)\, dx\, dy$

31. $\displaystyle\int_0^1 \int_0^{\sqrt{1-x^2}} f(x, y)\, dy\, dx$

33. $\displaystyle\int_0^1 \int_{\arcsin y}^{\pi y/2} f(x, y)\, dx\, dy$

35. $\displaystyle\int_0^2 \int_0^{y/2} x\sqrt{1 + y^3}\, dx\, dy = 13/18$

37. $\displaystyle\int_0^3 \int_0^{x^2} \sin(x^3)\, dy\, dx = (1 - \cos(27))/3$

39. $\displaystyle\int_0^1 \int_0^{y^2} \sqrt{1 + y^3}\, dx\, dy = (4\sqrt{2} - 2)/9$

41. $\displaystyle\int_0^1 \int_{x^2}^x \frac{\sin x}{x}\, dy\, dx = 1 - \sin 1$

Section 14.4

5. $3\pi/2$

7. 1/2

9. $\displaystyle\int_0^1 \int_0^{\pi/2} r^3\, d\theta\, dr = \pi/8$

11. $\displaystyle\int_0^{\pi/4} \int_0^{\tan\theta\sec\theta} r^3 \sin\theta \cos\theta\, dr\, d\theta = 1/24$

13. $\displaystyle\int_0^\pi \int_0^{2\cos\theta} r^3\, dr\, d\theta$

15. $\pi(e - 1)/4$

17. $\pi(\sqrt{2} - 1)/2$

19. $\ln(1 + \sqrt{2}) + \sqrt{2} - 1$

21. $2 + \pi$

23. $3\pi/4$

25. $2\pi(\ln 3 - \ln 2)$

27. 8/9

29. $\pi^5/20$

31. $3\pi/4$

33. $2 + \pi/4$

35. $\pi/3 + \sqrt{3}/2$

37. $\pi/3$

39. 32/9

41. $9\pi/2$

Section 14.5

1. (a) $\displaystyle\int_0^6 \int_0^{3-x/2} \int_0^{2-x/3-2y/3} 1\, dz\, dy\, dx$

(b) $\displaystyle\int_0^6 \int_0^{3-x/2} (2 - x/3 - 2y/3)\, dy\, dx$

(c) 6

3. (b) 1/30

5. (b) 1/12

7. (b) 9/4

9. (b) 7/60

11. (b) 1280/3

15. (b) $\pi/32$

Section 14.6

7. $r\cos\theta = 1$

9. $r^2 + z^2 = 9$

11. sphere of radius a centered at the origin; $x^2 + y^2 + z^2 = a^2$

13. cone; $z = \sqrt{x^2 + y^2}$

15. plane; $z = 3$

17. (a) $r = \sqrt{13}$; $\theta = \arctan(3/2)$; $z = 1$
   (b) $\rho = \sqrt{14}$; $\theta = \arctan(3/2)$; $\phi = \arctan(\sqrt{13})$

21. $\displaystyle\int_{\theta=0}^{\theta=2\pi} \int_{r=0}^{r=a} \int_{z=0}^{z=h} r\, dz\, dr\, d\theta = \pi a^2 h$

23. $1/3$

25. $16\pi/3$

27. $a^2 h^2 \pi/2$

29. $1/8$

31. 4π

35. $\pi(2 - \sqrt{2})/3$

37. $\displaystyle\int_0^{2\pi} \int_{\pi/2+\arcsin(3/5)}^{\pi} \int_{-3\sec\phi}^{5} \rho^2 \sin\phi\, d\rho\, d\phi\, d\theta$

39. (a) $\displaystyle\int_{-\sqrt{3}}^{\sqrt{3}} \int_{-\sqrt{3-x^2}}^{\sqrt{3-x^2}} \int_{\sqrt{x^2/3+y^2/3}}^{\sqrt{4-x^2-y^2}} y\, dz\, dy\, dx$

   (b) $\displaystyle\int_0^{2\pi} \int_0^{\sqrt{3}} \int_{r/\sqrt{3}}^{\sqrt{4-r^2}} r^2 \sin\theta\, dz\, dr\, d\theta$

   (c) $\displaystyle\int_0^{2} \int_0^{2\pi} \int_0^{\pi/3} \rho^3 \sin^2\phi \sin\theta\, d\phi\, d\theta\, d\rho$

Section 14.7

1. $2\sqrt{2}\pi$

3. (b) ≈ 6.68

5. (a) $\displaystyle\int_{-1}^{1} \int_{x^2}^{1} x^2 y\, dy\, dx = 4/21$

   (b) $\displaystyle\int_0^{1} \int_{-\sqrt{y}}^{\sqrt{y}} x^2 y\, dx\, dy = 4/21$

7. (b) $\displaystyle\pi \int_0^{1} (1-z)dz = \pi/2$

11. $2; 2$

13. $2; 2$

15. (a) $5/3$

17. (a) $2e^2 - 3e^{4/3} + 1$

19. $\displaystyle\int_1^{2} \int_0^{2} \int_0^{3} \left(\frac{uv+vw}{3u}\right) dw\, dv\, du$

Section 14.8

1. 1131 bushels; 1131/9 bushels/acre

3. $3\pi/2$

5. $1/6abc$

7. $2\pi/5$

9. π

11. $\pi/30$

13. $2/3$

15. $2/3$

17. $\displaystyle\int_0^{\pi/2} \int_0^{\pi/2} \int_0^{1} \rho^2 \sin\phi\, d\rho\, d\phi\, d\theta$

19. $3\pi/2$

23. $2/3$

25. $\displaystyle\int_0^{1/2} \int_0^{2\pi} \int_{\sqrt{3}r}^{\sqrt{1-r^2}} r\, dz\, d\theta\, dr$

27. 4π

29. $\pi/3$

31. (a) π

33. $5/8$

35. (a) $423/4$

   (b) 108

37. $24/5$

39. $256\pi/3$

41. $17\pi/6$

43. $2 + \pi/4$

45. $5\pi/2$

47. $-1/40$

CHAPTER 15

Section 15.1

1. (a) $\mathbf{p}(t) = t(1,2)/\sqrt{5}$
   (b) $\mathbf{p}(t) = (1,2) - t(1,2)/\sqrt{5}$
   (c) $\mathbf{p}(t) = (1,2) - t(400, 400)/\sqrt{32}$
   (d) $\mathbf{p}(t) = (5/9, 14/9) + t(4,4)/9$

3. (c) $\mathbf{v}(0) = (0, -3)$

5. $\mathbf{p}(t) = (\cos(t) + 0.4\cos(4t), \sin(t) + 0.4\sin(4t))$

7. $\mathbf{p}(t) = (\cos(4t) + 0.4\cos(t), \sin(4t) + 0.4\sin(t))$

9. (a) $\mathbf{v}(t) = (1 - \sin t, -\cos t)$
   (b) $s(t) = \sqrt{2 - 2\sin t}$; 8
   (c) $|\mathbf{a}(t)| = 1$

11. (a) $\mathbf{p}(t) = (1,2) + t(\sqrt{2}/2, \sqrt{2}/2)$, $0 \le t \le \sqrt{2}$; $\mathbf{q}(t) = (5,6) + t(\sqrt{2}/2, \sqrt{2}/2)$, $0 \le t \le \sqrt{2}$;
   (b) $(6,8)$; $(8,10)$; 2
   (c) $\mathbf{r}(t) = (6,8) + t(\sqrt{2}/2, \sqrt{2}/2)$, $0 \le t \le 4\sqrt{2}$

13. (a) $\mathbf{p}_{\text{long}} = (4\cos t, 4\sin t)$; $\mathbf{p}_{\text{medium}} = (2\cos(4t), 2\sin(4t))$; $\mathbf{p}_{\text{short}} = (\cos(8t), \sin(8t))$

(c) $\mathbf{v}(0) = (0, 20)$; $\mathbf{a}(0) = (-100, 0)$
(d) $\mathbf{v}(\pi/4) = (-2\sqrt{2}, 2\sqrt{2})$; $\mathbf{a}(\pi/4) = (-32 - 2\sqrt{2}, -2\sqrt{2})$

Section 15.2

3. (a) 3
   (b) $3/\sqrt{2}$
   (c) $3/\sqrt{2}$
   (d) $|b|/(1+m^2)$

5. (a) $\mathbf{q}(t) = (3\sin t, 3\cos t)$, $0 \le t \le 2\pi$
   (b) $\mathbf{q}(t) = (1, t)\left(1 + 1/\sqrt{1+t^2}\right)$
   (c) $\mathbf{q}(t) = (t+1)(\cos t, \sin t)$

7. (a) $\mathbf{p}(t) = t(2,4)$; $\mathbf{q}(t) = (2t, 4t^2)$; $\mathbf{r}(t) = (2t, 4t^4 - 4t^3 + 4t)$
   (b) $\mathbf{p}(t) = (\cos(\pi t), \sin(\pi t))$; $\mathbf{q}(t) = (1 - 2t, 0)$; $\mathbf{r}(t) = (t^2 - 2t^3 + \cos(\pi t) - t^2 \cos(\pi t), \sin(\pi t) - t^2 \sin(\pi t))$

Section 15.3

3. yes

5. 0

Section 15.4

1. (a) 1/4; none
   (b) $2; -2$

3. (a) $h(x) = -x^2 - 4x + 18$
   (b) $-3 \le x \le 3$

7. $x = y = 25$

9. (a) $(2, 3) = \lambda(4, 5)$ (b) no

11. (a) $(2, 3) = \lambda(a, b)$ holds if and only if $3a = 2b$

     (b) $f(x, y) = 13/2$ for every point on the line $4x + 6y = 13$

Section 15.5

3. $1 - e^{-9/2}$

5. $\sqrt{2 \ln 2}$

CHAPTER 16

Section 16.1

1. (b) 0   (c) 0

     (d) 0

3. positive   **5.** positive

7. 0   **9.** 0

11. 0   **13.** -2π

15. $ab/2$ **17.** 64

19. 0   **21.** 0

23. 70   **25.** 9

27. (a) 10   (b) 10

     (c) 10   (d) 10

29. (d) 0   **31.** 4

Section 16.2

1. zero   **3.** positive

5. negative   **7.** -17

9. 17   **11.** $h(x, y) = (x^2 + y^2)/2$

13. not a gradient   **15.** $h(x, y) = -\cos x + y$

17. 3   **19.** 0

21. $1 + \cos 1 - \cos 2$ **23.** 0

25. e **27.** (b) ab^2

29. (b) $h(x, y) = -ky$ **31.** must

33. must   **35.** might

37. must   **39.** cannot

41. might

43. (b) π; $-\pi$

     (c) 2π

     (d) no

Section 16.3

1. 0   **3.** π

5. $\pi/4 - 2/3$ **7.** -4

9. 4π **11.** -10

13. $-81/20$

15. (a) 2π (b) no

     (c) 2π (d) 0

17. 3π

Section 16.4

1. $(u, v, 5)$; $[-1, 2] \times [0, 4]$

3. $(u, v, \sqrt{u^2 + v^2}\,)$; $[-1, 1] \times [-1, 1]$

5. $(u, v) = (u \cos v, u \sin v, u)$; $[0, 1] \times [0, 2\pi]$

7. $\sqrt{14}$ **9.** $\sqrt{5}$

11. (a) unit sphere

13. (u, v, u^2); $[-2, 2] \times [-3, 2]$

15. $(u \cos v, u \sin v, u^2)$; $[0, 3] \times [0, 2\pi]$

17. $(u \cos v, u \sin v, 5)$; $[0, 2] \times [0, \pi/2]$

19. $(\sin u \cos v, \sin u \sin v, \cos u)$; $[0, \pi/4] \times [0, 2\pi]$

21. $(a \sin u \cos v, b \sin u \sin v, c \cos u)$; $[0, \pi] \times [0, 2\pi]$

23. $(a \cos u, b \sin u, v)$; $[0, 2\pi] \times [0, 2]$

Section 16.5

1. $\sqrt{4u^2 + 4v^2 + 1}$; $(-2u, -2v, 1)$

3. $\sqrt{4u^4 + u^2}$; $(-2u^2 \cos(v), -2u^2 \sin(v), u)$

5. $|\sin u|$; $(\sin^2(u) \cos(v), \sin^2(u) \sin(v), \cos(u) \sin(u))$

7. $2x + 2y - z = 2$ **9.** $2x + 2y - z = 2$

11. $x = 1$ **13.** 20

15. $\sqrt{14}\pi$ **17.** $(5\sqrt{5} - 1)\pi/6$

19. $3\sqrt{2}\pi$ **21.** $3\sqrt{29}$

23. $28\pi/3$

Section 16.6

1. $-\pi/2$ **3.** $\pi/2$

5. $-\pi/2$ **7.** $\pi/2$

9. (a) $\sqrt{3}/2$

     (b) $\mathbf{X}(u, v) = (u, v, 1 - u - v)$, where (u, v) lies in the triangle D with vertices at $(0, 0)$, $(1, 0)$, and $(0, 1)$

     (c) $\sqrt{3}/2$

     (d) $1/2$

     (e) $(a + b + c)/2$

     (f) $a + b + c = 0$

11. 3; $(0, 0, 0)$ **13.** 1; $(0, 0, 2)$

15. (a) $y \cos(xy)$

     (c) $\cos 1$; $\cos 1$; $-\cos 1$; $-\cos 1$

17. $0; (0, 0, 0)$

19. $2x; (0, 0, 0)$

9. 8π

11. 0

29. $\nabla \cdot (a\mathbf{f} + b\mathbf{g}) = a\nabla\mathbf{f} + b\nabla\mathbf{g}$

31. $\nabla \cdot (h\mathbf{f}) = h\nabla \cdot \mathbf{f} + (\nabla h) \cdot \mathbf{f}$

33. $\nabla \cdot (\nabla \times \mathbf{f}) = 0$

35. $\nabla \cdot (\nabla h \times \nabla j) = 0$

13. (a) 2π
  (b) 2π
  (c) 2π

15. (a) $(-2, -4, 1)$
  (b) $2x + 4y - z = 5$
  (c) ≈ 5.945
  (d) $\sqrt{21}$

Section 16.7

1. 0

3. $-\pi$

7. (b) 3

(c) 3

17. $3/5$

APPENDIX A

1. (a) $AB = \begin{bmatrix} 1 & 0 \\ 0 & 1 \end{bmatrix}$; $AD = \begin{bmatrix} 4 & 1 & 6 \\ 10 & 3 & 17 \end{bmatrix}$;

$AX = \begin{bmatrix} x + 2y \\ 3x + 5y \end{bmatrix}$; $UA = \begin{bmatrix} u + 3v & 2u + 5v \end{bmatrix}$;

$UX = \begin{bmatrix} ux + vy \end{bmatrix}$

(b) $BA = AB = I_2$; $XU = \begin{bmatrix} ux & xv \\ yu & vy \end{bmatrix}$

(c) yes — A and B

3. (a) $-1, -1, 1, 1, 1$

9. (b) yes, but zero matrix is the only example

INDEX

Lines in $\mathbb{R}^2$ (see page 649)

Line through $\mathbf{P_0} = (x_0, y_0)$ in direction of $\mathbf{v} = (a, b)$:

- **Vector form:** $\mathbf{X}(t) = \mathbf{P_0} + t\mathbf{v}$
- **Scalar form:** $x = x_0 + at; \quad y = y_0 + bt$
  (scalar form)

Lines in $\mathbb{R}^3$ (see page 681)

Line through $\mathbf{P_0} = (x_0, y_0, z_0)$ in direction of $\mathbf{v} = (a, b, c)$:

- **Vector form:** $\mathbf{X}(t) = \mathbf{P_0} + t\mathbf{v}$
- **Scalar form:** $x = x_0 + at; \quad y = y_0 + bt; \\ z = z_0 + ct$

Planes in $\mathbb{R}^3$ (see page 684)

Plane through $\mathbf{P_0} = (x_0, y_0, z_0)$, normal to $\mathbf{n} = (a, b, c)$:

- **Vector form:** $(\mathbf{X} - \mathbf{P_0}) \cdot \mathbf{n} = 0$
- **Scalar form:** $ax + by + cz = ax_0 + by_0 + cz_0$

Dot product (see page 668)

For vectors $\mathbf{X} = (x_1, x_2)$ and $\mathbf{Y} = (y_1, y_2)$ or
$\mathbf{X} = (x_1, x_2, x_3)$ and $\mathbf{Y} = (y_1, y_2, y_3)$:

$$\mathbf{X} \cdot \mathbf{Y} = x_1 y_1 + x_2 y_2 \quad \text{or} \quad \mathbf{X} \cdot \mathbf{Y} = x_1 y_1 + x_2 y_2 + x_3 y_3$$

Cross product (see page 688)

For vectors $\mathbf{X} = (x_1, x_2, x_3)$ and $\mathbf{Y} = (y_1, y_2, y_3)$:

$$\mathbf{X} \times \mathbf{Y} = (x_2 y_3 - x_3 y_2, \; x_3 y_1 - x_1 y_3, \; x_1 y_2 - x_2 y_1)$$

Arclength (see page 658)

For curve $x = x(t); \; y = y(t); \; a \leq t \leq b$:

$$\text{arclength} = \int_a^b \sqrt{x'(t)^2 + y'(t)^2} \, dt$$

For curve $y = f(x)$ for $a \leq x \leq b$:

$$\text{arclength} = \int_a^b \sqrt{1 + f'(x)^2} \, dx.$$

Determinants (see page A-6)

- **2-by-2 case:** $\det \begin{bmatrix} a & b \\ c & d \end{bmatrix} = ad - bc$

- **3-by-3 case:** $\det \begin{bmatrix} a & b & c \\ d & e & f \\ g & h & i \end{bmatrix}$

$$= aei - afh + bfg - bdi + cdh - ceg$$

Mass, center of mass
(see page 798)

For mass distributed over a plane region R with density ρ:

- **Calculating mass:** $\text{mass} = \displaystyle\iint_R \rho(x, y)\, dA$

- **Center of mass:** $\bar{x} = \dfrac{\iint_R x\, \rho(x, y)\, dA}{\text{mass}};$

  $\bar{y} = \dfrac{\iint_R y\, \rho(x, y)\, dA}{\text{mass}}.$

Divergence and curl of vector fields
(see page 914)

For $\mathbf{f} = (P, Q, R)$ a vector field on $\mathbb{R}^3$:

- **Divergence:** $\text{div}\,\mathbf{f} = P_x + Q_y + R_z$
- **Curl:** $\text{curl}\,\mathbf{f} = (R_y - Q_z,\; P_z - R_x,\; Q_x - P_y)$

Fundamental theorems   (see page 918)

- **Ordinary fundamental theorem:**   For f continuously differentiable on $[a, b]$:

$$\int_a^b f'(x)\, dx = f(b) - f(a)$$

- **Fundamental theorem for line integrals:**   For $f(x, y)$ continuously differentiable along a curve γ, starting at $\mathbf{a}$ and ending at $\mathbf{b}$:

$$\int_\gamma \nabla f \cdot d\mathbf{X} = f(\mathbf{b}) - f(\mathbf{a})$$

- **Green's theorem:**   For $\mathbf{f} = (P, Q)$ a vector field in $\mathbb{R}^2$, γ a closed curve (oriented counterclockwise), and D the region inside γ:

$$\iint_D (Q_x - P_y)\, dA = \int_\gamma P\, dx + Q\, dy$$

- **Stokes's theorem:**   For $\mathbf{f} = (P, Q, R)$ a vector field in $\mathbb{R}^3$, S a surface in $\mathbb{R}^3$ with unit normal $\mathbf{n}$ and closed boundary curve γ:

$$\iint_S (\text{curl}\,\mathbf{f}) \cdot \mathbf{n}\, dS = \pm \oint_\gamma \mathbf{f} \cdot d\mathbf{X}$$

- **Divergence theorem:**   For $\mathbf{f} = (P, Q, R)$ a vector field in $\mathbb{R}^3$, V a solid region in $\mathbb{R}^3$, bounded by a surface S with outward unit normal $\mathbf{n}$:

$$\iiint_V \text{div}\,\mathbf{f}\, dV = \iint_S \mathbf{f} \cdot \mathbf{n}\, dS$$